Lecture Notes in Computer Science 1335
Edited by G. Goos, J. Hartmanis and J. van Leeuwen

Springer
Berlin
Heidelberg
New York
Barcelona
Budapest
Hong Kong
London
Milan
Paris
Santa Clara
Singapore
Tokyo

Rolf H. Möhring (Ed.)

Graph-Theoretic Concepts in Computer Science

23rd International Workshop, WG'97
Berlin, Germany, June 18-20, 1997
Proceedings

Volume Editor

Rolf H. Möhring
Technische Universität Berlin, Fachbereich 3 Mathematik
Straße des 17. Juni 136, D-10623 Berlin, Germany
E-mail: moehring@math.tu-berlin.de

Cataloging-in-Publication data applied for

Die Deutsche Bibliothek - CIP-Einheitsaufnahme

Graph theoretic concepts in computer science : 23th international workshop ; proceedings / WG '97, Berlin, Germany, June 18 - 20, 1997. Rolf H. Möhring (ed.). - Berlin ; Heidelberg ; New York ; Barcelona ; Budapest ; Hong Kong ; London ; Milan ; Paris ; Santa Clara ; Singapore ; Tokyo : Springer, 1997
(Lecture notes in computer science ; Vol. 1335)
ISBN 3-540-63757-5

CR Subject Classification (1991): G.2.2, F.2, F.1.2-3, F.3-4, E.1

ISSN 0302-9743
ISBN 3-540-63757-5 Springer-Verlag Berlin Heidelberg New York

Typesetting: Camera-ready by author
SPIN 10647870 06/3142 – 5 4 3 2 1 0 Printed on acid-free paper

Preface

The 23rd International Workshop on Graph-Theoretic Concepts in Computer Science (WG'97) was held in Berlin, Germany, June 18–20, 1997, at the facilities offered by the Bildungszentrum am Müggelsee located at the shores of Lake Müggelsee in the green outskirts of Berlin. It was organized by the Algorithmic Discrete Mathematics group of the Technische Universität Berlin.

The workshop was attended by 72 participants from several countries. The location at Lake Müggelsee and the workshop infrastructure guaranteed a familiar atmosphere with on-site accommodation and intensive exchange of ideas.

The workshop on graph-theoretic concepts has a long tradition. Predecessors were organized at various places in central Europe. The workshop aims at uniting theory and practice by demonstrating how graph-theoretic concepts can be applied to various areas in computer science or by extracting new problems from applications. This workshop is thus a rarity in computer science, as it is neither purely theoretical, nor purely practical, nor completely oriented towards applications. It is instead a vertical cut through the different fields of computer science in which graphs and graph-theoretic concepts are applied. Based on this tradition, it is one of the oldest workshops with roots in the early years when computer science was established in Germany.

The program committee consisted of

G. Ausiello, Rome (I)
H. Bodlaender, Utrecht (NL)
M. Habib, Montpellier (F)
L. Kirousis, Patras (GR)
L. Kucera, Prague (CR)
A. Marchetti-Spaccamela, Rome (I)
E. Mayr, Munich (D)
R. Möhring, Berlin (D), chair
M. Nagl, Aachen (D)
H. Noltemeier, Würzburg (D)
O. Sykora, Bratislava (SK)
G. Tinhofer, Munich (D)
D. Wagner, Constance (D)
P. Widmayer, Zurich (CH)

The program committee selected 28 out of 42 papers after a careful refereeing process. This selection reflects several current research directions and up-to-date snapshots that are representative for the topic of the workshop. The present volume includes these papers, together with an abstract of the invited lecture by David P. Williamson (IBM-T. J. Watson Research Center).

We wish to thank the authors of submitted papers and the reviewers. Their contributions made this workshop possible. A list of reviewers is given on one of the following pages. Furthermore, I would like to express my gratitude to the sponsors of this workshop, namely the Deutsche Forschungsgemeinschaft, Elsevier

Science, Springer-Verlag, and the Technische Universität Berlin. Their support allowed us to finance parts of the travel expenses for some of the participants, as well as parts of the organizational cost.

Finally, my thanks go to my own staff for their support in organizing this workshop, in particular to Ewgenij Gawrilow, Stephan Hartmann, Ekkehard Köhler, and Sabine Marcus.

Berlin, August 1997 Rolf H. Möhring

The 23 WGs and Their Chairs

WG'75	U. Pape, Berlin
WG'76	H. Noltemeier, Göttingen
WG'77	M. Mühlbacher, Linz
WG'78	M. Nagl, H. J. Schneider, Schloss Feuerstein near Erlangen
WG'79	U. Pape, Berlin
WG'80	H. Noltemeier, Bad Honnef
WG'81	J. Mühlbacher, Linz
WG'82	H. J. Schneider, H. Göttler, Neunkirchen near Erlangen
WG'83	M. Nagl, J. Perl, Haus Ohrbeck near Osnabrück
WG'84	U. Pape, Berlin
WG'85	H. Noltemeier, Schloss Schwanenberg near Würzburg
WG'86	G. Tinhofer, G. Schmidt, Stift Bernried near München
WG'87	H. Göttler, H. J. Schneider, Schloss Banz near Bamberg
WG'88	J. van Leeuwen, Amsterdam
WG'89	M. Nagl, Schloss Rolduc near Aachen
WG'90	R. H. Möhring, Johannesstift Berlin
WG'91	G. Schmidt, R. Berghammer, Richterheim Fischbachau near München
WG'92	E. W. Mayr, W.-Kempf-Haus, Wiesbaden-Naurod
WG'93	J. van Leeuwen, Sports Center Papendal near Utrecht
WG'94	G. Tinhofer, E. W. Mayr, G. Schmidt, Herrsching near München
WG'95	M. Nagl, Haus Eich at Aachen
WG'96	G. Ausiello, A. Marchetti-Spaccamela, Foundation K. Adenauer at Cadenabbia
WG'97	R. H. Möhring, Bildungszentrum am Müggelsee, Berlin

List of Reviewers

S. Albers (Saarbrücken)
P. Alimonti (Rome)
G. Ausiello (Rome)
L. Babel (München)
H. L. Bodlaender (Utrecht)
R. Borndörfer (Berlin)
U. Brandes (Konstanz)
J. Chlebikova (Bratislava)
F. d'Amore (Rome)
B. de Fluiter (Utrecht)
J. Engelfriet (Utrecht)
M. Flammini (Rome)
D. Fotakis (Patras)
P. G. Franciosa (Rome)
R. Giaccio (Rome)
J. Gustedt (Berlin)
M. Habib (Montpellier)
D. Handke (Konstanz)
B. Haverkort (Aachen)
L. Kirousis (Patras)
E. Köhler (Berlin)
J. Kratochvíl (Czech Republic)
W. Kroell (Zürich)
L. Kucera (Prague)
A. Kurnier (Montpellier)
H. Leffmann (Konstanz)
S. Leonardi (Rome)
A. Liebers (Konstanz)
A. Marchetti-Spaccamela (Rome)
E. W. Mayr (München)
M. Müller-Hannemann (Berlin)
M. Nagl (Aachen)
T. Noll (Aachen)
H. Noltemeier (Würzburg)
L. Nourine (Montpellier)
W. Oberschelp (Aachen)
G. Pantziou (Patras)
C. Paul (Montpellier)
M. Schäffter (Berlin)
I. Schiermeyer (Cottbus)
K. Schlude (Zürich)
A. Schürr (Aachen)
A. S. Schulz (Berlin)
P. Schuurman (Eindhoven)
O. Spaniol (Aachen)
L. Stacho (Bratislava)
C. Stamm (Zürich)
F. Stork (Berlin)
O. Sykora (Bratislava)
R. Szelepcsenyi (Bratislava)
J. Richts (Aachen)
G. Tel (Utrecht)
D. Thilikos (Patras)
G. Tinhofer (München)
J. Verriet (Utrecht)
I. Vrto (Bratislava)
D. Wagner (Konstanz)
R. Wattenhofer (Zürich)
K. Weihe (Konstanz)
B. Westfechtel (Aachen)
P. Widmayer (Zürich)
M. Wolff (Köln)
G. M. Ziegler (Berlin)

Contents

Invited Lecture
Gadgets, Approximation, and Linear Programming: Improved Hardness Results for Cut and Satisfiability Problems*

David P. Williamson**

IBM T. J. Watson Research Labs
P. O. Box 218, Yorktown Heights, NY 10598, USA
dpw@watson.ibm.com

Abstract. The notion of a gadget is a central element in combinatorial reductions. Informally speaking, a gadget is a finite structure which converts a constraint of one optimization problem into constraints of a different one. Despite their central role, no uniform method has been developed to construct gadgets required for a given reduction. In fact till recently no formal definition seems to have been given. In a recent work Bellare, Goldreich, and Sudan presented a definition motivated by their work on non-approximability results. We use their definition and come up with a linear-programming based method for constructing gadgets. Using this new method we present a number of new (computer constructed) gadgets for reductions to and from MAX 3SAT, MAX 2SAT, MAX CUT, MAX DICUT, etc. The new gadgets improve hardness results for MAX CUT and MAX DICUT showing that approximating these problems to within a factor of 16/17 and 12/13 respectively are NP-hard. Interestingly, we can also use the improved reductions to present an improved approximation algorithm for MAX 3SAT which guarantees an approximation ratio of .801.

---

* This work appeared as: Luca Trevisan, Gregory B. Sorkin, Madhu Sudan, and David P. Williamson, "Gadgets, Approximation, and Linear Programming", in Proceedings of the Thirty-seventh Annual Symposium on Foundations of Computer Science (FOCS), Burlington, Vermont, 1996, pp. 617–626.

** Slides of this talk are available under
`http://www.research.ibm.com/people/w/williamson/Talks/gadgets.ps`.

Non-oblivious Local Search for MAX 2-CCSP with Application to MAX DICUT*

Paola Alimonti

Dipartimento di Informatica e Sistemistica, Università di Roma"la Sapienza", Italy and Dipartimento di Matematica, Università di Roma"Tor Vergata", Italy.
alimon@dis.uniroma1.it

Abstract. In this paper we give a fully dynamic 5/2-approximate algorithm for the class of Maximum binary conjunctive constraint satisfaction problem, and thus for the Maximum directed cut problem. The proposed algorithm is based on the non-oblivious local search technique and on a neighborhood mapping that allows to change either one item, or all the items in the current solution. The total time required to maintain 5/2-approximate solutions, while an arbitrary sequence of q constraint insertions and deletions is performed, is $O(m^2 + m \cdot q)$. This give $O(m)$ amortized time per update over a sequence of $\Omega(m)$ operations.

1 Introduction

Local search is a general technique that is extensively used to solve difficult combinatorial optimization problems. The general local search algorithm works starting from an initial feasible solution and repeatedly trying to improve the value of the solution searching for a better solution in its neighborhood.

In the standard local search technique the search for optimality is led by the objective function of the problem, and the neighborhood structure of the solutions is defined as the set of solutions obtained by changing a bounded number of items of the current solution.

Recently, the complexity, the computational power and the limits of local search in approximation has been studied from more theoretical points of view [7,11,17].

Successively, in [2,3], and independently in [18], a new local search technique, that generalizes the standard technique and enlarges the power of this general paradigm, has been presented. This new approach and the standard technique have been called, respectively, *non-oblivious* and *oblivious* local search [18].

Informally speaking, the non-oblivious technique differs from the oblivious one since the search for optimality is based on an auxiliary objective function, instead of being based on the objective function of the problem. In [1–3,18] it

* Work supported by: the CEE project ALCOM-IT ESPRIT LTR, project no. 20244, "Algorithms and Complexity in Information Technology"; the Italian Project "Algoritmi, Modelli di Calcolo e Strutture Informative," Ministero dell'Università e della Ricerca Scientifica e Tecnologica; Consiglio Nazionale delle Ricerche, Italy.

has been shown that, by means of the non-oblivious technique, it is possible both to achieve better performance ratio for some problems approximable by means of the oblivious techniques, and to approximate problems not approximable by means of the oblivious technique.

In several practical applications, such as interactive design processes, instances of optimization problems are allowed to dynamically change over time. In this situation, it will be desirable to compute a new solution without having to recompute it from scratch, and then to develop dynamic algorithms with a computational cost "competitive" with existing off-line algorithms [12,22].

Although most of the existing work on dynamic algorithms has been directed towards polynomially solvable problems (e.g [4–6,9,20]), recently some attention has been paid to dynamic approximations algorithms for NP-hard problems [16]. For such problems, the competitiveness of the dynamic algorithms can be measured both with respect to the quality of the obtained solution, and with respect to the computational cost of off-line algorithms.

In this work we give a fully dynamic 5/2-approximate algorithm for the class of Maximum binary conjunctive constraint satisfaction problems (MAX 2-CCSP), and thus for the Maximum directed cut problem (MAX DICUT). The proposed algorithm is based on the non-oblivious local search technique and on a neighborhood mapping that allows to change either one item, or all the items in the current solution. The total time required to maintain 5/2-approximate solutions while an arbitrary sequence of q constraint insertions and deletions is performed, is $O(m^2 + m \cdot q)$. This give $O(m)$ amortized time per operation over a sequence of $\Omega(m)$ constraint insertions and deletions.

Although mathematical programming algorithms for MAX 2-CCSP achieves better approximation ratio (namely 1.165 using semidefinite programming and 2 using linear programming [10,14,23]), our contribution remains interesting from various points of view.

First, we dynamically maintain approximate solutions, while the input instance is modified with an amortized computational cost that is remarkably better than the cost of the above mathematical programming algorithms. Indeed, the cost for solving a semidefinite program is $\tilde{O}(m \cdot n)$ [15] and the complexity of the linear programming algorithm, given in [23], is $\tilde{O}(m)$, where m is the number of constraints and n is the number of variables. Moreover, as far as we know, nothing better is known that recomputing solutions from scratch.

Then, by means of a simple technique, the non-oblivious local search technique, we have obtained a dynamic algorithm, that is the best known algorithm for MAX 2-CCSP based on a combinatorial approach. Such an approximation algorithm has guaranteed performance considerably better than that obtained by another simple combinatorial technique, that is the greedy technique [21]. Furthermore, it is well known that, although local search techniques do not always guarantee "good" worst case performances, they often allow to design useful heuristics [8,13,19].

Finally, we believe that, apart from the specific results achieved for MAX 2-CCSP problems, this work gives a contribution in the investigation of the computational power of the local search paradigm in the approximation of NP-hard optimization problems.

The remainder of the paper is organized as follows. In Section 2, we state the basic definitions. In Section 3, oblivious and non-oblivious local search techniques are introduced and previous results in the approximation of MAX 2-CCSP are reported. In Section 4.2 we give a fully dynamic non-oblivious 1-bounded local search 5/2-approximate algorithm for MAX 2-CCSP.

2 Definitions

A Maximum k-ary constraint satisfaction (MAX k-CSP) problem is defined by a set of variables, their associated domains and a set of k-ary constraints governing the assignment of values to variables.

MAX k-CSP

INSTANCE: Set Y of variables, set $C = \{c_1, \ldots, c_m)\}$ of k-ary constraints over Y, where $c_i = (X_i, P_i)$, X_i is a size k subset of Y, and $P_i : \{true, false\}^k \rightarrow \{true, false\}$ is a k-ary boolean predicate.
SOLUTION: Truth assignment of Y.
MEASURE: Cardinality of the set of constraints from C that are satisfied by the truth assignment.

A Maximum binary conjunctive constraint satisfaction (MAX 2-CCSP) problem is defined by making restrictions on the size and the structure of the allowed constraints.

MAX 2-CCSP

INSTANCE: Set Y of variables, set $C = \{c_1, \ldots, c_m)\}$ of binary conjunctive constraints over Y, where $c_i = (X_i, P_i)$, X_i is a size 2 subset of Y, and P_i is a conjunction of the literals in X_i.
SOLUTION: Truth assignment of Y.
MEASURE: Cardinality of the set of constraints from C that are satisfied by the truth assignment.

The class of all MAX k-CSP (MAX 2-CCSP) problems is denoted by the same name, as class MAX k-CSP (MAX 2-CCSP).

Let us consider the Maximum directed cut problem (MAX DICUT).

MAX DICUT

INSTANCE: Directed graph $G = (V, E)$.
SOLUTION: A partition of V into two parts: a red part P_R and a green part P_G.
MEASURE: Cardinality of the set of edges that are directed cut, i.e. edges with the first end point in P_R and the second end point in P_G.

It is easy to see that MAX DICUT belongs to the class MAX 2-CCSP. Indeed, it is the restriction of MAX 2-CCSP in which binary constraints are conjunctions of one positive literal and one negative literal.

Given a directed graph $G = (V, E)$, with $V = \{v_1, \ldots, v_n\}$ and $|E| = m$ we construct a collection $C =$ of m binary conjunctive constraints over the a Y of variables in the following way:

1. we associate to each vertex $v_i \in V$ a boolean variables $y_i \in Y$;
2. we associate to each directed edge $(v_i, v_j) \in E$, where v_i is the first end point of (v_i, v_j) and v_j is the second end point of (v_i, v_j), a binary conjunctive constraint in C of k literals $(\overline{y}_i \wedge y_j)$.

Now we have a correspondence between the values of the variables and the partition of the variable nodes—if the variable y_i is true then the node y_i is green and $\overline{y}_i$ is red, and if the variable y_i is false then the node y_i is red and $\overline{y}_i$ is green. Thus the number of cut edges in G will be the number of satisfied constraint of in the corresponding MAX 2-CCSP problem instance.

3 Preliminaries

3.1 Local Search and Approximation

To apply local search to obtain approximation algorithms in both the oblivious and the non-oblivious techniques some assumptions on the problems and on the neighborhood structures are made, that guarantee to develop polynomial time algorithms. In particular, according with [7,17], we have the following: 1) given an instance, all solutions have size polynomially bounded in the instance size, and some solution can be produced in polynomial time; 2) given an instance and a solution, the value of the solution is computable in polynomial time; 3) given an instance and a solution, it is possible to determine whether such solution is a local optimum, and, if not, to determine a better solution in its neighborhood in polynomial time.

3.2 Neighborhood Structure

The notion of neighborhood structure is intrinsically related with the computational power of local search algorithms. In particular, both the computational cost for finding local optima and the quality of local optima with respect to the global optima strictly depend on the superimposed neighborhood structure.

Taking into account that for most optimization problems a feasible solution may be represented as a set of items, most local search algorithms make use of a neighborhood mapping defined as the set of solutions obtained by changing a bounded number of items of a solution, that is the set of solutions with bounded Hamming distance from the current solution.

In this work we make use of a slight, but powerful extension of this definition of bounded neighborhood mapping that includes the solution of maximum Hamming distance from the current solution.

Given a set $Y = \{y_1, \dots, y_n\}$ of boolean variables we denote a truth assignment for Y by the set T of literals that are truth in such an assignment. Namely, for any variable $y \in Y$: if y is true under T, then the literal y belongs to the set T and, if y is false under T, then the literal $\overline{y}$ belongs to the set T.

We have the following definition.

Definition 1. Given a set $Y = \{y_1, \dots, y_n\}$ of boolean variables, and a truth assignment $T = \{p_1, \dots, p_n\}$ for Y, a *h-bounded neighborhood mapping* N_T for T is the set of truth assignments such that, either the value of up to h variable is changed, or the value of all variables is changed

$$N_T = \{T' = \{p'_1, \dots, p'_n\} : (p'_i = y_i \vee p'_i = \overline{y}_i) \wedge |T - T'| \leq h\} \cup \{\overline{p}_1, \dots, \overline{p}_n\}.$$

In particular, our algorithms make use of a 1-bounded neighborhood structure

Notice that, by including in the neighborhood mapping N_T of a solution its complement the size of N_T is not changed, and then such neighborhood structures are feasible for polynomial time local search algorithms.

3.3 Oblivious Local Search

In the oblivious technique the search for optimality is led by the objective function of the problem, and thus, for a MAX k-CSP problem, it is expressed in terms of the number of satisfied constraints.

In [1–3,18] it has been shown that:

1. the oblivious technique approximates MAX 2-CCSP within 4 by making use of the neighborhood structure in Definition 1;
2. if we do not include in the neighborhood mapping the complement of the solutions, MAX 2-CCSP cannot be approximated by the oblivious technique, whatever bounded neighborhood size we allow[1].

3.4 Non-Oblivious Local Search

In [2,3,18] a new and more powerful local search technique has been introduced: the non-oblivious technique. The difference between oblivious and non-oblivious local search consists on the fact that, while in the oblivious technique the search for optimality is based on the objective function of the problem, in the non-oblivious technique this search is led by an auxiliary objective function.

The reason why non-oblivious local search is more powerful than the oblivious one is due to the fact that the introduced objective function better represents the features of the problems, and in the search for optimality is allowed to move to solutions of decreasing real value.

By means of the non-oblivious technique it is possible:

[1] Moreover, such a negative result also holds for MAX DICUT. In [1] it has been shown that, the formulation of MAX DICUT as a MAX 2-CCSP problem gives rise to instances that exhibit to have the same behavior of the general MAX 2-CCSP instances.

1. to achieve better performance ratio for some problems approximable by means of the oblivious technique;
2. to approximate problems not approximable by means of the oblivious technique. In particular, in [1], it has been shown that MAX 2-CCSP is approximated within 3 by a non-oblivious local search algorithm by making use of a 1-bounded neighborhood structure that do not include the complement of the solutions.

4 Local Search Approximation for MAX 2-CCSP

In this section, we first show that, by making use of the neighborhood structure in Definition 1, the non-oblivious local search technique approximates MAX 2-CCSP within 5/2 in time $O(m^2)$, where m is the number of constraints in the MAX 2-CCSP instance.

Then, we give a fully dynamic algorithm for maintaining 5/2-approximate solutions of MAX 2-CCSP problems when both, insertions and deletions of constraints, are performed. The proposed algorithm is based on a powerful property of our local search techniques, that allows to efficiently update the current solutions in order to maintaining local optimality, when the input instance is "slowly" modified. In particular, our algorithms requires $O(m^2 + m \cdot q)$ total time for any sequence of q insertions and deletions of constraints, where m is the maximum number of constraints in C. Thus, the amortized cost per update is $O(m)$ over an arbitrary sequence of $\Omega(m)$ operations.

4.1 Non-Oblivious Local Search for MAX 2-CCSP

Consider a collection C of m binary conjunctive constraints over a set Y of n boolean variables. Let $T = A \cup B$ and $T^* = A \cup \overline{B}$ be a non-oblivious 1-bounded local optimum and a global optimum respectively, where (A, B) is a partition of T and $\overline{B} = \{\overline{b} | b \in B\}$. In the following a (,b,$\overline{b}$) is a literal in A ($\overline{A}$,B,$\overline{B}$).

Let C_j be the set of constraints where j literals are false in T and m_j be the cardinality of the set C_j. Let $|p|_j$ ($|\overline{p}|_j$) be the number of occurrences of the literal p ($\overline{p}$) in the set C_j ($0 \leq j \leq 2$).

The non-oblivious objective function for MAX 2-CCSP is

$$W = 4 \cdot m_0 + \cdot m_1$$

Now let m_j^* be the cardinality of the set of constraint in C_j satisfied by T^*, $m^* = \sum m_j^*$.

Then consider the set C_0 of constraints that are satisfied by T. We partition C_0 into three subsets $C_0^{a,a}$, $C_0^{a,b}$ and $C_0^{b,b}$ in the following way. Let $C_0^{a,a}$ be the set of constraints in C_0 such that both literals are in A, $C_0^{a,b}$ be the set of constraints in C_0 such that one literal is A and one literal is in B and $C_0^{b,b}$ be the set of constraints in C_0 such that that both literals are in B. Say $m_0^{a,a}$, $m_0^{a,b}$ and $m_0^{b,b}$ the cardinality of these sets.

Similarly for the sets C_1 and C_2, we construct the sets $C_1^{a,\overline{a}}$, $C_1^{a,\overline{b}}$, $C_1^{b,\overline{a}}$, $C_1^{b,\overline{b}}$, $C_2^{\overline{a},\overline{a}}$, $C_2^{\overline{a},\overline{b}}$ and $C_2^{\overline{b},\overline{b}}$ and say $m_1^{a,\overline{a}}$, $m_1^{a,\overline{b}}$, $m_1^{b,\overline{a}}$, $m_1^{b,\overline{b}}$, $m_2^{\overline{a},\overline{a}}$, $m_2^{\overline{a},\overline{b}}$ and $m_2^{\overline{b},\overline{b}}$ their cardinality.

The following equalities hold:

- $C_0 = C_0^{a,a} \cup C_0^{a,b} \cup C_0^{b,b}$
  $m_0 = m_0^{a,a} + m_0^{a,b} + m_0^{b,b}$
  $m_0^{a,a} = m_0^*$
  $\sum_a |a|_0 = 2 \cdot m_0^{a,a} + m_0^{a,b} = 2 \cdot m_0^* + m_0^{a,b}$
  $\sum_b |b|_0 = 2 \cdot m_0^{b,b} + m_0^{a,b}$
- $C_1 = C_1^{a,\overline{a}} \cup C_1^{a,\overline{b}} \cup C_1^{b,\overline{a}} \cup C_1^{b,\overline{b}}$
  $m_1 = m_1^{a,\overline{a}} + m_1^{a,\overline{b}} + m_1^{b,\overline{a}} + m_1^{b,\overline{b}}$
  $m_1^{a,\overline{b}} = m_1^*$.
  $\sum_a |a|_1 = m_1^{a,\overline{a}} + m_1^{a,\overline{b}} = m_1^{a,\overline{a}} + m_1^*$.
  $\sum_a |\overline{a}|_1 = m_1^{a,\overline{a}} + m_1^{b,\overline{a}}$.
  $\sum_b |b|_1 = m_1^{b,\overline{b}} + m_1^{b,\overline{a}}$.
  $\sum_b |\overline{b}|_1 = m_1^{b,\overline{b}} + m_1^{a,\overline{b}} = m_1^{b,\overline{b}} + m_1^*$.
- $C_2 = C_2^{\overline{a},\overline{a}} \cup C_2^{\overline{a},\overline{b}} \cup C_2^{\overline{b},\overline{b}}$
  $m_2 = m_2^{\overline{a},\overline{a}} + m_2^{\overline{a},\overline{b}} + m_2^{\overline{b},\overline{b}}$
  $m_2^{\overline{b},\overline{b}} = m_2^*$.
  $\sum_a |\overline{a}|_2 = 2 \cdot m_2^{\overline{a},\overline{a}} + m_2^{\overline{a},\overline{b}}$
  $\sum_b |\overline{b}|_2 = 2 \cdot m_2^{\overline{b},\overline{b}} + m_2^{\overline{a},\overline{b}} = 2 \cdot m_0^* + m_2^{\overline{a},\overline{b}}$

Theorem 2. *Let C be a collection of m binary conjunctive constraints over a set Y of boolean variables and T be an independently obtained truth assignment for Y. If T is a non-oblivious 1-bounded local optimum, then it satisfies at least $(5/2) \cdot m^*$ constraints.*

Proof. Since $T = A \cup B$ is also a non-oblivious 1-bounded local optimum no literal p exists such that $\Delta W(T - \{p\} \cup \overline{p}) = -3 \cdot |p|_0 + 3 \cdot |\overline{p}|_1 - |p|_1 + |\overline{p}|_2 > 0$. Then

$$3 \cdot |p|_0 + |p|_1 \geq 3 \cdot |\overline{p}|_1 + |\overline{p}|_2 \tag{1}$$

Therefore

$$3 \cdot \sum b_0 + \sum b_1 \geq 3 \cdot \sum \overline{b}_1 + \sum \overline{b}_2$$

$$6 \cdot m_0^{b,b} + 3 \cdot m_0^{a,b} + m_1^{b,\overline{a}} \geq 3 \cdot m_1^* + 2 \cdot m_1^{b,\overline{b}} + 2 \cdot m_2^* + m_2^{\overline{a},\overline{b}} \tag{2}$$

From equation 1

$$3 \cdot \sum a_0 + \sum a_1 \geq 3 \cdot \sum \overline{a}_1 + \cdot \sum \overline{a}_2$$

$$6 \cdot m_0^* + 3 \cdot m_0^{a,b} + m_1^* \geq 3 \cdot m_1^{b,\overline{a}} + 2 \cdot m_1^{a,\overline{a}} + 2 \cdot m_2^{\overline{a},\overline{a}}] + m_2^{\overline{a},\overline{b}} \qquad (3)$$

Therefore, by adding equation 2 and equation 3 divided by 3

$$6 \cdot m_0^{b,b} + 4 \cdot m_0^{a,b} + m_0^* + \geq \frac{*}{3} \cdot m_1^* + 2 \cdot m_2^* + 2 \cdot m_1^{b,\overline{b}} + \frac{2}{3} \cdot m_1^{a,\overline{a}} + \frac{2}{3} \cdot m_2^{\overline{a},\overline{a}} + \frac{4}{3} \cdot m_2^{\overline{a},\overline{b}}$$

Since T is a non-oblivious 1-bounded local optimum $m_0 = m_0^{a,a} + m_0^{a,b} + m_0^{b,b} \geq m_2 \geq m_2^*$. Therefore

$$\frac{20}{3} \cdot m_0^{b,b} + \frac{14}{3} \cdot m_0^{a,b} + \frac{8}{3} \cdot m_0^* \geq \frac{8}{3} \cdot m_1^* + \frac{8}{3} \cdot m_2^* + 2 \cdot m_1^{b,\overline{b}} + \frac{2}{3} \cdot m_1^{a,\overline{a}} + \frac{2}{3} \cdot m_2^{\overline{a},\overline{a}} + \frac{4}{3} \cdot m_2^{\overline{a},\overline{b}}$$

$$\frac{20}{3} \cdot m_0^{b,b} + \frac{14}{3} \cdot m_0^{a,b} + \frac{16}{3} \cdot m_0^* \geq \frac{8}{3} \cdot m_0^* + \frac{8}{3} \cdot m_1^* + \frac{8}{3} \cdot m_2^*$$

$$\frac{20}{3} \cdot m_0 \geq \frac{8}{3} \cdot m^*$$

$$m_0 \geq \frac{2}{5} \cdot m^*$$

Following the previous result we propose a local search algorithm for MAX 2-CCSP problems, which determines a truth assignment satisfying at least $(5/2) \cdot m^*$ constraints, starting from an independently obtained initial solution (see fig. 1).

Theorem 3. *A non-oblivious 1-bounded local search algorithm approximates* MAX 2-CCSP *within* 5/2 *in time* $O(m^2)$.

Proof. It follows immediately from theorem 2 that the algorithm approximates MAX 2-CCSP within 5/2. To show that it runs in time $O(m^2)$ first observe that, with a proper implementation of the algorithm, the cost of each local search steps is $O(m)$. Moreover, it is easy to see that, since at each step the value of the non-oblivious objective function $W = 4 \cdot m_0 + m_1$ increases, the number of step is $O(m)$.

Corollary 4. *A non-oblivious 1-bounded local search algorithm approximates* MAX DICUT *within* 5/2 *in time* $O(m^2)$.

Finally, we will show that the approximation ratio 5/2 achieved by the non-oblivious 1-bounded local search algorithm is tight. Indeed, there exists an instance of MAX 2-CCSP for which the algorithm calculates a solution with size exactly $2/5 \cdot m^*$ (see table 1).

Notice that the instance shown in table 1 is actually MAX DICUT instance, and, thus, that the ratio 5/2 is also tight for this problem.

```
Algorithm 1
Input: C = {c_1, ..., c_m}, T
Output: T
BEGIN
   i ← 1
   while (ΔW(p_i) ≤ 0) ∧ (i ≤ n) do i ← i + 1
   end while
   if i ≤ n then
         T ← T − {p_i} ∪ {p̄_i}
         aux ← ∅
         for each C_j (1 ≤ j ≤ 2) do
               for each c ∈ C_j − aux do
                     if p_i ∈ c then
                        aux ← aux ∪ {c}
                        C_j ← C_j − {c}
                        C_{j+1} ← C_{j+1} ∪ {c}
                     end if
                     if p̄_i ∈ c then
                        aux ← aux ∪ {c}
                        C_j ← C_j − {c}
                        C_{j−1} ← C_{j−1} ∪ {c}
                     end if
               end for each
         end for each
   end if
end.
```

Fig. 1. Non-oblivious 1-bounded local search algorithm for MAX 2-CCSP

4.2 Fully Dynamic Maintenance of Local Optima

Fully dynamic approximation algorithms for constraints satisfaction problems require to process a sequence of insertion of insertions and deletions of constraints. Since in this situation algorithms have no advance knowledge of future changes in the set of constraints, it may be necessary to modify the truth assignment in order to maintain solutions of guaranteed quality.

In this section we show that the non-oblivious local search paradigm with a 1-bounded neighborhood structure can be used for designing an efficient fully dynamic (5/2)-approximate algorithm for MAX 2-CCSP.

In the following we describe the insert and delete operations. In both cases we assume that, given a MAX 2-CCSP instance, before an insert (Delete) operation, there is an existing 1-bounded non-oblivious locally optimal solution T of the current set C of constraints (i.e. not including the effect of such an operation).

Insert(c) - To insert a new constraint c into C, first check if some variables in c is a new one and in that case insert new variables in Y and update T choosing for

C_0	C_1	C_2
$(\overline{y}_1, y_2)$	$(\overline{y}_{26}, y_9)$ $(\overline{y}_{10}, y_{25})$	$(\overline{y}_4, y_1)$
$(\overline{y}_3, y_4)$	$(\overline{y}_{26}, y_{11})$ $(\overline{y}_{12}, y_{25})$	$(\overline{y}_6, y_1)$
$(\overline{y}_5, y_6)$	$(\overline{y}_{26}, y_{13})$ $(\overline{y}_{14}, y_{25})$	$(\overline{y}_8, y_1)$
$(\overline{y}_7, y_8)$	$(\overline{y}_{28}, y_9)$ $(\overline{y}_{10}, y_{27})$	$(\overline{y}_2, y_3)$
$(\overline{y}_9, y_{10})$	$(\overline{y}_{28}, y_{15})$ $(\overline{y}_{16}, y_{27})$	$(\overline{y}_6, y_3)$
$(\overline{y}_{11}, y_{12})$	$(\overline{y}_{28}, y_{17})$ $(\overline{y}_{18}, y_{27})$	$(\overline{y}_8, y_3)$
$(\overline{y}_{13}, y_{14})$	$(\overline{y}_{30}, y_{19})$ $(\overline{y}_{20}, y_{29})$	$(\overline{y}_2, y_5)$
$(\overline{y}_{15}, y_{16})$	$(\overline{y}_{30}, y_{21})$ $(\overline{y}_{22}, y_{29})$	$(\overline{y}_4, y_5)$
$(\overline{y}_{17}, y_{18})$	$(\overline{y}_{30}, y_{23})$ $(\overline{y}_{24}, y_{29})$	$(\overline{y}_8, y_5)$
$(\overline{y}_{19}, y_{20})$	$(\overline{y}_9, y_{26})$ $(\overline{y}_{25}, y_{10})$	$(\overline{y}_2, y_7)$
$(\overline{y}_{21}, y_{22})$	$(\overline{y}_9, y_{28})$ $(\overline{y}_{27}, y_{10})$	$(\overline{y}_4, y_7)$
$(\overline{y}_{23}, y_{24})$	$(\overline{y}_9, y_{30})$ $(\overline{y}_{29}, y_{10})$	$(\overline{y}_6, y_7)$

Table 1. Let $C = C_0 \cup C_1 \cup C_2$. $T = \{\overline{y}_1, y_2, \overline{y}_3, y_4, \overline{y}_5, y_6, \overline{y}_7, y_8, \overline{y}_9, y_{10}, \overline{y}_{11}, y_{12}, \overline{y}_{13}, y_{14}, \overline{y}_{15}, y_{16}, \overline{y}_{17}, y_{18}, \overline{y}_{19}, y_{20}, \overline{y}_{21}, y_{22}, \overline{y}_{23}, y_{24}, y_{25}, \overline{y}_{26}, y_{27}, \overline{y}_{28}, y_{29}, \overline{y}_{30}\}$ is a non-oblivious 1-bounded local optimum, the number of satisfied constraints (or edges in the directed cut) is equal to 12 and m^* is equal to 30 (e.g. $T^* = \{y_1, \overline{y}_2, y_3, \overline{y}_4, y_5, \overline{y}_6, y_7, \overline{y}_8, y_9, \overline{y}_{10}, y_{11}, \overline{y}_{12}, y_{13}, \overline{y}_{14}, y_{15}, \overline{y}_{16}, y_{17}, \overline{y}_{18}, y_{19}, \overline{y}_{20}, y_{21}, \overline{y}_{22}, y_{23}, \overline{y}_{24}, y_{25}, \overline{y}_{26}, y_{27}, \overline{y}_{28}, y_{29}, \overline{y}_{30}\}$).

any new variable the value that make true the corresponding literal in c. Then, run Algorithm 1.

Delete(c) - To delete a constraint c from C, first check if some variable in Y do not occur in any constraints in C, but c, then remove it from Y and update T. Then, run Algorithm 1.

Notice, that the insertion of a satisfied constraint and the deletion of a constraint with both literals false in current solution do not effect the local optimality of such a solution, and thus can be performed in $O(1)$. In the other cases the solution may not be a non-oblivious 1-bounded local optimum anymore, requiring $O(m^2)$ to be update by Algorithm 1.

In the following of this section we will show that, although the worst case computational complexity is $O(m^2)$ per operation, the total cost of an arbitrary sequence of q operations over the set of constraints takes $O(m^2 + m \cdot q)$. Thus, the amortized cost per update is $O(m)$ over an arbitrary sequence of $\Omega(m)$ operations.

Theorem 5. *Let C be a* MAX 2-CCSP *problem. The total cost for maintaining a non-oblivious 1-bounded local optimal solution for C while an arbitrary sequence of q operations is performed is $O(m^2 + m \cdot q)$, where m is the maximum number of constraints in C.*

Proof. Given a MAX 2-CCSP problem C, consider an arbitrary sequence of q operations over C. We will show that the total number of changes of the value of the non-oblivious objective function performed in order to maintain local optimality is $O(m+q)$. Then, since the cost of each local search step is performed in $O(m)$,the thesis follows

Let W_i be a non-oblivious 1-bounded local optimal value of the solution before the i)-th operation $(1 \leq i \leq q)$, and W_{q+1} be the non-oblivious 1-bounded local optimal value of the solution after the execution of the sequence of operations. Of course, we have $W_i \leq 4 \cdot m$ $(1 \leq i \leq q+1)$.

Every time an operation is performed and the current solution is not a non-oblivious 1-bounded local optimum anymore, Algorithm 1 is run.

Say $\Delta W_i \geq 0$ the variation of the value of the non-oblivious objective function due to the application of Algorithm 1 after performing the i-th operation $(1 \leq i \leq q)$.

Notice that when a constraint c is deleted an additional variation can be produced if either one literal or both literals in c are true. Therefore, as far as the non-oblivious objective function is concerned, we have

$$W_i + \Delta W_i - 4 \leq W_{i+1} \qquad (1 \leq i \leq q) \tag{4}$$

It follows that

$$W_1 + \sum_{k=1}^{q} \Delta W_k - 4 \cdot q \leq W_{q+1}$$

Hence, in view of of the fact that $W_{q+1} \leq 4 \cdot m$

$$\Delta W = \sum_{k=1}^{q} \Delta W_k \leq 4 \cdot (m+q)$$

That concludes the proof.

Notice that, in the incremental case, we obtain the same time bound of the off-line algorithm, when performed on a set of m constraints.

With the above result we are able to state the following theorem.

Theorem 6. *The amortized cost for maintaining a non-oblivious 1-bounded local optimal solution of a* MAX 2-CCSP *problem per update is $O(m)$ over an arbitrary sequence of $\Omega(m)$ operations.*

Then, in view of Theorem 2, the above results on fully dynamic maintenance of non-oblivious 1-bounded local optima can be used for designing a fully dynamic approximation algorithm for these problems.

Theorem 7. *Let C be a* MAX 2-CCSP *problem. The total cost for maintaining a 5/2-approximate solution for C while an arbitrary sequence of q insert and delete operations is performed is $O(m^2 + m \cdot q)$, where m is the maximum number of constraints in C. Moreover, the amortized cost per update is $O(m)$ over an arbitrary sequence of $\Omega(m)$ operations.*

Acknowledgments: We would like to thank Giorgio Ausiello for his helpful comments and suggestions. Also many thanks to Daniele Frigioni for a careful reading of earlier drafts.

References

1. P. Alimonti, Non-Oblivious Local Search for Graph and Hypergraph Coloring Problems, *21st International Workshop on Graph-Theoretic Concepts in Computer Science*, Lecture Notes in Computer Science 1017, Springer Verlag, 167-180, 1995.
2. P. Alimonti, New Local Search Approximation Techniques for Maximum Generalized Satisfiability Problems, *Information Processing Letters* 57, 151-158, 1996.
3. P. Alimonti, and R. Ferroni, Algorithms for the Maximum Generalized Satisfiability Problem, *Rapporto Tecnico*, RAP 01.93, Dipartimento di Informatica e Sistemistica, Università degli Studi di Roma "la Sapienza", 1993.
4. P. Alimonti, S. Leonardi, A. Marchetti Spaccamela, Average Case Analysis of Fully Dynamic Connectivity for Directed Graphs, *RAIRO Journal on Theoretical Informatics and Applications* 30, 4, 305-318, 1996.
5. G. Ausiello, and G.F. Italiano, On-Line Algorithms for Polynomially Solvable Satisfiability Problems, *Journal of Logic Programming*, 10, 69-90, 1991.
6. G.Ausiello, G.F.Italiano, A.Marchetti-Spaccamela, U.Nanni, Incremental algorithms for minimal length paths, *J. of Algorithms*, 12 , 615-638, 1991.
7. G. Ausiello, and M. Protasi, Local Search, Reducibility and Approximability of NP Optimization Problems, *Information Processing Letters* 54, 73–79, 1995.
8. R.Battiti, M. Protasi, Reactive local search for the maximum clique problem. *Tech. Rep. TR-95-052*, International Computer Science Institute, Berkeley, CA, 1995.
9. D.Eppstein, Z.Galil, G.F.Italiano, A.Nissenzweig, Sparsification - a technique for speeding up dynamic graph algorithms, *Proc. 33rd Annual Symp. on Foundations of Computer Science*, 1992.
10. U. Feige, M. Goemans, Approximating the value of the two prover proof system with applications to MAX 2SAT and MAX DICUT, *Proceedings of the 3rd Israeli Symposium on Theory of Computing and Systems*, 182-189, 1995.
11. S.T. Fischer, The solution Sets of Local Search Problems, *PhD Thesis*, Department of Computer Science, University of Amsterdam, Amsterdam, 1995.
12. G.N. Frederickson, Data Structure for On-Line Updating of Minimum Spanning Tree with Applications, *SIAM J. Comput.*, 14, 781-798, 1985.
13. F. Glover, Tabu search, Part I, *ORSA Journal of Computing*, 1, 190-206, 1989.
14. M. Goemans, and D.P. Williamson, .878-Approximation Algorithms for MAX CUT and MAX 2SAT, *Proc. of the 35th Annual IEEE Conference on Foundations of Computer Science*, 1994.
15. P. Klein, H. Lu,i Efficient Approximation Algorithms for Semidefinite Programming Arising from MAX CUT and COLORING, *Proc. 28th ACM Symposium on Theory of Computinga*, 1996, 1996.
16. Z. Ivkovic, E.L.Lloyd, Fully Dynamic Maintenance of Vertex Cover, *Proc. of the 19th International Workshop on Graph-Theoretic Concept in Computer Science*, LNCS 790, 99-111, 1993.
17. D.S. Johnson, C.H. Papadimitriou, and M. Yannakakis, How Easy Is Local Search?, *Journal of Computer and System Sciences*, 37, 79-100, 1988.
18. S. Khanna, R. Motwani, M. Sudan, and U. Vazirani, On Syntactic versus Computational Views of Approximability, *Proc. of the 35th Annual IEEE Conference on Foundations of Computer Science*, 1994.
19. S. Kirkpatrick, C. Gelat, and M. Vecchi, Optimization by simulated annealing, *Science*, 220, 671-680, 1983.

20. J.A.La Poutré, J.van Leeuwen, Maintenance of transitive closure and transitive reduction of graphs, *Proc Work. on Graph Theoretic concepts in Comp. Sci.*, LNCS 314, Springer Verlag, Berlin, 106-120, 1988.
21. C. Papadimitriou, and M. Yannakakis, Optimization, Approximation, and Complexity Classes, *Journal of Computer and System Sciences*, 43, 425-440, 1991.
22. R.E.Tarjan, Amortized Computational Complexity, *SIAM J.Alg.Disc.Math.*, 6,306-318, 1985.
23. L.Trevisan, Positive linear programming, parallel approximation and PCP's, *Proceedings 4th Annual European Symposium on Algorithms*, 1996.

On the Number of Simple Cycles in Planar Graphs

Helmut Alt, Ulrich Fuchs, and Klaus Kriegel

Institut für Informatik, Freie Universität Berlin, Takustr. 9, D-14195 Berlin,
E-mail: *name*@inf.fu-berlin.de

Abstract. Let $C(G)$ denote the number of simple cycles of a graph G and let $C(n)$ be the maximum of $C(G)$ over all planar graphs with n nodes. We present a lower bound on $C(n)$ constructing graphs with at least 2.27^n cycles. Applying some probabilistic arguments we prove an upper bound of 3.37^n.
We also discuss this question restricted to the subclasses of grid graphs, bipartite graphs, and of 3–colorable triangulated graphs.

1 The Problem

The question addressed in this note came up in connection with problems in the theory of VLSI–layout and the computation of rectilinear Steiner trees [4], but is also a very natural one in graph theory:

> ***How many different simple cycles can there be in a planar graph with n nodes?***

Let us call this number $C(n)$. A very simple upper bound can be obtained by the following observation: By Eulers formula, a planar embedding of a graph with n nodes can have at most $2n - 4$ faces. Any simple cycle encloses a subset of these faces and is identified by this subset. Consequently,

$$C(n) \leq 2^{2n-4} = \frac{1}{16} \cdot 4^n \tag{1}$$

Observe that this upper bound was established making use of the dual version of our problem: *How many simply connected regions can be obtained by unifying countries in a planar map?* A special case of this question was studied by Blum and Hewitt [1]. Their aim was to estimate the number of simply connected patterns in an $k \times k$-grid. Thus, the construction in [1] proved a first lower bound on $C(n)$.

In particular the following set of simple cycles is constructed (see Figure 1): The leftmost column of the $k \times k$-grid belongs to every cycle, then in rows 1 and 2 we walk to the right and for each step rightwards we have the choice between row 1 and 2. We insert vertical steps between the horizontal ones, if necessary. In the same way we use rows 3 and 4 to walk back to the left. This procedure continues until, in a snakelike fashion, we have traversed the whole

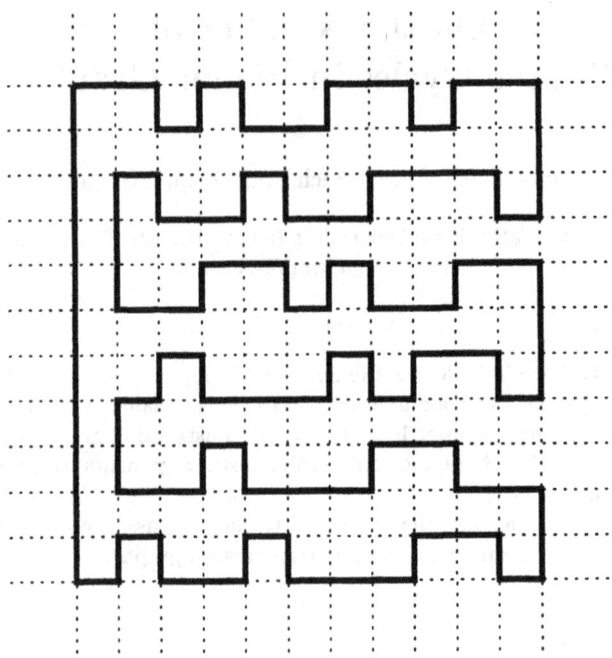

Fig. 1. Blum and Hewitt's construction

grid and are back at the lower end of the leftmost column. For any two rows, we have $k-2$ choices, so altogether there are $2^{k/2\cdot(k-2)}$ different cycles of this kind. Consequently, since the graph has $n = k^2$ nodes,

$$C(n) \geq \sqrt{2}^{\,n-2\sqrt{n}}$$

for infinitely many $n \in \mathbb{N}$.

This lower bound can be improved in an obvious way: Triangulating all tiles of the grid by a diagonal one has three choices for any step: row 1, row 2 or the diagonal. Thus we get

$$C(n) \geq \sqrt{3}^{\,n-2\sqrt{n}} \tag{2}$$

for infinitely many $n \in \mathbb{N}$.

So $C(n)$ is exponential in n and by (1) and (2) the basis of the exponent lies somewhere between $\sqrt{3}$ and 4. In this paper we will close this gap to some extend but not completely. In a recent paper Ding [2] gave a minor based characterization of graph classes, such that the number of cycles is polynominal. In fact, one can observe that our exponetial lower bound examples correspond to one of the forbidden minor types in a very natural way.

2 Lower Bounds

In order to get a better lower bound on $C(n)$ we first show the following lemma:

Lemma 1. *Let G be a planar graph with k nodes and a, b be two nodes of G incident to the outer face for some planar embedding of G. Let s be the number of simple paths between a and b. Then*

$$C(n) \geq \left(\sqrt[k-1]{s}\right)^n$$

for infinitely many $n \in \mathbb{N}$.

Proof. Consider the planar embedding where a, b are incident to the outer face. Then for any $t \in \mathbb{N}, t \geq 3$ we can construct a graph G_t with $n = t(k-1)$ nodes, as shown in Figure 2.

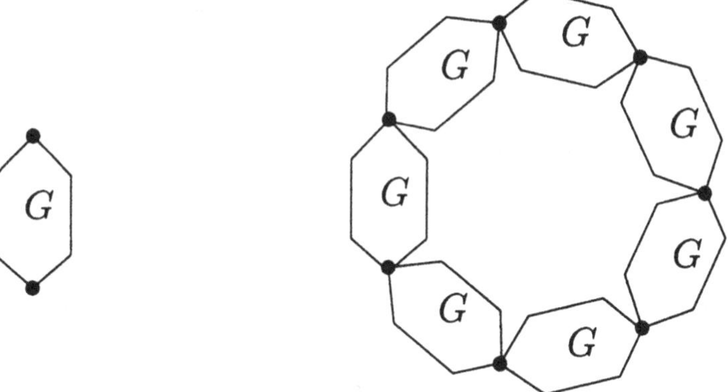

Fig. 2. The Graph G_t

Since there are s simple paths from a to b, there are at least s^t simple cycles within G_t. Consequently,

$$C(n) \geq s^t = s^{n/(k-1)} = \left(\sqrt[k-1]{s}\right)^n \qquad \square$$

Using Lemma 1 we obtain nontrivial lower bounds already for very simple graphs G, in fact the previous bound $\sqrt{2}$ is achieved already when G is a triangle, see Figure 3. The best lower bound on $C(n)$ we have, is obtained by applying Lemma 1 to the graph shown in Figure 4. Observe, that it is sufficient for a and b to be incident to some face, since it can be turned into the outer face by a homeomorphism. It also makes sense to triangulate the graph, since adding edges can only increase the lower bound. So without loss of generality a, b are the endpoints of an edge. For the graph G of Figure 4 we determined the number of simple paths between endpoints of edges by computer and found the maximum

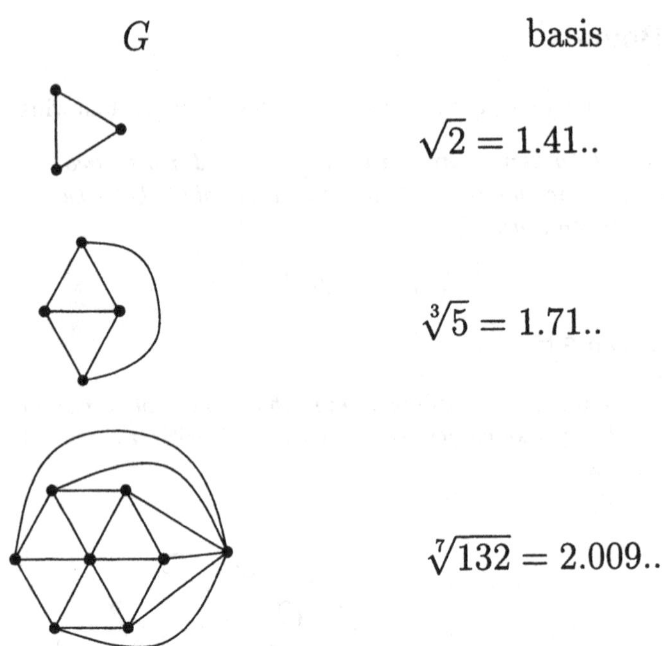

Fig. 3. Lower bounds obtained by simple graphs

$s = 4{,}359{,}234{,}132$ between the vertices a, b indicated in the picture. For the basis of the exponent we obtain $\sqrt[27]{s} = 2.275\ldots$, so we have as a lower bound for $C(n)$:

Theorem 2. *$C(n) \geq \alpha^n$, where $\alpha = \sqrt[27]{4{,}359{,}234{,}132} = 2.275\ldots$ for infinitely many $n \in \mathbb{N}$.* □

Our computer experiments indicated that the value of α can probably be increased by considering larger and larger graphs of the form of the graph G in Figure 4. We stopped with G, because of the excessive computation time. For G we needed 11 hours on a SPARC-Station-10/512 computer to determine s.

The method to obtain non-trivial upper bounds, that will be presented in the next section, gives better results for 3-colorable graphs. Hence we also had a look on the lower bounds for this class of graphs. A triangulated graph is 3–colorable iff the degree of each node is even. This was already mentioned in [5], proofs and more details can be found in [6] and [7]. We counted (36 hours of computation time) in a graph with 29 nodes, all of degree four or six, 8,133,124,602 paths between two adjacent vertices. Observe, that if G has only nodes of even degree and if t is three times a power of 2 the Graph G_t can be triangulated such that all nodes have even degree. Evaluating $\sqrt[28]{8{,}133{,}124{,}602}$ we get:

Theorem 3. *For infinitely many natural numbers n there is a 3-colorable planar graph with $\Omega(2.259^n)$ simple cycles.* □

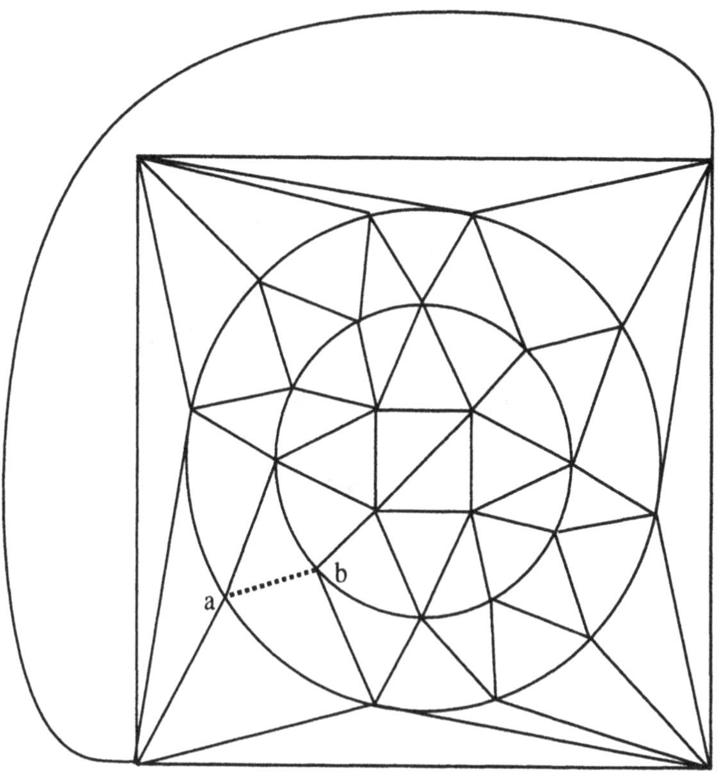

Fig. 4. Graph G yielding a bound of 2.275 ...

We also tested some bipartite i.e. 2–colorable graphs. For a graph with 36 nodes (more or less a 6×6-grid) we found 38,855,881 paths between two adjacent ones. For even t G_t remains bipartite and hence we get:

Theorem 4. *For infinitely many natural numbers n there is a bipartite planar graph with* $\Omega(1.647^n)$ *simple cycles.* □

3 Upper bounds

In order to get a better upper bound than $O(4^n)$ for $C(n)$ we use a probabilistic argument. Suppose we have some planar embedding for a graph $G = (V, E)$ and we color the faces randomly under uniform distribution with 2 colors ('black' and 'white'). If we are 'lucky' the total black area is simply connected, i.e. it has no holes and its interior is connected. Then the boundary of this area is a simple cycle (see Figure 5). Let us call such a coloring 'valid'. We will determine an upper bound on the probability of this event. Let us again assume that G is triangulated, clearly any upper bound on $C(n)$ under this assumption holds for

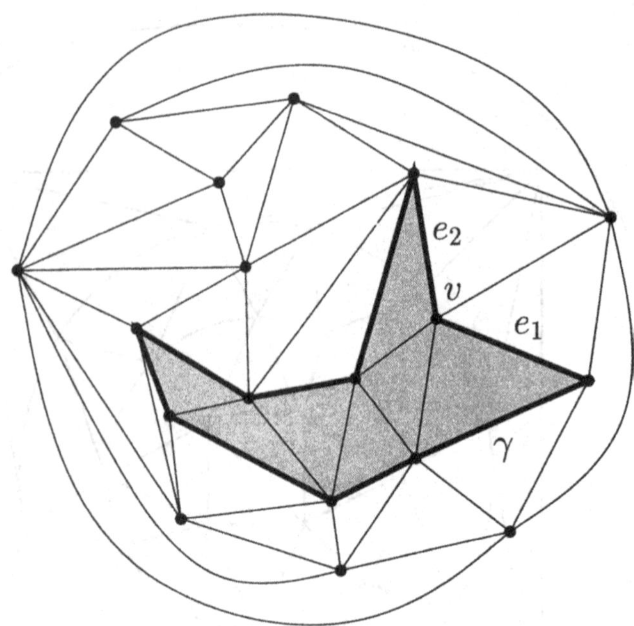

Fig. 5. The total black area is simply connected and identifies a simple cycle γ

general planar graphs as well. Consider some node v of G and inspect its incident faces in clockwise order (see Figure 5). Assume that there are both, black and white faces incident to v. In the 'valid' case it is necessary that the black faces start at some edge e_1, form a connected sequence and end at some edge e_2. Let us call v 'good' if this is the case or if all faces incident to v have the same color. The number of possible colorings of the incident faces making some node v of degree d good is

$$2 + d(d-1),$$

namely, coloring all faces black or all white or selecting edges e_1 and e_2 as described before and coloring black exactly the faces between e_1 and e_2. Therefore, the probability that v is good is

$$\frac{d(d-1)+2}{2^d}.$$

An upper bound for the probability that a coloring is valid, is the probability p_{good} that all nodes are good, since the latter condition is necessary for the former. If the goodness of all nodes were independent events we would obtain this probability by the product

$$\prod_{v \in V} \frac{d(v)(d(v)-1)+2}{2^{d(v)}} \tag{3}$$

where $d(v)$ is the degree of a node v. The denominator of this product equals

$$2^{\sum_{v\in V} d(v)} = 2^{6n-12},$$

since $\sum_{v\in v} d(v)$ equals twice the number of edges. The enumerator in (3) is maximal under the constraint $\sum_{v\in V} d(v) = 6n - 12$ if all $d(v)$ are equal. To see this, consider the function $f(d) = d(d-1)+2$. Straightforward calculations yield for $0 < \varepsilon < d$

$$\begin{aligned} f(d+\varepsilon)f(d-\varepsilon) &= (f(d))^2 + \varepsilon^2(\varepsilon^2 + 3 - 2d^2 + 2d) \\ &< (f(d))^2 + \varepsilon^2(-d^2 + 2d - 1 + 4) \\ &= (f(d))^2 + \varepsilon^2(4 - (d-1)^2) \\ &\le (f(d))^2, \end{aligned}$$

since G is triangulated and therefore $d \ge 3$. So, whenever two $d(v)$ in (3) differ by some amount 2ε we can increase the value by increasing one by ε and decreasing the other by ε, thus, making them equal. Consequently, the expression in (3) is maximized by letting $d(v) = \frac{6n-12}{n} \le 6$, so we have

$$\prod_{v\in V} \frac{d(v)(d(v)-1)+2}{2^{d(v)}} \le 2^{12}\left(\frac{6\cdot 5+2}{2^6}\right)^n = 2^{12}\left(\frac{1}{2}\right)^n. \tag{4}$$

However, this is not an upper bound on the probability of obtaining a valid coloring, since the events of the nodes being good are not independent. Let A_u denote the event that the vertex v is good for a random coloring of the faces, then it is not hard to check that the events A_u and A_v are not independent for adjacent vertices u and v. However, if we choose an independent set of vertices V, then the events $A_v(v \in V)$ are mutually independent because the sets of faces surrounding the vertices of V are disjoint. Note that this is only true for triangulated planar graphs. By the four–color–theorem the nodes of G can be partitioned into 4 independent sets $V_1, \ldots, V_4$. Let V_i be that one which minimizes

$$\prod_{v\in V_i} \frac{d(v)(d(v)-1)+2}{2^{d(v)}}.$$

So by (4) we have

$$p_{\text{good}} \le \prod_{v\in V_i} \frac{d(v)(d(v)-1)+2}{2^{d(v)}} \le \sqrt[4]{2^{12}\left(\frac{1}{2}\right)^n} = 8\left(\sqrt[4]{\frac{1}{2}}\right)^n.$$

Since there are $2n-4$ faces, there are 2^{2n-4} colorings of the faces. This implies the following upper bound on the number of cycles.

Theorem 5.

$$C(n) \le p_{\text{good}} \cdot \frac{1}{16} \cdot 4^n \le \frac{1}{2}\left(\sqrt[4]{1/2}\cdot 4\right)^n$$

and thus the basis is at most $\sqrt[4]{1/2} \cdot 4 = 3.363\ldots$.

The general upper bound can be improved for 3–colorable triangulated planar graphs in an obvious way. Since there are only three color classes we will find one such that

$$\prod_{v \in V_i} \frac{d(v)(d(v)-1)+2}{2^{d(v)}} \leq \sqrt[3]{2^{12}\left(\frac{1}{2}\right)}^n = 16 \cdot \sqrt[3]{1/2}^n .$$

Corollary 6. *The number of simple cycles in a 3–colorable triangulated planar graph is bounded by*

$$16\sqrt[3]{1/2}^n \cdot 2^{2n-4} = \left(\sqrt[3]{1/2} \cdot 4\right)^n \leq 3.175^n. \qquad \square$$

Now one could expect a further improvement of the upper bound for 2–colorable (i. e. bipartite) graphs. In fact we will obtain a much better bound, but we cannot argue as before, for the following reason: A bipartite graph is not triangulated and therefore the events A_v are no longer independent for vertices of the same color. On the other hand bipartite graphs have at most $n-2$ faces and thus the number of cycles can be at most 2^{n-2}. Hence, our aim is to improve this bound by probabilistic arguments.

Analogously to the general case we may assume that all faces of the graph G are cycles of length four. Then by Euler's formula we have n vertices, $2n-4$ edges and $n-2$ faces. Again, let A_v denote the event that a random coloring of the faces is good at v. Multiplying the probabilities of all events A_v we get

$$\prod_{v \in V} \frac{2+d(v)(d(v)-1)}{2^{d(v)}} \leq \frac{2+4\cdot 3}{2^{4n-8}} = 2^8 \cdot \left(\frac{7}{8}\right)^n .$$

Now we want to partition V in such a way that each corresponding subproduct represents mutually independent events A_v. Let $V = V_a \cup V_b$ be the partition of V given by the bipartiteness of G. We define auxiliary graphs G_a (G_b) with vertex sets V_a (V_b) and edge sets E_a (E_b) consisting of all face diagonals between vertices from V_a (V_b). Both graphs are planar. So we can color V_a with the colors $\{1, 2, 3, 4\}$ and V_b with the colors $\{5, 6, 7, 8\}$. It follows that for two vertices v, u of the same color the sets of the faces surrounding them are disjoint and thus the events A_v and A_u are independent. Consequently, there is a color class V_i such that

$$\prod_{v \in V} \frac{2+d(v)(d(v)-1)}{2^{d(v)}} \leq \sqrt[8]{2^8 \cdot \left(\frac{7}{8}\right)}^n = 2 \cdot \sqrt[8]{\frac{7}{8}}^n .$$

Taking into account that there are 2^{n-2} face colorings we get the following

Theorem 7. *The number of simple cycles in a bipartite planar graph is bounded by*

$$2 \cdot \sqrt[8]{\frac{7}{8}}^n \cdot 2^{n-2} \leq \left(\sqrt[8]{\frac{7}{8}} \cdot 2\right)^n \leq 1.967^n. \qquad \square$$

For grid graphs one can get a slightly better bound because in this case G_a and G_b are bipartite, which implies a partition of V into four color classes.

Corollary 8. *The number of cycles in an $m \times k$-grid($n = m \cdot k$) is bounded by*

$$4 \cdot \left(\sqrt[4]{\frac{7}{8}}\right)^n \cdot 2^{n-(m+k-1)} < \sqrt[4]{14}^{\,n} \leq 1.935^n. \qquad \square$$

4 Conclusions

The following table summarizes the lower and upper bounds on the maximal number $C(n)$ of cycles in planar graphs on n vertices:

graph class	lower bound	upper bound
general planar	2.275^n	3.364^n
3–colorable, triangulated	2.259^n	3.175^n
bipartite	1.647^n	1.967^n
grid graphs	1.414^n	1.935^n

The remaining challenge in this context is of course to further close the gaps between the lower and upper bounds. As already mentioned one can obtain minor improvements of the lower bounds counting the number of ab–paths for larger graphs. So in the meanwhile we could establish a 2.281^n lower bound based on a graph with 29 nodes. A similar lower bound of 2.279^n has been proved recently by Chrobak and Eppstein [3]. Their method does not require an excessive path counting but instead they have to compute the eigenvalues of a 6×6–matrix.

An other way to get minor improvements of the lower bounds consists in triangulating the inner and the outer face of the graph G_t constructed in the proof of Lemma 1. However, we could not improve the basis of the lower bound much more than by the factor of 1.001.

It would be interesting to prove an upper bound for not necessarily triangulated 3–colorable graphs which is better than the general one. Note that the additional assumption that the graphs have to be triangulated was necessary to get 3 sets of mutually independent events. On the other hand there are 3–colorable graphs for which no 3–colorable triangulations exist.

There is a natural approach to improve the upper bounds. In fact our previous upper bounds estimate the number of face colorings which are good in every vertex. In the dual version this means that we have bounds on the number of families of disjoint cycles in a graph. So one could try to estimate the probability that a random family of disjoint cycles in a graph consists of exactly one cycle. Consequently one improves the previous upper bound by multiplying it with the probability. So far we did not succeed in this way.

Finally we remark that it could be helpful to have an upper bound on the number of Hamiltonian cycles in planar graphs. Indeed, a bound of the form α^n

would imply $C(n) \leq (\alpha+1)^n$ applying the bound to each induced subgraph with k vertices $k = 1, \ldots, n$ and making use of the binomial formula.

Acknowledgement

We would like to thank Frank Hoffmann for several valuable hints concerning this research.

References

1. M. Blum, C. Hewitt, *Automata on a 2-dimensional tape*, 8th IEEE Conf. on SWAT, 1967, 155–160.
2. G. Ding, *Bounding the Number of Circuits of a Graph*, Combinatorica 16(3), 1996, 331–341.
3. D. Eppstein, *Pesonal Communications*, 1997
4. U. Fössmeier, M. Kaufmann, *On Exact Solutions for the Rectilinear Steiner Problem, Part I: Theoretical Results*, Technical Report, Computer Science Institute, Universität Tübingen, 1996.
5. P. J. Heawood, *Map Color Theorem*, Quarterly Journal of Pure and Applied Mathematics 24, 1890, 332–338.
6. T. R. Jensen, B. Toft, *Graph Coloring Problems*, Wiley, New York, 1995.
7. R. Steinberg, *The State of the Three Color Problem*, Annals of Discrete Mathematics 55, 1993, 211–248.

On the Separable-Homogeneous Decomposition of Graphs

Extended Abstract

Luitpold Babel[1]* and Stephan Olariu[2]**

[1] Institut für Mathematik, Technische Universität München, D-80290 München, Germany
[2] Department of Computer Science, Old Dominion University, Norfolk, VA 23529, U.S.A.

Abstract. We introduce a new decomposition scheme for arbitrary graphs which extends both the well-known modular and the homogeneous decomposition. It is based on a previously known structure theorem which decomposes a graph into its P_4-connected components and on a new decomposition theorem for P_4-connected graphs. As a final result we obtain a tree representation for arbitrary graphs which is unique up to isomorphism.

1 Introduction

It is a time-honored paradigm to model problems in communications, VLSI design, database design, network protocol design, and other areas of computer science and engineering, by graphs in the hope that the resulting graph problems can be solved fast. A powerful tool for obtaining efficient solutions to graph problems is *divide-and-conquer*, one of whose incarnations is *graph decomposition.*

In turn, an increasingly appealing approach to graph decomposition involves associating with the graph at hand G a rooted tree $T(G)$ whose leaves are subgraphs of G (e.g. vertices, edges, cliques, stable sets, cutsets) and whose internal nodes correspond to certain prescribed graph operations. In applications, it is most desirable that the corresponding tree representation be *unique* and that it be obtained *efficiently*, in time polynomial in the size of the graph G.

Tree representations satisfying these conditions are important, in particular, to solve the graph isomorphism problem. If the leaves of $T(G)$ can be tested for isomorphism in polynomial time, then the graph isomorphism problem can be solved efficiently for G since it reduces to labeled tree isomorphism. Unique tree representations have been obtained for several classes of graphs, among others for cographs [4], hook-up graphs [10], transitive series-parallel digraphs [11], interval

* This author was supported by the Deutsche Forschungsgemeinschaft (DFG)
** This author was supported in part by NSF grant CCR-95-22093 and by ONR grant N00014-97-1-0526

graphs [3], P_4-reducible graphs [6], P_4-extendible graphs [7], P_4-sparse graphs [8] and (q,t)-graphs [1].

The *modular decomposition*, also known as *substitution* decomposition, is a well-investigated type of graph decomposition which has applications in many areas of discrete mathematics. It allows to solve efficiently a wide class of combinatorial optimization problems. For a comprehensive review see [14] and [15]. Very recently, the modular decomposition has been extended to the *homogeneous decomposition* [9]. The main contribution of this paper is to introduce a new type of decomposition which extends the modular and the homogeneous decomposition in the sense that it goes further in decomposing graphs which are prime with respect to both decompositions.

Our decomposition scheme is based, on the one hand, on a known structure theorem for arbitrary graphs [9] which decomposes a graph into its P_4-connected components and, on the other hand, on a new decomposition theorem for P_4-connected graphs. As with the modular and the homogeneous decomposition, we obtain a tree representation for arbitrary graphs which is unique up to isomorphism and which can be obtained in polynomial time.

The concept of P_4-connectedness has been introduced by B. Jamison and S. Olariu in [9] as a generalization of the usual connectedness of graphs. By means of the above-mentioned structure theorem, it suggests a unique tree representation for arbitrary graphs. The leaves of this tree are the P_4-connected components of the graph. A detailed study of the structure of P_4-connected graphs will allow us to further decompose the leaves and to establish a unique tree representation for the P_4-connected components.

The paper is organized as follows. Section 2 presents the terminology and summarizes previous results about P_4-connectedness. In Section 3 we introduce an extension procedure that, starting with a P_4 in an arbitrary graph, tries to add one vertex after the other in such a way that the graph is P_4-connected in each step. We characterize the graphs for which this procedure can be applied successfully. In Section 4 we investigate the concept of separable-homogeneous sets, which is one of the main ingredients in our decomposition scheme, and present the decomposition theorem for P_4-connected graphs. Section 5 describes the separable-homogeneous decomposition along with the associated tree representation.

2 Terminology and previous results

All graphs in this paper are finite, with no loops nor multiple edges. In addition to standard graph-theoretical terminology, compatible with [2], we need some new terms that we are about to define.

Let $G = (V, E)$ be a graph with vertex-set V and edge-set E. For a vertex v of G, $N(v)$ denotes the set of all neighbors of v. If $U \subseteq V$ then $G(U)$ stands for the graph induced by U. Occasionally, to simplify the exposition, we shall blur the distinction between sets of vertices and the subgraphs they induce, using the same notation for both. $E(U)$ denotes the set of all edges joining vertices from

U. The complement of G is denoted by $\overline{G}$. A *clique* is a set of pairwise adjacent vertices, a *stable set* is a set of pairwise nonadjacent vertices. G is termed a *split* graph if its vertices can be partitioned into a clique and a stable set.

A vertex v is said to be *totally adjacent* to U if all vertices of U are neighbors of v. If v has no neighbors in U then v is *nonadjacent* to U. A subset H of V with $1 < |H| < |V|$ is termed a *homogeneous set* if each vertex outside H is either totally adjacent or nonadjacent to H. A homogeneous set H is *maximal* if no other homogeneous set properly contains H. The graph obtained from G by shrinking every maximal homogeneous set to one single vertex is called the *characteristic graph* of G.

As usual, we let P_k stand for the chordless path with k vertices and $k-1$ edges. In a P_4 $uvwx$ consisting of vertices u, v, w, x and edges uv, vw, wx we refer to v and w as the *midpoints* whereas u and x are called the *endpoints*.

Adapting the terminology of [9], a graph $G = (V, E)$ is *P_4-connected*, or *p-connected* for short, if for every partition of V into nonempty disjoint sets V_1 and V_2 there exists a *crossing* P_4, that is, a P_4 containing vertices from both V_1 and V_2. The *p-connected components* of a graph are the maximal induced subgraphs which are p-connected. Clearly, a p-connected component consists either of one single vertex or of at least four vertices. Vertices which are not contained in a p-connected component with at least four vertices are also called *weak vertices*. Note that the p-connected components of a graph G are closed under complementation and are connected subgraphs of G and $\overline{G}$. Moreover, it is easy to see that each graph has a unique partition into p-connected components and weak vertices.

A p-connected graph is termed *separable* if its vertex-set V can be partitioned into two nonempty disjoint sets V_1 and V_2 in such a way that each crossing P_4 has its midpoints in V_1 and its endpoints in V_2. We say that (V_1, V_2) is a *separation* of G. It is obvious that the complement of a separable p-connected graph is also separable. The partition (V_1, V_2) of G becomes (V_2, V_1) in $\overline{G}$. We recall some basic properties of separable p-connected graphs which are due to [9].

Theorem 2.1

(a) *Every separable p-connected graph G has a unique separation (V_1, V_2). Furthermore, every vertex of G belongs to a crossing P_4 with respect to (V_1, V_2).*
(b) *A p-connected graph is separable if and only if its characteristic graph is a split graph.* □

The introduction and study of separable p-connected graphs is justified by the following general structure theorem which has been found by B. Jamison and S. Olariu [9].

Theorem 2.2 *(Structure Theorem) For an arbitrary graph G exactly one of the following conditions is satisfied:*

(1) G is disconnected;
(2) $\overline{G}$ is disconnected;

(3) There is a unique proper separable p-connected component S of G with a partition (S_1, S_2) such that every vertex outside S is adjacent to all vertices in S_1 and to no vertex in S_2;

(4) G is p-connected. □

A graph $G = (V, E)$ is *minimally p-connected* if G is p-connected and, for all $v \in V$, $G - v$ is not p-connected. A graph is called a *spider* if its vertex-set V admits a partition into disjoint sets V_1 and V_2 such that the following two conditions are satisfied:

- $|V_1| = |V_2| \geq 2$, V_1 is a clique and V_2 is a stable set;
- there exists a bijection $f : V_2 \rightarrow V_1$ such that either

$$N(v) = \{f(v)\} \quad \text{for all vertices } v \in V_2 \quad \text{or}$$
$$N(v) = V_1 - \{f(v)\} \quad \text{for all vertices } v \in V_2.$$

The smallest spider is the P_4, which occasionally is referred to as the *trivial* spider. If G has more than four vertices then, if the first of the two alternatives holds, G is said to be a *thin spider*, otherwise a *thick spider*. Obviously, the complement of a thin spider is a thick spider and vice versa. Furthermore, it is an easy observation that spiders are separable p-connected graphs with separation (V_1, V_2). As pointed out in [1], spiders play a crucial role in the theory of p-connectedness:

Theorem 2.3 *A graph is a spider if and only if it is minimally p-connected.* □

3 Partner addition

Let X be the vertex-set of some P_4 in G. A vertex v outside X is said to have a *partner* in X if v together with three vertices from X induces a P_4 (occasionally, we shall say that v has a partner in a set U if v has a partner in a P_4 contained in U). It is straightforward to verify the following simple observation.

Observation 3.1 *v has a partner in X unless v is totally adjacent to X, nonadjacent to X or adjacent exactly to the midpoints of X.* □

Given a subset U of V, we denote by $T(U)$ and $I(U)$ the set of vertices in $V - U$ which are totally adjacent and nonadjacent to U, respectively. If U is separable p-connected with separation (U_1, U_2) then $M(U)$ denotes the set of vertices which are adjacent exactly to the vertices from U_1 (note that, by virtue of Theorem 2.1, the set $M(U)$ is well defined). Hence, a vertex v has a partner in X if and only if $v \notin T(X) \cup I(X) \cup M(X)$ holds. More generally, we have the following result.

Theorem 3.2 *Let $G = (V, E)$ be an arbitrary graph and let U be a proper subset of V such that $G(U)$ is p-connected. For every vertex v in $V - U$ the following statements are equivalent:*

(1) $G(U \cup \{v\})$ *is* p*-connected;*
(2) v *does not belong to* $T(U) \cup I(U)$ *and, if* $G(U)$ *is separable, also not to* $M(U)$*;*
(3) There is a set X *of vertices in* U *such that* X *induces a* P_4 *and* v *has a partner in* X*.*

Proof. (1) $\Rightarrow$ (2). Due to the assumption, $G(U)$ is p-connected. If v is totally adjacent or nonadjacent to U, then we are in case (1) or case (2) of the Structure Theorem which implies that $G(U \cup \{v\})$ is not p-connected. Similarly, if $G(U)$ is separable with separation (U_1, U_2) and if v is adjacent exactly to the vertices of U_1, then we are in case (3) and, again, $G(U \cup \{v\})$ is not p-connected.

(2) $\Rightarrow$ (3). Denote by W_1 the set of neighbors of v in U and let $W_2 := U - W_1$. Since $v \notin T(U) \cup I(U)$, v is adjacent but not totally adjacent to U, thus both sets W_1 and W_2 are nonempty. If U is separable then, since v does not belong to $M(U)$, the separation of U is different from (W_1, W_2). For that reason there is a crossing P_4, say with vertex-set X, between W_1 and W_2 which does not have both midpoints in W_1 and both endpoints in W_2, i.e. $v \notin T(X) \cup I(X) \cup M(X)$ holds. Now Observation 3.1 implies that v has a partner in X.

(3) $\Rightarrow$ (1). Let U_1, U_2 be an arbitrary partition of $U \cup \{v\}$ and assume, without loss of generality, that $v \in U_1$. If $U_1 - \{v\} \neq \emptyset$ then $U_1 - \{v\}, U_2$ is a partition of U and, since $G(U)$ is p-connected, there is a crossing P_4 between these sets. This P_4 is also a crossing P_4 for the partition U_1, U_2. Let $U_1 = \{v\}$. Due to the assumption, there is a P_4 in U with vertex-set X such that v together with three vertices from X induces a P_4. Obviously, this P_4 is crossing between U_1 and U_2. This shows that $G(U \cup \{v\})$ is p-connected. □

This result motivates to study the following extension procedure which starts with the vertex-set of a P_4 and adds a vertex whenever it has a partner in this set.

Procedure PARTNER ADDITION($G; X$)

Input: The vertex-set X of some P_4 in an arbitrary graph G.
Output: The vertex-set U of a p-connected subgraph of G.

```
begin
    Let U := X;
    while there is a vertex v ∈ V − U
    which has a partner in some P4 in G(U)
        do U := U ∪ {v};
end.
```

Theorem 3.2 immediately implies that the set U, which is generated by the above procedure, induces a p-connected graph. If $U \neq V$ then there is no vertex v outside U which has a partner in some P_4 in $G(U)$. This means that $G(U \cup \{v\})$

is not p-connected and each vertex $v \in V - U$ belongs either to one of the sets $T(U)$ and $I(U)$ or, if U is separable, to $M(U)$. If U is not separable or if $M(U)$ is empty, then U is homogeneous. Otherwise, we are in the situation where U is separable and $M(U)$ is nonempty. We shall call a set U with these properties a *separable-homogeneous set*[3]. In other words, U is separable-homogeneous if and only if U is separable p-connected and $V = U \cup T(U) \cup I(U) \cup M(U)$ holds with $M(U) \neq \emptyset$. Using this notation we obtain:

Lemma 3.3 *The procedure PARTNER ADDITION stops with a set U which induces a p-connected graph. If U is a proper subset of V then U is either homogeneous or separable-homogeneous.* □

We shall say that a P_4 with vertex-set X *extends* to U *by partner addition* if the above procedure, starting with X, provides U. Our next aim is to characterize the graphs which contain a P_4 that extends to the whole vertex-set V.

Clearly, a nontrivial spider does not contain a P_4 that extends to V. If G is a thin spider then the removal of all edges in the clique disconnects the graph leaving at least three connected components. This observation motivates the following definition. A graph G is *thin-spider-like* if

- G is separable p-connected with separation (V_1, V_2);
- the removal of all edges in V_1 disconnects the graph leaving at least three connected components.

A graph is *thick-spider-like* if its complement is thin-spider-like. The importance of spider-like graphs is exhibited in the following statement.

Theorem 3.4 *A graph G contains a P_4 that extends to V by partner addition if and only if G is p-connected and G is not spider-like.*

Proof. See the full version of the paper. □

4 Decomposing p-connected graphs

Let G be an arbitrary graph and let S be a separable-homogeneous set in G. We say that G^* results from G by *shrinking* S to a P_4 if S is replaced by a P_4 in the obvious way, i.e. a vertex v in G^* is either totally adjacent, nonadjacent or adjacent to the midpoints of the P_4, according to whether v belongs to $T(S)$, $I(S)$ or $M(S)$ in G. It is an important feature that shrinking a homogeneous set to a single vertex or shrinking a separable-homogeneous set to a P_4 preserves p-connectedness of a graph.

Theorem 4.1 *Let G be a p-connected graph and let H and S be a homogeneous respectively a separable-homogeneous set in G. Then the following holds:*

[3] The concept of separable-homogeneous sets generalizes the concept of regular sets which was introduced in [12] in the context of P_4-sparse graphs

(a) The graph obtained from G by shrinking H to a single vertex is p-connected;
(b) The graph obtained from G by shrinking S to a P_4 is p-connected.

Proof. *(a)* Let G^* denote the graph obtained from G by shrinking H to a single vertex h^* and let V_1^*, V_2^* be an arbitrary partition of the vertex-set of G^*. We can assume, without loss of generality, that $h^* \in V_2^*$. Consider the partition V_1^*, $V_2^* - \{h^*\} \cup H$ of the vertices in G. Since G is p-connected there is a P_4, say with vertex-set X, which is crossing with respect to this partition.

If X has no vertices in H then V must be present in G^*, too. If X has vertices in H then it contains exactly one vertex from H. Since H is homogeneous, the set X^* obtained from X by removing the unique vertex in H and replacing it by h^* still induces a P_4 in G^*. This shows that there is a P_4 which is crossing between V_1^* and V_2^*, thus G^* is p-connected.

(b) Let G^* denote the graph obtained from G by shrinking S to a P_4 with vertex-set S^* and let again V_1^*, V_2^* be an arbitrary partition of the vertices of G^*. Clearly, if S^* is not completely included in one of the two sets of the partition, then S^* is a crossing P_4 and we are done. Therefore we can assume, without loss of generality, that $S^* \subseteq V_2^*$. Consider the partition V_1^*, $(V_2^* - S^*) \cup S$ of the vertices in G and let X denote the vertex-set of a crossing P_4.

It is easy to see that X has at most two vertices in S. If X has one vertex in S then the set X^* obtained by removing the unique vertex in S and replacing it by a midpoint from S^* still induces a P_4. If X has precisely two vertices in S then one of them belongs to S_1 and the other one to S_2 (with (S_1, S_2) denoting the separation of S). If these two vertices are adjacent then replace them in X by two adjacent endpoints and midpoints from S^*, if they are nonadjacent then replace them by two nonadjacent endpoints and midpoints from S^*. Again we obtain a crossing P_4 between V_1^* and V_2^*. This implies that G^* is p-connected. □

We shall now investigate properties of separable-homogeneous sets in arbitrary, not necessarily p-connected graphs.

Theorem 4.2 *Let G be an arbitrary graph and S, S' be separable-homogeneous sets in G with nonempty intersection such that no set contains the other. Then the following statements hold:*

(a) $S \cup S'$ induces a spider-like graph;
(b) If $S \cup S' \neq V$ then $S \cup S'$ is homogeneous or separable-homogeneous.

Proof. See the full version of the paper. □

The previous theorem elucidates the way in which the separable-homogeneous sets relate to each other in an arbitrary graph. These results can be extended to reveal the interaction of maximal separable-homogeneous sets in a p-connected graph.

Corollary 4.3 *Let G be a p-connected graph which contains no homogeneous set and which is not spider-like. Then any two distinct maximal separable-homogeneous sets are disjoint.*

Proof. Let S and S' be distinct maximal separable-homogeneous sets with $S \cap S' \neq \emptyset$. By Theorem 4.2 (a), the graph induced by $S \cup S'$ is spider-like. Since G is not a spider-like graph, $S \cup S' \neq V$ holds. By Theorem 4.2 (b), $S \cup S'$ is either homogeneous or separable-homogeneous. Since G is assumed to be without homogeneous sets, we conclude that $S \cup S'$ is separable-homogeneous. This, however, contradicts to the maximality of S and S'. □

We are now in a position to state a decomposition theorem for p-connected graphs which lays the foundation for the separable-homogeneous decomposition of general graphs. For this purpose, call a graph *prime* if it contains no homogeneous set and no proper separable-homogeneous set, i.e. no separable-homogeneous set with more than four vertices.

Theorem 4.4 *(Decomposition Theorem) Let G be a p-connected graph. Then exactly one of the following statements is satisfied:*

(1) G is thin-spider-like;
(2) G is thick-spider-like;
(3) There is a maximal prime p-connected subgraph Y of G and a unique partition P of V such that for each $U \in P$ either
  - *$|U| = 1$ and $|U \cap Y| = 1$ holds or*
  - *U is homogeneous and $|U \cap Y| = 1$ holds or*
  - *U is separable-homogeneous and $U \cap Y$ induces a P_4.*

Proof. Let G be a spider-like graph with separation (V_1, V_2). It is easy to realize that, if $G - E(V_1)$ has at least three connected components, then $\overline{G} - E(V_2)$ is connected. Thus, G is either thin-spider-like or thick-spider-like.

In order to prove the remainder of the statement we first observe that, in a p-connected graph, any two maximal homogeneous sets are disjoint. Let H and H' be distinct maximal homogeneous sets with nonempty intersection. Then either $H \cup H'$ is a homogeneous set or $H \cup H' = V$ holds. In the first case H and H' are not maximal, in the latter case it is easy to verify that G or $\overline{G}$ is disconnected, thus G not p-connected.

Next, we show that in an arbitrary graph any two homogeneous and separable-homogeneous sets do not overlap. Let H be a homogeneous and S a separable-homogeneous set such that no one contains the other. Assume that $W = H \cap S \neq \emptyset$. Since S induces a p-connected graph there is a crossing P_4 $uvwx$ between $S - W$ and W. Note that W is a homogeneous set in $G(S)$, therefore exactly one vertex of the P_4 belongs to W. We can assume, without loss of generality, that u belongs to W. Let further y be a vertex from $H - W$. Since v is adjacent to u and u belongs to H we conclude that v is totally adjacent to H and, in particular, adjacent to y. This implies that y belongs to $M(S) \cup T(S)$ and, therefore, y is also adjacent to w. This provides a contradiction since now w must be totally adjacent to H and, in particular, must be adjacent to u.

Now the conclusion follows immediately from Theorem 4.1, Corollary 4.3, and the above observations. □

5 The separable-homogeneous decomposition

The first step towards our decomposition scheme for arbitrary graphs is the *primeval decomposition* which has been introduced and investigated in [9]. Motivated by the Structure Theorem, it provides a unique tree representation for arbitrary graphs, where the leaves of the tree correspond to the p-connected components and weak vertices of the graph. In order to give an outline of the primeval decomposition we have to define a number of graph operations.

Let $G_1 = (V_1, E_1)$ and $G_2 = (V_2, E_2)$ be disjoint graphs. The *disjoint union* and the *disjoint sum* of G_1 and G_2 are the graphs which result, respectively, from the operations

- $G_1 \circledcirc G_2 = (V_1 \cup V_2, E_1 \cup E_2)$ and
- $G_1 ① G_2 = (V_1 \cup V_2, E_1 \cup E_2 \cup \{xy \,|\, x \in V_1, y \in V_2\})$.

Obviously, operations ⓪ and ① reflect the first two cases of the Structure Theorem. Let $G_1 = (V_1, E_1)$ be separable p-connected with separation (V_1^1, V_1^2) and $G_2 = (V_2, E_2)$ be an arbitrary graph disjoint from G_1. The third case of the Structure Theorem is reflected by the operation

- $G_1 ② G_2 = (V_1 \cup V_2, E_1 \cup E_2 \cup \{xy \,|\, x \in V_1^1, y \in V_2\})$.

As pointed out in [9], each graph can be obtained uniquely from its p-connected components and its weak vertices by a finite sequence of operations ⓪ , ① and ② . Furthermore, the Structure Theorem in a quite natural way suggests a tree representation for arbitrary graphs which is unique up to isomorphism. The tree associated with a graph G is called the *primeval tree* of G. The internal nodes of the tree are labeled by integers $i \in \{0, 1, 2\}$, where an i-node means that the graph which is associated to the subtree with this node as a root is obtained from the graphs corresponding to its children by an ⓘ operation. The leaves of the tree are the p-connected components of G.

The primeval decomposition lays the foundation of the *homogeneous decomposition* [9], which additionally involves the homogeneous sets of the graph. Given the primeval tree, it constructs a new tree representation by introducing a graph operation which, loosely speaking, replaces homogeneous sets by single vertices (this operation will also occur in our forthcoming decomposition). The homogeneous decomposition properly extends the *modular* or *substitution decomposition* [14], a well-investigated and extremely useful technique to decompose a graph G into certain subgraphs, called *modules*. A module M is a set of vertices in G which cannot be distinguished from vertices in $V - M$, i.e. each vertex outside M is either totally adjacent or nonadjacent to M. In particular, the graph itself and each single vertex is considered to be a module. In this sense, homogeneous sets are precisely the nontrivial modules of a graph. The result of the modular decomposition is a tree that describes the submodules of G.

By virtue of Theorem 4.4 we are now able to go substantially further and decompose graphs which are prime with respect to the modular and to the homogeneous decomposition. For this purpose, we now introduce several graph operations which are meant to explain the decomposition of spider-like graphs and which reflect the substitution of homogeneous and separable-homogeneous sets by single vertices and P_4s, respectively.

In a thin-spider-like graph G, the removal of the edges in the first set of the associated separation leaves at least three connected components. Let G_i, $i = 1, 2, \ldots, t$, denote the subgraphs of G which are induced by the vertex-sets of these components. Note that, since the characteristic graph of G is a split graph, the characteristic graph of each subgraph G_i is a split graph, too. The reverse of the decomposition of G into the subgraphs $G_1, \ldots, G_t$ is reflected by the following operation.

Let $G_i = (V_i, E_i)$, $i = 1, 2, \ldots, t$, denote disjoint graphs with $t \geq 3$ and let $V_i = V_i^1 \cup V_i^2$. The graph $G = (V, E)$ is said to arise from $G_1, \ldots, G_t$ by a ③ operation if

- $V = \bigcup_{i=1}^t V_i$ and
- $E = \bigcup_{i=1}^t E_i \cup \{xy \,|\, x \in V_i^1, y \in V_j^1, 1 \leq i < j \leq t\}$.

Similarly, $G = (V, E)$ arises from $G_1, \ldots, G_t$ by a ④ operation if

- $V = \bigcup_{i=1}^t V_i$ and
- $E = \bigcup_{i=1}^t E_i \cup \{xy \,|\, x \in V_i^2, y \in V_j, 1 \leq i < j \leq t\}$.

Clearly, a thin-spider-like graph results from a ③ operation applied to certain induced subgraphs, a thick-spider-like graph results from a ④ operation.

The reverse of shrinking homogeneous sets to single vertices and separable-homogeneous sets to P_4s is established by the following operation. Let $G_0 = (V_0 \cup \{y_1, y_2, \ldots, y_s\} \cup X_{s+1} \cup \ldots \cup X_t, E_0)$ be a graph such that each of the sets X_j induces a P_4. Let further $G_i = (V_i, E_i)$, $i = 1, \ldots, s$, be arbitrary and $G_j = (V_j, E_j)$ be separable p-connected graphs with separation (V_j^1, V_j^2), $j = s+1, \ldots, t$. The graph $G = (V, E)$ arises from $G_0, G_1, \ldots, G_t$ by means of a ⑤ operation if G is obtained by replacing every vertex y_i in G_0 by the graph G_i and each set X_j by the separable p-connected graph G_j in the obvious way, i.e.

- $V = \bigcup_{i=0}^t V_i$ and
- $E = \bigcup_{i=0}^t E_i - E' \cup E'' \cup E'''$,

where E' denotes the edges in G_0 which are incident to a vertex y_i or to a vertex from a set X_j, the set E'' arises by joining each vertex in V_i to every neighbor of y_i, and E''' arises by joining each vertex from V_j to every vertex which is totally adjacent to X_j and every vertex from V_j^1 to every vertex which is adjacent precisely to the midpoints of the P_4 induced by X_j.

It is now easy to verify that all graphs are constructible from certain atomic subgraphs by means of the operations defined above. More precisely, we state the following result (the proof is straightforward and therefore omitted).

Theorem 5.1 *Every graph can be obtained uniquely from prime p-connected subgraphs by a finite sequence of operations* ⓪ ,① ,...,⑤ . □

Theorems 2.2 and 4.4 in a quite natural way suggest a tree representation for arbitrary graphs which is unique up to isomorphism. The tree $T(G)$ belonging to a graph G will be called the *separable-homogeneous tree* of G. The internal nodes of $T(G)$ are labeled with integers $i \in \{0, 1, \ldots, 5\}$, where an i-node means that the subgraph associated with this node as a root is constructed from the subgraphs associated with its children by an ⓘ operation. The leaves of the tree are the prime p-connected subgraphs of G.

The following recursive procedure describes the formal construction of the separable-homogeneous tree of an arbitrary graph G.

Procedure BUILD SEPARABLE-HOMOGENEOUS TREE(G)

Input: An arbitrary graph $G = (V, E)$.
Output: The separable-homogeneous tree $T(G)$ associated with G.

```
begin
    if |V| = 1 then
            return the tree T having G as its unique vertex;
    else if G is disconnected then begin
            let G_1,...,G_t be the connected components of G;
            let T_1,...,T_t be the associated trees rooted at r_1,...,r_t;
            return the tree T(G) obtained by adding r_1,...,r_t
            as children of a 0-node;
            end
    else if \overline{G} is disconnected then begin
            let \overline{G}_1,...,\overline{G}_t be the connected components of \overline{G};
            let T_1,...,T_t be the associated trees rooted at r_1,...,r_t;
            return the tree T(G) obtained by adding r_1,...,r_t
            as children of a 1-node;
            end
    else if G is not p-connected then begin
            let G_1, G_2 be such that G = G_1 ② G_2;
            let T_1, T_2 be the associated trees rooted at r_1, r_2;
            return the tree T(G) obtained by adding r_1, r_2
            as children of a 2-node;
            end
    else if G is thin-spider-like then begin
            let G_1,...,G_t be the graphs induced by
            the connected components of G - E(V_1);
            let T_1,...,T_t be the associated trees rooted at r_1,...,r_t;
            return the tree T(G) obtained by adding r_1,...,r_t
            as children of a 3-node;
            {Comment: it is assumed that each vertex of G stores
            to which set of the separation (V_1, V_2) it belongs}
            end
```

```
    else if G is thick-spider-like then begin
            let G_1,...,G_t be the graphs induced by
            the connected components of \overline{G} - E(V_2);
            let T_1,...,T_t be the associated trees rooted at r_1,...,r_t;
            return the tree T(G) obtained by adding r_1,...,r_t
            as children of a 4-node;
            {Comment: it is assumed that each vertex of G stores
            to which set of the separation (V_1, V_2) it belongs}
            end
    else begin let Y be the maximal prime p-connected subgraph of G
            obtained by shrinking maximal homogeneous sets to single vertices
            and maximal separable-homogeneous sets to P_4s;
            let α be the tree having Y as its unique vertex;
            let H_1,...,H_s be the maximal homogeneous sets
            and S_{s+1},...,S_t the maximal separable-homogeneous sets of G;
            let T_1,...,T_t be the trees associated
            with H_1,...,H_s,S_{s+1},...S_t rooted at r_1,...,r_t;
            return the tree T(G) obtained by adding α,r_1,...,r_t
            as children of a 5-node;
            {Comment: it is assumed that the root r_i of the
            tree associated with H_i resp. S_i stores y_i resp. X_i}
            end
end.
```

With the previous considerations it is quite obvious that the separable-homogeneous tree $T(G)$ which is associated with a graph G is unique up to labeled tree isomorphism. It remains to show that the separable-homogeneous tree of an arbitrary graph can be constructed in polynomial time. Due to [9], the primeval tree can be obtained in polynomial time. Furthermore, it is well known that homogeneous sets can be found very efficiently [5, 13]. In order to check whether a p-connected graph is spider-like we shrink the maximal homogeneous sets to single vertices, test whether the resulting graph is a split graph and determine the number of connected components in the graph (or in the complement) after having removed the edges of the clique. Finally, in order to find all separable-homogeneous sets, we apply the procedure PARTNER ADDITION starting with each P_4 of the graph. Obviously, this all can be done in time polynomial in the size of the graph.

References

1. Babel L., Olariu, S.: On the isomorphism of graphs with few P_4s. Graph-Theoretic Concepts in Computer Science, 21th International Workshop, WG'95, Lecture Notes in Computer Science **1017** Springer, Berlin, 1995, 24–36
2. Bondy J.A., Murty, U.S.R.: Graph Theory with Applications. North-Holland, Amsterdam, 1976

3. Booth, K.S., Lueker, G.S.: A linear time algorithm for deciding interval graph isomorphism. Journal of the ACM **26** (1979) 183–195
4. Corneil, D.G., Lerchs, H., Stewart Burlingham, L.: Complement reducible graphs. Discrete Applied Mathematics **3** (1981) 163–174
5. Cournier, A., Habib, M.: A new linear time algorithm for modular decomposition. Trees in Algebra and Programming, Lecture Notes in Computer Science **787** Springer, 1994, 68–84
6. Jamison, B., Olariu, S.: P_4-reducible graphs, a class of uniquely tree representable graphs. Studies in Applied Mathematics **81** (1989) 79–87
7. Jamison, B., Olariu, S.: On a unique tree representation for P_4-extendible graphs. Discrete Applied Mathematics **34** (1991) 151–164
8. Jamison, B., Olariu, S.: A unique tree representation for P_4-sparse graphs. Discrete Applied Mathematics **35** (1992) 115–129
9. Jamison, B., Olariu, S.: p-components and the homogeneous decomposition of graphs. SIAM Journal of Discrete Mathematics **8** (1995) 448–463
10. Klawe, M.M., Corneil, D.G., Proskurowski, A.: Isomorphism testing in hook-up graphs. SIAM Journal on Algebraic and Discrete Methods **3** (1982) 260–274
11. Lawler, E.L.: Graphical algorithms and their complexity. Math. Center Tracts **81** (1976) 3–32
12. Lin, R., Olariu, S.: A fast parallel algorithm to recognize P_4-sparse graphs. Submitted
13. McConnell, R., Spinrad, J.: Linear-time modular decomposition and efficient transitive orientation of comparability graphs. Fifth Annual ACM-SIAM Symposium of Discrete Algorithms, 1994, 536–545
14. Möhring, R.H.: Algorithmic aspects of comparability graphs and interval graphs. Graphs and Orders, Dordrecht, Holland, 1985
15. Möhring, R.H., Rademacher, F.J.: Substitution decomposition and connections with combinatorial optimization. Annals of Discrete Mathematics **19** (1984) 257–356

Pseudo-Hamiltonian Graphs

Luitpold Babel[1]* and Gerhard J. Woeginger[2]**

[1] Institut für Mathematik, TU München, D-80290 München, Germany
[2] Institut für Mathematik B, TU Graz, Steyrergasse 30, A-8010 Graz, Austria

Abstract. A pseudo-h-hamiltonian cycle in a graph is a closed walk that visits every vertex exactly h times. We present a variety of combinatorial and algorithmic results on pseudo-h-hamiltonian cycles: First, we show that deciding whether a graph is pseudo-h-hamiltonian is NP-complete for any given $h \geq 1$. Surprisingly, deciding whether there exists an $h \geq 1$ such that the graph is pseudo-h-hamiltonian, can be done in polynomial time. We also present sufficient conditions for pseudo-h-hamiltonicity that are based on stable sets and on toughness. Moreover, we investigate the computational complexity of finding pseudo-h-hamiltonian cycles on special graph classes like bipartite graphs, split graphs, planar graphs, cocomparability graphs; in doing this, we establish a precise separating line between easy and difficult cases of this problem.

1 Introduction

For an integer $h \geq 1$, we shall say that an undirected graph $G = (V, E)$ is *pseudo-h-hamiltonian* if there exists a circular sequence of $h \cdot |V|$ vertices such that

- every vertex of G appears precisely h times in the sequence, and
- any two consecutive vertices in the sequence are adjacent in G.

A sequence with these properties will be termed a *pseudo-h-hamiltonian cycle*. In this sense, *pseudo-1-hamiltonian* corresponds to the standard notion *hamiltonian*, and a *pseudo-1-hamiltonian cycle* is just a *hamiltonian cycle*. The *pseudo-hamiltonicity number* $\mathrm{ph}(G)$ of the graph G, is the smallest integer $h \geq 1$ for which G is pseudo-h-hamiltonian; in case no such h exists, $\mathrm{ph}(G) = \infty$. A graph G with finite $\mathrm{ph}(G)$ is called *pseudo-hamiltonian*. Pseudo-h-hamiltonicity is a non-trivial graph property. E.g. for every $h \geq 2$, the graph G_h that results from glueing together h triangles at one of their vertices, is pseudo-h-hamiltonian but it is not pseudo-$(h-1)$-hamiltonian.

Results of this paper. The problem of deciding whether a given graph is hamiltonian is NP-complete. Hence, it is not surprising at all that for each fixed value of $h \geq 1$, the problem of deciding whether $\mathrm{ph}(G) \leq h$ holds for a given

* This author was supported by the Deutsche Forschungsgemeinschaft (DFG)
** This author acknowledges support by the Start-program Y43-MAT of the Austrian Ministery of Science

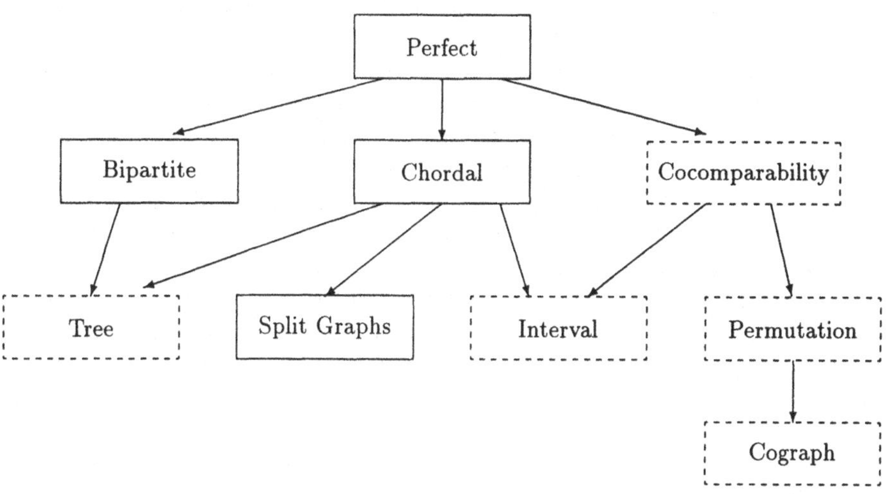

Fig. 1. Complexity results for some of the treated graph classes. NP-complete problems have a solid frame, polynomially solvable problems have a dashed frame.

graph G is also NP-complete. However, if we just ask whether $\mathrm{ph}(G) < \infty$, i.e. whether there exists some value of h for which G is pseudo-h-hamiltonian, then we can answer this question in polynomial time (and this is perhaps surprising). This polynomial time result is based on the close relationship of pseudo-hamiltonian graphs with *regularizable* graphs (cf. Section 2).

We also provide a nice and simple characterization of pseudo-hamiltonian graphs that is based on the stable sets of vertices of the graph. We show that every pseudo-hamiltonian graph G must be $1/\mathrm{ph}(G)$-tough, and that every 1-tough graph is pseudo-hamiltonian. We prove that the square of a connected graph is always pseudo-hamiltonian. For d-regular graphs with $d \geq 3$, we derive a tight result of the following form: There exists a threshold $\tau(d)$ such that for $h < \tau(d)$, it is NP-complete to decide whether a d-regular graph is pseudo-h-hamiltonian, whereas for every $h \geq \tau(d)$, a d-regular graph automatically is pseudo-h-hamiltonian. Hence, the computational complexity of deciding pseudo-h-hamiltonicity of regular graphs jumps at $\tau(d)$ from trivial immediately to NP-complete.

Finally, we will investigate the computational complexity of computing $\mathrm{ph}(G)$ on many well-known special graph classes, like bipartite graphs, split graphs, partial k-trees, interval graphs, planar graphs etc. Figure 1 summarizes some of our results together with some of their implications for special graph classes. Directed arcs represent containment of the lower graph class in the upper graph class. For classes with a solid frame, the computation of $\mathrm{ph}(G)$ is NP-complete, and for classes with a dashed frame, this problem is polynomial time solvable (for exact definitions of all these graph classes cf. Johnson [10]). Note that the results for trees, bipartite graphs, split graphs and cocomparability graphs imply all the other results in Figure 1.

Organization of the paper. Section 2 investigates the connections between pseudo-hamiltonicity and regularizable graphs, and it states several general complexity results. Section 3 relates pseudo-hamiltonicity to stable sets, to connectivity and to toughness. Section 4 derives the complexity threshold for d-regular graphs, and Section 5 deals with squares of graphs. Finally, Section 6 collects the complexity results for the special graph classes.

Notation and conventions. Throughout this paper, we only consider undirected graphs. All graphs have at least three vertices. For convenience we often write $G - W$ instead of $G(V - W)$, the graph that results from removing the vertices in W together with all incident edges from G. For a set $W \subseteq V$, we denote by $N(W)$ the set of all vertices outside W which are adjacent to vertices from W. A stable set is a set of pairwise non-adjacent vertices. A stable set S is maximal if there is no stable set S' which properly contains S. The stability number $\alpha(G)$ is the size of a largest stable set in G.

2 Complexity Aspects of Pseudo-Hamiltonicity

In this section, we give several characterizations of pseudo-hamiltonian graphs that are based on regularizable graphs. These characterizations imply that one can decide in polynomial time whether $\mathrm{ph}(G) < \infty$. On the other hand, we will show that for every fixed integer $h \geq 1$ it is NP-complete to decide whether $\mathrm{ph}(G) \leq h$.

A graph $G = (V, E)$ is called *regularizable* (see Berge [2, 3]), if for each edge $e \in E$ there is a positive integer $m(e)$ such that the multigraph which arises from G by replacing every edge e by $m(e)$ parallel edges is a regular graph. A useful characterization of regularizable graphs can be found in Berge [2].

Proposition 2.1 *(Berge [2])*
A connected graph $G = (V, E)$ is regularizable if and only if at least one of the following two statements holds

(a) G is elementary bipartite
(i.e. G is bipartite, connected and every edge of G appears in a perfect matching);

(b) G is 2-bicritical
(i.e. $|N(S)| > |S|$ holds for every stable set $S \subsetneq V$). □

Regularizable graphs are related to pseudo-hamiltonian graphs as follows.

Lemma 2.2 *A graph G is pseudo-hamiltonian if and only if G has a connected spanning regularizable subgraph.*

Proof. (Only if). Clearly, in a pseudo-h-hamiltonian cycle (considered as a multigraph) each vertex has degree $2h$. Hence, the *skeleton* of a pseudo-h-hamiltonian cycle (that is, the simple graph arising from replacing parallel edges by simple edges) of a graph G constitutes a regularizable subgraph of G which, additionally, is connected and contains all the vertices of G.

(If). Conversely, assume that a graph G has a connected spanning regularizable subgraph H. Let H^* denote the associated regular multigraph, say of degree $2h$ (if the degree of the regular multigraph is odd, multiply every number $m(e)$ by two). Clearly, H^* has an Eulerian cycle. This Eulerian cycle corresponds to a pseudo-h-hamiltonian cycle in G. □

A graph has a *perfect 2-matching* if one can assign weights 0, 1 or 2 to its edges in such a way that for each vertex, the sum of the weights of the incident edges is equal to 2. The following characterization of regularizable graphs can be found in the book by Lovász and Plummer [11].

Proposition 2.3 *(Lovász and Plummer [11])*
A graph $G = (V, E)$ is regularizable if and only if for each edge $e \in E$ there exists a perfect 2-matching of G in which e has weight 1 *or* 2. □

Proposition 2.3 has several important consequences.

Corollary 2.4 *(i) For any integer h with $1 \leq h < \mathrm{ph}(G)$, graph G does not possess a pseudo-h-hamiltonian cycle. (ii) For any integer $h \geq \mathrm{ph}(G)$, graph G does possess a pseudo-h-hamiltonian cycle.*

Proof. Statement (i) trivially follows from the definition of $\mathrm{ph}(G)$. In order to prove (ii), we show that if a graph has a pseudo-h-hamiltonian cycle then it also has a pseudo-$(h+1)$-hamiltonian cycle: Let C be a pseudo-h-hamiltonian cycle in G. Then the skeleton of C is regularizable, and consequently possesses a perfect 2-matching. If one adds this perfect 2-matching to the $2h$-regular multigraph that corresponds to C, one gets a $(2h+2)$-regular multigraph that corresponds to a pseudo-$(h+1)$-hamiltonian cycle. □

Proposition 2.3 together with Lemma 2.2 also allows us to construct an algorithm to decide efficiently whether a graph is pseudo-hamiltonian (or, equivalently, to decide whether a graph has a connected spanning regularizable subgraph). The algorithm runs through all the edges of the graph and deletes all those edges which do not allow a perfect 2-matching with the desired property. If the remaining graph is disconnected then G is not pseudo-hamiltonian. Otherwise, one obtains a connected spanning regularizable subgraph of G, i.e. G is pseudo-hamiltonian.

Algorithm PSEUDO-HAMILTON(G)

1. UNCHECKED:= E; E^*:= E;
2. While UNCHECKED $\neq \emptyset$ **do**
  Pick an arbitrary edge e $\in$UNCHECKED;
  Check whether the graph (V, E^*) possesses a perfect
    2-matching in which the edge e has weight 1 or 2;
  If there is no such perfect 2-matching
    then $E^* := E^* - \{e\}$;
  UNCHECKED:= UNCHECKED$-\{e\}$;
3. If the graph (V, E^*) is connected
  then return 'yes' **else return** 'no'.

Since perfect 2-matchings can be found in polynomial time (cf. Lovász and Plummer [11]), the whole algorithm can be implemented to run in polynomial time.

Theorem 2.5 *It can be decided in polynomial time, whether* $\text{ph}(G) < \infty$ *holds for a given graph* G. □

In strong contrast to Theorem 2.5, it is NP-complete to compute $\text{ph}(G)$ exactly.

Theorem 2.6 *For every fixed value* $h \geq 1$, *the problem of deciding whether* $\text{ph}(G) \leq h$ *holds for a given graph* G *is NP-complete.*

Proof. The proof is done by a transformation from the NP-complete hamiltonian cycle problem (for details we refer to the full version of the paper). □

Question 2.7 *What can be said about approximating* $\text{ph}(G)$*? Can one always find in polynomial time a, say, pseudo-*$2\text{ph}(G)$*-hamiltonian cycle?*

3 Stable Sets, Connectivity and Toughness

This section discusses the relationship of pseudo-hamiltonicity with the structure of stable subsets, with the connectivity of a graph, and with the toughness of a graph. First, consider the following two conditions (C1) and (C2) on a graph $G = (V, E)$.

(C1) $|N(S)| \geq |S|$ holds for every maximal stable set $S \subseteq V$.
(C2) $|N(S)| > |S|$ holds for every non-maximal stable set $S \subseteq V$.

Lemma 3.1 *If a graph* $G = (V, E)$ *is pseudo-hamiltonian, then it fulfills the conditions (C1) and (C2).*

Proof. Consider a pseudo-h-hamiltonian cycle C and let S be a stable set in G. Every vertex from S appears h times in C. Since S is stable, each vertex from S must be followed by a vertex from $N(S)$. Hence the set $N(S)$ is visited at least $h \cdot |S|$ times. Since each vertex from $N(S)$ also appears h times in C we obtain

$$|N(S)| \geq |S|.$$

Now assume that $|N(S)| = |S|$. Then vertices from S and from $N(S)$ must alternate in C, and it is not possible to visit any vertex from $V - S - N(S)$. This implies that $V = S \cup N(S)$, or equivalently, that S is a maximal stable set. □

Corollary 3.2 *If $G = (V, E)$ is pseudo-hamiltonian then the following holds:*

(a) G has no end-vertices.
(b) $\alpha(G) \leq \frac{1}{2}|V|$. □

We can use the results on regularizable graphs (cf. Section 2) in order to show that, for a connected graph, the conditions (C1) and (C2) are also sufficient for the existence of a pseudo-hamiltonian cycle.

Lemma 3.3 *If a connected graph $G = (V, E)$ fulfills conditions (C1) and (C2), then it is pseudo-hamiltonian.*

Proof. If $|N(S)| > |S|$ holds for every stable set $S \subseteq V$ then G is 2-bicritical and, by Proposition 2.1, also regularizable. Since G is connected, Lemma 2.2 implies that in this case G is pseudo-hamiltonian.

Otherwise, there exists a stable set S with $|N(S)| = |S|$. Then by condition (C1), S is maximal and $V = S \cup N(S)$ holds. Let H denote the spanning subgraph of G which arises from deleting all edges between vertices from $N(S)$. We show that H is elementary bipartite. Then, again by Proposition 2.1, the subgraph H is regularizable and, since H is also connected, Lemma 2.2 implies that G is pseudo-hamiltonian.

By construction, the graph H is bipartite. H is connected, since otherwise we can easily find a proper subset $S' \subset S$ with $|N(S')| \leq |S'|$ in contradiction to the assumption. Let (s, t) be an arbitrary edge in H with $s \in S$. In $H - \{s, t\}$ we have $|N(S')| \geq |S'|$ for each set $S' \subseteq S - \{s\}$ (note that S' is not maximal stable in G). It is well known that this condition implies the existence of a perfect matching in $H - \{s, t\}$. Hence there is a perfect matching in H containing the edge (s, t). □

Every hamiltonian graph must be 2-connected. However, it is easy to see that this is not a necessary condition for a graph to be pseudo-h-hamiltonian for some $h \geq 2$. On the other side one may ask whether there exists a number k such that every k-connected graph is also pseudo-hamiltonian. The following example shows that this is not true in general.

Example 3.4 *Consider the complete bipartite graph $K_{k+1,k}$, i.e. the graph consisting of two stable sets S and S' of cardinality $k+1$ and k, respectively, where* *any two vertices from S and S' are adjacent. The reader may convince himself that each pair of vertices is joined by k vertex-disjoint paths. Hence, the graph is k-connected. However, since $|N(S)| = k < k+1 = |S|$, we conclude from Lemma 3.1 that the graph is not pseudo-hamiltonian.*

Chvátal [5] defines the *toughness* $t(G)$ of a graph G (where G is not a complete graph) by

$$t(G) = \min_W \frac{|W|}{c(G-W)},$$

where W is a cutset of G and $c(G-W)$ denotes the number of connected components of the graph $G-W$. It is well known that a hamiltonian graph has toughness at least 1. As an extension of this result we obtain:

Lemma 3.5 *If G is pseudo-h-hamiltonian, then $t(G) \geq \frac{1}{h}$.*

Proof. Let W^* be a cutset of G with $t(G) = |W^*|/c(G-W^*)$. Each path between two vertices of different connected components of $G - W^*$ contains vertices from W^*. Hence, in a pseudo-h-hamiltonian cycle of G there appears at least $c(G-W^*)$ times a vertex from W^*, i.e. each vertex from W^* appears at least $c(G-W^*)/|W^*|$ times. This implies $h \geq 1/t(G)$ and the correctness of the claim. □

It is known (cf. Chvátal [5]) that there are graphs with toughness 1 which are not hamiltonian. Similarly, the converse of Lemma 3.5 is not always true for $h \geq 2$. The complete bipartite graph $K_{3,2}$ has toughness $t(K_{3,2}) = 2/3 \geq 1/h$. However, as argued in Example 3.4 above, this graph is not pseudo-h-hamiltonian.

Another sufficient condition for pseudo-hamiltonicity relies on the toughness of the graph.

Lemma 3.6 *(i) Any graph G with $t(G) \geq 1$ is pseudo-hamiltonian. (ii) For every $\varepsilon > 0$, there exists a graph G with $t(G) \geq 1 - \varepsilon$ that is not pseudo-hamiltonian.*

Proof. Consider a graph G with toughness at least 1. Clearly, G is connected. We will show that G fulfills the conditions (C1) and (C2), and then Lemma 3.3 implies statement (i).

Let S be a maximal stable set in G and assume that $|N(S)| < |S|$ holds. With $W := N(S)$, we obtain $c(G-W) > |W|$ as the vertices of S form the connected components of $G-W$. Hence $t(G) < 1$, in contradiction to the assumption.

Let S be a non-maximal stable set in G and assume that $|N(S)| \leq |S|$. Define again $W := N(S)$. Then the vertices of S are again connected components of $G - W$, and since S is not maximal there is at least one further component. Hence $c(G-W) > |W|$ holds, which implies that $t(G) < 1$.

In order to prove (ii), consider the complete bipartite graphs $K_{k+1,k}$ from Example 3.4: $K_{k+1,k}$ has toughness $k/(k+1)$. As k tends to infinity, this expression tends to one. □

4 Regular Graphs

In this section, we discuss the problem of deciding whether a given d-regular graph possesses a pseudo-h-hamiltonian cycle. We will show that for every d, there is a precise threshold for h where the computational complexity of recognizing pseudo-h-hamiltonian d-regular graphs jumps from NP-complete to trivial.

Lemma 4.1 *(i) For odd $d \geq 3$, every connected d-regular graph G fulfills* $\text{ph}(G) \leq d$. *(ii) For even $d \geq 4$, every connected d-regular graph G fulfills* $\text{ph}(G) \leq d/2$.

Proof. For even d, graph G itself is Eulerian and the Eulerian cycle yields a pseudo-$d/2$-hamiltonian cycle. For odd d, the multigraph that contains two copies of every edge in G is Eulerian and thus yields a pseudo-d-hamiltonian cycle. □

Lemma 4.2 *(i) For odd $d \geq 3$, it is NP-complete to decide whether* $\text{ph}(G) \leq d-1$ *holds for a d-regular graph G. (ii) For even $d \geq 4$, it is NP-complete to decide whether* $\text{ph}(G) \leq d/2-1$ *holds for a d-regular graph G.*

Proof. We only prove (i). The proof of (ii) can be done by analogous (somewhat tedious) arguments.

For every odd $d \geq 3$, the proof of (i) is based on the following auxiliary graph H_d: H_d has $2d-1$ vertices that are divided into three parts X, Y and Z. Part X consists of a single vertex x, parts $Y = \{y_1, \ldots, y_{d-1}\}$ and $Z = \{z_1, \ldots, z_{d-1}\}$ both contain $d-1$ vertices. There is an edge between x and every vertex in Y, and there is an edge between every vertex in Y and every vertex in Z. We add a perfect matching between the vertices in Z such that z_1 and z_2 are matched with each other. This completes the description of H_d. Note that in H_d, vertex x has degree $d-1$ and all vertices in $Y \cup Z$ have degree d. Moreover, we will use the following connected multigraph $M(H_d)$: $M(H_d)$ has the same vertex set as H_d. Vertex x is connected by a single edge to y_1 and y_2, respectively, and by two edges to each vertex in $Y - \{y_1, y_2\}$. For $1 \leq j \leq 2$, y_j is connected by $2d-3$ edges to z_j, and for $3 \leq j \leq d-1$, y_j is connected by $2d-4$ edges to z_j. Finally, there is one edge that connects z_1 to z_2, and there are two copies of every other edge in the matching over Z. Note that in the resulting graph $M(H_d)$, vertex x has degree $2d-4$ and all vertices in $Y \cup Z$ have degree $2d-2$.

The NP-completeness proof for result (i) is done by a reduction from the NP-complete hamiltonian cycle problem in cubic graphs (cf. Garey and Johnson [9]). Consider an instance $G' = (V', E')$ of this problem, and construct a d-regular graph $G = (V, E)$ from G' as follows:

- For every $v \in V'$, introduce a corresponding vertex v^* in V. Moreover, introduce $d-3$ pairwise disjoint copies of H_d. The x-vertex of every such copy is connected to v^*.
- For every edge $(u, v) \in E'$, introduce two new vertices $a_{u,v}$ and $a_{v,u}$ together with the three edges $(u^*, a_{u,v})$, $(a_{u,v}, a_{v,u})$ and $(a_{v,u}, v^*)$, i.e. the vertices $a_{u,v}$ and $a_{v,u}$ essentially subdivide the original edge (u, v) into three sub-edges.

- For every new vertex $a_{u,v}$, create $d-2$ pairwise disjoint copies of H_d and connect the x-vertex of every copy to $a_{u,v}$.

It is easy to verify that the resulting graph G is d-regular (since in H_d, vertex x has degree $d-1$ and all other vertices have degree d). We claim that G possesses a pseudo-$(d-1)$-hamiltonian cycle if and only if G' possesses a hamiltonian cycle.

(If). Assume that G' possesses a hamiltonian cycle. Construct from this hamiltonian cycle a $(2d-2)$-regular multigraph M^* as follows: For every copy of H_d in G, introduce the corresponding edges of $M(H_d)$ in M^*, together with two edges that connect the x-vertex to that vertex to which the copy has been attached. For every edge (u,v) that is used by the hamiltonian cycle, introduce the three edges $(u^*, a_{u,v})$, $(a_{u,v}, a_{v,u})$ and $(a_{v,u}, v^*)$ in M^*. For every edge (u,v) that is not used by the hamiltonian cycle, introduce two copies of $(u^*, a_{u,v})$ and two copies of $(a_{v,u}, v^*)$ in M^*. The resulting multigraph is $(2d-2)$-regular, is connected (as it simulates the hamiltonian cycle in G'), and it is spanning. Hence, the corresponding Eulerian cycle in G yields a pseudo-$(d-1)$-hamiltonian cycle for G.

(Only if). Now assume that G possesses a pseudo-$(d-1)$-hamiltonian cycle C. Then the edges that are traversed by C form a $(2d-2)$-regular connected multigraph M^C. For every copy of H_d in G, the cycle C traverses the edge that connects the x-vertex to the vertex to which the copy has been attached, at least twice and an even number of times. Hence, for every edge $(u,v) \in E'$ the vertex $a_{v,u}$ in M^C is connected by at least $2d-4$ edges to the x-vertices of the attached copies of H_d, and there remain only two edges that can connect $a_{v,u}$ to the rest of the graph. With this it is easy to verify that there remain only two possibilities how the cycle C may traverse the three edges $(u^*, a_{u,v})$, $(a_{u,v}, a_{v,u})$ and $(a_{v,u}, v^*)$ that correspond to some edge $(u,v) \in E'$ in the original graph: Either all three edges are traversed exactly once (this is called a traversal of *type-1*), or $(u^*, a_{u,v})$ and $(a_{v,u}, v^*)$ are both traversed exactly twice and $(a_{u,v}, a_{v,u})$ is not used at all (this is called a traversal of *type-2*). Note that traversals of type-1 do not help in connecting the corresponding vertices u^* and v^* to each other.

Now consider for some vertex v^* the types of traversals of the three incident edges that lead to vertices a_{uv}. If exactly one or all three of them are traversed by traversals of type-1, the cycle C visits and leaves an odd number of times the component that consists of v^* and all copies of H_d that are attached to v^*; this clearly is impossible. If none of these edges is traversed by a traversal of type-1, then the component is isolated from the rest of the graph; this contradicts the connectivity of M^C. Hence, exactly two of these edges are traversed by traversals of type-1. Consequently, the edges $(u,v) \in E'$ whose corresponding edges in G are traversed by traversals of type-1 form a hamiltonian cycle in G'. □

In fact, for any odd d and $1 \leq h \leq d-1$ it is NP-complete to decide whether a d-regular graph is pseudo-h-hamiltonian, and for any even d and $1 \leq h \leq d/2-1$ it is NP-complete to decide whether a d-regular graph is pseudo-h-hamiltonian. The proofs for these statements can be done analogously to the proof of Lemma 4.2. The main difference is that one has to work with slightly more complicated auxiliary graphs H_d, and that the copies of these auxiliary graphs that are attached to the same vertex are connected to each other in an appropriate way. Moreover, for even d the reduction is done from the NP-complete hamiltonian cycle problem in 4-regular graphs.

Theorem 4.3 *For every $d \geq 3$, there exists a threshold $\tau(d)$ with the following properties: For every $1 \leq h < \tau(d)$, it is NP-complete to decide whether a given d-regular graph fulfills* $\mathrm{ph}(G) \leq h$, *and for every $h \geq \tau(d)$,* every *connected d-regular graph fulfills* $\mathrm{ph}(G) \leq h$. □

5 The Square of a Graph

The *square* G^2 of a graph $G = (V, E)$ is the graph with vertex set V, where two distinct vertices are joined whenever in G the distance between them is at most 2. A famous result of Fleischner [8] states that the square of every 2-connected graph is hamiltonian.

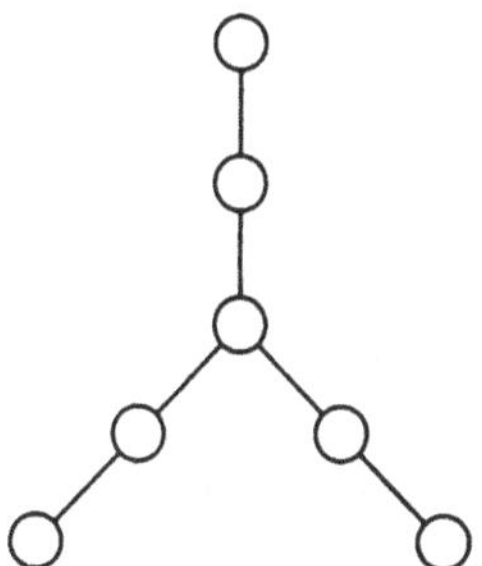

Fig. 2. A graph with non-hamiltonian square.

The square of a *connected* graph is not necessarily hamiltonian: see e.g. the graph in Figure 2. However, the reader may convince himself that the square of the graph in Figure 2 is pseudo-2-hamiltonian.

Lemma 5.1 *The square of a connected graph is pseudo-hamiltonian.*

Proof. Let $G = (V, E)$ be a connected graph. We will show that G^2 fulfills conditions (C1) and (C2). Combining this statement with the statement of Lemma 3.3 then yields that G^2 is pseudo-hamiltonian.

Let S be a stable set in G^2. Since G results from G^2 by deleting certain edges, S is also stable in G. In G, we have $N(s) \cap N(s') = \emptyset$ for all $s, s' \in S$, since otherwise s and s' would be adjacent in G^2. Since G is connected (and nontrivial), we have $N(s) \neq \emptyset$ for all $s \in S$. Therefore, $|N(S)| \geq |S|$ holds in G and since G^2 is a supergraph of G, the same inequality is true in G^2.

Now assume that there is a stable set S in G^2 that fulfills $|N(S)| = |S|$. Then, in G, we must have $|N(s)| = 1$ for each $s \in S$. Let $S = \{s_1, s_2, \ldots, s_r\}$ and denote by t_i the unique neighbor of s_i in G. Then in G, $N(S) = \{t_1, t_2, \ldots, t_r\}$ holds. We claim that $S \cup N(S) = V$: Otherwise, since G is connected, at least one of the vertices t_i must be adjacent to some vertex $u \notin S \cup N(S)$. But then s_i is connected to u in G^2, and this contradicts the assumption that $|N(S)| = |S|$ in G^2. Hence, $S \cup N(S) = V$ must hold, and S is a maximal stable set. □

Lemma 5.2 *It is NP-complete to decide whether the square of a given graph is pseudo-2-hamiltonian.*

Proof. The proof is done by a reduction from the NP-complete hamiltonian cycle problem in bipartite graphs (cf. Garey, Johnson [9]). For details see the full version of the paper. □

6 Special Graph Classes

In this section, we show that deciding whether a graph is pseudo-h-hamiltonian is NP-complete even for some very restricted graph classes with a strong combinatorial structure. Moreover, we present polynomial time algorithms for other classes of structured graphs.

6.1 Trees and Planar Graphs

By Corollary 3.2.(a), a pseudo-hamiltonian graph cannot have any vertices of degree one. Hence, $\mathrm{ph}(T) = \infty$ for any tree T.

If we start the construction in the proof of Theorem 2.6 with a planar graph G', then the constructed graph G is also planar. This yields that for every $h \geq 1$ it is NP-complete to decide whether a planar graph is pseudo-h-hamiltonian.

6.2 Partial k-trees

The class of *partial k-trees* is a well-known generalization of ordinary trees (see e.g. the survey articles by Bodlaender [4] and by van Leeuwen [12]). It is known that series-parallel graphs and outerplanar graphs are partial 2-trees and that Halin graphs are partial 3-trees. Large classes of algorithmic problems can be solved in polynomial time on partial k-trees if k is constant. Essentially, each graph problem that is expressible in the Monadic Second Order Logic (MSOL) is solvable in linear time on partial k-trees with constant k (cf. e.g. Arnborg, Lagergren, Seese [1]).

Lemma 6.1 *For every $h \geq 1$ and for every $k \geq 1$, it can be decided in linear time whether a given partial k-tree is pseudo-h-hamiltonian.*

Proof. We only show the statement for $h = 2$; the other cases can be settled analogously. For a given graph $G = (V, E)$, the property of having a connected 4-regular submultigraph can be expressed in MSOL as follows:

1. There exist three pairwise disjoint subsets E_1, E_2 and E_3 of E
2. Every vertex is either incident to (i) four edges from E_1, or to (ii) two edges from E_1 and one edge from E_2, or to (iii) one edge from E_1 and one edge from E_3, or to (iv) two edges from E_2
3. There does not exist a partition of the vertex set V into two non-empty sets V_1 and V_2, such that none of the edges in $E_1 \cup E_2 \cup E_3$ connects V_1 to V_2.

Intuitively speaking, the edges in E_1 (E_2, E_3) occur once (twice, thrice) in the submultigraph. The second condition then takes care of the 4-regularity, and the third condition ensures that the submultigraph is connected. □

6.3 Bipartite Graphs and Split Graphs

Lemma 6.2 *For every integer $h \geq 1$, it is NP-complete to decide whether a bipartite graph is pseudo-h-hamiltonian.*

Proof. It is NP-complete to decide whether a bipartite graph G' is hamiltonian (cf. Garey and Johnson [9]). Consider a bipartite graph $G' = (V', E')$ with bipartition $V' = V_1' \cup V_2'$, and construct from G' another bipartite graph G as follows. For every vertex $v \in V'$, introduce two vertices ℓ_v and r_v in V together with auxiliary vertices a_v^i and b_v^i, $i = 1, \ldots, 2h-2$. In E, there are the edges (ℓ_v, r_v) together with the edges (ℓ_v, a_v^i), (a_v^i, b_v^i), and (b_v^i, r_v) for $i = 1, \ldots, 2h-2$. Moreover, for every edge $(u, v) \in E'$ with $u \in V_1'$ and $v \in V_2'$, we introduce the two edges (ℓ_u, r_v) and (ℓ_v, r_u).

It can be verified that the resulting graph G is also bipartite. Moreover, G possesses a pseudo-h-hamiltonian cycle if and only if G' possesses a hamiltonian cycle. □

A *split graph* is a graph whose vertex set can be partitioned into two parts such that the subgraph induced by the first part is a clique and the subgraph induced by the second part is a stable set.

Corollary 6.3 *For every integer $h \geq 1$, it is NP-complete to decide whether a split graph is pseudo-h-hamiltonian.*

Proof. In the NP-completeness proof for bipartite graphs in Lemma 6.2, both classes in the bipartition of the constructed graph G are of equal cardinality. Transform G into a split graph G^* by adding all edges between vertices in one part of the bipartition. It is easy to see that a pseudo-h-hamiltonian cycle in G^* can never use these added edges, and hence G^* is pseudo-h-hamiltonian if and only if G' is hamiltonian. □

6.4 Cocomparability Graphs

A *comparability graph* is a graph $G = (V, E)$ whose edges are exactly the comparable pairs in a partial order on V. The complementary graph is called a *cocomparability graph*. The class of cocomparability graphs properly contains all cographs, permutation graphs and interval graphs.

Lemma 6.4 *For every integer $h \geq 1$, it can be decided in polynomial time whether a cocomparability graph is pseudo-h-hamiltonian.*

Proof. It is known that a hamiltonian cycle in a cocomparability graph can be found in polynomial time (cf. Deogun and Steiner [7]). Given a cocomparability graph $G = (V, E)$, we construct another cocomparability graph $G' = (V', E')$ as follows. V' contains the vertices in V together with $(h-1)|V|$ new vertices. For every vertex $v \in V$ there are $h-1$ new vertices that are called v^i, where $i = 2, \ldots, h$. For simplicity of notation, let $v^1 := v$. If (u, v) is an edge in E then all edges (u^i, v^j) with $i, j = 1, \ldots, h$ belong to E' (roughly spoken, G' arises from G by replacing each vertex by a stable set of h vertices). It is easy to see that G' is again a cocomparability graph. We show that G has a pseudo-h-hamiltonian cycle if and only if G' has a hamiltonian cycle.

(If). Assume that G' possesses a hamiltonian cycle. We obtain a pseudo-h-hamiltonian cycle in G if each vertex v^i, $i = 2, \ldots, h$, is replaced by the corresponding vertex v.

(Only if). Now assume that G possesses a pseudo-h-hamiltonian cycle C. Each vertex of G appears h times in C. For each $v \in V$ replace $h-1$ copies of v in C by $v^2, \ldots, v^h$. This yields a *2-factor* of G', i.e. a subgraph of G' such that each vertex has degree 2. If the 2-factor is a cycle then we have a hamiltonian cycle in G' and we are done. Otherwise the 2-factor is a disjoint union of cycles. In this case the following principle allows to reduce the number of cycles: Let C_1 and C_2 denote two disjoint cycles such that v^i belongs to C_1 and v^j belongs to C_2 (it is straightforward to see that such cycles must exist). Let further x be the predecessor of v^i in C_1 and y the predecessor of v^j in C_2. Replace the edges (x, v^i) and (y, v^j) by (x, v^j) and (y, v^i). One obtains a new cycle that contains all vertices from C_1 and C_2. Repeatedly merging cycles in this way finally provides the desired hamiltonian cycle in G'. □

We leave it as an open problem to determine the complexity of computing the pseudo-hamiltonicity number of *asteroidal triple-free* graphs, AT-free graphs for short (cf. Corneil, Olariu, and Stewart [6]). Note that for an AT-free graph G, the graph G' that is constructed in the proof of Lemma 6.4 above is also AT-free. However, the complexity of finding a hamiltonian cycle in AT-free graphs is currently unknown.

References

1. Arnborg, S., Lagergren, J., Seese, D.: Easy problems for tree-decomposable graphs. J. Algorithms **12** (1991) 308–340
2. Berge, C.: Regularizable graphs I. Discrete Mathematics **23** (1978) 85–89
3. Berge, C.: Regularizable graphs II. Discrete Mathematics **23** (1978) 91–95
4. Bodlaender, H.L.: Some classes of graphs with bounded treewidth. Bulletin of the EATCS **36** (1988) 116–126
5. Chvátal, V.: Tough graphs and hamiltonian circuits. Discrete Mathematics **5** (1973) 215–228

6. Corneil, D.G., Olariu, S., Stewart, L.: Asteroidal triple-free graphs. Proceedings of the 19th International Workshop on Graph-Theoretic Concepts in Computer Science WG'93, Springer Verlag, LNCS **790**, 1994, 211–224
7. Deogun. J.S., Steiner, G.: Polynomial algorithms for hamiltonian cycles in cocomparability graphs. SIAM J. Computing **23** (1994) 520–552
8. Fleischner, H.: The square of every two-connected graph is hamiltonian. J. Combinatorial Theory B **16** (1974) 29–34
9. Garey. M.R., Johnson, D.S.: Computers and intractability, A guide to the theory of NP-completeness. Freeman, San Francisco, 1979
10. Johnson, D.S.: The NP-completeness column: an ongoing guide. J. Algorithms **6** (1985) 434–451
11. Lovász, L., Plummer, M.D.: Matching theory. Annals of Discrete Math. **29**, North-Holland, 1986
12. van Leeuwen, J.: Graph algorithms, in: Handbook of Theoretical Computer Science, A: Algorithms and Complexity Theory, 527–631, North Holland, Amsterdam, 1990

Acyclic Orientations for Deadlock Prevention in Interconnection Networks* (extended abstract)

Jean-Claude Bermond[1], Miriam Di Ianni[2], Michele Flammini[3,1] and Stephane Perennes[4,1]

[1] Project SLOOP I3S-CNRS/INRIA/Université de Nice–Sophia Antipolis, 930 route des Colles, F–06903 Sophia-Antipolis Cedex, France. E-mail: bermond@unice.fr

[2] Istituto di Elettronica, University of Perugia, via G. Duranti 1/A, I-06123 Perugia, Italy. E-mail: diianni@istel.ing.unipg.it

[3] Dipartimento di Matematica Pura ed Applicata, University of L'Aquila, via Vetoio loc.Coppito, I-67100 L'Aquila, Italy. E-mail: flammini@univaq.it

[4] Dept. Math. and Computer Science - TU Delft, Makelweg 4, 26 28 CD Delft, The Netherlands. E-mail: perennes@math.tudelft.fr

Abstract. In this paper we extend some of the computational results presented in [6] on finding an acyclic orientation of a graph which minimizes the maximum number of changes of orientations along the paths connecting a given subset of source-destination couples. The corresponding value is called rank of the set of paths. Besides its theoretical interest, the topic has also practical applications. In fact, the existence of a rank r acyclic orientation for a graph implies the existence of a deadlock-free routing strategy for the corresponding network which uses at most r buffers per vertex.

We first show that the problem of minimizing the rank is NP-hard if all shortest paths between the couples of vertices wishing to communicate have to be represented and even not approximable within an error in $\mathbf{O}(k^{1-\epsilon})$ for any $\epsilon > 0$, where k is the number of source-destination couples wishing to communicate, if only one shortest path between each couple has to be represented.

We then improve some of the known lower and upper bounds on the rank of all possible shortest paths between any couple of vertices for particular topologies, such as grids and hypercubes, and we find tight results for tori.

Keywords: computational and structural complexity, graph theory, parallel algorithms, routing and communication in interconnection networks

* Work supported by the EU TMR Research Training Grant N. ERBFMBICT960861, the EU ESPRIT Long Term Research Project ALCOM-IT under contract N. 20244, the French action RUMEUR of the GDR PRS and the Italian MURST 40% project "Algoritmi, Modelli di Calcolo e Strutture Informative".

1 Introduction

In this paper we investigate a graph theoretical combinatorial problem, naturally arising from communication issues in interconnection networks. Namely, given a graph and a set of paths connecting a subset of source-destination couples of vertices, we want to determine an acyclic orientation of the graph which minimizes the maximum number of changes of orientations along the dipaths.

Practical applications of this problem concern the design of deadlock-free routing strategies which use a low number of buffers. Deadlocks arise due to limited buffer availability and since messages are allowed to request buffers while holding others. They are network configurations in which no message can be delivered because of cyclic waitings and influence not only the efficiency of the routing strategy but also its correctness. A central issue in the design of deadlock-free routing algorithms is limiting the number of buffers in the vertices necessary to guarantee its deadlock-freedom property.

Several techniques have been developed to design deadlock-free routing strategies in which deadlocks are avoided by ordering the buffers and allowing each message to use them in a monotonically increasing fashion ([12, 11, 14, 5, 1, 8, 7, 13, 2, 3, 4, 9] among the others). This idea results in the generation of a directed acyclic resource dependencies graph (DAG).

A DAG-based method has been introduced in [11, 15] and furtherly studied in [6], in which the ordering in the set of buffers of each vertex is based on the idea of acyclic orientations of a graph. Informally, an acyclic orientation of a graph G is a directed acyclic graph $\vec{G}$ obtained by orienting the edges of G. The buffers contained in each vertex are then partitioned into a suitable number of classes and a message using buffers of class i moves to buffers of class $i+1$ every time two consecutive traversed edges cause a change of orientation, i.e. whenever exactly one of them is crossed according to the direction in $\vec{G}$. Such rule guarantees the acyclicity of the resource dependencies graph and the number of buffers per vertex yielded by the acyclic orientation is given by the minimum number of classes which is sufficient to implement the above mechanism (in case we associate one buffer per class). The method thus defined was introduced in [6] and is equivalent to the *"Peaks and Valleys"* scheme presented in [11]. However, in this last paper, the author did not give results for specific network topologies. A more general definition can be found in [15], together with some results on ring networks.

In [6] the authors first proved the hardness of devising optimal acyclic orientations if only one shortest path between each couple of vertices wishing to communicate must be represented. Then, they provided some lower and upper bounds on the required number of buffers for particular network topologies.

In this paper we improve the above mentioned hardness result by showing that approximating the number of buffers yielded by an optimum acyclic orientation is NP-hard even within an error $\mathbf{O}(k^{1-\epsilon})$ for any $\epsilon > 0$, where k is the number of source-destination couples wishing to communicate. Moreover, we prove the hardness of finding an optimal acyclic orientation when all shortest paths between each couple of vertices wishing to communicate have to be represented. We then improve the known lower bounds for tori, grids and hypercubes, and we give new upper bounds for tori and grids. There is still a little gap left between lower and upper bounds for grids and hypercubes, while the results for tori are tight.

The paper is organized as follows: in section 2 we give the basic notation and definitions we use throughout the paper; in section 3 we show the hardness results; in sections 4 we give new lower and upper bounds for specific network topologies, and finally in section 5 we discuss some conclusive remarks and we address some open problems.

2 Definitions

In this section we give the necessary notation and definitions to be used throughout the paper.

G will always denote a digraph. Since we consider networks in which two processors can communicate in both directions, in the sequel we always refer to symmetric digraphs obtained from a network by replacing each communication link or edge $\{u, v\}$ between two processors u and v with the two arcs (u, v) and (v, u).

Definition 1. An acyclic orientation of a digraph $G = (V, E)$ is an acyclic digraph $\vec{G} = (V, \vec{E})$ such that $\vec{E} \subseteq E$. We say that two consecutive arcs (u, v) and (v, w) in E cause a change of orientation if exactly one of them belongs to $\vec{E}$.

Definition 2. Let $\vec{G} = (V, \vec{E})$ be an acyclic orientation of $G = (V, E)$. Given a dipath $P = \langle u_1, u_2, \ldots, u_h \rangle$ in G, let c be the number of changes of orientation caused by all the pairs of consecutive arcs along P.

We define the rank $r(P, \vec{G})$ of P with respect to $\vec{G}$ as $r(P, \vec{G}) = c + 1$ if $(u_1, u_2) \in \vec{E}$ and $r(P, \vec{G}) = c + 2$ if $(u_1, u_2) \notin \vec{E}$.

Given a set $\mathcal{P}$ of dipaths in G, the rank of $\mathcal{P}$ with respect to $\vec{G}$ is defined as $r(\mathcal{P}, \vec{G}) = \max_{P \in \mathcal{P}} r(P, \vec{G})$.

Finally, the rank of $\mathcal{P}$ is $r_G(\mathcal{P}) = \min_{\vec{G}} r(\mathcal{P}, \vec{G})$.

Informally, if a dipath P has rank r, then P can be expressed as the concatenation of r directed subpaths $P_1, \ldots, P_r$ such that for each i, $1 \leq i \leq r$, P_i is a dipath in $\vec{G}$ if i is odd and P_i is a dipath in the opposite orientation of $\vec{G}$ if i is even.

Due to efficiency requirements, messages are generally routed along shortest dipaths which connect the sender to the destination. Then, for the sake of brevity, if a set of dipaths $\mathcal{P}$ includes *all* shortest dipaths connecting any couple of vertices in the network, then we denote $r(\mathcal{P}, \vec{G})$ and $r_G(\mathcal{P})$ respectively as $r(\vec{G})$ and r_G.

In packet routing let us denote as s_u the number of buffers assigned by the routing scheme to vertex u. Then, the importance of acyclic orientations is stated by the following classical theorem (see [11] for a formally equivalent theorem and [15] for a more general statement).

Theorem 3. *Given a network G, an acyclic orientation $\vec{G}$ of G and the set of dipaths $\mathcal{P}$ there exists a deadlock free packet routing scheme for G which routes messages along the dipaths in $\mathcal{P}$ and such that $s_u \leq r(\mathcal{P}, \vec{G})$, for each vertex u.*

3 Finding minimal acyclic orientations

In many applications not all pairs of vertices need to exchange messages with each other. Thus, it is worthwhile to specify a set $R = \{(s_1, t_1), \ldots, (s_k, t_k)\} \subseteq V^2$ of *communication requests* denoting the couples of vertices wishing to communicate.

Given a network G, a set of communication requests R, a set $\mathcal{P}$ of dipaths connecting all pairs in R and an integer $k > 0$, we now consider the problem of deciding if $r_G(\mathcal{P}) \leq k$.

Unfortunately, in [6] it has been proved that the problem of deciding if

$$\min\{r_G(\mathcal{P}) : \mathcal{P} \text{ includes exactly one shortest dipath for each couple in } R\} \leq 2$$

is NP-hard. We now extend this result to the set $\mathcal{P}$ containing all shortest dipaths connecting each couple in R.

Theorem 4. *Given a graph G, a set R of communication requests and the set $\mathcal{P}$ of all shortest dipaths connecting each couple in R, it is* NP*-hard to decide if $r_G(\mathcal{P}) \leq 5$.*

Proof. Consider the 3-SAT problem: given a boolean function f in conjunctive normal form in which each clause contains exactly three literals, decide if there exists a truth assignment satisfying f. We will provide a polynomial-time reduction which associates to an instance of 3-SAT a network G and a set of communication requests R such that there exists a truth assignment for f if and only if $r_G(P_R) \leq 5$, where P_R is the set containing all shortest dipaths between each couple in R. Then the assertion will follow from the NP-completeness of 3-SAT [10].

We say that an orientation $\vec{G}$ is *acceptable* for $< G, R >$ if $r(P_R, \vec{G}) \leq 5$. Notice that if $\vec{G}$ is acceptable then for any request $(s_i, t_i) \in R$ the dipath from s_i to t_i can have at most 4 changes of orientation.

Let $f = c_1 \wedge \ldots \wedge c_m$ be a formula in conjunctive normal form defined on the set of variables $X = \{x_1 \ldots x_n\}$ such that each clause contains three literals. The corresponding network G is constructed as follows.

We associate to each variable x_i 10 columns grouped two by two: the network is built from a set of $10n$ *columns* divided into 5 blocks of $2n$ columns each. We will denote these columns as $C_b(i)$ with $1 \leq i \leq n$, $1 \leq b \leq 5$ and $C \in \{P, Q\}$. $P_b(i)$ (resp. $Q_b(i)$) is the column of type P (resp. Q) belonging to block b and associated to variable x_i. Columns of type P will be said *constrained* and the ones of type Q *free*. C will denote a generic column. *Columns* are parallel vertical dipaths of length L (which will be specified later) and the set of vertices of a column C is $\{C.x \mid 0 \leq x < L\}$. For a given vertex $v = C.x$, we say that x and C are respectively the *coordinate* and the column of v. The edges of column C join vertices $C.x$ and $C.(x+1)$ for $0 \leq x < L-1$. For reasons that will be explained in a few lines, columns are divided into $40n+1$ horizontal slices of thickness S. More formally, the slice s $(0 \leq s < 40n+1)$ is the subgraph induced by the vertices of coordinate $x \in [sS, (s+1)S-1]$ and so $L = (40n+1)S$. We will denote by *atom* $A_{s,b}$ the subset of the vertices in slice s and block b.

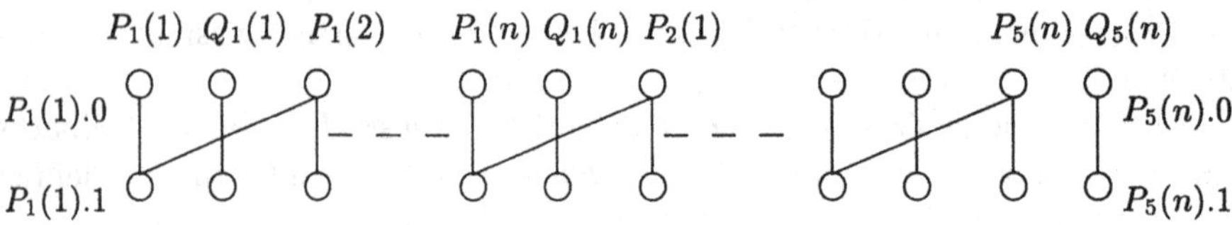

Fig. 1. The dipath constructed in each slice

We say that $\vec{G}$ is *uniform* on column C and slice s if C is uniformly oriented downward or upward in the slice s, that is, either the dipath from $C.(sS)$ to $C.((s+1)S-1)$ belongs to $\vec{G}$ or the dipath from $C.((s+1)S-1)$ to $C.(sS)$ belongs to $\vec{G}$. Similarly, $\vec{G}$ is *uniform* on slice s (resp. on atom $A_{s,b}$) if $\vec{G}$ is uniform on each column of slice s (resp. of atom $A_{s,b}$). We will say that $\vec{G}$ is *strongly uniform downward* (resp. *upward*) on atom $A_{s,b}$ if $\vec{G}$ is uniform on $A_{s,b}$ and furthermore all columns of type P (the constrained ones) are oriented downward (resp. upward). Again, this means that all the dipaths from $P_b(i).(sS)$ to $P_b(i).((s+1)S-1)$ belong to $\vec{G}$ (resp. all the dipaths from $P_b(i).((s+1)S-1)$ to $P_b(i).(sS)$ belong to $\vec{G}$), for $1 \leq i \leq n$.

Now, we put in the set of requests the pairs formed by the initial and the terminal vertices of each column (all the couples $(C.0, C.(L-1))$). Then, if $\vec{G}$ is acceptable for this set of request, $\vec{G}$ is uniform on at least one slice. Indeed, since the unique shortest dipath from $C.0$ to $C.(L-1)$ is the column C itself, we know that on each column there are at most 4 changes of orientation. Since the total number of columns is $10n$ and each column may contribute to the non uniformity of at most 4 slices, the maximum number of non uniform slices is $40n$. Hence, there must exists a uniform slice, because the total number of slices is $40n+1$.

We now add some new edges and requests so that any acceptable orientation $\vec{G}$ has to be strongly uniform on an atom $A_{s,b}$. In each slice we perform the same construction as described in the following. We refer to the coordinate of a vertex in slice s by its offset from the coordinate of the initial vertex of the slice $s_0 = sS$, so in what follows the vertex $C.(sS + x)$ will be simply denoted as $C.x$.

The edges needed to complete the following dipath are added (see figure 1).

$$\begin{array}{rl} T = < & P_1(1).0,\ P_1(1).1,\ P_1(2).0,\ P_1(2).1,\ P_1(3).0,\ \ldots,\ P_1(n).1,\\ & P_2(1).0,\ P_2(1).1,\ P_2(2).0,\ P_2(2).1,\ P_2(3).0,\ \ldots,\ P_2(n).1,\\ & P_3(1).0,\ P_3(1).1,\ P_3(2).0,\ P_3(2).1,\ P_3(3).0,\ \ldots,\ P_3(n).1,\\ & P_4(1).0,\ P_4(1).1,\ P_4(2).0,\ P_4(2).1,\ P_4(3).0,\ \ldots,\ P_4(n).1,\\ & P_5(1).0,\ P_5(1).1,\ P_5(2).0,\ P_5(2).1,\ P_5(3).0,\ \ldots,\ P_5(n).1 > \end{array}$$

The communication request $(P_1(1).0, P_5(n).1)$ is added to the set of requests.

Let us consider now a slice s_0, such that $\vec{G}$ is uniform on s_0. Since the dipath from $P_1(1).0$ to $P_5(n).1$ has at most four orientation changes, $\vec{G}$ is necessarily such that for some b_0, $1 \leq b_0 \leq 5$, all columns $Q_{b_0}(i)$, $1 \leq i \leq n$, have the same orientation in slice s_0. Thus $\vec{G}$ is strongly uniform downward or upward in the atom A_{s_0,b_0}.

The remaining and the key part of our construction is devoted to show the requests (and the shortest dipaths) associated to the clauses of f in such a way that f is satisfiable if and only if there exists an acceptable orientation $\vec{G}$ for a strongly uniform block.

To this aim, we add edges and requests for each atom and each clause in the same way. We first split the vertices of a column C in slice s as follows. On each column and for each slice s we reserve h_0 vertices (namely vertices $C.x$ with $x \in [sS, sS + h_0 - 1]$) for the path T defined above, h vertices per clause c_k (namely vertices $C.x$ with $x \in [sS + h_0 + kh, sS + h_0 + (k+1)h - 1]$) and h_0 vertices (namely vertices $C.x$ with $x \in [sS + h_0 + mh, sS + h_0 + mh + h_0 - 1]$) at the end of the atom to separate it from the next one. Thus, $S = mh + 2h_0$. The two parameters h_0 and h will be adjusted later in such a way that the dipaths that we consider in the proof are unique shortest dipaths. In order to have simpler notations, for the clause c_k in a generic atom $A_{s,b}$ we will denote the vertex $C_b(i).sS + kh + h_0 + x$ by $C_b(i).x$.

Let $c_k = l_{j_1} \vee l_{j_2} \vee l_{j_3}$, with $j_1 < j_2 < j_3$, where l_{j_u} is either x_{j_u} or $\overline{x_{j_u}}$. The vertex E is defined as $Q(j_3).1$ if $l_{j_3} = x_3$, or as $Q(j_3).0$ if $l_{j_3} = \overline{x_3}$. The communication request $(P(j_1).0, E)$ is added to the set of requests and the edges necessary to build the following dipaths are added to G (see also figure 2):

- $< P(j_1).0,\ P(j_1).1,\ Q(j_1).0,\ Q(j_1).1,\ P(j_2).0 >$ if $l_{j_1} = x_{j_1}$
  $< P(j_1).0,\ P(j_1).1,\ Q(j_1).1,\ Q(j_1).0,\ P(j_2).0 >$ if $l_{j_1} = \overline{x_{j_1}}$

- $< P(j_2).0,\ P(j_2).1,\ Q(j_2).0,\ Q(j_2).1,\ P(j_3).0 >$ if $l_{j_2} = x_{j_2}$
  $< P(j_2).0,\ P(j_2).1,\ Q(j_2).1,\ Q(j_2).0,\ P(j_3).0 >$ if $l_{j_2} = \overline{x_{j_2}}$
- $< P(j_3).0,\ P(j_3).1,\ Q(j_3).0,\ Q(j_3).1 >$ if $l_{j_3} = x_{j_3}$
  $< P(j_3).0,\ P(j_3).1,\ Q(j_3).1,\ Q(j_3).0 >$ if $l_{j_3} = \overline{x_{j_3}}$

Consider now an acceptable orientation $\vec{G}$ for the set of communication requests $\{(C.0, C.(L-1))\}$ for every column C and a strongly uniform atom $A_{s_0 b_0}$ for $\vec{G}$. It is possible to associate to $\vec{G}$ a truth assignment for X as follows. In slice s_0, all columns $P_{b_0}(i)$ are oriented downward (resp. upward) if $\vec{G}$ is strongly uniform downward (resp. upward) in the atom, and each column $Q_{b_0}(i)$ can be independently oriented downward or upward. If the orientations of $P_{b_0}(i)$ and $Q_{b_0}(i)$ are identical (resp. opposite) we will associate to x_i the value true (resp. false).

Notice that, if the truth assignment associated to the strongly uniform atom A_{s_0,b_0} is such that the clause c_k is false, then the dipath from $P(j_1).0$ to E when $\vec{G}$ is strongly uniform downward or upward in the atom has to use the orientations $\vec{G}\overleftarrow{G}\vec{G}\overleftarrow{G}\vec{G}\overleftarrow{G}$, where $\overleftarrow{G}$ is the reversal of $\vec{G}$. Hence, such a dipath has at least 5 changes of orientation, i.e. rank at least 6, and $\vec{G}$ cannot be acceptable.

Thus, if $\vec{G}$ is acceptable then f is satisfiable.

To complete the proof we must provide an acceptable orientation $\vec{G}$ when f is satisfiable. To this aim, we choose a truth assignment for variables x_i satisfying f and we define $\vec{G}$ as follows: constrained columns are directed downward, free columns are directed according to the truth assignment (as previously shown), horizontal arcs are directed from left to right. More formally:

all columns of $P_b(i)$ such that $1 \leq b \leq 5$ and $1 \leq i \leq n$ are directed downward (that is arc $(P_b(i).x, P_b(i).(x+1))$ is in $\vec{G}$).

if x_i is true all columns $Q_b(i)$ such that $1 \leq i \leq n$ are directed downward $((Q_b(i).x, Q_b(i).(x+1)) \in \vec{G})$, otherwise they are directed upward $((Q_b(i).(x+1), Q_b(i).x) \in \vec{G})$.

if there is an edge between two vertices $C_b(i).x$ and $C_{b'}(i').x'$ with $b < b'$ or $b = b'$ and $i < i'$, then the arc $(C_b(i).x, C_{b'}(i').x')$ belongs to $\vec{G}$.

if there is an edge between two vertices $P_b(i).x$ and $Q_b(i).x'$, then the arc $(P_b(i).x, Q_b(i).x')$ belongs to $\vec{G}$.

Such an orientation is clearly acyclic (any dipath in $\vec{G}$ either stays on a column and goes upward or downward, or it goes strictly from left to right). Since all the clauses are true under the chosen truth assignment and consequently each of them contains at least one true literal, one can check that the dipaths associated to clauses have rank at most 5. All the other requests are fulfilled with no change of orientation, thus we have constructed an acceptable orientation for the graph G.

In order to complete the proof it suffices to observe that by choosing $h_0 \geq 2n$ and $h \geq 10$ all the considered dipaths are the (unique) shortest ones. This leads to $S = 10m + 4n, L = (40n + 1)(10m + 4n)$ and to a total number of vertices in the graph equal to $10nL = 10n(40n + 1)(10m + 4n)$.

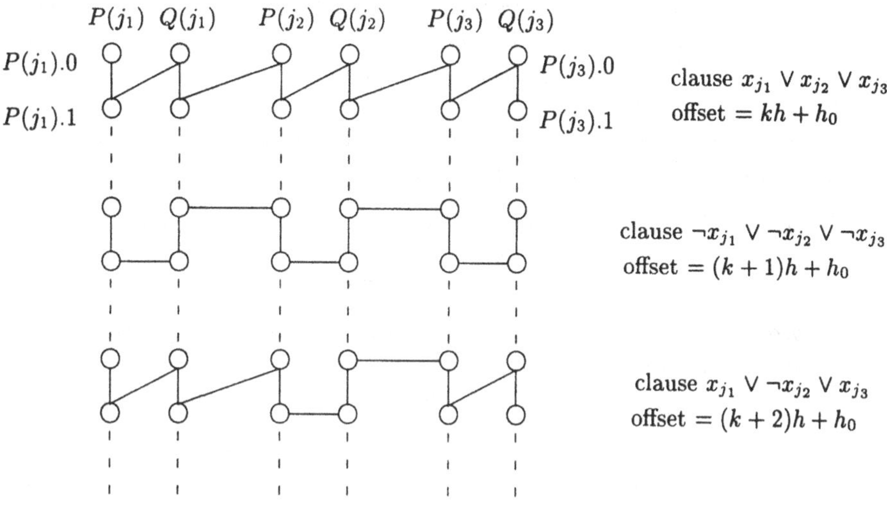

Fig. 2. Sample of 3 out of 8 possible clauses.

We now turn our attention on the possibility of devising polynomial time algorithms that are able to find approximate solutions, that is, solutions whose sizes have constant approximation error with respect to the optimal ones. The formal definition of the approximation error of a minimization problem Π is defined as $\frac{m(S_A)}{m(S^*)}$, where $m(S^*)$ is the size of an optimum solution S^* and $m(S_A)$ is the approximate solution computed by some algorithm A. A problem is said to be ϵ-approximable if a polynomial time algorithm A exists such that the approximation error is never greater than ϵ.

The technique used in the previous theorem can be exploited to prove a stronger result.

Theorem 5. *Given a graph G and a set of communication requests R in G, it is* NP*-hard to approximate the*

$$\min\{r_G(\mathcal{P}) : \mathcal{P} \text{ includes exactly one shortest dipath for each couple in } R\}$$

within an error in $\mathbf{O}(k^\epsilon)$ *for any* $\epsilon < 1$, *where* $k = |R|$.

The previous results motivate us in the next section to look for minimal schemes for some classes of graph which are widely used in distributed and parallel systems.

4 Bounds for fixed topologies

In this section we consider only the case where $R = V \times V$ and $\mathcal{P}$ includes all shortest dipaths connecting every couple of vertices and we provide new bounds on r_G (i.e. $r_G(\mathcal{P})$) for tori, grids and hypercubes, which are classical interconnection networks.

The Cartesian sum $G_1 \Box G_2$ of two graphs $G_1 = (V_1, E_1)$ and $G_2 = (V_2, E_2)$, often called Cartesian product or box product, is the graph whose vertices are the pairs (x_1, x_2) where x_1 is a vertex of G_1 and x_2 is a vertex of G_2. Two vertices (x_1, x_2) and (y_1, y_2) of $G_1 \Box G_2$ are adjacent if and only if $x_1 = y_1$ and (x_2, y_2) is an arc of G_2 or $x_2 = y_2$ and (x_1, y_1) is an arc of G_1. Many graphs can be defined in terms of Cartesian sum of simpler graphs:

- the *hypercube* H_d can be recursively defined from K_2 (the complete graph on two vertices) by $H_d = K_2 \Box H_{d-1} = \underbrace{K_2 \Box \ldots \Box K_2}_{d \text{ times}}$;
- the *toroidal mesh* or *torus* $T_{p\times q}$ is the Cartesian sum $C_p \Box C_q$ (where C_h is the cycle with h vertices);
- the *mesh* or *grid* $G_{p\times q}$ is the Cartesian sum $P_p \Box P_q$ (P_h being the path of h vertices).

Before starting the analysis for fixed topologies, let us remark one of the key properties of orientations related to the traversability of a cycle. Let C be a 4-cycle consisting of the arcs e_0, e_1, e_2, e_3. As any orientation $\vec{G}$ is acyclic, in the subgraph induced by the cycle C there is at least one sink and one source. So, if we consider any four dipaths of length 2: P_0, P_1, P_2, P_3, where P_i contains arcs e_i and $e_{(i+1) \bmod 4}$, at least two of them have one change of orientation in the cycle.

We first consider torus networks. The vertices of $T_{p\times q}$ will be denoted as (i,j) with $i \in Z_p$, $j \in Z_q$. Vertex (i,j) is joined to vertices $(i+1,j)$ and $(i-1,j)$ by horizontal arcs and to vertices $(i, j+1)$ and $(i, j-1)$ by vertical arcs.

Theorem 6. *Let $p \geq q$, then $r_{T_{p\times q}} \geq \lfloor \frac{q}{2} \rfloor + 2$.*

Proof. Let $p' = \lfloor \frac{p}{2} \rfloor$, $q' = \lfloor \frac{q}{2} \rfloor$ and $N = pq$ (the number of vertices).

Consider first the case $p' = q'$. Let $\mathcal{P}_s$ be the subset of the set of all shortest dipaths $\mathcal{P}$ constituted by the following $8N$ "staircase dipaths": for each vertex (i,j) we associate 8 shortest dipaths of length the diameter $D = p' + q' = 2q'$ where arcs alternate in directions. Such dipaths are of the form $(e_1, f_1, e_2, f_2, \ldots, e_{q'}, f_{q'})$ where the e_i's are all horizontal (resp. vertical) arcs and all the f_i's vertical (resp. horizontal). These dipaths join vertex (i,j) to vertices $(i \mp p', j \mp q')$.

Notice that if a dipath from (i,j) to (i',j') belongs to $\mathcal{P}_s$ then the opposite dipath from (i',j') to (i,j) also belongs to $\mathcal{P}_s$.

Due to the symmetry of the torus, each of the 8 dipaths of length 2 of any 4-cycle belongs to the same number $2(2q'-1)$ of dipaths in $\mathcal{P}_s$. So, for any acyclic orientation $\vec{T}_{p\times q}$, the N cycles of length 4 yield globally a total of $4N(2q'-1)$ changes of orientation over the $8N$ dipaths in $\mathcal{P}_s$.

Therefore, either one dipath of $\mathcal{P}_s$ has at least $q'+1$ changes or $4N$ dipaths in $\mathcal{P}_s$ have exactly q' changes of orientation and the remaining $4N$ dipaths of $\mathcal{P}_s$ have $q'-1$ changes. If there is a dipath P with $q'+1$ changes, then by definition of rank $r_{T_{p\times q}}(\mathcal{P}) \geq r(\mathcal{P}, \vec{T}_{p\times q}) \geq q'+2$ and we have proven the lower bound, so let us suppose that the second condition holds.

In this case assume by contradiction that $r_{T_{p\times q}}(\mathcal{P}) \leq q'+1$. Since there are as many dipaths in $\mathcal{P}_s$ starting with an arc in $\vec{T}_{p\times q}$ than with an arc not in $\vec{T}_{p\times q}$, this means that all the dipaths starting with an arc not in $\vec{T}_{p\times q}$ have $q'-1$ changes of orientation (otherwise they would have rank $q'+2$) and all the dipaths starting with an arc in $\vec{T}_{p\times q}$ have q' changes.

In this case, all the dipaths in $\mathcal{P}_s$ should have the last (vertical arc) in $\vec{T}_{p\times q}$ if q' is even and not in $\vec{T}_{p\times q}$ if q' is odd, but this is impossible since for for any i and j there are dipaths in $\mathcal{P}_s$ ending with arc $((i,j),(i,j+1))$ and dipaths ending with arc $((i,j+1),(i,j))$.

Suppose now $p' > q'$. We use a similar technique, but now we take the set of dipaths $\mathcal{P}_s$ as the $4N$ shortest dipaths of length $2q'+1$ $(\leq D)$ starting at any vertex with a horizontal arc and where arcs alternate (so the last one is horizontal). These dipaths join vertex (i,j) to vertices $(i \mp (q'+1)', j \mp q')$. The total number of changes yielded by the N 4-cycles is now $2N(2q')$ for the $4N$ dipaths of $\mathcal{P}_s$. Therefore, either one dipath in $\mathcal{P}_s$ has at least $q'+1$ changes of orientation, or all dipaths of $\mathcal{P}_s$ have q' changes. If there is a dipath with $q'+1$ changes we have proven the lower bound, otherwise all dipaths in $\mathcal{P}_s$ starting with an arc not in $\vec{T}_{p\times q}$ (one half of the total) have rank at least $q'+2$.

For tori we have an upper bound of the same order.

Theorem 7. *Let $p \geq q$, then $r_{T_{p\times q}} \leq \lceil \frac{q}{2} \rceil + 4$.*

Sketch of Proof. It suffices to consider the acyclic orientation such that all vertical arcs are oriented from (i,j) to $(i,j+1)$ for $0 \leq j \leq n-2$ and from $(i,0)$ to $(i,n-1)$. Horizontal arcs are oriented if j is even from (i,j) to $(i+1,j)$ for $0 \leq i \leq n-2$ and from $(0,j)$ to $(n-1,j)$, while if j is odd from $(i+1,j)$ to (i,j) for $0 \leq i \leq n-2$ and from $(n-1,j)$ to $(0,j)$.

For this acyclic orientation we can check that any shortest dipath has rank at most $\lceil \frac{q}{2} \rceil + 4$.

Also better bounds can be determined for grid networks. In [6] it has been proved that $r_{G_{q\times q}} \geq \lceil \frac{q-1}{3} \rceil$. This lower bound can be easily improved to $\lceil \frac{q}{2} \rceil$ using a similar proof as for tori. However, we have been able to obtain a better value which we conjecture to give the right order.

Theorem 8. *Let $p \geq q$, then $r_{G_{p\times q}} \geq \lceil (2-\sqrt{2})q \rceil - 1$.*

Proof. Consider only the $q \times q$ subgrid $G_{q\times q}$ of $G_{p\times q}$ induced by vertices (i,j) such that $0 \le i \le q-1$ and $0 \le j \le q-1$. Let α be a fixed number such that $\frac{q-1}{2} \le \alpha \le q-1$. The sets of shortest dipaths considered will consist of two disjoint sets $\mathcal{P}_1$ and $\mathcal{P}_2$. $\mathcal{P}_1$ contains the 2α dipaths from $(0,0)$ to $(q-1,q-1)$ constituted by a sequence of horizontal (resp. vertical) arcs till a given vertex $(j,0)$ (resp. $(0,j)$), where $1 \le j \le \alpha$, then followed by arcs alternating in direction starting with a vertical (resp. horizontal) arc, then by a vertical (resp. horizontal) dipath from $(q-1,q-1-j)$ (resp. $(q-1-j,q-1)$) to $(q-1,q-1)$. We will call such dipaths "almost staircase". $\mathcal{P}_2$ consists of the 2α "almost staircase" shortest dipaths from $(0,q-1)$ to $(q-1,0)$ constituted by an horizontal (resp. vertical) dipath till $(j,q-1)$ (resp. $(0,q-1-j)$) with $1 \le j \le \alpha$, then a staircase dipath and finally a vertical (resp. horizontal) one.

Any 4-cycle will be said to be "inner" if it consists of the four vertices $(i,j),(i,j+1),(i+1,j+1)$ and $(i+1,j)$ where $q-\alpha-1 \le i+j \le q+\alpha-3$, $i-j \le \alpha-1$, $j-i \le \alpha-1$. Hence, the total number of inner cycles is $c = (q-1)^2 - 2(q-\alpha-1)(q-\alpha)$.

Notice that, for each inner cycle there are exactly 2 dipaths of $\mathcal{P}_1$ using respectively the arcs $(i,j)(i+1,j)(i+1,j+1)$, $(i,j)(i,j+1)(i+1,j+1)$ and 2 dipaths of $\mathcal{P}_2$ using the arcs $(i,j+1)(i,j)(i+1,j)$ and $(i,j+1)(i+1,j+1)(i+1,j)$. By the remark on the acyclicity of the orientations, at least two of these dipaths must change orientation inside the cycle. Hence, the c inner cycles yield globally a total of at least $2c$ changes of orientation over all the dipaths of $\mathcal{P}_1 \cup \mathcal{P}_2$.

Since $|\mathcal{P}_1 \cup \mathcal{P}_2| = 4\alpha$, one dipath $P \in \mathcal{P}_1 \cup \mathcal{P}_2$ has at least $\frac{c}{2\alpha} = \frac{1}{2\alpha}(-(q-1)^2 + 2(2q-1)\alpha - 2\alpha^2)$ changes of orientation.

A simple derivation shows that $\frac{c}{2\alpha}$ is the maximum for $\alpha = \sqrt{\frac{q^2-1}{2}}$. For this value of α, it gives $\frac{c}{2\alpha} = 2(q-1) - \sqrt{2(q^2-1)}$. Since we are considering only integers one can show that $\frac{c}{2\alpha} \ge (2-\sqrt{2})q - 2$. So the dipath P has rank at least $\lceil (2-\sqrt{2})q \rceil - 1$.

We conjecture that this lower bound is asymptotically the right order for the value $r_{G_{q\times q}}$. Till now we have been able to design a simple construction yielding rank $\frac{2q}{3} + o(q)$ and a slightly more complicated one of order $\frac{3q}{5} + o(q)$. According to our method, we conjecture the existence of a recursive solution of order $\frac{aq}{b}$ for any fraction $a/b \ge 2 - \sqrt{2}$ and for q large enough.

Consider now an hypercube H_d. In [6] it has been proved that $r_{H_d} \ge \lceil r \cdot (d+1) \rceil$, where $r = 1 - \frac{d}{2(d-1)}$. By using a proof similar to the above reasoning for the 4-cycles, this lower bound can be improved as follows.

Theorem 9. $r_{H_d} \ge \lceil \frac{d+1}{2} \rceil$.

Notice that the above bound is within a multiplicative factor of one half far from the trivial $d+1$ upper bound, i.e. the general one given for any network as the diameter plus one (see [6]).

As the shown results for tori, grids and hypercubes suggest, even for particular cases the task of determining tight bounds is not trivial. Anyway, in all the above cases it is possible to give acyclic orientations of rank at most twice the optimal one.

5 Conclusions and open problems

In this paper we have investigated the problem of finding acyclic orientations for communication networks in order to prevent deadlock configurations.

In particular, new results have been presented both from a theoretical computational complexity point of view and from a practical one by providing concrete bounds on deadlock free routing schemes for specific topologies.

One of the main questions left open in this paper is whether or not the problem of minimizing the number of buffers yielded by the acyclic orientations can be approximated in polynomial time when all shortest dipaths between each communication request must be represented.

Concerning the topology dependent results, while tight bounds have been determined for tori, it would be worthwhile to establish the exact order for $q \times q$ grids (we conjecture a value of order $(2-\sqrt{2})q$) and hypercubes of dimension d (we conjecture order d).

Finally, it would be worth to extend the known results to more general classes of networks.

References

1. B. Awerbuch, S. Kutten, and D. Peleg. Efficient deadlock-free routing. In *10th Annual ACM Symposium on Principles of Distributed Computing (PODC)*, pages 177–188, Montreal, Canada, 1991.
2. P.E. Berman, L. Gravano, G.D. Pifarré, and J.L.C. Sanz. Adaptive deadlock and livelock-free routing with all minimal paths in torus networks. In *4th Symposium on Parallel Algorithms and Architectures (SPAA)*, pages 3–12, June 1992.
3. J.C. Bermond and M. Syska. Routage wormhole et canaux virtuel. In M. Cosnard M. Nivat and Y. Robert, editors, *Algorithmique Parallèle*, pages 149–158. Masson, 1992.
4. Robert Cypher and Luis Gravano. Requirements for deadlock-free, adaptive packet routing. In 11*th Annual ACM Symposium on Principles of Distributed Computing (PODC)*, pages 25–33, 1992.
5. W. J. Dally and C. L. Seitz. Deadlock-free message routing in multiprocessor interconnection networks. *IEEE Trans. Comp.*, C-36, N.5:547–553, May 1987.
6. M. Di Ianni, M. Flammini, R. Flammini, and S. Salomone. Systolic acyclic orientations for deadlock prevention. In *2nd Colloquium on Structural Information and Communication Complexity (SIROCCO)*, pages 1–12. Carleton University Press, 1995.
7. J. Duato. Deadlock-free adaptive routing algorithms for multicomputers: evaluation of a new algorithm. In *3rd IEEE Symposium on Parallel and Distributed Processing*, 1991.
8. J. Duato. On the design of deadlock-free adaptive routing algorithms for multicomputers: theoretical aspects. In *2nd European Conference on Distributed Memory*

Computing, volume 487 of *Lecture Notes in Computer Science*, pages 234–243. Springer-Verlag, 1991.

9. E. Fleury and P. Fraigniaud. Deadlocks in adaptive wormhole routing. Research Report, Laboratoire de l'Informatique du Parallélisme, LIP, École Normale Supérieure de Lyon, 69364 Lyon Cedex 07, France, March 1994.
10. M.R. Garey and D.S. Johnson. *Computers and Intractability. A Guide to the Theory of NP-completeness.* W.H. Freeman, 1977.
11. K.D. Gunther. Prevention of deadlock in packet-switched data transport system. *IEEE Trans. on Commun.*, COM-29:512–514, May 1981.
12. P.M. Merlin and P.J. Schweitzer. Deadlock avoidance in store-and-forward networks: Store and forward deadlock. *IEEE Trans. on Commun.*, COM-28:345–352, March 1980.
13. G.D. Pifarré, L. Gravano, S.A. Felperin, and J.L.C. Sanz. Fully-adaptive minimal deadlock-free packet routing in hypercube, meshes, and other networks. In *3rd Symposium on Parallel Algorithms and Architectures (SPAA)*, pages 278–290, June 1991.
14. A.G. Ranade. How to emulate shared memory. In *Foundation of Computer Science*, pages 185–194, 1985.
15. Gerard Tel. *Introduction to Distributed Algorithms.* Cambridge University Press, Cambridge, U.K., 1994.

Weak-Order Extensions of an Order*

Karell Bertet[1], Jens Gustedt[2], and Michel Morvan[1]

[1] LIAFA/IBP - Université Denis Diderot Paris 7 - Case 7014 - 2, place Jussieu - 75256 Paris Cedex 05 - France. Email: {bertet,morvan}@litp.ibp.fr

[2] Technische Universität Berlin, Sekr. MA 6-1, D-10623 Berlin, Germany. Email: gustedt@math.tu-berlin.de. Supported by IFP *Digitale Filter*.

Abstract In this paper, at first we describe a graph representing all the weak-order extensions of a partially ordered set and an algorithm generating them. Then we present a graph representing all of the minimal weak-order extensions of a partially ordered set, and implying a generation algorithm. Finally, we prove that the number of weak-order extensions of a partially ordered set is a comparability invariant, whereas the number of minimal weak-order extensions of a partially ordered set is not a comparability invariant.

1 Introduction and Motivations

In this paper, we are interested in the algorithmic and structural study of extensions of a partially ordered set, **orders** for short. The extensions are restricted to a certain class of orders. A lot of previous works deals with studies of restricted extensions classes:

- The linear extensions (extensions which are total orders) of an order are in one-to-one correspondence with the maximal chains of the lattice of the antichains of the order [2].
- The minimal interval extensions of an order are in one-to-one correspondence with the maximal chains of the lattices of the maximal antichains of the order [8].
- The MacNeille completion of an order studied in [3, 9] is an extension of an order belonging to the class of lattices.
- Series-parallel orders are used as extensions of an order to resolve scheduling problems [11].

Among these classes, exist particular extensions which are the extensions of an order obtained by only adding some comparabilities in the order, as the linear extensions or the minimal interval extensions. We are interested in these extensions, especially the weak-order extensions of an order. Informally, a weak-order is an order composed of a set of complete bipartite orders one above an other. Weak-order extensions are suited for the scheduling of tasks [5, 6]: consider a partial order of tasks, a weak-order extension of this order is a scheduling of the

* This work was supported by the PROCOPE Program

tasks over processes or machine in the time. In this way, Lamport's work on time-stamping in [10] can be seen as on-line computing of a particular weak-order extension of the causal order associated to a distributed execution.

In Sect. 3 we present a one-to-one correspondence between all the weak-order extensions of an order and all the paths from the unique source to the unique sink of a certain graph. This result is related to the similar characterization of linear and minimal interval extensions cited above. We use this characterization to develop an efficient generation algorithm.

Sect. 4 deals with the minimal weak-order extensions of an order. We first present a one-to-one correspondence between all the minimal weak-order extensions of an order and all the paths from any source to any sink of a particular graph. This graph is not a suborder of the above graph since the minimal weak-order extensions of an order are not directly implied from the weak-order extensions as we illustrate with an example. We also present an efficient generation algorithm implied from this graph.

The notion of comparability invariance is fundamental in the study of orders [4, 7, 8]. It is based on the notion of a comparability graph associated to any order. A comparability graph of an order is the undirected graph obtained by deleting the direction on the edges of the order. A parameter of an order is a comparability invariant if it has the same value on any other order having the same comparability graph. Almost all classical parameters on orders are comparability invariants. For example, the number of linear extensions, the dimension, the jump number, the number of the minimal interval extensions are comparability invariants. On the other hand, the number of the interval extensions of an order is not a comparability invariant.

Surprisingly, the inverse statement than for the interval extensions holds for the weak-order extensions: the number of weak-order extensions is a comparability invariant whereas the number of minimal weak-order extensions is not a comparability invariant, as we show in Sect. 5.

2 Definitions and Notations

A *partially ordered set* $P = (X, \leq_P)$ is a reflexive, antisymmetric and transitive binary relation on a set X. Instead of partially ordered set, we often talk about an *order*. We represent an order by a diagram (Hasse diagram) where $x <_P y$ if and only if there is a sequence of connected lines moving upwards from x to y.

Two distinct elements x and y are said to be *comparable* if $x \leq_P y$ or $y \leq_P x$. Otherwise, they are said *incomparable*, denoted by $x \parallel_P y$. We say that y *covers* x, $x \prec_P y$, iff $x <_P y$ and there is no z such that $x <_P z <_P y$.

We define the following sets for P, for an element x of P, and for a subset A

of P:

$$\begin{aligned}
\mathit{Max}(P) &= \{x \in X \mid \text{ for all } y \in X, y \not>_P x\} \\
\mathit{Min}(P) &= \{x \in X \mid \text{ for all } y \in X, x \not<_P y\} \\
\mathit{Ideal}(A) &= \{y \in X \mid y \leq_P x, \text{ for some } x \in A\} \\
\mathit{Pred}(x) &= \{y \in X \mid y <_P x\} \\
\mathit{Pred}(A) &= \bigcup_{x \in A} \mathit{Pred}(x) \\
\mathit{Succ}(x) &= \{y \in X \mid y >_P x\} \\
\mathit{Succ}(A) &= \bigcup_{x \in A} \mathit{Succ}(x)
\end{aligned}$$

A subset A of X is called an *antichain* (resp. *chain*) of P if it contains only pairwise incomparable (resp. comparable) elements. We denote by A_P the set of all antichains of P. A subset A of X is a *maximal antichain* (resp. chain) if it is maximal under inclusion.

$A(P)$ is the order on A_P defined as follows: $A \leq_{A(P)} B$ iff for all x in A, there is y in B such that $x \leq_P y$. It is well known that $A(P)$ equipped with that order is a distributive lattice. By $AM(P)$, we denote the suborder of $A(P)$ restricted to the maximal antichains of P. $AM(P)$ is a lattice, but in general it is not distributive.

The ordering on P is a *weak-order* iff it does not contain the order $2 \bigoplus 1$ as a suborder. Here, $2 \bigoplus 1$ denotes the union of a singleton and a chain composed of two elements. An other characterization of a weak-order P is that $AM(P)$ is a total order such that for all $A \neq B$ in $AM(P)$, $A \cap B = \emptyset$. This allows us to represent a weak-order by a sequence of antichains $A_0, \ldots, A_n$ with $A_i <_{A(P)} A_{i+1}$.

A *directed graph* $G = (X, E)$ is given by a set X of elements or *nodes*, and a subset $E \subseteq X \times X$, the *arcs*. A subset $x_1, \ldots, x_n$ of X such that $(x_i, x_{i+1}) \in E$ for $i < n$ is called a *path* from x_1 to x_n. An node $x \in X$ such that for all $y \in X$ there is no arc (x, y) is called a *sink*. If there is no arc (y, x), x is called a *source*.

3 Weak-Order Extensions of an Order

In this section, we define a graph which represents all weak-orders extensions of an order P. It gives rise to a one-to-one correspondence between certain paths and all the weak-orders extensions of the P. From this graph we define an efficient generation algorithm.

An order $Q = (X, \leq_Q)$ is an *extension* of an order $P = (X, \leq_P)$ if and only if for all x and y in X, $x \leq_P y$ implies $x \leq_Q y$. Then we say that P is a *reduction* of Q. If P is not a weak-order, it clearly admits weak-order extensions.

Definition 3.1. Let $P = (X, \leq_P)$ be an order. We define the directed graph $\mathit{WE}(P) = (A_P, E_{we})$ as follows. For $A \neq B$ two antichains of P, $(A, B) \in E_{we}$

iff the following two conditions are satisfied:

$$A \subseteq \mathit{Ideal}(B) \tag{1}$$

$$B \setminus A = \mathit{Ideal}(B) \setminus \mathit{Ideal}(A). \tag{2}$$

The binary relation induced by E_{we} is an anti-reflexive and antisymmetric relation. So, the reflexo-transitive closure of $WE(P)$ is $A(P)$.

Since $A(P)$ admits a minimal element which is $\emptyset$ and a maximal element which is $Max(P)$, the same holds for $WE(P)$ which admits a unique source and a unique sink.

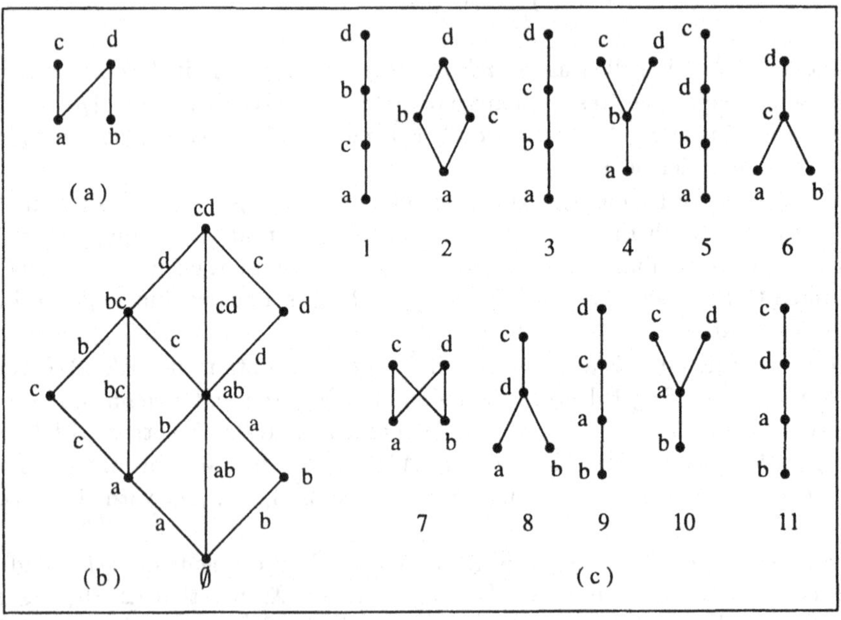

Figure 1. Weak-Order Extensions of an Order

Let P be the order in (a) in Fig. 1. The directed graph $WE(P)$ is given in (b), with arcs labeled with the difference between the two corresponding vertices (direction of the arcs is from bottom to top). The eleven weak-order extensions of P are represented in (c). All these orders are represented by their Hasse diagram. We see that there is a correspondence between the labeled arcs of $WE(P)$ and the weak-order extensions of P, and that $WE(P)$ admits a unique source, and a unique sink. This correspondence is such that $WE(P)$ represents all the weak-orders extensions of P as follows:

Theorem 3.2. *There is a one-to-one correspondence between all the paths of $WE(P)$ from the unique source to the unique sink and all the weak-orders extensions of P.*

For a sketch of a proof let us just describe the mapping. Let $A_0, \ldots, A_n$ be a path of $WE(P)$ from the source to the sink. Then $A_1 \backslash A_0, \ldots, A_i \backslash A_{i-1}, \ldots, A_n \backslash A_{n-1}$ are the maximal antichains of a weak-order extension of P.

The definition of $WE(P)$ can easily be modified as follows: For $A \neq B$ two antichains of P, $(A, B) \in E_{we}$ iff the following two conditions are satisfied:

$$A \subseteq Ideal(B) \tag{3}$$

$$B \setminus A \subseteq Min(P \setminus Ideal(A)) \tag{4}$$

This new definition gives us a way to compute all the weak-orders extensions of an order: Consider that we have $Min(P \setminus Ideal(A_i))$, for a path $A_0, \ldots, A_n$ of $WE(P)$ from the source to the sink. Then we compute $B_{i+1} = A_{i+1} \setminus A_i$ by choosing a subset of this set, and

$$Min\,(P \setminus Ideal\,(A_{i+1})) = Min\,(P \setminus (Ideal\,(A_i) \cup B_{i+1}))\,. \tag{5}$$

At the beginning, $A_0 = \emptyset$ and $Min(P \backslash Ideal(\emptyset)) = Min(P)$. So we may conclude that

$$A_{i+1} = Max\,(Ideal\,(A_i) \cup B_{i+1})\,. \tag{6}$$

Algorithm 1: Weak-Order Extensions of an Order

Input: The arrays *Succ* and *Pred* for an order P reduced transitively

Output: The weak-orders extensions of P and the number of weak-order extensions of P

```
begin
    let L be an inverse linear extension of P;
    for x in L such that x not visited do
        y = x;
        while |Pred(y)| = 1 and |Succ(Pred(y))| = 1 do
            y = Pred(y);
            EndChain(y) = x;
            mark y visited;
    nbext = Find1(Min(P));
    print "there are" nbext "weak-order extensions";
end
```

Algorithm 1 computes all the weak-orders extensions of an order by using the recursive function *Find1* that is presented in Algorithm 2. It distinguishes the special case that $Min(P \setminus Ideal(A_i))$ contains only one element. This allows to amortize the complexity as it is done in the following theorem. Here m is the number of comparabilities of the transitive reduction of P, n_w is the number of weak-order extensions of P, and Δ is the maximum number of immediate successors of the elements of P.

Theorem 3.3. *Algorithm 1 computes all the weak-order extensions of an order P, and requires $O(m)$ space and $O(n_w\Delta + m)$ time.*

The main idea of the proof is to amortize the work that is done for an individual extension by distributing the cost of a call to *Find1* to the subsequent recursive calls that are issued by this call.

Algorithm 2: The Function *Find1*

```
Input: X a subset
Output: The weak-orders extensions of P \ (Ideal(X) ∪ X) and the num-
        ber of weak-order extensions of P \ (Ideal(X) ∪ X)
begin
    nbext = 0;
    if X = ∅ then
        print "End of a weak-order extension";
        return 1;
    if |X| = 1 then
        if EndChain(X) exist then
            print "X → EndChain(X)";
            X = Succ(EndChain(X));
        else
            print "X";
            X = Succ(X);
    foreach B ⊆ X, B ≠ ∅ do
        print "B";
        X = Min((X \ B) ∪ Succ(B));
        nbext+ = Find1(X);
    return nbext;
end
```

It is also possible to obtain a better time complexity by increasing the space complexity if we explicitly compute $WE(P)$.

Algorithm 1 entirely computes all the weak-order extensions of P, but these extensions have common parts which are computed several times. With the knowledge of parts of these extensions already computed during the execution of the algorithm, we can avoid this. If we consider $WE(P)$, at each step i we can compute the corresponding node A_i of $WE(P)$ which is $Max(Ideal(A_i) \cup B_i)$. If this node already exists in $WE(P)$ then the corresponding part of the path is already computed and vice-versa.

Algorithm 3 describes the function *Find2* which is a modified version of *Find1* that enables us to compute $WE(P)$ in addition. Initially, it is called as $Find2(Min(P), \emptyset)$. Let n' be the number of elements of $WE(P)$, m' be the

Algorithm 3: The Function *Find2*

```
Input: Y a subset, A an antichain of P
Output: The arrays Succ for WE(P)
begin
    if Y = ∅ then
        return ;
    foreach B ⊆ Y, B ≠ ∅ do
        A' = Max(Ideal(A) ∪ B);
        Succ(A) = Succ(A) ∪ A';
        if A' not visited then
            Y' = Min((Y \ B) ∪ Succ(B));
            Find2(Y', A');
        return;
end
```

number of comparabilities of $WE(P)$ transitively reduced, and w be the width of P, that is the maximum size of an antichain of P. Then we have:

Theorem 3.4. *Algorithm 3 computes WE(P) and uses a space of $O(wn' + m')$ and a time of $O(m'w \log n)$.* □

Then, to compute all the weak-order extensions of P, we have to visit all the paths of $WE(P)$ from the source to the sink. So, we have:

Corollary 3.5. *By Algorithm 3, it is possible to compute all the weak-order extensions of an order P in $O(m + m'w \log n + n_w)$ time and $O(wn' + m')$ space.*

4 Minimal Weak-Order Extensions of an Order

Now, we characterize the minimal weak-order extensions of an order. Then we present a one-to-one correspondence between all the minimal weak-order extensions of an order and certain paths of a graph and we use this correspondence to develop an efficient generation algorithm.

A weak-order extension Q of P is a *minimal weak-order extension* of P if there is no weak-order extension Q' of P such that Q is an extension of Q'. Informally, a minimal weak-order extension of P is a weak-order extension of P which is as close as possible to P.

The main part of Algorithms 1 and 3 was to choose a subset B of Y, and to delete this subset from Y. In this way, all the weak-orders extensions of an order have been computed. We easily could add conditions to this choice, as e.g the size of the chosen subset. But if we want all the minimal weak-order extensions of P, there are no obvious local conditions that only involve the subset B chosen in Y at each step. The way to define and to compute them is not directly inherited from the general case.

We have the following characterization of a minimal weak-order extension of an order:

Lemma 4.1. *Let $P = (X, \leq_P)$ be an order. Let $Q = A_0, \dots, A_n$ be a weak-order extension of P. The two following properties are equivalent:*

1. *Q is a minimal weak-order extension of P.*
2. *For all A_i such that $i < n$, there are x in A_i, y in A_{i+1} such that $x \prec_P y$.*

□

In the same way as for the weak-order extensions of P, our goal is now to define a directed graph such that there is a one-to-one correspondence between certain paths of this graph and the minimal weak-order extensions of P. Suppose we choose *WE*(P) restricted to the paths from the source to the sink such that the corresponding weak-order extension verify Lemma 4.1. Let us demonstrate by Fig. 2 that this graph contains paths that do not correspond to a minimal weak-order extension of P.

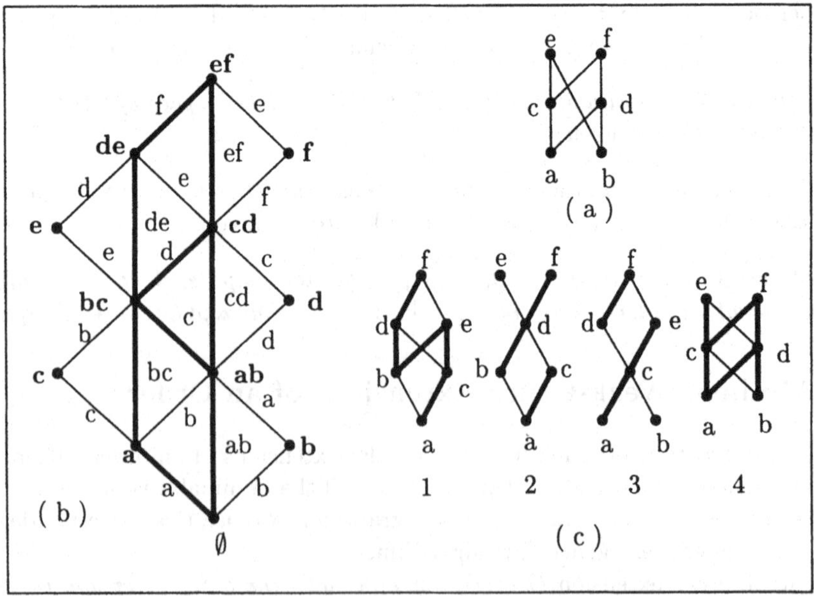

Figure 2. Creation of Wrong Paths

Let P be the order in (a) and *WE*(P) in (b) of Fig. 2. The minimal weak-order extensions of P are represented in (c), with covering relations as required for Lemma 4.1 in bold. The corresponding paths of *WE*(P) are given in bold, too. Indeed the subgraph induced by these paths contains 5 paths from $\emptyset$ to ef instead of 4: ab, c, d, ef does not correspond to a minimal weak-order extension of P because

$$\{c,d\} \setminus \{b,c\} \cup \{b,c\} \setminus \{a,b\} = \{c,d\} \tag{7}$$

is an antichain of P.

To avoid this, we use the linegraph transformation: We replace a node of $WE(P)$ belonging to a valid path by an arc, and an arc of $WE(P)$ belonging to a valid path by a node associated with one of the extremities of this arc. Then we differentiate nodes with the same label.

Let us now give a formal definition of this directed graph representing the minimal weak-order extensions of an order:

Definition 4.2. Let $P = (X, \leq_P)$ be an order. We define $WE_m(P) = (\mathcal{X}, \mathcal{E})$, with $\mathcal{X}$ the set of pairs of antichains of P, as follows. For $A_0, \ldots, A_n$ a path of $WE(P)$ from the source to the sink, and $B_i = A_i \setminus A_{i-1}$ with $0 < i < n$ such that $B_i \cup B_{i+1}$ is not an antichain of P include the following objects into WE_m:

$$\begin{array}{ll} (A_i, B_i) \in \mathcal{X} & \text{for all } 0 < i \leq n \\ ((A_i, B_i), (A_{i+1}, B_{i+1})) \in \mathcal{E} & \text{for all } i \text{ with } 0 < i < n \end{array}$$

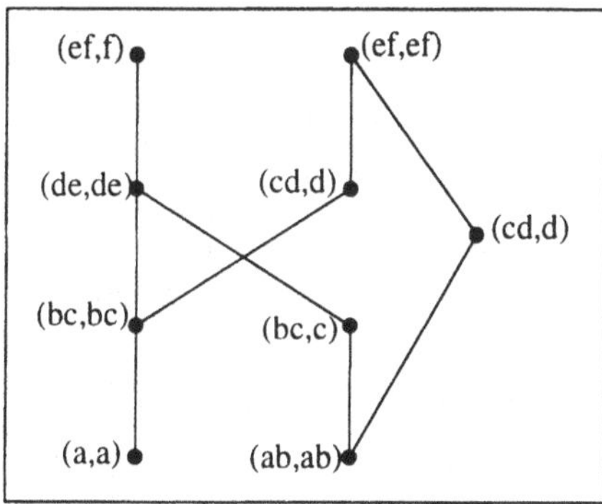

Figure 3. Minimal Weak-Order Extensions of an Order

Let P be the order in (a) of Fig. 2; the directed graph $WE_m(P)$ for P is shown in Fig. 3; the corresponding minimal weak-order extensions are given in (c) of Fig. 2.

This graph represents the minimal weak-order extensions of an order as follows:

Lemma 4.3. *There is a one-to-one mapping between the paths of* $WE_m(P)$ *from any source to any sink and the minimal weak-orders extensions of* P. □

Now, our goal is to compute all the minimal weak-order extensions of P. Consider the function *Find1* in Algorithm 2 and the function *Find2* in Algorithm 3. The principal step in these functions is to choose a subset B in Y, Y being the current set of minimal elements. Suppose the last set that was chosen is called B', then computing a minimal weak-order extension of P consists in choosing a subset B in Y with respect to Lemma 4.1 which is equivalent to the two following conditions:

C1: There are $x' \in B'$ and $x \in B$ with $x' <_P x$.
C2: The deletion of B introduces a new minimal element x''.

These conditions are clearly necessary. They are also sufficient since there always is a trivial choice possible in each step: choosing the whole set of minimal elements.

Let $New = Min(Succ(Y))$. Then x'' is in *New* and we may choose a new subset B of Y such that

$$Z = Min((Y \setminus B) \cup Succ(B)) \cap New \neq \emptyset. \tag{8}$$

Then, to verify condition C1, we have to choose a subset B of Y such that B contains x as above. This can easily be done with a set *Mark* containing all these x added in Y in the previous recursive call. In this way, we have to choose B such that

$$B \cap \mathit{Mark} \neq \emptyset. \tag{9}$$

So, conditions C1 and C2 are detailed by Eq. 9 and 8, resp.

Every valid pair (B, Z) contains a pair (i, j) where $i \in Mark \cap B$ and $j \in Z$. Then if we assume that *Mark* and *New* are totally ordered, we can associate to each such pair (that is each B) the lexicographically minimal such pair (i, j).

Using that fact, it is possible to avoid the generation of invalid subsets of Y and to enumerate all valid B in time asymptotically smaller than n^2 times the number of valid B. A more subtle analysis leads to a better bound but has to be omitted for the sake of brevity. We can state the following result.

Proposition 4.4. *It is possible to compute all the minimal weak-order extensions of an order with an amortized complexity which requires* $O(m)$ *space and* $O(n_{wm}n^2\Delta + m)$ *time.* □

In a second approach we compute $WE_m(P)$ in order to avoid computing parts of minimal extensions several times as in Algorithm 3 of Sect. 3, and we compute all the minimal weak-orders extensions with a visit of $WE_m(P)$. So, we have the following complexity, where n'' is the number of elements of $WE_m(P)$ and m'' is the number of comparabilities of the transitive reduction of $WE_m(P)$:

Proposition 4.5. *It is possible to compute all the minimal weak-orders extensions of an order in space* $O(wn'' + m'')$ *and in time* $O(m''w^2\Delta + n_{wm})$. □

5 Comparability Invariants

Here, we will prove that the number n_w of weak-order extensions and the number n_{wm} of *minimal* weak-order extensions behave quite differently with respect to the property of being a comparability invariant. Indeed, whereas the first is such an invariant, the second is *not*, as will be shown by an example.

The *comparability graph* of an order P is the undirected graph obtained from P -seen as a directed graph- by deleting the direction of the arcs. A parameter of an order is *comparability invariant* if whenever two orders P and Q have isomorphic comparability graphs, the value of the parameter is the same for P and Q.

The *reversed order* $P^d = (X, \leq_{P^d})$ of P is defined by $x \leq_{P^d} y$ iff $y \leq_P x$.

Definition 5.1 (Substitution). Let $P = (X, \leq_P)$ and $M = (Y, \leq_M)$ be two orders such that $X \cap Y = \emptyset$. Let a be in X. $P_a^M = (X \setminus \{a\} \cup Y, \leq_{P_a^M})$, the *substitution* of a by Q, is defined by $x \leq_{P_a^M} y$ iff one of the following cases holds:

$$\begin{aligned} x, y \in X &\text{ and } x \leq_P y \\ x, y \in Y &\text{ and } x \leq_M y \\ x \in X, y \in Y &\text{ and } x \leq_P a \\ y \in X, x \in Y &\text{ and } a \leq_P y. \end{aligned}$$

Below we will use the following theorem:

Theorem 5.2. *[7] A parameter α of finite orders is a comparability invariant iff for every pair of finite orders P and M, $W = M^d$, and every element a of P,*

$$\alpha\left(P_a^M\right) = \alpha\left(P_a^W\right). \tag{10}$$

Let us first demonstrate that n_{wm} is not a comparability invariant with the counter example represented in Fig. 4. Let P and M be the orders in (a) and (b) respectively and $W = M^d$. Then $WE_m\left(P_X^M\right)$ is represented in (c) and $WE_m\left(P_X^W\right)$ in (d). Clearly, P_X^M admits seven minimal weak-order extensions, and P_X^W eleven.

Theorem 5.3. *n_w is a comparability invariant.*

Proof. We prove this by giving a one-to-one correspondence between the weak-order extensions of P_a^M and the weak-order extensions of P_a^W, for any P, M, $W = M^d$ and $a \in P$.

Let $Q = C_0, \ldots, C_m$ be a weak-order extension of P_a^M. Clearly, the suborder of Q induced by the elements of P is a weak-order extension of $P \setminus \{a\}$, and the suborder of Q induced by the elements of M is a weak-order extension of M. Let $Q_P = B_0, \ldots, B_{n'}$ and $Q_M = A_0, \ldots, A_n$ be these two suborders respectively. We define the mapping $turn_M = Q' = C'_0, \ldots, C'_m$ for the weak-order extensions of P_a^M such that for each $j < m$:

$$C'_j = \begin{cases} (C_j \setminus A_i) \cup A_{n-i} & \text{for some } i \leq n \text{ such that } A_i \subseteq C_j \\ C_j & \text{if there is no such } i. \end{cases} \tag{11}$$

□

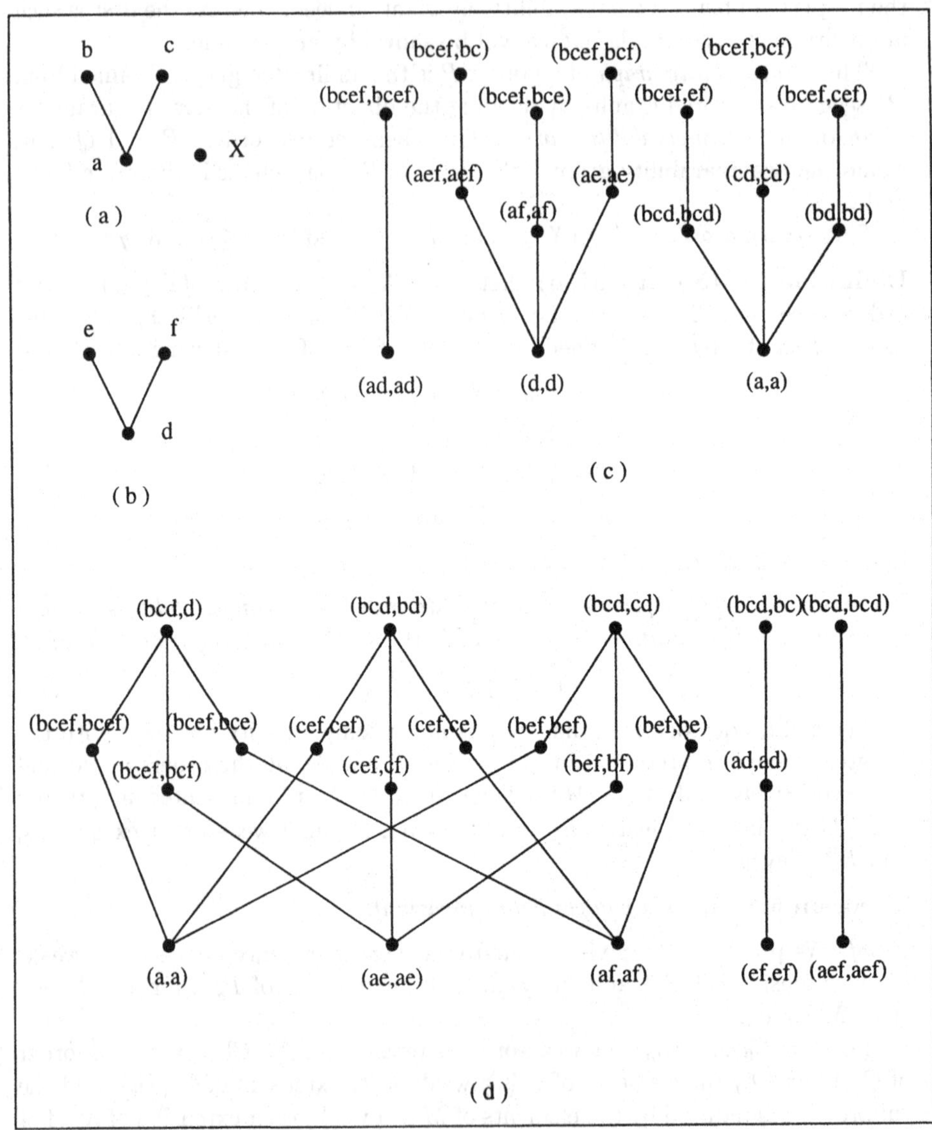

Figure 4. A Counter Example

References

1. BIRKHOFF, G. *Lattice Theory*, 3rd ed. American Math. Soc., Providence, RI, 1967.
2. BONNET, R., AND POUZET, M. Extensions et stratifications d'ensembles dispersés. *C. R. Acad. Sci. 268* (1969), 1512–1515.
3. BORDAT, J. Calcul pratique du treillis de gallois d'une correspondance. *Math. Sci. Hun. 96* (1986), 31–47.
4. BOUCHITTÉ, V., AND HABIB, M. The calculation of invariants for partially ordered sets. In *Algorithms and Order* (1989), I. Rival, Ed., Kluwer Acad. Publ., Dordrecht, pp. 231–279.
5. CHRETIENNE, P., AND CARLIER, J. *Problèmes d'Ordonnancement.* Masson, 1988.
6. CONWAY, R., MAXWELL, W., AND MILLER, L. *Theory of Scheduling.* Addison-Wesley, 1967.
7. DREESEN, B., POGUNTKE, W., AND WINKLER, P. Comparability invariance of the fixed point property. *Order 2* (1985), 269–274.
8. HABIB, M., MORVAN, M., POUZET, M., AND RAMPON, J. Extensions intervallaires minimales. *C. R. Acad. Sci. 313* (1991), 893–898.
9. JOURDAN, G.-V., RAMPON, J.-X., AND JARD, C. Computing on-line the lattice of maximal antichains of posets. *Order 11*, 3 (1994), 197–210.
10. LAMPORT, L. Time, clocks, and the ordering of events in a distributed system. *Communications of the ACM 21*, 7 (1978), 558–565.
11. MÖHRING, R. H., AND RADERMACHER, F. J. The order-theoretic approach to scheduling: The deterministic case. In *Advances in Project Scheduling* (1989), R. Slowinski and J. Weglarz, Eds., Elsevier Science Publishers B.V., Amsterdam, pp. 29–66.

An Upper Bound for the Maximum Cut Mean Value

Alberto Bertoni, Paola Campadelli*, and Roberto Posenato**

Dipartimento di Scienze dell'Informazione
Università degli Studi di Milano
Via Comelico 39
I-20135 Milano - Italy

Abstract Let $MaxCut(G)$ be the value of the maximum cut of a graph G. Let $f(x,n)$ be the expectation of $MaxCut(G)/xn$ for random graphs with n vertices and xn edges and let $r(x,n)$ be the expectation of $MaxCut(G)/xn$ for random $2x$-regular graphs with n vertices.
We prove, for sufficiently large x:
1. $\lim_{n\to\infty} f(x,n) \leq \frac{1}{2} + \sqrt{\frac{\ln 2}{2x}}$,
2. $\lim_{n\to\infty} r(x,n) \leq \frac{1}{2} + \frac{1}{\sqrt{x}} + \frac{1}{2}\frac{\ln x}{x}$.

1 Introduction

An experimental evaluation of the quality of approximate solutions of optimization problems requires the exact solution and, for **NP**-hard problems, this is possible only for small size instances. This difficulty can be overcome if the problem instances are chosen at random and a closed form for the expectation of the optimum solution value is known.

This technique has been successfully applied to the maximum clique problem; in this case a result due to Matula [5] gives an accurate estimate of the size of the maximum clique in a random graph when the number of vertices is sufficiently large. More precisely, for random graphs of density δ, the expected value $M(n,\delta)$ of the maximum clique is $M(n,\delta) \sim 2\log_{\frac{1}{\delta}} n$ as $n \to \infty$.

It is interesting to obtain similar results for other combinatorial optimization problems. In this perspective we consider the maximum cut problem, Max Cut, which, besides its theoretical importance, has applications in circuit layout design and statistical physics [1].

The Max Cut problem requires, given an undirected graph $G = (V, E)$, to find the set S of vertices that maximizes the cardinality of $Cut(S)$, that is the cardinality of the set of the edges with one endpoint in S and the other in $\overline{S}$.

Max Cut is one of the Karp's original **NP**-complete problems [6] and it is solvable in polynomial time for some special classes of graphs (e.g. if the

Acknowledgements: This work has been supported by grant CT 93.05230.ST74 of the Consiglio Nazionale delle Ricerche (CNR).

* campadelli@dsi.unimi.it
** posenato@dsi.unimi.it

graphs are planar [4]). As efficient algorithms are unlike to exist for **NP**-hard maximization problems, a typical approach to solving them consists in finding an ϵ-*approximation algorithm*, that is, a polynomial-time algorithm that delivers a solution of value at least ϵ times the optimal one. For several years, some 0.5-approximation algorithms for MAX CUT have been known [8] and, only recently, has an 0.878-approximation algorithm been proposed [3]. On the other hand, it is known that there exists no 0.941-approximation algorithm [9].

In this paper we study the expected MAX CUT value both for random graphs (Sect. 2) and for random regular graphs (Sect. 3). In particular let $f(x,n)$ be the expectation of $MaxCut(G)/xn$ for random graphs with n vertices and xn edges and let $r(x,n)$ be the expectation of $MaxCut(G)/xn$ for random $2x$-regular graphs with n vertices. We show, for sufficiently large x:

1. $\lim_{n\to\infty} f(x,n) \le \frac{1}{2} + \sqrt{\frac{\ln 2}{2x}}$,
2. $\lim_{n\to\infty} r(x,n) \le \frac{1}{2} + \frac{1}{\sqrt{x}} + \frac{1}{2}\frac{\ln x}{x}$.

The first result is obtained by elementary counting methods, while the second one is a direct consequence of results obtained in the area of spectral theory of graphs.

2 Expected Max Cut Value for Random Graphs

We want to estimate the expectation of the maximum cut for particular classes of graphs.

Given a graph (V,E) and a set $S \subset V$, we consider the adjacency matrix $(w_{i,j})$ of the graph and the characteristic vector (x_k^S) of S:

$$w_{i,j} = \begin{cases} 1 & \text{if } \{i,j\} \in E; \\ 0 & \text{otherwise;} \end{cases}$$

and

$$x_k^S = \begin{cases} 1 & \text{if } k \in S; \\ 0 & \text{otherwise.} \end{cases}$$

The cardinality of the cut induced by the set S is:

$$|Cut(S)| = \sum_{\substack{1\le i\le |V| \\ 1\le j\le |V|}} w_{i,j} x_i^S (1 - x_j^S) \ .$$

We state an upper bound on the expectation of the maximum cut for random graphs with a fixed number of vertices and edges.
If $|V| = n$, let us consider the probability space $\langle \mathcal{G}_{n,\mathfrak{e}}, \mathcal{U}\rangle$ where $\mathcal{U}$ is the uniform distribution and $\mathcal{G}_{n,\mathfrak{e}}$ is the class of graphs (V,E) with $|E| = \mathfrak{e} = xn$, where $x > 0$ is fixed. Furthermore, we consider the following functions:

$$g(n,\mathfrak{e}) = |\mathcal{G}_{n,\mathfrak{e}}|;$$
$$g(n,\mathfrak{e},h) = |\{G \mid G \in \mathcal{G}_{n,\mathfrak{e}}, MaxCut(G) \ge h\}| \ .$$

An upper bound on the expectation of the maximum cut can be obtained by an estimation of $g(n,\mathfrak{e})$ and $g(n,\mathfrak{e},h)$, up to a polynomial factor in n (see Th. 2.2). This estimation problem can be simplified considering the triples of the following type:

$$T = \langle (V_1, E_1), (V_2, E_2), H \rangle$$

where (V_1, E_1) and (V_2, E_2) are two graphs such that $V_1 \subseteq V$ and $V_2 = \overline{V_1}$, $H \subseteq V_1 \times V_2$ is a set of edges with an end point in V_1 and the other one in V_2 and such that $|E_1|+|E_2|+|H| = xn = \mathfrak{e}$. Let $\mathcal{T}_{n,\mathfrak{e}}$ the set of such triples. We call *vertex* and *edge* of T a generic element $k \in V$ and a generic pair $(i,j) \in E_1 \cup E_2 \cup H$, respectively.

Each graph $G = (V, E)$ with $MaxCut(G) = h$ can be associated at least to a triple $\langle (V_1, E_1), (V_2, E_2), H \rangle$ where $|H| = h,\ \ V_1 \cup V_2 = V,\ \ E_1 \cup E_2 \cup H = E$. Therefore, calling $t(n,\mathfrak{e},h) = |\{T \mid T \in \mathcal{T}_{n,\mathfrak{e}}, |H| \geq h\}|$, we have:

$$g(n,\mathfrak{e},h) \leq t(n,\mathfrak{e},h) \ .$$

Consider the following functions:

$$\begin{aligned} t_1(n,\mathfrak{e},n_1,e_1,h_1) &= |\{T \mid T \in \mathcal{T}_{n,\mathfrak{e}}, |V_1| = n_1, |E_1| = e_1, |H| = h_1\}| \\ m(n,\mathfrak{e},h) &= \max_{n_1,e_1,h_1 \geq h} t_1(n,\mathfrak{e},h_1,n_1,e_1) \ . \end{aligned}$$

We observe that

$$\begin{aligned} t(n,\mathfrak{e},h) &= \sum_{n_1,e_1,h_1 \geq h} t_1(n,\mathfrak{e},h_1,n_1,e_1); \\ &\leq n^5 \cdot m(n,\mathfrak{e},h) \ . \end{aligned}$$

Therefore

$$m(n,\mathfrak{e},h) \leq t(n,\mathfrak{e},h) \leq n^5 \cdot m(n,\mathfrak{e},h) \ .$$

We can then establish an estimate for $m(n,\mathfrak{e},h)$ instead of $t(n,\mathfrak{e},h)$.

Theorem 1. *Given the class of graphs $\mathcal{G}_{n,\mathfrak{e}}$, such that $\mathfrak{e} = xn$ ($x > 0$ fixed),*

$$m(n,\mathfrak{e},h) = \binom{n}{\frac{n}{2}} \binom{\binom{\frac{n}{2}}{2}}{\frac{\mathfrak{e}-h}{2}}^2 \binom{\frac{n^2}{4}}{h} n^{O(1)},$$

when $h \geq \mathfrak{e}/2$.

Proof. From the definition of $t_1()$ it follows that

$$t_1(n,\mathfrak{e},n_1,e_1,h_1) = \binom{n}{n_1} \binom{\binom{n_1}{2}}{e_1} \binom{\binom{n_2}{2}}{e_2} \binom{n_1 n_2}{h_1},$$

where $n_1 + n_2 = n$ and $e_1 + e_2 + h_1 = \mathfrak{e}$.

Let

$$f(n,\mathfrak{e},n_1,e_1,h_1) = \binom{\binom{n_1}{2}}{e_1} \binom{\binom{n_2}{2}}{e_2} \binom{n_1 n_2}{h_1},$$

with

$$\begin{array}{ll} n_1 = x_1 n & \text{where } 0 < x_1 < 1; \\ n_2 = x_2 n & \text{where } x_2 = 1 - x_1; \\ e_1 = y_1 \mathfrak{e} & \text{where } 0 \leq y_1 < 1; \\ e_2 = y_2 \mathfrak{e} & \text{where } 0 \leq y_2 < 1; \\ h_1 = z_1 \mathfrak{e} & \text{where } 0 < z_1 \leq 1 \text{ is fixed.} \end{array}$$

If $n, k \to +\infty$ and $k = \Theta(n^{1/2})$, by means of the Stirling's approximations we obtain:

$$\ln \binom{n}{k} = k \ln \frac{n}{k} + k + \Theta(\ln n) \ .$$

Therefore:

$$\begin{aligned} \ln f(n, \mathfrak{e}, n_1, e_1, h_1) = \mathfrak{e} \ln \frac{n^2}{\mathfrak{e}} &+ \mathfrak{e}(1 + 2y_1 \ln x_1 + 2y_2 \ln x_2 + z_1 \ln x_1 + z_1 \ln x_2 \\ &- y_1 \ln 2y_1 - y_2 \ln 2y_2 - z_1 \ln z_1) + \Theta(\ln n) \end{aligned}$$

In order to maximize $\ln f(n, \mathfrak{e}, h_1, n_1, e_1)$:

$$\begin{cases} \frac{\partial f()}{\partial x_1} = \frac{2y_1}{x_1} - \frac{2y_2}{x_2} + \frac{z_1}{x_1} - \frac{z_1}{x_2} = 0 \\ \\ \frac{\partial f()}{\partial y_1} = 2 \ln x_1 - 2 \ln x_2 - \ln 2y_1 + \ln 2y_2 = 0 \end{cases}$$

$$\begin{cases} \frac{1}{x_1}(1 + y_1 - y_2) - \frac{1}{x_2}(1 - y_1 + y_2) = 0 \\ \left(\frac{x_1}{x_2}\right)^2 = \frac{y_1}{y_2} \\ x_1 + x_2 = 1 \\ y_1 + y_2 = 1 - z_1 \end{cases}$$

Let $\phi = x_1/x_2$ and $a = 1 - z_1$. The solutions of the system are given by the solutions of the following equation:

$$(1 - a)\phi^3 - (1 + a)\phi^2 + (1 + a)\phi - 1 + a = 0$$

that, when $h_1 \geq \mathfrak{e}/2$, has only one real solution $\phi = 1$.
Therefore $x_1 = x_2 = 1/2$ and $y_1 = y_2 = (1 - z_1)/2$; as a consequence $n_1 = n_2 = n/2$ and $e_1 = e_2 = (\mathfrak{e} - h)/2$.
Since the binomial coefficient $\binom{n}{n_1}$ is maximum for $n_1 = n/2$, the function $t_1() = \binom{n}{n_1} f()$ assumes the maximum value for $n_1 = n_2 = n/2$ and $e_1 = e_2 = (\mathfrak{e} - h)/2$. □

Let $f(x, n)$ be the expectation of the ratio between the maximum cut and the number of edges in a random graphs with n vertices. Let be $f(x) = \lim_{n \to \infty} f(x, n)$.

Theorem 2. *Given the class of graphs $\mathcal{G}_{n,\mathfrak{e}}$, such that $\mathfrak{e} = xn$ ($x > 0$ fixed),*

$$f(x) \leq \frac{1}{2} + \sqrt{\frac{\ln 2}{2x}}$$

Proof. Let us denote by $P(n, \mathfrak{e}, h)$ the probability that a graph of n vertices and $\mathfrak{e}$ edges has maximum cut h at least, with $h \geq \mathfrak{e}/2$:

$$P(n, \mathfrak{e}, h) = \frac{g(n, \mathfrak{e}, h)}{g(n, \mathfrak{e})},$$

where

$$g(n, \mathfrak{e}) = \binom{\binom{n}{2}}{\mathfrak{e}}$$

is the cardinality of the class $\mathcal{G}_{n,\mathfrak{e}}$.
Then

$$P(n, \mathfrak{e}, h) \leq e^{\frac{n}{\lg e}(1+x(-1+\mathcal{H}(z)))+O(\ln n)},$$

where $\mathcal{H}(z) = -(1-z)\lg(1-z) - z\lg z$ is the entropy function, $h = z\mathfrak{e}$ with $\frac{1}{2} \leq z \leq 1$.
In fact

$$\begin{aligned}
\ln P(n, \mathfrak{e}, h) &= \ln g(n, \mathfrak{e}, h) - \ln g(n, \mathfrak{e}) \\
&\leq \ln m(n, \mathfrak{e}, h) - \ln g(n, \mathfrak{e}) + O(\ln n) \\
&\leq \frac{n}{\lg e}(1 + x(-1 + \mathcal{H}(z))) + O(\ln n),
\end{aligned}$$

since

$$\ln m(n, \mathfrak{e}, h) = \frac{n}{\lg e}\mathcal{H}(\frac{1}{2}) + \mathfrak{e}\ln\frac{n^2}{\mathfrak{e}} + \mathfrak{e}(1 - 2\ln 2 + \frac{1}{\lg e}\mathcal{H}(z)) + O(\ln n),$$

$$\ln g(n, \mathfrak{e}) = \mathfrak{e}\ln\frac{n^2}{\mathfrak{e}} + \mathfrak{e}(1 + \ln 2) + O(\ln n) \ .$$

If $(1 + x(-1 + \mathcal{H}(z))) < 0$ then $P(n, \mathfrak{e}, h) \to 0$ as $n \to \infty$. By means of the Taylor's approximation in the point 1/2,

$$\mathcal{H}(z) = 1 - \frac{2}{\ln 2}(z - \frac{1}{2})^2 + \varepsilon(z),$$

where $\varepsilon(z) < 0$ when $z \neq 1/2$. Thus, if $z = \frac{1}{2} + \sqrt{\frac{\ln 2}{2x}}$

$$P(n, \mathfrak{e}, h) \leq e^{nx\varepsilon(z)+O(\ln n)} = o(1) \ .$$

Therefore

$$Prob\left(\frac{MaxCut(G)}{\mathfrak{e}} \leq \frac{1}{2} + \sqrt{\frac{\ln 2}{2x}}\right) \sim 1 \ .$$

That implies that the expectation of the maximum cut is no more than $|E|(\frac{1}{2} + \sqrt{\frac{\ln 2}{2x}})$. □

3 Expected Max Cut Value for Random Regular Graphs

Let $r(x,n)$ be the expectation of the ratio between the maximum cut and the number of edges in a random $2x$-regular graphs with n vertices, that is for graphs in which every vertex has $2x$ incident edges. Let be $r(x) = \lim_{n\to\infty} r(x,n)$.

Theorem 3. *For sufficiently large x, it holds that*

$$r(x) \leq \frac{1}{2} + \frac{1}{\sqrt{x}} + \frac{1}{2}\frac{\ln x}{x} \ .$$

Proof. As a direct application of a result in [7], it is possible to bound the maximum cut of a graph in term of the second eigenvalue of the adjacency matrix of the graph. In fact, given a d-regular graph G and denoting with λ_2 the second eigenvalue (in decreasing order), it holds:

$$MaxCut(G) \leq \left(\frac{1}{4}d + \frac{1}{4}\lambda_2\right) n \ .$$

In [2] the expectation of the second eigenvalue for random d-regular graphs is estimated. In particular, the second eigenvalue magnitude is no more than $2\sqrt{2d-1} + 2\ln d + C'$ with probability $1 - n^{-C}$, for suitable constants C and C' and sufficiently large n.
Therefore,

$$Prob\left(\frac{MaxCut(G)}{\mathfrak{e}} \leq \frac{1}{2} + \frac{1}{\sqrt{x}} + \frac{1}{2}\frac{\ln x}{x} + \frac{C'}{x}\right) \geq 1 - n^{-C}.$$

This implies that, for sufficiently large x,

$$r(x) \leq \frac{1}{2} + \frac{1}{\sqrt{x}} + \frac{1}{2}\frac{\ln x}{x} \ .$$

□

References

1. Francisco Barahona, M. Jünger M. Grötschel, and G. Reinelt. An application of combinatorial optimization to statistical physics and circuit layout design. *Operations Research*, (36):493–513, 1988.
2. Joel Friedman. On the second eigenvalue and random walks in random d-regular graphs. *Combinatorica*, 11(4):331–365, 1991.
3. Michel X. Goemans and David P. Williamson. .878-approximation algorithms for MAX CUT and MAX 2SAT. In *Proceedings of the 26th ACM Symposium on the Theory of Computation*, 1994.
4. F. Hadlock. Finding a maximum cut of a planar graph in polynomial time. *SIAM Journal on Computing*, 4(3):221–225, September 1975.
5. Arun Jagota. Approximating maximum clique with a Hopfield network. *IEEE Transactions on Neural Networks*, 6(3):724–735, May 1995.

6. Richard M. Karp. Reducibility among combinatorial problems. In R. E. Miller and J. W. Thatcher, editors, *Complexity of Computer Computations*, pages 85–103, New York, 1972. Plenum Press.
7. J. Komlos and R. Paturi. Effect of conectivity in associative memory models. Technical Report CS88-131, University of California. San Diego, August 1988.
8. Sartaj Sahni and Teofilo Gonzalez. P-Complete Approximation Problems. *Journal of the ACM*, 23(3):555–565, July 1976.
9. Johan Håstad. Some optimal in-approximability results. November 1996.

$\mathcal{NP}$–Completeness Results for Minimum Planar Spanners*

Ulrik Brandes and Dagmar Handke

Universität Konstanz, Fakultät für Mathematik und Informatik,
78457 Konstanz, Germany.
email: {Ulrik.Brandes,Dagmar.Handke}@uni-konstanz.de

Abstract. For any fixed parameter $t \geq 1$, a *t–spanner* of a graph G is a spanning subgraph in which the distance between every pair of vertices is at most t times their distance in G. A *minimum* t–spanner is a t–spanner with minimum total edge weight or, in unweighted graphs, minimum number of edges. In this paper, we prove the $\mathcal{NP}$–hardness of finding minimum t–spanners for planar weighted graphs and digraphs if $t \geq 3$, and for planar unweighted graphs and digraphs if $t \geq 5$. We thus extend results on that problem to the interesting case where the instances are known to be planar. We also introduce the related problem of finding minimum *planar* t–spanners and establish its $\mathcal{NP}$–hardness for similar fixed values of t.

1 Introduction

A *t–spanner* of a graph G is a spanning subgraph S in which the distance between every pair of vertices is at most t times their distance in G. The main idea of this concept is to find a subgraph of a given graph G that is sparse, but still guarantees a so–called *stretch factor* on the vertex–to–vertex distances of G. The stretch factor will be bounded by a constant independent of the size of G (i.e. in $\mathcal{O}(1)$). Observe that the minimum spanning tree does not necessarily meet this specification. Consider, for example, the complete graph K_n with vertices $1, 2, \ldots, n$ and unit edge weights. Then the simple path $1, 2, \ldots, n$ forms a minimum spanning tree yielding a stretch factor of $t = n - 1$.

The concept of spanners has been introduced by Peleg and Ullman in [PU87], where they used spanners to synchronize asynchronous networks. One of many other applications for spanners are communication networks, where one is interested in finding a sparse subnetwork that nevertheless guarantees constant delay factors. A survey of some results on the existence and efficient constructibility of (sparse) spanners is given in [PS89]. Further results and discussions concerning t–spanners and variants thereof can be found in [Soa92].

In most applications the sparseness of a spanner is crucial. The problem of finding t–spanners with a minimum number of edges has been shown to be $\mathcal{NP}$–hard for most values of t by Cai in [Cai94]. Therefore subsequent efforts have

* Research partially supported by DFG under grant Wa 654/10-1.

concentrated on finding spanners that are maybe not minimum, but sufficiently sparse (see for example [ADD+93]).

Several authors considered variants of t-spanners. In [CC95], Cai and Corneil deal with *tree t-spanners* (i.e. t-spanners that are trees) and also examine the complexity status of the corresponding decision problem. Liestman and Shermer introduced the notion of *additive* spanners, which employ an additive instead of multiplicative stretch function on the distances [LS93].

Here we consider spanners in *planar* graphs (either weighted or unweighted, directed or undirected), i.e. we restrict the set of input instances. We thereby (partially) settle a question raised in [Cai94]. We also introduce the notion of *planar t-spanners*. These are subgraphs, which in addition to being t-spanners are planar, no matter whether the original graph is planar or not.

This paper is organized as follows: After introducing some basic notation and the examined problems, our results of $\mathcal{NP}$-completeness are stated in Sect. 2. Proofs of these in unweighted, weighted, and directed graphs make up for Sects. 3, 4, and 5, respectively.

2 Problems and Results

In what follows $G = (V, E; w)$ (respectively $G = (V, A; w)$) denotes a simple, weighted undirected (directed) graph with vertex set V, edge set E (arc set A), and edge weights $w : E \to \mathbb{R}^+$ ($w : A \to \mathbb{R}^+$). If all edges have unit weight, i.e. all weights are equal to 1, the graph is said to be *unweighted.* A directed graph (*digraph*) is said to be an *oriented* graph, if it does not contain a cycle of two arcs. For simplicity, we will use the terminology for undirected graphs throughout most of this paper. The terms are naturally extended to digraphs. Since spanners of each connected component can be determined independently, we only consider connected graphs. The *length* of a path is the sum of the weights of its edges. The *distance* between two vertices u and v in G, i.e. the length of the shortest (directed) path, is denoted by $d_G(u, v)$. A t-spanner is defined as follows:

Definition 1 (t-spanner). *For any real valued parameter $t \geq 1$, a spanning subgraph $S = (V, E'; w)$ with $E' \subseteq E$ is a* t-spanner *of an edge-weighted graph $G = (V, E; w)$, if $d_S(u, v) \leq t \cdot d_G(u, v)$ for all $u, v \in V$.*

The parameter t is called *stretch factor*. We say that an edge $e \in E$ is *covered* (by an edge $f \in S$), if in S there exists a path of length at most $t \cdot w(e)$ (and containing f) that connects the endpoints of e.

In order to prove that a given spanning subgraph is a t-spanner, we do not have to consider all pairwise distances of the vertices. It is sufficient to only look at edges of the original graph that are not part of the spanning subgraph.

Lemma 2 ([CC95]). *Let $S = (V, E'; w)$ be a spanning subgraph of a weighted graph $G = (V, E; w)$. Then S is a t-spanner of G if and only if $d_S(u, v) \leq t \cdot w(u, v)$ for every edge $\{u, v\} \in E \setminus E'$.*

A t–spanner is called a *minimum* t–spanner of a weighted graph G, if it has minimum total edge weight among all t–spanners of G. The corresponding decision problem is defined as follows:

Minimum t–Spanner Problem (MinS$_t$)
Given: A graph G with associated (positive real valued) edge weights and a positive real value W.
Problem: Does G contain a t–spanner with total edge weight at most W?

Obviously, for an unweighted graph, the only 1–spanner is the graph itself. For a weighted graph, Hakimi and Yau [HY64] proved that there is a unique 1–spanner with a minimal number of edges. From [CC95] we know that this must also be the unique minimum 1–spanner, and that it can be determined in polynomial time. The $\mathcal{NP}$–completeness of MinS$_t$ for general graphs has been established by Cai [Cai94] for $t \geq 2$ in directed and undirected, and $t \geq 3$ in oriented graphs, even if they are unweighted. From the transformation used in [CC95] to prove the $\mathcal{NP}$–completeness of the Tree t–Spanner Problem it can be seen that MinS$_t$ is also $\mathcal{NP}$–complete for $1 < t < 2$.

Here we will show that the problem remains $\mathcal{NP}$–complete for most values of t when G or S are restricted to be planar[1]. In particular, we prove the following theorems.

Theorem 3. *For any fixed integer $t \geq 5$, MinS$_t$ is $\mathcal{NP}$–complete for undirected, planar, biconnected graphs with unit edge weights.*

Theorem 4. *For any fixed integer $t \geq 3$, MinS$_t$ is $\mathcal{NP}$–complete for undirected, planar, biconnected graphs with edge weights equal to 1 or 2.*

Theorem 5. *For any fixed integer $t \geq 5$ ($t \geq 3$), MinS$_t$ is $\mathcal{NP}$–complete for unweighted (weighted) planar oriented graphs.*

The proofs of the theorems are given in the next three sections. All three of them are transformations from the Planar Satisfiability Problem with three literals in each clause, and they can be viewed as modifications of each other. Therefore we treat the unweighted, undirected case in detail, and outline the changes necessary for the other cases.

Note that in unweighted graphs every t–spanner is also a $\lfloor t \rfloor$–spanner, while there is no such correspondence in weighted graphs, even if all edges have integer weights. At the end of Sect. 4 it will be easy to see how our construction can be adjusted to allow arbitrary real values of $t \geq 3$ in the weighted case. Since the above results are valid for more specific instances, the following corollary is then obtained immediately.

Corollary 6.
1. *For any fixed real valued $t \geq 5$, MinS$_t$ is $\mathcal{NP}$–complete for unweighted planar graphs, planar oriented graphs, and planar digraphs.*

[1] Note that planarity of G implies planarity of S, while the converse is not true.

2. *For any fixed real valued $t \geq 3$, $MinS_t$ is $\mathcal{NP}$–complete for weighted planar graphs, planar oriented graphs, and planar digraphs.*

We now introduce a new variant of general t–spanners, for which similar results are implied by the above theorems.

Definition 7 (planar t–spanner). *For any real valued parameter $t \geq 1$, a spanning subgraph $S = (V, E'; w)$ with $E' \subseteq E$ is a* planar t-spanner *of a weighted graph $G = (V, E; w)$, if $d_S(u, v) \leq t \cdot d_G(u, v)$ for all $u, v \in V$, and S is planar.*

We now give the decision formulation of the corresponding minimization problem for planar t–spanners.

Minimum Planar t–Spanner Problem (MinPS$_t$)
Given: A graph G with associated (positive real valued) edge weights and a positive real value W.
Problem: Does G contain a planar t–spanner with total edge weight at most W?

As noted above, the only 1–spanner of an unweighted graph is the graph itself. Therefore MinPS$_1$ is in $\mathcal{P}$ for unweighted graphs, because planarity can be tested in linear time (cf. [HT74,BL76]). On the other hand, it is $\mathcal{NP}$–complete to decide whether an unweighted graph contains a tree t–spanner, i.e. a t–spanner which is a tree, if $t \geq 4$ [CC95]. Observe that spanning trees are planar spanning subgraphs with the least possible number of edges. Together with Theorems 3 and 5 we have the following consequences:

Corollary 8.
1. *For any fixed real valued $t \geq 4$, $MinPS_t$ is $\mathcal{NP}$–complete for unweighted graphs.*
2. *For any fixed real valued $t \geq 5$, $MinPS_t$ is $\mathcal{NP}$–complete for unweighted graphs, oriented graphs, and digraphs, even if they are planar.*

For weighted graphs the situation is different. The graph itself need not be the only 1–spanner. But, as mentioned above, the unique 1–spanner with a minimal number of edges also is the unique minimum 1–spanner, and can be determined in polynomial time. Since all edge weights are positive, and every subgraph of a planar graph is planar, a minimum planar 1–spanner has a minimal number of edges. Therefore a minimum planar 1–spanner would have to be identical to the minimum 1–spanner, and we can conclude that MinPS$_1$ is also in $\mathcal{P}$ for weighted graphs by testing the minimum 1–spanner for planarity.

In [CC95] the $\mathcal{NP}$–completeness of the Tree t–Spanner Problem for $t > 1$ in weighted graphs is proven. By a close look at the transformation used there and by an appropriate choice of the bound on the total weight of a planar t–spanner, the proof can be modified to show the $\mathcal{NP}$–completeness of MinPS$_t$ for $t > 1$ in weighted, undirected graphs. We omit the details and combine this observation with Theorems 4 and 5.

Table 1. The complexity status of MinS_t and MinPS_t in undirected graphs

t	**MinS_t**, general graphs [Cai94,CC95]		**MinPS_t**, general graphs		**Min(P)S_t**, planar graphs	
	unweighted	weighted	unweighted	weighted	unweighted	weighted
1	$\mathcal{P}$	$\mathcal{P}$	$\mathcal{P}$	$\mathcal{P}$	$\mathcal{P}$	$\mathcal{P}$
(1,2)	$\mathcal{P}$	$\mathcal{NPC}$	$\mathcal{P}$	$\mathcal{NPC}$	$\mathcal{P}$	?
[2,3)	$\mathcal{NPC}$	$\mathcal{NPC}$	?	$\mathcal{NPC}$	?	?
[3,4)	$\mathcal{NPC}$	$\mathcal{NPC}$	?	$\mathcal{NPC}$	?	$\mathcal{NPC}$
[4,5)	$\mathcal{NPC}$	$\mathcal{NPC}$	$\mathcal{NPC}$	$\mathcal{NPC}$	?	$\mathcal{NPC}$
$[5,\infty)$	$\mathcal{NPC}$	$\mathcal{NPC}$	$\mathcal{NPC}$	$\mathcal{NPC}$	$\mathcal{NPC}$	$\mathcal{NPC}$

Corollary 9.

1. *For any fixed real valued $t > 1$, $MinPS_t$ is $\mathcal{NP}$-complete for weighted graphs.*
2. *For any fixed real valued $t \geq 3$, $MinPS_t$ is $\mathcal{NP}$-complete for weighted graphs, oriented graphs, and digraphs, even if they are planar and the edge weights are restricted to be equal to 1 or 2.*

Table 1 summarizes the results for the complexity status of the problems considered in this paper in undirected graphs. We give the complexity status for MinS_t with arbitrary input instances (as shown in [Cai94] and [CC95]), for MinPS_t with arbitrary input instances, and for both problems with planar input instances[2]. The results are listed for both the weighted and the unweighted case. A "?" indicates that the complexity status is unknown.

3 MinS_t for Unweighted, Planar Graphs

In this section we prove Theorem 3, so all graphs are unweighted and planar. The other theorems are proven along the same lines and therefore this proof is described in detail first. Part of the proof modifies ideas of [Cai94].

Let $t \geq 5$ be an arbitrary fixed integer. Clearly MinS_t is in $\mathcal{NP}$, since the test whether a spanning subgraph S is a t-spanner can be done in polynomial time. By Lemma 2, we just have to check the (at most linear number of) edges of G that do not belong to S. To show the $\mathcal{NP}$-completeness, we transform the Planar 3-Satisfiability Problem to MinS_t. For this, given an instance of the Planar 3-Satisfiability Problem, we construct a planar graph G, choose a weight W, and then show the equivalence of both problems for these instances.

For the construction we use the fact that we can force edges to be in every minimum t-spanner by adding some additional edges. So this section first introduces the Planar 3-Satisfiability Problem and the concept of forcing, then gives the reduction.

[2] Observe that MinS_t and MinPS_t are the same for planar instances.

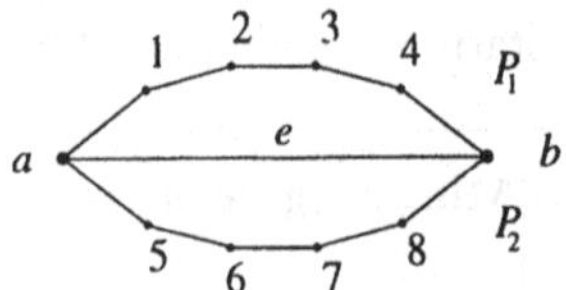

Fig. 1. Forcing edge $\{a, b\}$ into a 5–spanner

3.1 The Planar 3–Satisfiability Problem

The Planar 3–Satisfiability Problem is a variant of the 3–Satisfiability Problem with the additional restriction that the underlying bipartite graph, where clause vertices are connected to variable vertices if the corresponding variables appear within the clause, is planar.

Planar 3–Satisfiability Problem (P3SAT)
Given: A set U of variables, and a collection C of clauses over U with $|c| = 3$ for all $c \in C$. Furthermore the bipartite graph $G = (V, E)$ where $V = U \cup C$ and $E = \{\{x, c\} : x \text{ or } \overline{x} \text{ occurs in } c\}$ is planar.
Problem: Is there a satisfying truth assignment for C?

The $\mathcal{NP}$–completeness proof for this problem can be found in [Man83]. We use the planarity of the underlying graph of P3SAT to construct a planar graph in which we can easily determine the minimum t–spanner.

3.2 Forcing Edges into a Minimum t–Spanner

We can force an edge into a spanner by adding auxiliary edges to the given graph such that every minimum t–spanner of the new graph contains this edge. This concept has appeared in [Cai94] and will be used extensively.

Lemma 10 ([Cai94]). *Let e be an arbitrary edge of an unweighted graph G, and let G' be the graph constructed from G by adding two distinct paths P_1 and P_2 of length t (all internal vertices of P_1 and P_2 are new vertices) between the two ends of e. Then for any minimum t-spanner S of G', edge e belongs to S.*

The two auxiliary paths P_1 and P_2 are called *forcing paths*, edge e is called *forced edge*. A *forced l–component* is a simple path of length l consisting of l forced edges together with their forcing paths. For an example of a forced edge $e = \{a, b\}$ with $t = 5$, see Fig. 1. A minimum t–spanner of this graph contains exactly $2 \cdot (t - 1) + 1 = 2t - 1 = 11$ edges: edge $\{a, b\}$ and $t - 1 = 4$ edges from the forcing paths $(a, 1, 2, 3, 4, b)$ and $(a, 5, 6, 7, 8, b)$ each.

3.3 Construction of the Instance

We start with the planar, embedded underlying graph of the given instance (U, C) of P3SAT and extend the variable and clause vertices to form *variable components* and *clause components*. Then these components are combined to form *truth assignment testing components* which reflect the relationship between the satisfiability of a clause and the existence of a minimum t–spanner. As a last step we choose the bound on the number of edges in the Minimum t–Spanner Problem.

The Variable Components. The key idea behind the variable component is that it is planar and each of its possible minimum t–spanners reflects exactly one truth assignment for the variable. For every variable $x \in U$ we construct a variable component T_x as follows. Let k be the number of (positive and negative) occurrences of the variable x in all clauses.

1. Create a central vertex x^*.
2. For each occurrence of x in a clause c create a block of four new vertices $x_1^{(c)}$, $\overline{x}_1^{(c)}$, $x_2^{(c)}$, and $\overline{x}_2^{(c)}$, thus yielding $4k$ so–called *literal vertices* in total. Within each block, the vertices are positioned in this order, and the blocks are arranged circularly around x^* according to the embedding of the underlying graph of the instance of P3SAT.
3. Connect each pair of neighboring literal vertices by a forced $(t-1)$–component such that a circle of $4k$ forced $(t-1)$–components is formed altogether.
4. Connect x^* with all literal vertices by an edge, called *literal edge*. An edge $\{x_i^{(c)}, x^*\}$ is called *positive* literal edge, an edge $\{\overline{x}_i^{(c)}, x^*\}$ is called *negative* literal edge.
5. Create new auxiliary vertices between all pairs of neighboring literal edges, i.e. in total $4k$ auxiliary vertices. Connect each of these by an edge with x^* (called *auxiliary edge*) and by two distinct forced $(t-1)$–components with its neighboring literal vertices. Their literal edges are then called *associated* literal edges of the auxiliary edge and vice versa.

Figure 2 illustrates this construction. For readability a symbolic representation is used later on when larger portions of the graph are drawn. Now, a minimum t–spanner can contain only *consistent* literal edges:

Lemma 11. *Any minimum t–spanner of a variable component T_x contains either all positive or all negative literal edges.*

Proof. Let S be an arbitrary minimum t–spanner of T_x. Then S contains all forced edges and $t-1$ edges from each forcing path. Observe that these edges together with either all $2k$ positive or all $2k$ negative literal edges form a t–spanner. Thus S can contain at most $2k$ edges out of the $8k$ literal and auxiliary edges.

By construction of the variable component, both associated auxiliary edges and both neighboring negative (resp. positive) literal edges are covered by a positive (resp. negative) literal edge in S. But, by an auxiliary edge in S, only the associated literal edges are covered.

Now assume that S contains an auxiliary edge. Then S also contains either the next auxiliary edge, too, or the next not associated literal edge. In total, this

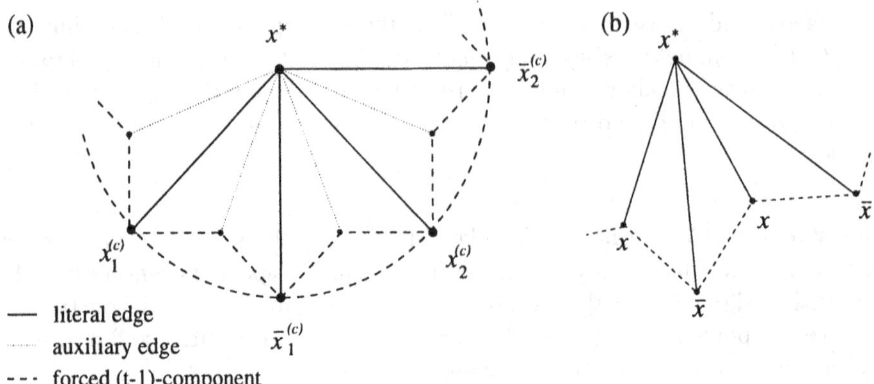

Fig. 2. (a) Part of the variable component T_x for the variable x occurring in clause c, (b) its symbolic representation

leads to more than $2k$ additional edges and thus contradicts the minimality of S. Similarly, assume that S contains two inconsistent literal edges. Then there must be at least one auxiliary edge belonging to S or more than $2k$ literal edges to cover all other edges. Again, this contradicts the minimality of S. Thus S contains exactly every other literal edge. □

With this lemma it can easily be deduced that the number of edges of each minimum t–spanner of T_x is $4 \cdot 3k \cdot (t-1) \cdot (2t-1) + 2k$.

The Clause Components. The clause component for each clause $c \in C$ is basically a quadrilateral consisting of four *clause vertices* 1,2,3, and 4, where the sides are formed by distinct forced $(t-2)$–components. Vertices 1 and 3 are additionally connected by an edge, called *clause edge*. See Fig. 3(a) for an example[3]. Observe that any minimum t–spanner for $t \geq 5$ of such an isolated clause component must contain the clause edge.

The Truth Assignment Testing Components. Now we combine the clause components with the variable components according to the given clauses by identifying vertices. Three sides of the quadrilateral in the clause component each correspond to one of the literals in the corresponding clause. (The fourth side is used to make the arguments symmetrical.) The endpoints of each such side of the quadrilateral are thus identified with the two corresponding literal vertices of the corresponding block in the variable component: if clause c contains the positive literal we use the positive literal vertices $x_i^{(c)}$, and $\overline{x}_i^{(c)}$ otherwise. See Fig. 3(b) for an example.

[3] Our construction is a bit more complex than actually needed in the unweighted case, but will not have to be changed much when being modified for the weighted and the directed case.

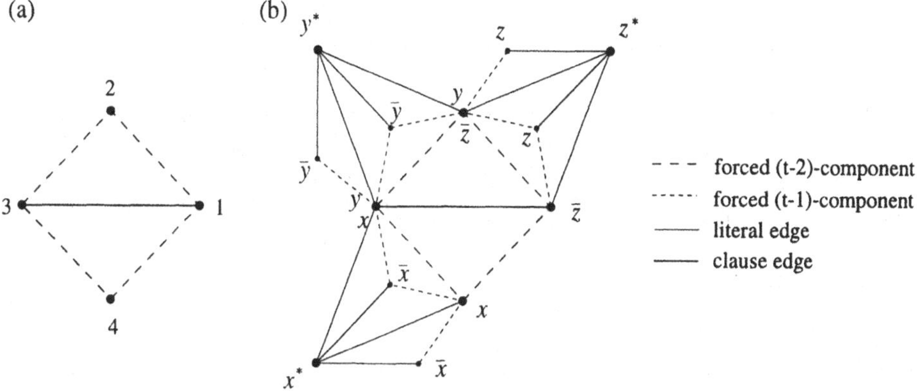

Fig. 3. (a) Clause component, (b) the truth assignment testing component for clause $c = x \vee y \vee \bar{z}$ using the symbolic representation for relevant blocks of the variable components

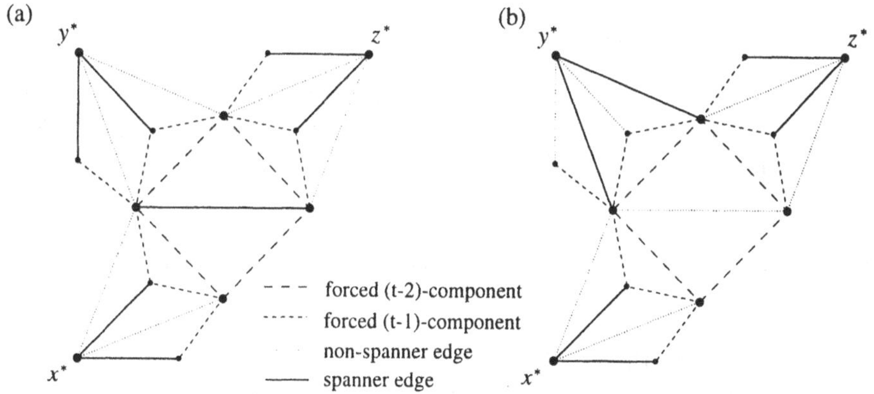

Fig. 4. The minimum t–spanner in the truth assignment testing component

Note that the combination of the variable components with the clause components does not affect the validity of Lemma 11. We now have the following lemma:

Lemma 12. *For any fixed integer $t \geq 5$, a minimum t–spanner S of a truth assignment testing component contains the clause edge if and only if S contains no pair of consistent literal edges that is incident to the clause edge.*

Proof. If S does not contain a pair of literal edges that is incident to the clause edge (cf. Fig. 4(a)), then every path connecting the endpoints of the clause edge in S either uses the clause edge or has length at least $2(t-2) > t$, if $t \geq 5$.

For the other direction see Fig. 4(b). Assume that S contains a pair of incident consistent literal edges. Then this provides a shortcut for one of the forced $(t-2)$–

components, and thus there is a path of length $2 + (t - 2) = t$ in S connecting the endpoints of the clause edge. Hence the clause edge is covered. □

Thus the number of edges in a minimum t–spanner of such a truth assignment testing component reflects the truth value of the corresponding clause. This completes the construction of the graph. All isolated components are planar, and since we start from an instance of P3SAT, the whole graph is planar. It is also easily seen that the instance is biconnected, and can be constructed in polynomial time.

Choice of W. We set W, the bound on the number of edges in a t–spanner, to

$$W = 6m + 36m(2t - 1)(t - 1) + 4m(2t - 1)(t - 2),$$

where m is the number of clauses of the instance of P3SAT.

3.4 Equivalence of the Problems

In this subsection, let (U, C) be an instance of P3SAT, and (G, W) the instance for MinS_t constructed as described above. We will show that there is a satisfying truth assignment for (U, C), if and only if G has a t–spanner with at most W edges.

Lemma 13. *If the set of clauses C of (U, C) is satisfiable, then there exists a planar t–spanner of G with at most W edges.*

Proof. Suppose that the set of clauses C is satisfiable, and let θ be a satisfying truth assignment. From this we construct the subgraph S of G as follows:

1. S contains all forced edges.
2. S contains $t - 1$ arbitrarily chosen edges from each forcing path.
3. For each variable $x \in U$, S contains all positive literal edges if $\theta(x)$ is *true*, and all negative literal edges otherwise.

By this construction, S is trivially a spanning subgraph. The number of edges W' in S computes as follows. S contains

- $3 \cdot m \cdot 3 \cdot 4 \cdot (t - 1)$ forced edges from the variable components (overall number of variable occurrences is $3m$),
- $3 \cdot m \cdot 3 \cdot 4 \cdot (t - 1) \cdot 2(t - 1)$ edges from the forcing paths of the variable components,
- $4 \cdot m \cdot (t - 2)$ forced edges from the clause components,
- $4 \cdot m \cdot (t - 2) \cdot 2(t - 1)$ edges from the forcing paths of the clause components, and
- $3 \cdot m \cdot 2$ literal edges.

Hence S contains exactly $W' = 6m + 36m(2t-1)(t-1) + 4m(2t-1)(t-2) = W$ edges.

It remains to show that S is a t–spanner of G. According to Lemma 2, we only have to show that for every edge not contained in S, there exists a path of length at most t connecting the endpoints of that edge. This is obvious for the variable components. For the clause edges observe that, since θ is a satisfying truth assignment, there is at least one literal in each clause that is true. Due to the construction of S we thus have at least one incident pair of literal edges in each clause component. From Lemma 12 it follows that S is a t–spanner. □

To show the opposite direction we need another lemma:

Lemma 14. *Any minimum t-spanner S of G contains at least W edges.*

Proof. Any t–spanner S of G must contain all forced edges and $t-1$ edges from each forcing path. By Lemma 11, S contains at least either all positive or all negative literal edges for each variable component. This sums up to W. □

Lemma 15. *If G has a t-spanner with at most W edges, then there exists a satisfying truth assignment for (U, C).*

Proof. Suppose S is a t–spanner of G with at most W edges. Then by Lemma 14 S is a minimum t–spanner and contains exactly W edges. All forced edges and the according number of edges from the forcing paths must be in S. Hence there remain only $6m$ further edges which can only be consistent literal edges (by Lemma 11). Thus we can uniquely define a truth assignment θ by setting, for each $x \in U$, $\theta(x) = \mathit{true}$, if S contains the positive literal edges of T_x, and $\theta(x) = \mathit{false}$ otherwise.

Since S is a t–spanner and S contains no clause edge it follows from Lemma 12 that there is at least one incident pair of literal edges for every clause edge. Hence θ satisfies all clauses. □

This completes the proof of Theorem 3.

4 MinS$_t$ for Weighted, Planar Graphs

We will now prove Theorem 4. Again, we transform an instance of P3SAT to an instance of MinS$_t$ by extending variable and clause vertices to appropriate components. The fact that we are now allowed to also assign edge weights of value 2 will be exploited to lower the bound on t, thus yielding a stronger result than in the unweighted case.

The variable components remain the same with all edges having unit edge weight, and the results about minimum t–spanners for these components keep valid (Lemma 11). The clause components again consist of four clause vertices, but now three sides of the quadrilateral remain unconnected. Only one side is connected by two consecutive forced $(t-1)$–components with unit edge weights.

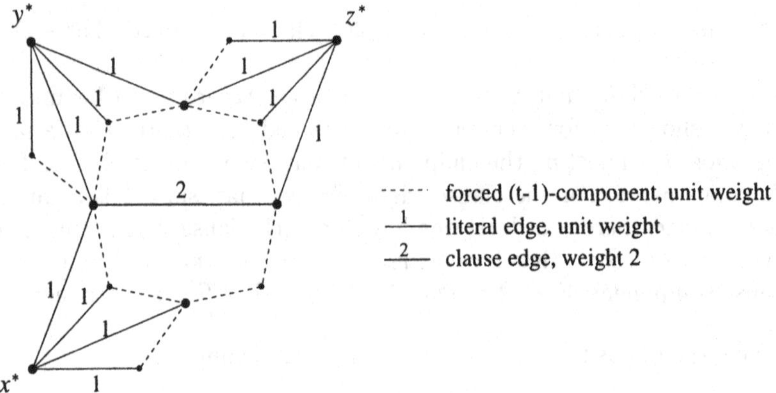

Fig. 5. The truth assignment testing component in the weighted case

As before, we have one clause edge, now having edge weight 2. We combine the components to form the truth assignment testing components as we did in the unweighted case by identifying the corresponding vertices (see Fig. 5 for an example). Note that every edge in the so–constructed instance has unit edge weight, except for the clause edges which are assigned a weight of 2.

To make use of the proof structure from Sect. 3, we provide the following lemma (cf. Lemma 12).

Lemma 16. *For any fixed integer $t \geq 3$, a minimum t–spanner S of a truth assignment testing component contains the clause edge if and only if S contains no pair of literal edges incident to the clause edge.*

Proof. If S does not contain a pair of literal edges that is incident to the clause edge, then every path connecting the endpoints of the clause edge in S either uses the clause edge or has length at least $4(t-1) > 2t$, if $t \geq 3$.

Assume that S contains a pair of incident consistent literal edges. Then we can combine these edges with the two forced $(t-1)$–components of the neighboring side of the quadrilateral to form a path of length $2 + 2(t-1) = 2t$ connecting the endpoints of the clause edge. Hence in this case the clause edge is covered. □

It is easily seen that the constructed graph is again planar and biconnected. By choosing $W = 6m + 36m(t-1)(2t-1) + 2m(t-1)(2t-1)$ the arguments of the previous section can be repeated to complete the proof of Theorem 4.

Corollary 6 states that Theorem 4 can be generalized to allow real values of $t \geq 3$. This can be seen by using forced $(\lfloor t \rfloor - 1)$–components in the construction described above. All results about minimum t–spanners then keep valid.

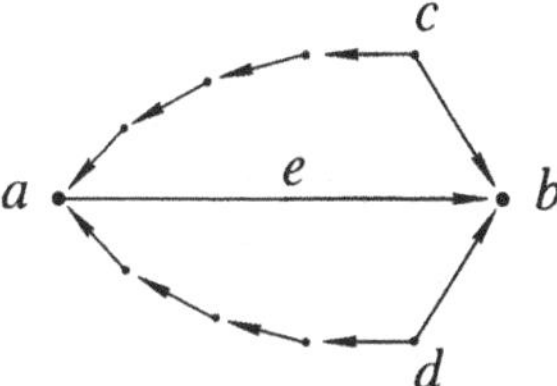

Fig. 6. Forcing arc (a, b) into a 5–spanner

5 MinS$_t$ for Planar Digraphs

In this section we turn to Theorem 5. To show the $\mathcal{NP}$–completeness for digraphs, we again use a modification of the reduction of the previous sections. Here we only give the details of the construction for the unweighted case, since the weighted case is then straightforward from what has been established so far.

Forcing Arcs into a Minimum t–Spanner. Similar to the undirected case, an arc (a, b) of a digraph can be forced to be in every minimum t–spanner as described in [Cai94]. For this purpose we create two new vertices c and d, add two arcs (c, b) and (d, b), and then add two distinct directed paths of length $t-1$ from c to a and from d to a, respectively. Then a minimum t–spanner of this component consists of arc (a, b) and all arcs of the paths of length $t-1$. Figure 6 shows an example for $t = 5$.

The Variable Components. A directed variable component consists of a central vertex x^*, and four *literal vertices* $x_1^{(c)}$, $\overline{x}_1^{(c)}$, $x_2^{(c)}$, and $\overline{x}_2^{(c)}$ together with four *literal arcs* for each positive or negative occurrence of variable x in a clause c. The orientation of the literal arcs depends on what the connection to the clause components will be like (see below). In the following, $\langle x, y\rangle$ stands for exactly one of the arcs (x, y) and (y, x), which will never be present at the same time. We add the following components and auxiliary vertices or arcs:

- All pairs of neighboring literal vertices $x_i^{(c)}$ and $\overline{x}_i^{(c)}$ are connected by two distinct directed forced $(t-1)$–components, one in either direction. Their literal arcs will be directed both either from or to x^*. If they are both directed from (resp. to) x^*, add an auxiliary vertex a_i and an auxiliary arc (x^*, a_i) (resp. (a_i, x^*)). Also connect a_i with $x_i^{(c)}$ and $\overline{x}_i^{(c)}$ by two distinct directed forced $(t-1)$–components directed from the literal vertices to a_i (from a_i to the literal vertices).
- Between the other pairs of neighboring literal arcs add an auxiliary arc $\langle x_i^{(c)}, \overline{x}_i^{(c)}\rangle$, such that $\langle x_i^{(c)}, x^*\rangle, \langle x^*, \overline{x}_i^{(c)}\rangle, \langle\overline{x}_i^{(c)}, x\rangle$ do not form a directed cycle. Additionally connect $x_i^{(c)}$ and $\overline{x}_i^{(c)}$ with x^* by directed forced $(t-1)$–components parallel to their corresponding literal arcs.

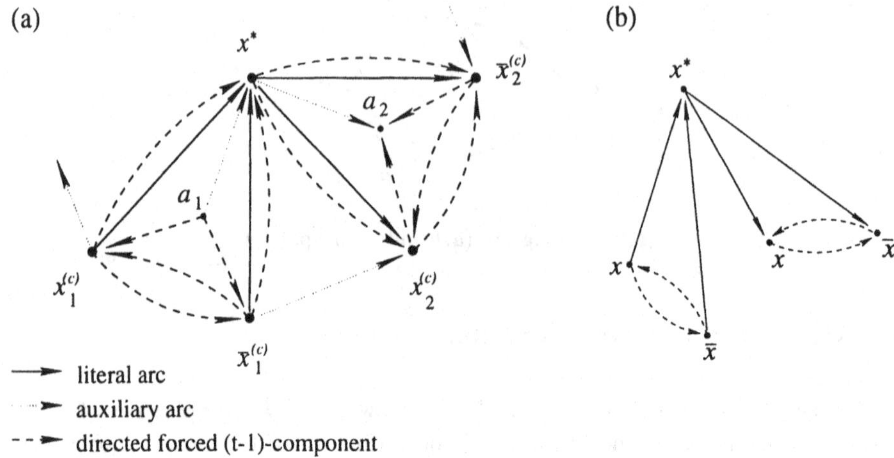

Fig. 7. (a) Part of the directed variable component for the variable x occurring in clause c, (b) its symbolic representation

Figure 7 gives an example of a directed variable component and its symbolic representation. As in the undirected case this construction guarantees that every minimum t–spanner of such a variable component only contains consistent literal arcs (cf. Lemma 11). This can be seen as follows. All positive (or negative, respectively) literal arcs together with the appropriate arcs from the forced arcs form a t–spanner. By the construction of the auxiliary arcs at least every other literal arc has to be included into a t–spanner. No other auxiliary arc is covered by an auxiliary arc in the t–spanner.

The Clause Components. We define clause components analogously to the undirected case, where the clause arc and the forced $(t-2)$-components are oriented such that they start and end at the same vertices of the quadrilateral.

The Truth Assignment Testing Components. Again we combine the variable and clause components by identifying the corresponding vertices. According to the choice of the orientation of the clause arc, the corresponding literal arcs are now oriented such that the literal arcs together with one of the directed forced $(t-2)$-components of the clause component form a directed path parallel to the clause arc. The remaining literal arcs of the variable component are oriented such that pairs $\langle x_i^{(c)}, x^* \rangle$ and $\langle \overline{x}_i^{(c)}, x^* \rangle$ of corresponding inconsistent literal arcs are directed likewise from or to x^*. Figure 8 shows an example of such a directed truth assignment testing component.

This completes the construction for the directed, unweighted case. It is easily seen that the graph is planar and oriented. Choosing W as in the undirected case ($W = 6m + 36m(t-1)(2t-1) + 4m(t-2)(2t-1)$), the proof of the equivalence of P3SAT and MinS$_t$ is straightforward as before.

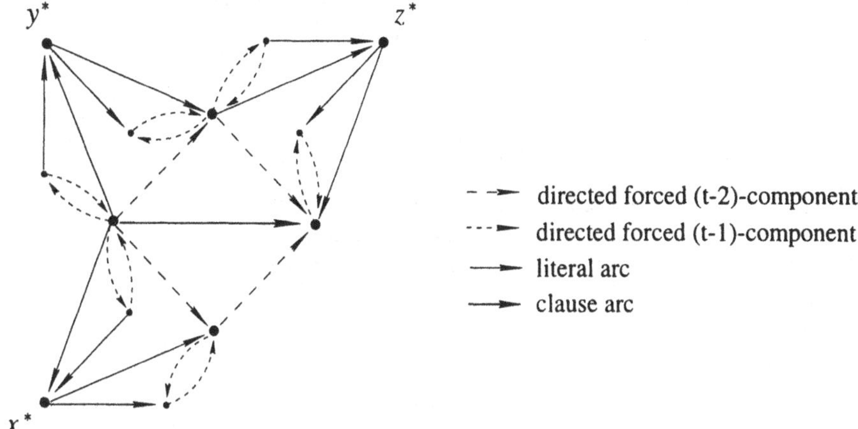

Fig. 8. The directed truth assignment testing component for unweighted digraphs

Weighted Digraphs. In the weighted, directed case the same variable components (with unit arc weights) are used. The clause components are the ones from the weighted, undirected case, and orientations are determined analogously.

References

[ADD+93] Ingo Althöfer, Gautam Das, David Dobkin, Deborah Joseph, and José Soares. On sparse spanners of weighted graphs. *Discrete Comput. Geom.*, 9:81–100, 1993.

[BL76] Kellogg S. Booth and George S. Lueker. Testing for the consecutive ones property, interval graphs, and planarity using PQ-tree algorithms. *J. of Computer and System Sciences*, 13:335–379, 1976.

[Cai94] Leizhen Cai. $\mathcal{NP}$-completeness of minimum spanner problems. *Discrete Applied Math.*, 48:187–194, 1994.

[CC95] Leizhen Cai and D.G. Corneil. Tree spanners. *SIAM J. Discrete Math.*, 8(3):359–387, 1995.

[HT74] John E. Hopcroft and Robert E. Tarjan. Efficient planarity testing. *J. of the Association for Computing Machinery*, 21:549–568, 1974.

[HY64] S.L. Hakimi and S.-T. Yau. Distance matrix of a graph and its realizability. *Q. J. Mech. appl. Math.*, 22:305–317, 1964.

[LS93] A.L. Liestman and Thomas Shermer. Grid spanners. *Networks*, 23:123–133, 1993.

[Man83] Anthony Mansfield. Determining the thickness of graphs is $\mathcal{NP}$-hard. *Math. Proc. Camb. Phil. Soc.*, 93:9–23, 1983.

[PS89] David Peleg and Alejandro A. Schäffer. Graph spanners. *J. of Graph Theory*, 13:99–116, 1989.

[PU87] David Peleg and Jeffrey D. Ullman. An optimal synchronizer for the hypercube. In *Proceedings 1987 6th ACM Symposium on Principles of Dist. Comp., Vancouver*, pages 77–85, 1987.

[Soa92] José Soares. Graph spanners: a survey. *Congressus Numerantium*, 89:225–238, 1992.

Computing the Independence Number of Dense Triangle-Free Graphs

Stephan Brandt

FB Mathematik & Informatik, WE 2
Freie Universität Berlin
Arnimallee 3, 14195 Berlin, Germany
e-mail: brandt@math.fu-berlin.de

Abstract. Computing the independence number of a graph remains $\mathcal{NP}$-hard, even restricted to the class of triangle-free graphs. So the question arises, whether this remains valid if the minimum degree is required to be large. While in general graphs this problem remains $\mathcal{NP}$-hard even within the class of graphs with minimum degree $\delta > (1-\varepsilon)n$, the situation is different for triangle-free graphs. It will be shown that for triangle-free graphs with $\delta > n/3$ the independence number can be computed as fast as matrix multiplication, while within the class of triangle-free graphs with $\delta > (1-\varepsilon)n/4$ the problem is already $\mathcal{NP}$-hard.

1 Introduction and Main Results

An early $\mathcal{NP}$-completeness result due to Poljak [20] says that computing the independence number remains $\mathcal{NP}$-hard, even if one restricts to the class of triangle-free graphs. Here we consider triangle-free graphs with large minimum degree and investigate the complexity of computing their independence number. It turns out that we can compute all maximum independent sets of a triangle-free graph of order n with minimum degree $\delta > n/3$ as fast as matrix multiplication (the current record for matrix multiplication is time $\mathcal{O}(n^{2.377})$ [9]).

Theorem 1. *Let G be a triangle-free graph of order n. If $\delta(G) > n/3$ then all maximum independent sets of G can be computed asymptotically in time needed to square its adjacency matrix.*

We will give a simple, efficient and practical algorithm for this problem based on the fact, that the maximum independent sets of such a graph are the neighborhoods of vertices of maximum degree of an easily computable supergraph. On the other hand, for any $\varepsilon > 0$ it remains $\mathcal{NP}$-hard to compute the independence number within the class of triangle-free graphs with $\delta > (1-\varepsilon)n/4$. In fact we prove a slightly stronger result.

Theorem 2. *For every pair of real numbers $c, \varepsilon > 0$ it is $\mathcal{NP}$-hard to compute the independence number within the class of triangle-free graphs with minimum degree $\delta > n/4 - cn^{\varepsilon}$.*

This result is shown by reducing INDEPENDENT SET in general graphs to INDEPENDENT SET in triangle-free graphs with large degree.

Recent interest in the complexity of problems in dense graphs (i.e. graphs with $\Omega(n^2)$ edges, or, more restrictive, with minimum degree $\Omega(n)$) arose from the observation that denseness can help to make hard problems more tractable. Arora, Karger and Karpinski [1] proved that there are polynomial time approximation schemes for Max-$\mathcal{SNP}$-hard optimization problems like MAX-CUT in graphs with $\Omega(n^2)$ edges. This is impossible in general graphs unless $\mathcal{P} = \mathcal{NP}$ [2] (see [3] for a thorough introduction into approximation complexity).

Other problems even become polynomial time solvable in dense graphs. Edwards [12] proved that k-COLORABILITY is polynomial time solvable in graphs with minimum degree at least cn for every $c > (k-3)/(k-2)$, so, in particular, 3-COLORABILITY is polynomial time solvable in graphs with linear degree. HAMILTONIAN CYCLE can trivially be decided in constant time in the class of graphs with minimum degree $\delta \geq n/2$, since the answer is always YES, by Dirac's famous result [10]. Few years ago, Häggkvist [14] proved by standard techniques that HAMILTONIAN CYCLE remains $\mathcal{NP}$-complete within the class of (general) graphs with minimum degree $\delta > (1-\varepsilon)n/2$. Anyway, INDEPENDENT SET remains $\mathcal{NP}$-complete for general graphs with minimum degree $\delta > (1-\varepsilon)n$.

If we restrict to dense triangle-free graphs the situation is different. Recently, Brandt [6] proved that the circumference, i.e. the length of a longest cycle, of a triangle-free graph with minimum degree $\delta > n/3$ can be computed in time $\mathcal{O}(n^{2.5})$ based on the matching algorithm of Micali and Vazirani [17]. Conversely, Veldman [22] proved that HAMILTONIAN CYCLE remains $\mathcal{NP}$-complete even within the class of bipartite graphs with minimum degree $\delta > (1-\varepsilon)n/4$.

It was shown in [6] that the circumference of triangle-free graphs with $\delta > n/3$ can be calculated directly from the independence number. So the question arose whether it is easy to compute this parameter as well. We will show that this is indeed the case, and, as a by-product, we get a slightly improved running time of $\mathcal{O}(n^{2.377})$ for computing the circumference compared to $\mathcal{O}(n^{2.5})$ from the matching-based approach. This corresponds to many results obtained by different motivations, indicating that there is not much structural freedom for triangle-free graphs with $\delta > n/3$ (see [7] for further results and references and for some historical remarks).

2 The Polynomial Algorithm

We consider undirected graphs without loops and multiple edges and we assume that the reader is familiar with basic graph theoretic concepts (see, e.g. [23]) and with standard computational complexity notions (see, e.g. [13]). A *maximum* independent set is a subset of the vertex set of maximal cardinality where no two vertices are joined by an edge, while a *maximal* independent set is one to which no further vertex may be added. A *maximal* triangle-free graph is a triangle-free graph to which no edge may be added without producing a triangle. Note that for graphs with $n \geq 3$ vertices being maximal triangle-free is equivalent to having diameter 2.

We need two steps to show that finding all maximum independent sets of a triangle-free graph G with minimum degree $\delta > n/3$ can be performed as fast as matrix multiplication. In the first step we show that such a graph G has a uniquely determined maximal triangle-free supergraph, the triangle-free closure $\mathrm{cl}_\Delta(G)$, which has the same maximum independent sets as G. In the second step we show that in a maximal triangle-free graph with $\delta > n/3$ every maximum independent set is the neighborhood of a vertex.

Let us start with the unique maximal triangle-free supergraph of such a graph. Closure concepts for graphs turned out to be powerful tools in hamiltonian graph theory. The first of them was introduced by Bondy and Chvátal [5], followed by many generalizations and variants, i.e. a recent striking variant due to Ryjáček [21] linking longest cycles in claw-free graphs to longest cycles in line graphs of triangle-free graphs.

Here we introduce a closure concept for triangle-free graphs with a slightly different flavour. We successively add edges to a triangle-free graph that do not create a triangle until no further edge can be added (i.e. the resulting graph is maximal triangle-free). In other words, we add edges joining vertices at distance at least 3 until a diameter 2 graph remains. In general, the result is far from being unique but, as we will see, for $\delta \geq n/3$ the resulting graph $\mathrm{cl}_\Delta(G)$ is uniquely determined. Closely related to the closure is the ***dist-2-graph***. For a given graph $G = (V, E)$ the edge set E_2 of the dist-2-graph $G_2 = (V, E_2)$ consists of the pairs of vertices which are at distance 2 in G. Note that a maximal triangle-free graph is the complement of its dist-2-graph. We will now characterize the triangle-free graphs with unique closure (in the labeled sense) via their dist-2-graphs.

Theorem 3. *Let G be a triangle-free graph. Then $\mathrm{cl}_\Delta(G)$ is uniquely determined if and only if $\overline{G_2}$ is triangle-free. In this case, $\mathrm{cl}_\Delta(G) = \overline{G_2}$.*

Proof. All edges appearing in some triangle-free supergraph of G are those joining vertices at distance 1 or at distance ≥ 3 in G, i.e. exactly those of $\overline{G_2}$. So if $\overline{G_2}$ is triangle-free we obtain that $\mathrm{cl}_\Delta(G) = \overline{G_2}$ is unique. Conversely, if $\overline{G_2}$ contains a triangle x, y, z, then at least two of its edges, say xy, xz, are not in G. So G has a maximal triangle-free supergraph G' containing xy and yz and another one $G'' \neq G'$ containing xz. □

The dist-2-graph will be important for the algorithm. But first we show that for $\delta(G) \geq n/3$ the closure $\mathrm{cl}_\Delta(G)$ is indeed unique.

Corollary 4. *Let G be a triangle-free graph of order n. If $\delta(G) \geq n/3$ then $\mathrm{cl}_\Delta(G)$ is uniquely determined.*

Proof. Suppose that $\overline{G_2}$ has a triangle x, y, z with $xy, xz \notin E(G)$. Since no pair of these vertices has distance 2 in G, the neighborhoods $N_G(x)$, $N_G(y)$, $N_G(z)$ are disjoint. Moreover, x is contained in none of the sets, since x has distance ≥ 3 to y and z, so $d_G(x) + d_G(y) + d_G(z) \leq n - 1$. □

Next we show that large independent sets of G are independent sets in $\mathrm{cl}_\Delta(G)$.

Lemma 5. *Let G be a triangle-free graph of order n. If $\delta(G) \geq n/3$ then every independent set S of G with $|S| > n/3$ is an independent set in $\mathrm{cl}_\Delta(G)$.*

Proof. Suppose that there is an independent set S with $|S| > n/3$ containing vertices u, v, which are joined by an edge in $\mathrm{cl}(G)$. Then u and v cannot have a common neighbor. So $n - |S| \geq d_G(u) + d_G(v) \geq 2n/3$, implying $|S| \leq n/3$, a contradiction. □

In triangle-free graphs the neighborhood of every vertex forms an independent set. Pach [19] characterized the maximal triangle-free graphs where every maximal independent set S is the neighborhood of a vertex, i.e. $S = N(v)$ for some vertex v. Jin [15] proved that every maximal triangle-free graph with $\delta > 10n/29$ has this property. There are maximal triangle-free graphs with minimum degree δ for $n/3 < \delta \leq 10n/29$, containing maximal independent sets which are not the neighborhood of a vertex. Our central observation is that *maximum* independent sets are still neighborhoods of vertices for $\delta > n/3$.

Theorem 6. *Let G be a maximal triangle-free graph of order n. If $\delta > n/3$ then every maximum independent set is the neighborhood of a vertex of G.*

Proof. Suppose that S is a maximum independent set which is not contained in the neighborhood of a vertex. Let S' be a minimal subset of S which is not contained in the neighborhood of a vertex. Since G has diameter 2 we get $s := |S'| \geq 3$ and, by the minimality of S', for every vertex u_i of S $(1 \leq i \leq s)$ there is a vertex $w_i \in V(G)$ which is adjacent to all vertices of S' except u_i. Since G has diameter 2, for every pair of vertices u_i, w_i there is a common neighbor v_i, which cannot be adjacent to any further vertex u_j, w_j $(j \neq i)$. Since $\delta > n/3$, there must be a vertex x which is adjacent to more than s vertices in $\bigcup_{1 \leq i \leq s}\{u_i, v_i, w_i\}$. The indices may be chosen such that $u_1, w_1, v_2, \ldots, v_s$ are the neighbors of x.

Now suppose that $s \geq 4$. Consider the subgraph H spanned by the 9 vertices $\bigcup_{2 \leq i \leq 4}\{u_i, v_i, w_i\}$. Since $\delta > n/3$ there must be a vertex y of G being adjacent to 4 vertices of H which we may assume to be v_2, v_3, u_4, w_4. Since G is triangle-free no vertex of G can be adjacent to more than four vertices in the subgraph induced by $\{x, v_2, v_3, y\} \cup \bigcup_{1 \leq i \leq 4}\{u_i, w_i\}$ and hence $\delta(G) \leq n/3$, a contradiction (see Fig. 1).

So we may assume $s = 3$. Let T_0 be the set of vertices of G having no neighbor in S' and T_2 be the vertices having exactly two neighbors. Since $\delta > n/3$ and no vertex has 3 neighbors in S', we have $|T_2| > |T_0|$. Since any two vertices of T_2 have a common neighbor in S', the set T_2 is an independent set of larger cardinality than $S \subseteq T_0$, the final contradiction. □

For every integer $s \geq 2$ there is a maximal triangle-free graph of order $6s + 3$ with minimum degree $2s + 1$, maximum degree $2s + 2$ and independence number

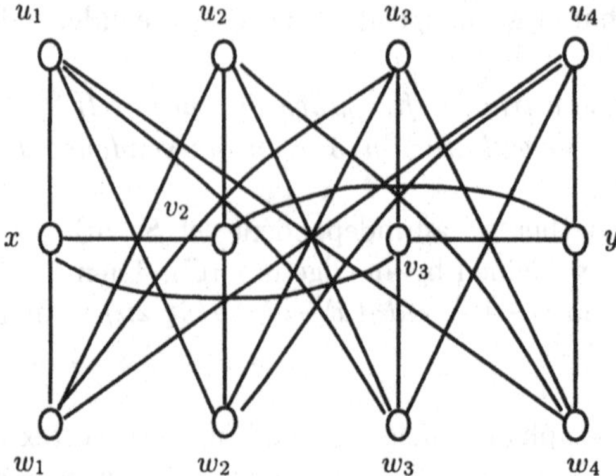

Fig. 1. The resulting subgraph of order 12.

$3s$ (see Fig. 2). So if we lower the degree bound to $\delta \geq n/3$ no maximum independent set needs to be the neighborhood of a vertex any more.

Based on the previous three results we can show that the following simple algorithm $\mathcal{A}$ computes all maximum independent sets of a triangle-free graph G with $\delta(G) > n/3$:

Input: adjacency matrix $A(G)$

Output: all maximum independent sets of G

- Compute the square $A^2(G) = (a_{ij}^{(2)})_{n \times n}$;
- For each row of $A^2(G)$ with largest number of 0's output the list of column indices of the 0 entries.

In order to prove our main Theorem 1 we have to prove that algorithm $\mathcal{A}$ computes all maximum independent sets of G and that its running time is asymptotically equal to squaring the adjacency matrix.

Proof (of Theorem 1). Let $B = (b_{ij})_{n \times n}$ be the matrix obtained from $A^2(G)$ by setting

$$b_{ij} = \begin{cases} 1 & : \quad a_{ij}^{(2)} = 0, \\ 0 & : \quad a_{ij}^{(2)} > 0. \end{cases}$$

Since $a_{ij}^{(2)}$ is the number of walks of length 2 joining the vertex v_i to v_j we have $a_{ij}^{(2)} = 0$ iff $\text{dist}(v_i, v_j) \geq 3$ or $\text{dist}(v_i, v_j) = 1$, using that G is triangle-free. So B is the adjacency matrix of $\overline{G_2}$ and by Theorem 3 and Corollary 4 we get $B = A(\text{cl}_\Delta(G))$.

Since the independence number $\alpha(G) \geq \delta(G) > n/3$, every maximum independent set of G is an independent set in $\text{cl}_\Delta(G)$ by Lemma 5. By Theorem 6 every maximum independent set of $\text{cl}(G)$, and therefore also of G, is the neighborhood of a vertex in $\text{cl}(G)$. So, indeed, $\mathcal{A}$ computes all maximum independent sets of G.

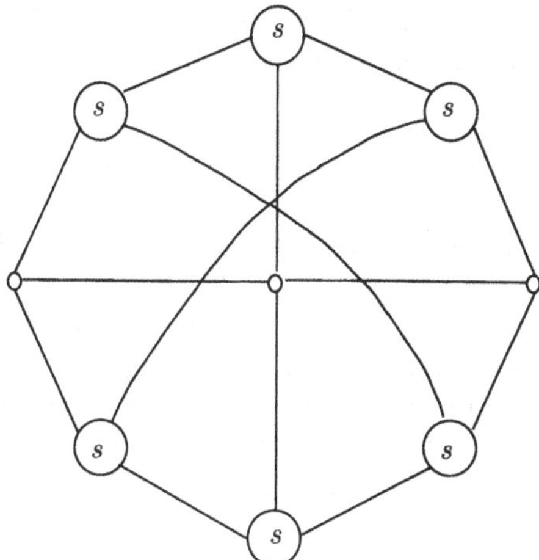

Fig. 2. A graph derived from the Petersen graph with one edge contracted (the big circles represent an independent set of s vertices, and the edges a complete bipartite subgraph joining two vertex sets).

For the running time, the matrix multiplication needed to compute $A^2(G)$ is the crucial step. Currently, there is an upper bound of $\mathcal{O}(n^{2.377})$ [9] for fast matrix multiplication and a lower bound of $\Omega(n^2)$. The other step can be performed by tracing $A^2(G)$ twice in time $\mathcal{O}(n^2)$. □

Note that, in terms of boolean matrix multiplication, $B = \neg(A(G) \wedge A(G))$, where $\neg$ is the elementwise negation. Anyway, the upper and lower bounds for the running time of boolean matrix multiplication are the same as for standard matrix multiplication.

Finally, given a triangle-free graph G, then the decision problem whether G is maximal triangle-free can clearly be solved in time $\mathcal{O}(n^{2.377})$ as well. Even the problem whether there is a unique maximal triangle-free supergraph can be solved in time $\mathcal{O}(n^{2.377})$, since the closure is unique if and only if $\overline{G_2}$ is triangle-free which is the case if and only if the trace of B^3 is 0 (see also [18] for computing cliques via fast matrix multiplication).

3 $\mathcal{NP}$-Completeness

In order to show that computing the independence number is $\mathcal{NP}$-hard even within a class of relatively dense triangle-free graphs we employ Poljak's original construction [20], giving a polynomial time reduction of INDEPENDENT SET in general graphs to INDEPENDENT SET in triangle-free graphs. Since the resulting graph has minimum degree at most 2, we have to modify it to a graph with large minimum degree later. Poljak offered his result without proof, so, for completeness, we include the simple but non-trivial proof here.

Lemma 7 (Poljak [20]). *Let G be a graph and G^{**} be the graph obtained from G by subdividing every edge of G exactly twice then the independence number $\alpha(G^{**}) = |E(G)| + \alpha(G)$.*

Proof. Let U be a maximum independent vertex set of G. For every edge e of G choose a vertex of G^{**} subdividing e which is not in the neighborhood of a vertex in U to obtain the vertex set W. Then $U \cup W$ is independent in G^{**} so $\alpha(G^{**}) \geq |E(G)| + \alpha(G)$.

Let U^{**} be an independent set of G^{**}. For any two vertices x and y in U^{**} which are the endvertices of an edge in G replace y by its subdivision neighbor on the edge xy in G^{**}. Performing this operation successively, we terminate with an independent set of the same cardinality as U^{**}, consisting of an independent vertex set of G and for each edge e of G at most one vertex subdividing e. Hence $\alpha(G^{**}) \leq |E(G)| + \alpha(G)$. □

Note that a graph G^{**} obtained by subdividing every edge of a graph G twice is triangle-free and has a 3-coloring which is easy to find: Color each vertex of G green and for every edge color one subdivision vertex yellow and the other red. We will now construct a big graph containing G^{**} as a little spot. This graph will have minimum degree almost $n/4$ and we can compute the independence number of G from the independence number of the big graph. The backbone of the big graph is a sequence of regular triangle-free graphs with chromatic number 4 and both degree and independence number equal to $n/4$. The existence of such graphs for every order $n \equiv 0 \pmod{96}$ was verified by the author in [8] by analyzing a product construction due to Bauer, van den Heuvel and Schmeichel [4] more carefully. Note that the chromatic number satisfies $\chi \geq n/\alpha$ for every graph (since coloring is partitioning into independent sets) and $\alpha \geq \Delta \geq \delta$ (Δ being the maximum degree) holds for triangle-free graphs, so these graphs are extremal, satisfying all three inequalities with equality.

Theorem 8 (Brandt [8]). *For every pair of integers k, s ($k \geq 2$, $s \geq 1$) there is a k-colorable n/k-regular triangle-free graph $H_{k,s}$ of order $n = 2^{k-2}k!s$ with independence number n/k.*

The graphs $H_{k,1}$ are easily computed by iterating the product construction of [4]. We get $H_{2,1} = K_2$ and $H_{3,1}$ is the graph depicted in Fig. 1 and $H_{4,1}$ is (in the terminology of [4]) the graph obtained by layering the graph $H_{3,1}$ four times. The graph $H_{k,s}$ is simply the lexicographic product $H_{k,1}[\overline{K}_s]$ (for the definition see e.g. [23, p. 370]). Given a graph G and real numbers $c, \varepsilon > 0$ we construct a graph $G_{c,\varepsilon}$ in the following way:

Let p be the order of G^{**} then set $s = \lceil((p/4c)^{1/\varepsilon} - p + 1)/96\rceil$. Let $G_{c,\varepsilon}$ be the graph obtained from a copy of G^{**} and a copy of $H_{4,s}$ in the following way: Choose a 3-coloring of G^{**} and a 4-coloring of $H_{4,s}$ and for each color class S of G^{**} select a distinct color class S' of $H_{4,s}$ and add all edges between S and S'. Note that every color class in a 4-coloring of $H_{4,s}$ has exactly $24s$ vertices.

Proof (of Theorem 2). We will reduce INDEPENDENT SET in general graphs (which is a well-known $\mathcal{NP}$-complete problem, see [13]) to INDEPENDENT

SET in triangle-free graphs with minimum degree $\delta > n/4 - cn^{\varepsilon}$. For fixed constants c, ε we can construct $G_{c,\varepsilon}$ in time polynomial in the order $n_{c,\varepsilon}$ of $G_{c,\varepsilon}$ and therefore in polynomial time in the order of G as well. Note that $\delta(G_{c,\varepsilon}) > n_{c,\varepsilon}/4 - cn^{\varepsilon}_{c,\varepsilon}$ and $\alpha(G_{c,\varepsilon}) = \alpha(H_{4,s}) + \alpha(G^{**})$, so by Lemma 7 we have $\alpha(G) = \alpha(G_{c,\varepsilon}) - 24s - |E(G)|$. Assuming that we could compute $\alpha(G_{c,\varepsilon})$ in polynomial time in the order of $G_{c,\varepsilon}$ we could also compute $\alpha(G)$ in polynomial time in the order of G as well. □

The reduction showing that computing the independence number in general graphs with $\delta > (1-\varepsilon)n$ is $\mathcal{NP}$-hard is simply by considering the join $G \vee K_s$, whose independence number is just the independence number of G.

4 Final Remarks

The results leave a grey area in the degree range between $n/4$ and $n/3$, where the complexity of INDEPENDENT SET in triangle-free graphs remains unsolved. There seems to be a continuous loss of information in this degree range. E.g. it was shown in [6] that the possible range for the (polynomial time computable) vertex connectivity κ of a triangle-free graph with minimum degree δ is $\delta \geq \kappa \geq \min\{\delta, 4\delta - n\}$, so we start with $\kappa = \delta$ for $\delta \geq n/3$ and end with no information $\delta \geq \kappa \geq 0$ for $\delta \leq n/4$. Anyway, the author would guess that INDEPENDENT SET becomes $\mathcal{NP}$-complete within the class of triangle-free graphs with $\delta > (1-\varepsilon)n/3$. It should be mentioned that all the cited $\mathcal{NP}$-completeness results also hold for the slightly stronger requirement $\delta > n/\gamma - cn^{\varepsilon}$ for any constants $c, \varepsilon > 0$ (similar to the statement of Theorem 2), though most of them were only stated in the less technical form $\delta > (1-\varepsilon)n/\gamma$.

The situation seems to be somewhat more difficult concerning the chromatic number. It follows from a result of Jin [15] that computing the chromatic number of triangle-free graphs with $\delta > 10n/29$ have chromatic number either 2 or 3). Triangle-free 4-chromatic graphs with $\delta > n/3$ are known, but it seems not unlikely, that there are no 5-chromatic graphs (in contrast to a conjecture of Jin [16]), so, using Edwards' result [12] that 3-COLORABILITY can be decided in polynomial time in graphs with linear degree, this would lead to a polynomial time algorithm for the chromatic number. A construction of Erdős, Hajnal and Simonovits [11] with a similar flavor as our construction in Theorem 2 shows, that for every $\varepsilon > 0$ there are triangle-free graphs with $\delta > (1-\varepsilon)n/3$ with arbitrarily large chromatic number. This construction seems to indicate that computing the chromatic number within this range is already $\mathcal{NP}$-hard.

Acknowledgement: I thank Thomas Emden-Weinert for drawing my attention to Edwards' result [12].

References

1. S. Arora, D. Karger and M. Karpinski, Polynomial time approximation schemes for dense instances of $\mathcal{NP}$-hard problems, in: Proc. 27th ACM Symp. on Theory of Computing, 1995, pp. 284–293.

2. S. Arora, C. Lund, R. Motwani, M. Sudan and M. Szegedy, Proof verification and hardness of approximation problems, in: Proc. IEEE Foundations of Computer Science, 1992, pp. 14–23.
3. S. Arora and C. Lund, Hardness of approximations, in: Approximation algorithms for NP-hard problems (D. S. Hochbaum, ed.), PWS, Boston, 1995, pp. 399–446.
4. D. Bauer, J. van den Heuvel and E. Schmeichel, Toughness and triangle-free graphs, *J. Combin. Theory Ser. B* **65**, (1995), 208–221.
5. J. A. Bondy and V. Chvátal, A method in graph theory, *Discrete Math.* **15** (1976), 111–135.
6. S. Brandt, Cycles and paths in triangle-free graphs, in: The Mathematics of Paul Erdős (R. L. Graham and J. Nešetřil, eds.), Springer, 1996, pp. 32–42.
7. S. Brandt, On the structure of dense triangle-free graphs, submitted.
8. S. Brandt, Triangle-free graphs whose independence number equals the degree, manuscript.
9. D. Coppersmith and S. Winograd, Matrix multiplication via arithmetic progressions, *J. Symbolic Comput.* **9** (1990), 251–280.
10. G. A. Dirac, Some theorems on abstract graphs, *Proc. London Math. Soc. (3)* **2** (1952), 69–81.
11. P. Erdős and M. Simonovits, On a valence problem in extremal graph theory, *Discrete Math.* **5** (1972), 323–334.
12. K. Edwards, The complexity of colouring problems on dense graphs, *Theoretical Comp. Sci.* **43** (1986), 337–343.
13. M. R. Garey and D. S. Johnson, Computers and intractability: a guide to NP-completeness, W. H. Freeman, New York, 1979.
14. R. Häggkvist, On the structure of non-hamiltonian graphs I, *Comb., Prob. and Comp.* **1** (1992), 27–34.
15. G. Jin, Triangle-free graphs with high minimal degrees, *Comb. Prob. and Comp.* **2**, (1993), 479–490.
16. G. Jin, Triangle-free four-chromatic graphs, *Discrete Math.* **145** (1995), 151–170.
17. S. Micali and V. V. Vazirani, An $\mathcal{O}(V^{1/2}E)$ algorithm for finding maximum matching in general graphs, *Proc. 21st Ann. Symp. on Foundations of Computer Sc.* IEEE, New York (1980), 17–27.
18. J. Nešetřil and S. Poljak, On the complexity of the subgraph problem, *Comment. Math. Univ. Carolin.* **26** (1985), 415–419.
19. J. Pach, Graphs whose every independent set has a common neighbour, *Discrete Math.* **37** (1981), 217–228.
20. S. Poljak, A note on stable sets and colorings of graphs, *Comment. Math. Univ. Carolinae* **15** (1974), 307–309.
21. Z. Ryjáček, On a closure concept in claw-free graphs, to appear in J. Combin. Theory Ser. B.
22. H. J. Veldman, Personal communication, 1994.
23. D. B. West, Introduction to graph theory, Prentice Hall, 1996.

Algorithms for the Treewidth and Minimum Fill-in of HHD-Free Graphs

H. J. Broersma[1], E. Dahlhaus[2] and T. Kloks[1]*

[1] University of Twente
Faculty of Applied Mathematics
P.O.Box 217
7500 AE Enschede, the Netherlands
[2] Department of Computer Science
University of Bonn
Bonn, Germany

Abstract. A graph is HHD-free is it does not contain a house (i.e., the complement of P_5), a hole (a cycle of length at least 5) or a domino (the graph obtained from two 4-cycles by identifying an edge in one C_4 with an edge in the other C_4) as an induced subgraph. The MINIMUM FILL-IN problem is the problem of finding a chordal supergraph with the smallest possible number of edges. The TREEWIDTH problem is the problem of finding a chordal embedding of the graph with the smallest possible clique number. In this note we show that both problems are solvable in polynomial time for HHD-free graphs.

Keywords: graphs, algorithms, HHD-free graphs, treewidth, minimum fill-in.
MSC: 68R10.

1 Introduction

A graph is HHD-free if it does not contain a house (i.e., the complement of P_5), a hole (C_k for $k \geq 5$) or a domino (see Figure 1).

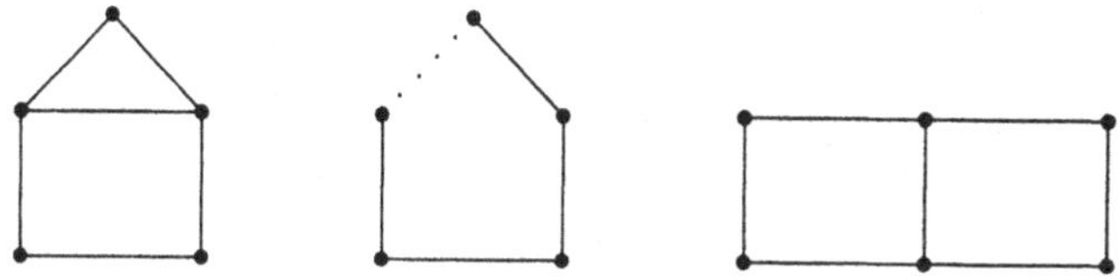

Fig. 1. 'House' (left), 'hole' (middle) and 'domino' (right)

* kloks@math.utwente.nl

Elimination and structural properties for HHD-free graphs were obtained in [8, 10]. For more information, the reader is referred to [17, 4].

A graph is chordal if it does not contain a chordless cycle of length at least four as an induced subgraph. A triangulation of a graph is a chordal supergraph with the same vertex set. Two triangulation problems have drawn much attention because of the large number of applications. The first is to find a triangulation of the graph such that the number of edges is minimum. This is called the MINIMUM FILL-IN problem. This problem is strongly related to Gaussian elimination of matrices. The second is called the TREEWIDTH problem. The objective in this case is to find a triangulation of a graph such that the clique number is as small as possible (the treewidth of the graph is the minimum clique number over all triangulations minus one). Both problems are NP-complete in general [21, 1], but polynomial time algorithms exist for many special graph classes such as cographs, circle and circular arc graphs, permutation graphs and, more generally, cocomparability graphs with bounded dimension, chordal bipartite graphs etc. [3, 12, 16, 15, 2, 20, 13, 5, 11, 18].

In this paper we show that the TREEWIDTH and the MINIMUM FILL-IN problem are solvable for HHD-free graphs.

Notice that adding an edge between two non adjacent vertices of a C_4 in an HHD-free graph, may introduce a new chordless cycle, and hence the resulting graph may no longer be HHD-free. This is illustrated for example by a graph consisting of a path and two non adjacent vertices that are adjacent to all vertices of the path. Joining the end vertices of the path by an edge would destroy the outer cycle (that was a C_4). But we get a cycle that consists of the path and the new edge. When we make the path long enough we get a cycle of length at least five. Hence, it is not clear whether a 'minimum C_4 destroying set of chords' leads to a chordal graph (note that all chordal graphs are HHD-free). If true, this could lead to a possible solution for the minimum fill-in problem by finding a minimum vertex cover in an auxiliary graph defined on the chords of the C_4's (if the VERTEX COVER problem can be solved for this auxiliary graph) (see [19, 5]).

Instead of taking this approach we only make some fairly easy observations for the minimal separators of an HHD-free graph, which enable us to use a 'standard' dynamic programming technique to solve the problem.

2 Preliminaries

We denote the number of vertices of a graph $G = (V, E)$ by n and the number of edges by m. For a vertex $x \in V$, $N(x)$ is the neighborhood of x and $N[x] = \{x\} \cup N(x)$ is the closed neighborhood of x.

If Ω is a set and $x \in \Omega$, then we write $\Omega - x$ instead of $\Omega \setminus \{x\}$. For a subset Q of vertices, we write $G[Q]$ for the graph induced by the vertices of Q. For a vertex x, we write $G - x$ instead of $G[V - x]$, and for a subset W of vertices we write $G - W$ for the graph $G[V - W]$.

A *hole* is an induced cycle of length at least five. The house, hole and domino are depicted in Figure 1.

Definition 1. A graph is *HHD-free* if it does not contain a house, hole or domino as an induced subgraph.

Definition 2. A graph is *chordal* if it does not contain an induced cycle of length more than three.

Definition 3. Let a and b be non adjacent vertices. A set S of vertices is a minimal a, b-separator if a and b are in different connected components of $G - S$ and there is no proper subset of S with the same property. A minimal separator is a set S of vertices for which there exist non adjacent vertices a and b such that S is a minimal a, b-separator.

For the following lemma, we refer to [7].

Lemma 4. *A set S of vertices is a minimal separator if and only if there exist two connected components C_1 and C_2 in $G - S$ such that every vertex of S has at least one neighbor in C_1 and at least one neighbor in C_2.*

Definition 5. Let S be a minimal separator and C a connected component of $G - S$ such that every vertex of S has a neighbor in C. Then S is *close to C*.

There exist many characterizations of chordal graphs. We use the characterization given by Dirac [6] using minimal separators.

Lemma 6. *A graph G is triangulated if and only if every minimal vertex separator induces a complete subgraph of G.*

Definition 7. A *triangulation* of a graph G is a graph H with the same vertex set as G such that G is a subgraph of H and H is chordal. A triangulation H of G is *minimal* if no proper subgraph of H is also a triangulation of G.

Definition 8. The *minimum fill-in* of a graph G, denoted by $\text{mfi}(G)$, is the minimum number of edges which are not edges of G, of a a triangulation of G. We write $\text{mfi}^*(G) = m + \text{mfi}(G)$ for the number of edges in a triangulation realizing the minimum fill-in. The *treewidth* of a graph G, $\text{tw}(G)$, is the minimum clique number of a triangulation of G minus one.

Remark. Notice that for the treewidth and minimum fill-in problem we only have to consider triangulations that are minmal.

For a proof of the following, see, e.g., [14].

Lemma 9. *Let H be a minimal triangulation of a graph G and let S be a minimal a, b-separator of H for non adjacent vertices a and b in H. Then S is also a minimal a, b-separator in G, and if C is the vertex set of a connected component of $H - S$ then C induces also a connected component in $G - S$.*

For a proof of the following corollary of Lemma 6 and Lemma 9, we refer to [15].

Corollary 10. *If G is a clique then the treewidth equals the number of vertices minus one. The minimum fill-in of a clique is zero. Assume G is not a clique. Then*

$$tw(G) = \min_S \max_C tw(H(S,C))$$

where the minimum is taken over all minimal separators S in G and the maximum is taken over all connected components C of $G - S$. For the minimum fill-in, we have:

$$mfi^*(G) = \min_S \binom{|S|}{2} + \sum_C \left(mfi^*(H(S,C)) - \binom{|S|}{2} \right)$$

2.1 Minimal separators in HHD-free graphs

Lemma 11. *Let S be a minimal separator in an HHD-free graph G. Let C be a connected component of $G - S$ such that S is close to C. For every pair of vertices x and y in S, there exists a vertex p in C adjacent to x and y.*

Proof. Assume x and y do not have a common neighbor in C. By lemma 4, there exists at least one other component C' of $G - S$ such that x and y have a neighbor in C'. Then either a house, a hole, or a domino must exist. □

Theorem 12. *Let S be a minimal separator in a HHD-free graph and let C be a connected component of $G - S$ such that S is close to C. Then there is a vertex p in C adjacent to all vertices of S.*

Proof. Consider two adjacent vertices p and q in C and assume they have private neighbors in S, p' and q' respectively (i.e., p' is not adjacent to q and q' is not adjacent to p). By Lemma 11 p' and q' have a common neighbor in some other connected component C'. This gives a house or an induced 5-cycle. Hence, since C is connected, there is a linear ordering by inclusion of $N(x) \cap S$ for the vertices x in C. Since every vertex of S has a neighbor in C a maximal element in this ordering must be adjacent to all vertices of S. □

Corollary 13. *Let G be an HHD-free graph and let S be a minimal separator in G. Then there exist non adjacent vertices p and q such that $S = N(p) \cap N(q)$. Hence, if G is HHD-free, then there are at most $O(n^2)$ different minimal separators in G.*

Proof. Since S is a minimal separator there exist connected components C and C' of $G - S$ such that S is close to both. By Lemma 12 there are vertices p and q in C and C' respectively such that $S \subseteq N(p) \cap N(q)$. The neighbors of p are contained in $S \cup C$ and the neighbors of q are contained in $S \cup C'$. This proves the corollary. □

3 From lumps to smaller lumps

Definition 14. Let S be a minimal separator and C a connected component of $G - S$. The pair (S, C) is called a *lump*. We write $H = H(S, C)$ for the graph obtained from $G[C \cup S]$ by adding edges such that S becomes a clique. The graph H is called a *realizer* of the lump.

Our algorithms for treewidth and minimum fill-in use dynamic programming on lumps. In this section we describe in detail how the minimum fill-in and treewidth of the realizer of a lump are expressed in the treewidth and minimum fill-in of smaller realizers.

In the first stage of the algorithm a list is made of all lumps (S, C) and this list is sorted according to $|S| + |C|$, the number of vertices of $G[S \cup C]$. For each lump (S, C), the treewidth and minimum fill-in of the realizer $H = H(S, C)$ is computed (in a way described hereafter). When this is completed the treewidth and minimum fill-in of G can be obtained using Corollary 10.

We describe in detail how the treewidth and minimum fill-in of a realizer is expressed in the treewidth and minimum fill-in of smaller realizers in the rest of this section.

Throughout this section, let S be a minimal separator of G, let C be a connected component of $G \setminus S$, and let $H = H(S, C)$ be the realizer of the lump (S, C).

Lemma 15. *Let $S^* \subseteq S$ be the set of vertices of S with a neighbor in C. Then S^* is a minimal separator in G close to C.*

For the treewidth of H, we have $tw(H) = \max(|S| - 1, tw(H^*))$ where $H^* = H(S^*, C)$, and for the minimum fill-in: $mfi^*(H) = \binom{|S|}{2} - \binom{|S^*|}{2} + mfi^*(H^*)$.

Proof. If $S^* = S$ there is nothing to prove.

We show that S^* is a minimal separator in G. C is a connected component of $G - S^*$ and every vertex of S^* has a neighbor in C. There exists at least one connected component C' different from C in $G - S$ such that every vertex of S has a neighbor in C'. Since $S^* \subseteq S$, C' is contained in connected component different from C in $G - S^*$. Using Lemma 4 this proves the lemma. □

Let Ω be the set of vertices in C which are adjacent to all vertices of S. Hence if S is close to C, by Theorem 12, $\Omega \neq \emptyset$.

We first consider minimal triangulations Q of H in which Ω is *not* a clique.

Lemma 16. *Assume $\Omega \neq \emptyset$. Let Q be a minimal triangulation of H such that Ω is not a clique in Q. Then there is a minimal separator S' of Q which is also a minimal separator of G with $S \subset S' \subseteq S \cup C$.*

Let $C_1, \ldots, C_t$ be the connected components of $H - S'$. If Q realizes the treewidth of H, then $tw(H) = \max_i tw(H_i)$ where $H_i = H(S', C_i)$. If Q realizes the minimum fill-in we have $mfi^*(H) = \binom{|S'|}{2} + \sum_i \left(mfi^*(H_i) - \binom{|S'|}{2}\right)$.

Proof. Let $p, q \in \Omega$ be non adjacent in Q. Clearly every minimal p, q-separator S' in Q contains S since p and q are in Ω. Since S' is a minimal separator in Q for vertices p and q in C, and since S is a clique in Q, $S' \subset S \cup C$. Since Q is a minimal triangulation of H, S' is a minimal p, q-separator in H. Since $S \subset S'$, G and H have the same set of edges, except between vertices which are both in S'. Hence S' is a minimal p, q-separator of G. The formulae follow from Corollary 10. □

Remark. Notice that the the number of vertices in each lump H_i in Lemma 16 is strictly less than the number of vertices in H, since p and q are not both contained in the same lump.

We now consider minimal triangulations of H in which Ω is a clique.

Lemma 17. *Assume $\Omega = C$ and let Q be a minimal triangulation Q of H such that Ω is a clique in Q.*

If Q realizes the treewidth of H, then $tw(H) = |C| + |S| - 1$. If Q realizes the minimum fill-in: $mfi^*(H) = \binom{|C \cup S|}{2}$.

Proof. Obvious, since Q is a clique. □

Before we continue we need the following crucial observation.

Theorem 18. *Assume $\Omega \neq \emptyset$ and $\Omega \neq C$. Let C^* be a connected component of $G[C - \Omega]$. Then at least one vertex of S does not have a neighbor in C^*.*

Proof. Consider the graph $G^* = G - \Omega$. Then G^* is HHD-free and C^* is a connected component of $G^* - S$. If every vertex of S has a neighbor in C^* then S would is a minimal separator of G^* close to C^*. But then, by Theorem 12, there is a vertex in C^* adjacent to all vertices of S, which is a contradiction. □

Lemma 19. *Assume $\Omega \neq \emptyset$ and $\Omega \neq C$. Let Q be a minimal triangulation of H such that Ω is a clique in Q. Let $C_1, \ldots, C_t$ be the connected components of $G[C - \Omega]$. Let S_i be the set of vertices in $S \cup \Omega$ with a neighbor in C_i in G. Then S_i is a minimal separator in G.*

If Q realizes the treewidth of H, $tw(H) = \max(|S \cup \Omega| - 1, \max_i tw(H_i))$, where $H_i = H(S_i, C_i)$. If Q realizes the minimum fill-in we find: $mfi^*(H) = \binom{|S \cup \Omega|}{2} + \sum_i \left(mfi^*(H_i) - \binom{|S_i|}{2}\right)$.

Proof. Since Q is a minimal triangulation C_i $(i = 1, \ldots, t)$ are the connected components of $Q[C - \Omega]$ and S_i is the set of vertices with a neighbor in C_i in Q. By Theorem 18 there exists a vertex $p_i \in S$ without a neighbor in C_i. Then for any vertex $q_i \in C_i$ the set S_i is a minimal p_i, q_i-separator in Q and hence also in H. We show that S_i is also a minimal separator in G. There exists a connected component C' other than C such that S is close to C'. The component C' is contained in a connected component C^* of $G - S_i$ different from C_i. Then $p_i \in C^*$. Hence every vertex of $S_i \cap \Omega$ is a neighbor of $p_i \in C^*$ and every vertex of $S_i \cap S$ has a neighbor in $C' \subset C^*$. It follows that H_i is a lump. □

Remark. Notice that the number of vertices of each realizer H_i is strictly less than the number of vertices in H since at least one vertex of S is not a vertex of H_i.

4 Algorithms for the treewidth and minimum fill-in

The algorithms we propose use dynamic programming on lumps to compute the treewidth and minimum fill-in of the realizers of these lumps. First a list of all lumps (S, C) is made and this list is sorted according to the number of vertices of $S \cup C$.

We use the adjacency matrix to test for adjacencies. Creating a list of all minimal separators can be performed in $O(n^3)$ time using Corollary 13. For a minimal separator S, the vertex set of each connected component of $G - S$ can be computed in $O(n + m)$ time. Hence, creating a list of all lumps takes $O(n^2(n + m))$ time. Furthermore, within the same time bound, for each lump (S, C) a minimal separator $S^* \subseteq S$ can be determined such that S^* is close to C.

Notice that the total size of the list of lumps is $O(n^3)$. Sorting this list according to the number of vertices in the lumps using a linear time sorting algorithm takes $O(n^3)$ time.

If for a lump (S, C) the separator S is not close to C, the treewidth and minimum fill-in are given by Lemma 15. Since S^* is known, the treewidth and minimum fill-in of $H(S, C)$ can be determined in constant time in this case.

Now consider a lump (S, C) such that S is close to C. Clearly a possible triangulation of the realizer $H = H(S, C)$ is to make a clique of $S \cup C$. In that case the the treewidth and minimum fill-in of that triangulation can easily be determined.

Determining the set Ω takes linear time (for each x, one only has to count the number of neighbors in C, and if this number is equal to the size of C then x belongs to Ω).

We first consider triangulations Q of H such that Ω is not a clique in Q. We use Lemma 16 to determine the minimum fill-in and treewidth in that case. Creating a list of minimal separator S^* with $S \subset S^* \subseteq S \cup C$ can be done in $O(n^3)$ time. For each such S^*, we determine the components $C_1, \ldots, C_t$ that partition $G[C - (S^* \setminus S)]$ in time $O(n)$ time. If this partition contains only one component we can ignore the choice of S^* since in that case S^* cannot be a minimal separator for two vertices in Ω. Otherwise update the minima for the treewidth and minimum fill-in of H according to the formulae given in Lemma 16.

Now we consider triangulations Q where Ω is a clique in Q. If $\Omega = C$, then updating the current minima for the minimum fill-in and treewidth is trivial according to Lemma 17. If $\Omega \neq C$ we use Lemma 19. In that case the components $C_1, \ldots, C_t$ of $G[C - \Omega]$ can be computed in linear time. It is easy to see that the sets $S_i \subset S \cup \Omega$ which have a neighbor in C_i can be determined in $O(n + m)$

time. Hence updating the minima for the treewidth and minimum fill-in of H in this case takes $O(n+m)$ time.

Theorem 20. *There exists an $O(n^6)$ time algorithm computing the treewidth and minimum fill-in of an HHD-free graph.*

5 Conclusions

In this note we presented a polynomial time algorithm to compute the treewidth and minimum fill-in for HHD-free graphs. We do not claim that our algorithm is a very practical one. Indeed we feel that it is possible to improve the time bounds for these algorithms by analyzing the structure of the minimal separators or, equivalently, the structure of the C_4's in more detail.

Another question which is left open, is whether a minimum cover of all C_4's by 'diagonals' gives a chordal graph. If this is the case, this could lead to a more efficient algorithm for the minimum fill-in of HHD-free graphs.

The class of HHD-free graphs is properly contained in that of the weakly chordal graphs. These are graphs without induced C_k or $\overline{C_k}$ for any $k \geq 5$. It is easy to see (by using the results on so called two-pairs [9]) that the number of minimal separators in this case is also at most $O(n^2)$. However, unitl now, the complexity of the treewidth and minimum fill-in problem for this graph class is unknown.

References

1. Arnborg, S., D. G. Corneil and A. Proskurowski, Complexity of finding embeddings in a k-tree, *SIAM J. Alg. Disc. Meth.* **8**, (1987), pp. 277–284.
2. Bodlaender, H., T. Kloks, D. Kratsch and H. Müller, Treewidth and minimum fill-in on *d*-trapezoid graphs, Technical report RUU-CS-1995-34, Utrecht University, The Netherlands, 1995.
3. Bodlaender, H. and R. Möhring, The pathwidth and treewidth of cographs, *SIAM Journal on Discrete Mathematics* *7* (1993), pp. 181-188.
4. Brandstädt, A., Special graph classes – A survey, Schriftenreihe des Fachbereichs Mathematik, SM-DU-199 (1991), Universität Duisburg Gesamthochschule.
5. Maw-Shang Chang, Algorithms for maximum matching and minimum fill-in on chordal bipartite graphs. ISAAC'96 (T. Asano et al. ed.), LLNCS 1178, pp. 146-155.
6. Dirac, G. A., On rigid circuit graphs, *Abh. Math. Sem. Univ. Hamburg* **25**, (1961), pp. 71–76.
7. Golumbic M. C., *Algorithmic graph theory and perfect graphs*, Academic Press, New York, 1980.
8. Hammer, P. L. and F. Maffray, Completely separable graphs, *Discrete Applied Mathematics* **27**, (1990), pp. 85–99.
9. Hayward, R., C. T. Hoang and F. Maffray, Optimizing weakly triangulated graphs, *Graphs and combinatorics* **5**, (1989), pp. 339–349.
10. Jamison, B. and S. Olariu, On the semi-perfect elimination, *Advances in Applied Mathematics* **9**, (1988), pp. 364–376.

11. Kloks, T., *Treewidth - Computations and Approximations*, Springer Verlag, Lecture Notes in Computer Science 842, (1994).
12. Kloks, T., Treewidth of circle graphs, *International Journal of Foundations of Computer Science* **7**, (1996), pp. 111–120.
13. Kloks, T. and D. Kratsch, Treewidth of chordal bipartite graphs, *J. of Algorithms* **19**, (1995), pp. 266–281.
14. Kloks, T., D. Kratsch and H. Müller, Approximating the bandwidth for AT-free graphs, *Proceedings of the Third Annual European Symposium on Algorithms (ESA'95)*, Springer-Verlag, Lecture Notes in Computer Science 979, (1995), pp. 434–447.
15. Kloks, T., D. Kratsch and J. Spinrad, Treewidth and pathwidth of cocomparability graphs of bounded dimension, Computing Science Notes, 93/46, Eindhoven University of Technology, Eindhoven, The Netherlands, (1993), to appear in *Order*.
16. Kloks, T., D. Kratsch and C. K. Wong, Minimum fill-in of circle and circular arc graphs, *Proceedings of the* 21th *International Symposium on Automata, Languages and Programming (ICALP'96)*, Springer-Verlag Lecture Notes in Computer Science 1113, (1996), pp. 256–267.
17. Olariu, S., Results on perfect graphs, PhD thesis, Scool of Computer Science, McGill University, Montreal, 1986.
18. Parra, A., Scheffler, P., How to use minimal separators for its chordal triangulation, *ICALP'95*, LLNCS 944, pp. 123-134.
19. Spinrad, J., A. Brandstädt and L. Stewart, Bipartite permutation graphs, *Discrete Applied mathematics* **18**, (1987), pp. 279–292.
20. Sundaram, R., K. Sher Singh and C. Pandu Rangan, Treewidth of circular arc graphs. To appear in *SIAM J. Disc. Math.*
21. Yannakakis, M., Computing the minimum fill-in is NP-complete, *SIAM J. Alg. Disc. Meth.* **2**, (1981), pp. 77–79.

Block Decomposition of Inheritance Hierarchies

Christian Capelle

LIRMM - UMR 9928 UM II/CNRS
161, rue Ada - 34 392 Montpellier Cedex 05 - France
tel : 33 4 67 41 85 41 - fax : 33 4 67 41 85 85
email:capelle@lirmm.fr

Abstract. Inheritance hierarchies play a central role in object oriented languages as in knowledge representation systems. These hierarchies are acyclic directed graphs representing the underline structure of objects. This paper is devoted to the study of efficient algorithms to decompose recursively an inheritance hierarchy into independent subgraphs which are inheritance hierarchies themselves. This process gives a tree called decomposition tree.
The decomposition proposed here is based on the concept of block which is an extension of the concept of h-module proposed by R. Ducournau and M. Habib [7]. M. Habib, M. Huchard and J. Spinrad [8] have presented a linear algorithm to decompose an inheritance hierarchy into h-modules. The algorithm proposed here to decompose an inheritance hierarchy into blocks generalizes the algorithm of Habib *et al.*. It computes a linear extension of the hierarchy such that the blocks are factors of the extension. This is a general technique applicable to different decompositions [2, 1]. The unicity of the block decomposition comes from a proposition showing the links between blocks and modules of the well known modular decomposition of directed graphs, and from the theorem of unicity of the modular decomposition.
While the cost to compute the block decomposition is greater than the h-module decomposition one, it allows a greater factorization of the information of inheritance represented by the hierarchy. This decomposition can be useful for graph drawing applications [6] and could also be used for hierarchy coding applications. Such a decomposition can also be seen as a tool to help object oriented languages programmer to "understand" their hierarchies. Some linearizations of object hierarchies can be defined from this decomposition.

1 Introduction

Multiple inheritance is a common feature of many object oriented systems. This implies an ordering of object sets according to an inheritance relation which is generally formalized by a directed acyclic graph. Such a hierarchy denotes an "is a" or "is a kind of" link between objects. The objects, classes of concepts (depending of the context: object oriented languages or knowledge representation) have attributes/properties/methods that are inherited through the edges of the hierarchy. So it induces a factorization of information.

The main drawback of multiple inheritance comes from the lack of visibility of the structure of the inheritance relation which is more complicated than in systems without multiple inheritance. Automated hierarchical representations of the inheritance hierarchies introduce the necessity to develop some decompositions based on partitioning techniques. Such a technique can improve the understanding of the organization of objects by the programmer/user.

Graph decompositions can take many different forms. In this paper decomposition is seen as a procedure partitioning recursively the edge set of a hierarchy in partial sub-hierarchies whose neighbourhood to the outside verifies some properties. In fact, the properties checked define the decomposition. The whole recursive process of decomposition of a given hierarchy is described by a tree whose root is the hierarchy itself and leaves are the edges of the transitive reduction of the hierarchy. Each internal node denotes a sub-hierarchy and its children describe its partition.

Habib, Huchard and Spinrad [8] have studied such a decomposition based on the concept of h-module. Informally, an h-module is a subset of the vertex set of the inheritance hierarchy which induces an inheritance sub-hierarchy, and such that all its vertices have strictly the same neighbourhood towards the outside according to the transitive closure of the hierarchy. The problem considered is the following: let v be some object of a hierarchy H, how can the sub-hierarchy $[v, 1_H]$ of H can be decomposed/partitioned in h-modules, where 1_H denote the greater element of the hierarchy. $[v, 1_H]$ is called the hierarchy of v.

In this paper we study a similar problem where a hierarchy is not partitioned in h-modules but in *blocks*. This concept of block is an extension of the concept of h-module. It induces a new decomposition called *block decomposition* which extends h-module decomposition. As an h-module, a block is a subset of the vertex set of the inheritance hierarchy which induces an inheritance sub-hierarchy, but the properties on the neighbourhood a block must verify are less strong.

This paper is organized as follows: first the concepts of *h-module* and *block* are presented as well as the results previously obtained by M. Habib, M. Huchard and J. Spinrad about h-module decomposition of hierarchies. In Sect. 3 the well known *modular decomposition theory* is briefly presented in the case of digraphs. This decomposition exploits the concept of module. The links between modules and blocks are highlighted. This allows to define formally a *block decomposition theorem.* The following sections deal with the algorithmic problem of the computation of blocks (or the block decomposition). In Sect. 4 the algorithmic perspectives given by the study of Sect. 3 are explored, then in Sect. 5 the skeleton of an algorithm to compute block decomposition is presented as well as the concept of *factorizing linear extension.* An algorithm to compute such a linear extension is proposed. Section 6 deals with the following problem: computing connected blocks from a factorizing linear extension. In the last section, it is shown how to compute the block decomposition from a factorizing permutation.

This paper does not contain the proofs of the results, neither the proofs of the algorithms. However these proofs are in [1]. This Ph'd thesis is available at the following URL: http://www.lirmm.fr/~capelle/publications.html. Unfortunately, this document is in french.

2 h-modules and Blocks of Inheritance Hierarchies

Except if it is explicitly stated, all the graphs considered in this paper are digraphs with vertex set X and edge set E ($|X| = n$ and $|E| = m$).

An **inheritance hierarchy** or an **inheritance graph** (for short in the following we will say a **hierarchy**) is a directed acyclic multi-graph $H = (X, E)$ with a **least element** denoted by 0_H, and a **greater element** denoted by 1_H. This is a formalization of the interval $[v, 1_H]$ presented above. The transitive closure H^{tc} of H induces a partial order. In the context of object oriented languages this order is called the **inheritance relation**. Figure 1 (a) gives an example of such a hierarchy. By understanding, edges are always drawn from the bottom to the top. In the algorithms the successors of a vertex v (drawn above v on the figure) are denoted by $\Gamma^+(v)$. They are the vertices from which v inherits directly. The set of vertices from which v inherits is the set of successors of v in the transitive closure of H. It is denoted by $\Gamma^{+tc}(v)$.

An **h-module**[1] is a subset M of the vertex set X of the hierarchy such that H_M (the subgraph of H induced by M) is a hierarchy, and any directed path from M to $X \setminus M$ meets the greater element of the h-module (denoted by 1_M) and any directed path from $X \setminus M$ to M meets its least element (denoted by 0_M).

A **block** is a subset B of X such that H_B is a hierarchy and any directed path from B to $X \setminus B$ meets the greater or the least element of the block and any directed path from $X \setminus B$ to B meets its least or its greater element. The set $B \setminus \{0_B, 1_B\}$ is called the **inside** of B.

A **connected block** is a block whose inside induces a connected subgraph of H. It will be see later that the concept of connected block is of great importance in algorithmic point of view.

Each edge, the empty set and the whole hierarchy are the **trivial blocks** of the hierarchy. A hierarchy is said **prime** if all its blocks are trivial.

A directed path from an h-module M to $X \setminus M$ must leave M by the greater element while for a block B a directed path leaving B directly from the least element is also allowed. In a symmetric way, a directed path from $X \setminus M$ to M must come into M by the least element, while for a block a directed path entering directly to the greater element is also accepted. Obviously, an h-module is a block, so the concept of block extends strictly the concept of h-module. Figure 1 (b) gives an example of a hierarchy with an h-module and a block. And the set $\{c, d, e, f, i\}$ is a block of the hierarchy of Fig. 1 (a).

Considering the inheritance relation induced by a hierarchy H, all the vertices of an h-module M have the same neighbourhood to the outside of M according to the inheritance relations, while it is not the case for vertices of a block.

[1] this concept was originally called *module* but we do not use this word to avoid any ambiguity with the concept of module in the modular decomposition theory

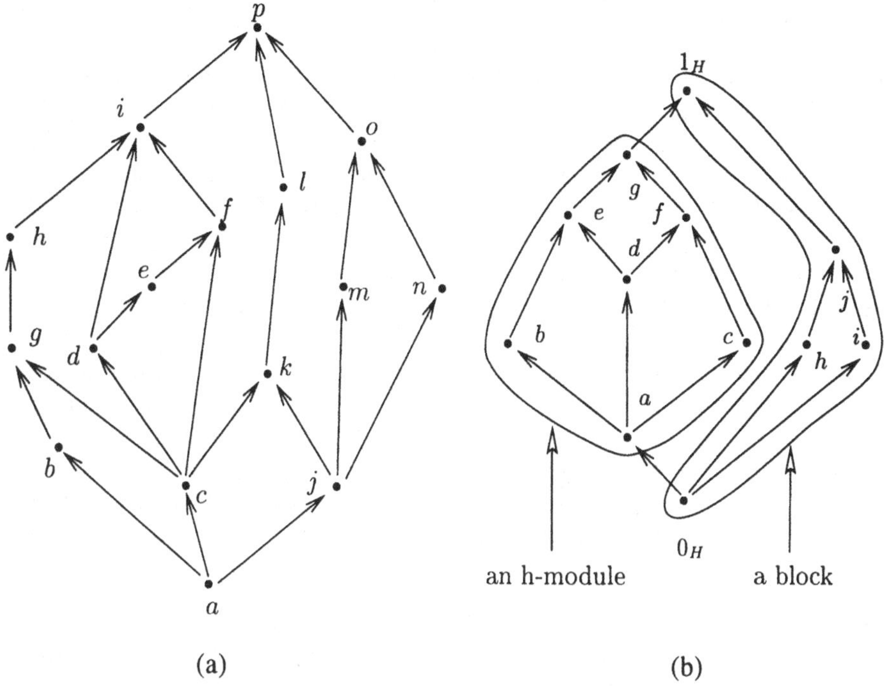

Fig. 1. Two examples of inheritance hierarchies

The vertices of the inside of the block B have the same neighbourhood to the top of the hierarchy than the greater element of the block, and the same neighbourhood to the bottom of the hierarchy than the least element. As it will be seen in the following section, h-module or block decompositions consist to partition a hierarchy in h-modules or in blocks. The quotient hierarchy describes a factorization of the information (w.r.t. the inheritance relation) contained in the hierarchy. For h-module decomposition, an h-module can be replaced by a single vertex representing it towards the rest of the hierarchy. For block decomposition, a block can be replaced by an edge (from its least to its greater element) labeled by the inside of the block. So no inheritance information is lost.

h-module decomposition: h-module decomposition has been studied by M. Huchard [10], M. Habib, M. Huchard and J. Spinrad [8]. There are two cases to consider to partition a hierarchy H: if H is not biconnected, it is partitioned according to its biconnected components which are h-modules. If the hierarchy is biconnected it is partitioned in proper inclusion maximal h-modules. The algorithm of Habib *et. al.* computes in linear time these h-modules called *principal h-modules*: (1) the vertex set V, (2) the inclusion maximal h-modules included in a biconnected h-module and (3) the biconnected components in a non biconnected h-module.

3 From Modular Decomposition to Block Decomposition

In this section we study the links between blocks and the well known *modules* of the *modular decomposition theory* (also called substitution decomposition). Let us start recalling some classical results about modular decomposition (see [12] for a general presentation of substitution decomposition and its applications).

A **module** M of a digraph $G = (X, E)$ is a subset of X such that for each $x, y \in M$, $z \in X \setminus M$, $xz \in E$ (*resp.* $zx \in E$) iff $yz \in E$ (*resp.* $zy \in E$). Each vertex of M has the same neighbourhood to the outside of M. The sets X, $v \in X$ and $\emptyset$ are the **trivial** modules of G. A graph is **prime** if it contains only trivial modules.

As every decomposition considered in this paper, modular decomposition is defined by a theorem (Theor. 1) describing how to partition a given digraph in modules according to its structural properties. In this theorem some unusual notations are used: G^a (the *asymmetric* part of G) is defined from G by $E(G^a) = \{xy \in E(G) \text{ such that } yx \notin E(G)\}$. G^{un}, the undirected graph associated to G, contains an edge xy iff xy or yx is an edge of the directed graph G. $\overline{G}$ is the complement of G.

Theorem 1 (modular decomposition of digraphs). *Let $G = (X, E)$ a digraph such that $|X| \geqslant 2$, then one and only one of the four following propositions is true:*

1. *G is not connected. It can be decomposed according to its connected components (parallel decomposition),*
2. *$\overline{G}$ is not connected. G can be decomposed according to the connected components of $\overline{G}$ (series decomposition),*
3. *G^a is connected, and $\overline{(G^a)^{un}}$ is non connected. G can be decomposed according to the connected components of $\overline{(G^a)^{un}}$ (order decomposition).*
4. *G is such that G, $\overline{G}$ and $\overline{(G^a)^{un}}$ are connected, then the partition P defined below defines the decomposition of G (prime decomposition):*
   There exists a subset Y of X such that $|Y| > 2$, and a uniquely defined partition of X, $P = \{p_1, \ldots p_k\}$ verifying:
   (a) *G_Y is a prime subgraph of G,*
   (b) *For each i, $0 < i \leqslant k$, p_i is a module of G,*
   (c) *For each i, $0 < i \leqslant k$, $|Y \cap p_i| = 1$.*

The vertex set partition defined by the preceding theorem allows to define a factorization of the adjacency information contained in a graph. By recursive application of the theorem a graph is decomposed completely. Graphs reduced to the vertices of the initial graph are the last level of this recursive process. This process is completely memorized by the **modular decomposition tree** of the graph. This tree is rooted by X and its leaves are the vertices of G. Any internal node is defined by a module of G and it is labeled *parallel*, *series*, *order* or *prime* according to the proposition of Theor. 1 which is applied. Its children are defined by the elements of its partition. Let us study the links between modules and blocks:

Proposition 2. *If B is a block of H, then $B \setminus \{0_B, 1_B\}$ is a module of H^{tc}.*

The converse of this proposition is not true, so it cannot be used in a useful way. The following proposition gives an equivalence between modules and blocks. To introduce it we must define some notations : let B be a subset of X, $E(B)$ is the set of edges of the subgraph of H induced by B. $L(H)$ is the line graph obtained from H. Its vertex set is the edge set of H and there is an an edge in $L(H)$ from xy to zt (where xy and zt are of course two edges of H) iff $y = z$.

Proposition 3. *B is a block of H if and only if $E(B)$ is a module of $L(H)^{tc}$.*

The four propositions of Theor. 1 can be considered with the digraph $L(H)^{tc}$. According to the structural properties of $L(H)^{tc}$, the theorem defines a partition of its vertex set. Using the preceding theorem, we can deduce a partition of the edge set of H in blocks according to structural properties of H. *parallel*, *order* and *prime* decompositions of $L(H)^{tc}$ can be expressed on H. However there is no *series* decomposition because $L(H)^{tc}$ is always connected. $L(H)^{tc}$ is not connected iff $H \setminus \{0_H, 1_H\}$ is not connected. The *parallel* decomposition induced on H is obtained partitioning it according to the connected components of $H \setminus \{0_H, 1_H\}$ adding them 0_H and 1_H as least and greater elements (so the hierarchy H is partitioned in hierarchies which are blocks of H). *order* decomposition is applied to $L(H)^{tc}$ iff H is not biconnected. This gives the *order* decomposition of H partitioning it in blocks according to its articulation points. If none of the two preceding cases applied, a *prime* decomposition of H can be deduced from *prime* decomposition of $L(H)^{tc}$. This decomposition corresponds to a partition of the edge set of H in proper blocks maximal for the inclusion order as defined in the following theorem:

Theorem 4 (block decomposition of inheritance hierarchies). *Let H be a hierarchy on vertex set X and edget set E such that $|X| \geq 3$, then one and only one of the following propositions is true:*

1. *$H \setminus \{0_H, 1_H\}$ is not connected and it can be decomposed in* parallel*,*
2. *H is not biconnected and it can be decomposed in* order*,*
3. *A prime decomposition can be applied to H: there exists a subset Y of X and a single partition of E $P = \{p_1, \ldots p_k\}$ $(k > 1)$ verifying:*
   (a) *For each i, $0 < i \leqslant k$, p_i is block of H maximal for the inclusion,*
   (b) *If for each i, $0 < i \leqslant k$, p_i is replaced by an edge from its lower to its greater element, the resulting hierarchy is prime with vertex set Y.*

As for modular decomposition, by recursive application of this partition theorem up to obtain hierarchies reduced to an edge, a **block decomposition tree** can be built. The whole hierarchy is associated to the root of the tree, while the edges are associated to the leaves. Any internal node is associated to a block labeled *parallel*, *order* or *prime* according to the proposition of Theor. 4 which can be applied to this block. The children of the nodes are associated to the elements of the partition (see Fig. 3 (a) for an example of such a block decomposition tree and an example of the different propositions of Theor. 4).

By equivalence with the modular decomposition stated by Theor. 3, Theor. 4 defines a *unique* decomposition tree up to isomorphism.

Let us remark that some blocks of H are not recognized by the process of decomposition. For instance the decomposition tree of a hierarchy defined by a simple n vertices directed path identifies only n blocks (the whole path and the $n-1$ edges) and the $O(n^2)$ proper sub-paths are not identified. The blocks corresponding to nodes of the decomposition tree are called **strong blocks**. The number of strong blocks is linear in the size of the hierarchy (the proof comes easily considering the number of nodes of the decomposition tree) while the total number of blocks can be exponential (for instance considering a hierarchy with two vertices 0 and 1, and m edges from 0 to 1, any part of the edge set is a block). However the non strong blocks can be obtained by union of some but not all of the children of a node of the decomposition tree. So the concept of strong blocks is more relevant than the general concept of block.

4 Computing Blocks Using Modular Decomposition

To deal with the algorithmic problem of computing blocks (or strong blocks) from a hierarchy a possible idea is to use Theor. 3. Such an algorithm can be separate in two parts : first, compute $L(H)^{tc}$, then compute modules of $L(H)^{tc}$.

On the one hand the best known algorithm to compute transitive closure has complexity $O(n^{2.376})$ [3] which represents a lower bound for our algorithm, on the other hand considering the hierarchy of Fig. 2 (a) with $|X| = n = (k+1)d + 2 = \theta(k.d)$ and $|U| = m = kd^2 + 2d = \theta(k.d^2)$ it can be shown that its line graph $L(H)$ has m vertices and $2d^2 + (k-1)d^3 = \theta(k.d^3)$ edges. The number of edges in $L(G)^{tc}$ is $\Omega(d^k)$. So, even if a linear time algorithm was used to compute modules of $L(G)^{tc}$ (for instance a generalization to digraphs of a linear modular decomposition algorithm [4, 11, 5]), the complexity of the whole process, expressed on the size of the initial hierarchy H, could not be good. So we have developed an algorithm computing blocks directly from H without considering $L(H)^{tc}$. The rest of this paper is devoted to its presentation.

5 Computing Blocks Using a Factorizing Linear Extension

As we have seen in the preceding section, the number of blocks of a hierarchy can be exponential, so the computation of all the blocks is not a good idea. Two more interesting problems are considered in this paper : the computation of connected blocks, and the computation of strong blocks (or the block decomposition tree).

The algorithms proposed here have two steps: first, to compute a linear extension of the hierarchy such that some blocks are factors of this extension, then to search for this extension to exhibit connected blocks or strong blocks (depending of the problem considered). Let us consider the first step of the algorithm.

In [1], we have studied the links between different decompositions and a new concept: a **factorizing permutation**. For each decomposition defined by a

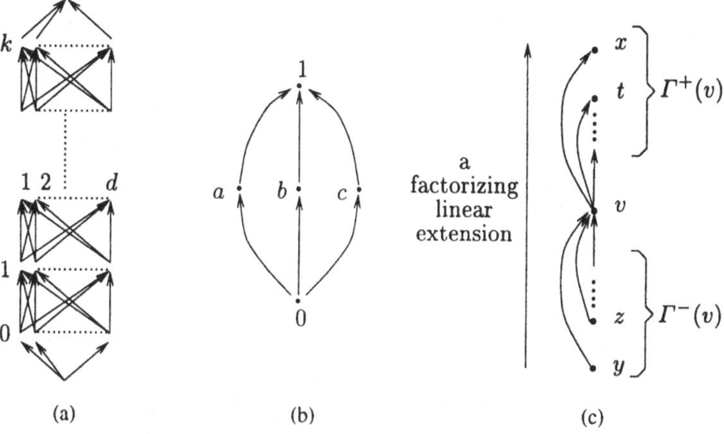

Fig. 2. (a)A hierarchy $H = (X, U)$ with $|X| = (k+1)d + 2$ and $|U| = kd^2 + 2d$. (b) There is no linear extension of this hierarchy such that all its strong blocks be factors of the extension. (c) Variables used by the `Connected-Blocks` algorithm.

partition theorem applied recursively, a factorizing permutation is a permutation such that all the set identified by the decomposition process (according to the decompositions, these sets are modules, h-modules, blocks ...) are factors of the permutation. In this context, a **permutation** is a total ordering (or an enumeration) of the vertex set of the graph. A **factor** of a permutation is a totally ordered set of successive elements of the permutation.

In the case of modular decomposition a factorizing permutation is a total ordering of the vertex set such that every strong module (the modules which correspond to nodes of the modular decomposition tree) is a factor of the permutation. It has been shown that for chordal graphs a factorizing permutation can be obtained from the graph using a modified version of the Lex-BFS algorithm (Lexicographic Breath First Search [13]) named cardinality-Lex-BFS. And a simple modular decomposition algorithm for chordal graph derives from this algorithmic result [9]. This result has been generalized by C. Capelle and M. Habib [2] proposing a linear algorithm computing the modular decomposition tree of any digraph from a factorizing permutation. This result shows the strong connections between decomposition tree and factorizing permutation.

If the digraphs to decompose are acyclic a factorizing permutation is generally defined as a linear extension. For h-module decomposition of hierarchies, R. Ducournau and M. Habib [7] have observed that any greedy deep first linear extension of a hierarchy is a factorizing permutation (the h-modules of the hierarchy are factors of such a linear extension). The linear h-module decomposition algorithm proposed by M. Habib, M. Huchard and J. Spinrad [8] is based on the search of such a linear extension.

In the case of the block decomposition, the problem is more complicated. Parallel blocks of a hierarchy share the same least and greater element so there is no linear extension such that all blocks (seen as a vertex set), or even strong blocks, are factors of the permutation. Moreover, considering the hierarchy of Fig. 2 (b), there no exists a linear extension such that for each bloc B, $B \setminus \{0_B, 1_B\}$ be a factor of a linear extension. Moreover, some algorithmic reasons (we want to compute connected blocks and strong blocks) make useful to consider the following definition of factorizing linear extensions for the block decomposition : a **factorizing linear extension** of a hierarchy H is a linear extension L such that for each connected block B of H, $B \setminus \{0_B, 1_B\}$ (the inside of B) is a factor of L.

The algorithm `Factorizing-Linear-Extension` (Algorithm 1) computes such a linear extension from the hierarchy.

The principle of the algorithm is the following: for each vertex v with more than one successor in H, its successors are linearly ordered ($lsucc(v)$ in the algorithm) in a way that if v is the least element of several connected blocks $\{B_1, \ldots, B_k\}$, the elements of $min(B_i \setminus \{v\})$ appears consecutively in the linear order. Then, a linear extension of H is computed, choosing the successors of v according to the linear order. The algorithm exploits the following property of the connected blocks $\{B_1, \ldots, B_k\}$ with the same least element v: for any two blocks B_i and B_j, for each $x, y \in min(B_i \setminus \{v\})$, $z \in min(B_j \setminus \{v\})$, the least common successor of x and y in the inheritance relation H^{tc}, is lower or equal to the least common successor of x and z. So the linear order on the successors in H of a vertex v is computed merging first the vertices whose the least common successor in H^{tc} is minimal.

This algorithm must consider the inheritance relation which is represented by H^{tc}. Its complexity in time and space is linear in the size of H^{tc}: $O(n + m^{tc})$ (with m^{tc} the size of the edge set of H^{tc}).

6 Connected Blocks Algorithm

The algorithm `Connected-Blocks` (Algorithm 2) computes the connected blocks of H from a factorizing linear extension with a complexity linear in the size of H and in the size of the set of connected blocks (there are at most $O(n^2)$ connected blocks). The principle of the algorithm is the following: a set F of subsets of X, candidate to be blocks, is maintained by the algorithm. They are created, modified, recognized as connected blocks, or suppressed from F. The algorithm consists of a search through the hierarchy according to a factorizing linear extension. The candidates in F have the following form: $(e, [i, j], s)$ where e and s are respectively the least and greater element of the candidate. The interval $[i, j]$ (an interval of L) is the inside of the candidate or it is equal to $[0, 0]$. A candidate to be block has a vertex set $[i, j] \cup \{e, s\}$ or $\{e, s\}$ if $i = j = 0$.

The algorithm is based on the following property of connected blocks. The vertex set is supposed to be renumbered according to a factorizing permutation. A subset B of X is a connected block if and only if either $B = \{1_B, 0_B\}$, either $|B| > 2$ and there exists i and j such that $B = [i, j] \cup \{1_B, 0_B\}$ with $G_{[i,j]}$ a connected graph and for each $x \in [i, j]$, $\Gamma^+(x) \subset [x, j] \cup \{1_B\}$ and $\Gamma^-(x) \subset [i, x] \cup \{0_B\}$.

Algorithm 1: Factorizing-Linear-Extension

```
Data   : a hierarchy H = (X, E) and H^tc its transitive closure
Result : L a block-factorizing linear extension of H.
begin
  Vertices are re-numbered according to a deep first linear extension of H
  For each vertex x, its successor in H^tc (Γ^{+tc}(x)) are sorted according to the
  minimal length path in H between x and its successors (in case of equality
  vertices are sorted according to their label)
  for v ∈ X such that |Γ^+(v)| > 1 do
    for i ⟵ 1 to |X| do
      D[i] ← ∅ { D[i] will contain the vertices of Γ^+(v) whose i is a
      descendant}
    i ← 0
    for x ∈ Γ^+(v) do
      y ← min(Γ^{+tc}(x)) { those whose distance to x is minimal first}
      D[y] ← D[y] ∪ {x}
      creation of the list number i containing the vertex x
      i ← i + 1
    nblists ← i
    while nblists > 1 do
      i ← min(j such that D[j] ≠ ∅)
      merge of all the lists containing at least an element of D[i]
      for x ∈ D[i] do
        j ← next(Γ^{+tc}(x))
        x is moved from D[i] to D[j]
    lsucc(v) ← the remaining list
  Computation of L, a greedy deep first linear extension of H such that the
  successors of a vertex x are chosen according to the total order lsucc(x)
end
```

More precisely, the algorithm works on a slightly different formulation of this characterization, but easier to algorithmically verify: a subset B of X ($|B| > 2$) is a connected block with least and greater element O_B and 1_B if and only if there exists i and j such that $B = [i,j] \cup \{1_B, 0_B\}$ with $G_{[i,j]}$ being a connected graph, and for each $v \in [i,j]$: $max(\Gamma^+(v)) \in [v,j] \cup \{1_B\}$ and $max(\Gamma^+(v) - max(\Gamma^+(v))) \leqslant j$, and $min(\Gamma^-(v)) \in [i,v] \cup \{0_B\}$ and $min(\Gamma^-(v) - min(\Gamma^-(v))) \geqslant i$.

In the algorithm, $max(\Gamma^+(v))$ is denoted by x, $max(\Gamma^+(v) - max(\Gamma^+(v)))$ is denoted by t, $min(\Gamma^-(v))$ is denoted by y, and $min(\Gamma^-(v) - min(\Gamma^-(v)))$ is denoted by z (see Fig. 2 (c)).

7 Block Decomposition Algorithm

The algorithm to compute block decomposition will not be presented in details. There is two steps in this algorithm: first, to compute the set of strong blocks of the hierarchy — those which correspond to the nodes of the decomposition tree, and then to compute the decomposition tree itself.

The algorithm to compute the strong blocks is a slightly modified version of the algorithm `Connected-Blocks`.

Some strong blocks are not connected, they can be easily defined. They are the (maximal sized) parallel union of connected blocks sharing the same least and greater element. They are associated to the parallel nodes of the decomposition tree. They can easily be computed from connected blocks.

Conversely, some connected blocks are not strong. The hierarchy induced by such a connected block B is a non biconnected hierarchy. Each biconnected component of H_B is a strong block. B is such that there exists a block B' with $|B \cap B'| = 1$ (the least element of B is the greater of B', or conversely). In other words a connected block is strong iff it is biconnected, or if it is a maximal "chain" of biconnected blocks. These two conditions are checked in an on-line way by the `Strong-Blocks` algorithm without recognizing all the connected blocks of H.

The decomposition tree corresponds to the inclusion order of strong blocks (seen as sets of arcs). To build the decomposition tree from the set of strong blocks, the following property is used: if two strong blocks B and B' are such that $B \subset B'$, then the inside of B is included in the inside of B'.

The inside of connected blocks are factors of a factorizing permutation, so if the vertex set of H is renumbered according to such a permutation the inside of connected blocks can be represented by intervals $[i,j]$ of $[1,n]$. Sorting them by inclusion can be done in linear time. The case of non connected strong blocks can be traited separately with no difficulty. This allows to obtain the decomposition tree of H. So, the block decomposition tree of H can be obtained from H using `Factorizing-Linear-Extension` and `Strong-Blocks` algorithms with an $O(n + m^{tc})$ complexity.

Figure 3 presents the block decomposition tree (a) and the h-module decomposition tree (b) of the hierarchy of Fig. 1 (b). It can be seen how the block

Algorithm 2: Connected-Blocks

```
Data   : a hierarchy H = (X, U) such that its vertex set is numbered according
         to a factorizing linear extension
Result : the set 𝓑 of connected blocks of H (including the edges of H).
𝓑 ← ∅, F ← ∅
for v ⟵ 2 to |X| do
    { initialization: }
    x, t, y and z are computed according to v (see Fig. 2 (c))
    for u ∈ Γ⁻(v) do
        { the edge uv is inserted in the list F of candidates: }
        add (u, [0, 0], v) to the head of F
    B = (e, [i, j], s) ← first element of F, or ∅ if F is empty
    while {v belongs to the inside of B or B is an edge: } v ∈ [i, j] or i = j = 0
    do
        if { B is an edge: } i = j = 0 then
            𝓑 ← 𝓑 ∪ {(e, [0, 0], v)}
            if Γ⁻(v) = {y} then
                { B grows "by the top": }
                i ← v, j ← max(t, v) and s ← x
            else
                (e, [i, j], s) is suppressed from F
        if (e, [i, j], s) has not been suppressed then
            if { B cannot become a block, i.e. v has a predecessor outside B: }
            (y < i and y ≠ e) or (z ≠ 0 and z < i) then
                (e, [i, j], s) is suppressed from F
            else
                if { B is a block: } j = v and Γ⁺(v) = {s} then
                    𝓑 ← 𝓑 ∪ {(e, [i, v], Γ⁺(v))}
                    if { 1_H has not be reach: } Γ⁺(x) ≠ ∅ then
                        { B grows "by the top": }
                        j ← x and s ← max(Γ⁺(x))
                    else (e, [i, j], s) is suppressed from F
                if (e, [i, j], s) has not been suppressed then
                    { B is updated using informations about the neighbourhood
                    of v: }
                    if t ≠ 0 then j ← max(j, t)
                    if x > j then
                        if x > s then  j ← max(j, s) and s ← x
                        else
                            if x ≠ s then j ← x
                    the new values of B = (e, [i, j], s) are updated in F
        B = (e, [i, j], s) ← next element of F or ∅ if the end of F is reach
```

decomposition allows a better factorization of the inheritance relation information than the h-module decomposition.

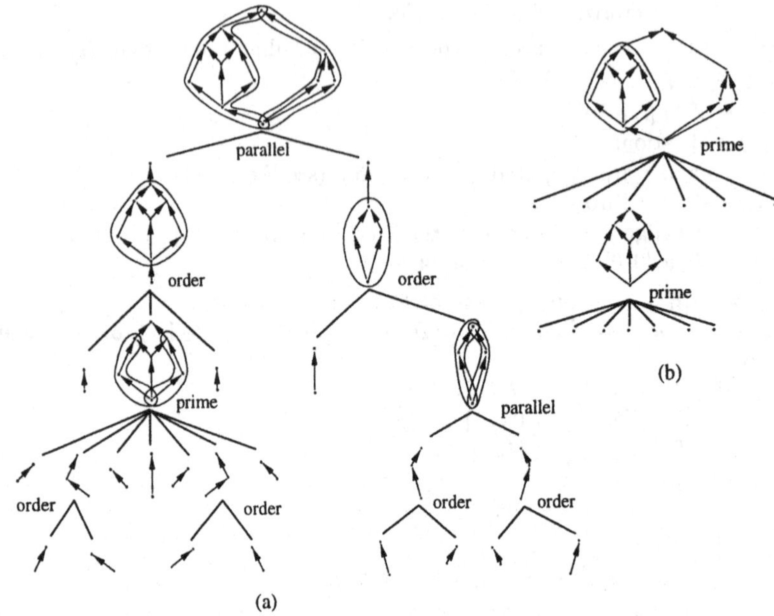

Fig. 3. (a) an example of block decomposition tree (the non trivial strong blocks are highlighted). (b) an example of h-module decomposition tree.

References

1. Christian Capelle. *Décompositions de Graphes et Permutations Factorisantes.* PhD thesis, Université Montpellier II, 161 rue Ada, 34392 Montpellier Cedex 5, France, January 1997.
2. Christian Capelle and Michel Habib. Graph decomposition and factorizing permutations. pages 132–143, Juin 1997. fith Israel Symposium on the Theory of Computing and Systems (ISTCS'97).
3. D. Coppersmith and S. Winograd. Matrix multiplication via arithmetic progressions. In *Proceedings of 19th Annual Symposium on the Theory of Computation*, pages 1–6, 1987.
4. A. Cournier and M. Habib. A new linear algorithm for modular decomposition. In S. Tison, editor, *Lectures notes in Computer Science, 787. Trees in Algebra and Programming-CAAP'94*, pages 68–84. Springer-Verlag, April 1994. 19th International Colloquium, Edinburgh, U.K., April 1994. Procedings.
5. Elias Dahlhaus, Jens Gustedt, and Ross M. McConnell. Efficient and practical modular decomposition. In *8th Annual ACM-SIAM Symposium On Discrete Algorithms (SODA)*, pages 26–35, January 1997.

6. Céline De Hadjetlache and Carine Escale. Dessin de hiérarchies d'héritage. Master's thesis, Université Montpellier II, June 1996. Mémoire de Stage de Recherche.
7. Roland Ducournau and Michel Habib. La multiplicité de l'héritage dans les langages à objects. *Technique et Science Informatique*, 8(1):41–62, 1989.
8. M. Habib, M. Huchard, and J. Spinrad. A linear algorithm to decompose inheritance graphs into modules. *Algorithmica*, (13):573–591, 1995.
9. Wen-Lian Hsu and Tze-Heng Ma. Substitution decomposition on chordal graphs and applications. In *Proceedings of the 2nd ACM-SIGSAM Internationnal Symposium on Symbolic and Algebraic Computation*, pages 52–60, 1991.
10. M. Huchard. *Sur quelques questions algorithmiques de l'héritage multiple.* PhD thesis, Université Montpellier II, 1992.
11. R. M. McConnell and J. Spinrad. Linear-time modular decomposition and efficient transitive orientation of undirected graphs, 1994. Proc. of the fifth Annual ACM-SIAM Symposium of Discrete Algorithms.
12. R. H. Möhring and F. J. Radermacher. Substitution decomposition for discrete structures and connections with combinatorial optimization. *Ann. Discrete math*, (19):257–356, 1984.
13. D.J. Rose, R.E. Tarjan, and G.S. Lueker. Algorithmic aspects of vertex elimination of graphs. *SIAM journal of computing*, 5:266–283, 1976.

Minimal Elimination Ordering Inside a Given Chordal Graph

Elias Dahlhaus

Dept. of Computer Science
University of Bonn
Bonn, Germany *
and
Department of Mathematics and Department of Computer Science,
University of Cologne,
Cologne, Germany **

Abstract. We consider the following problem, called *Relative Minimal Elimination Ordering*. Given a graph $G = (V, E)$ which is a subgraph of the chordal graph $G' = (V, E')$, compute an inclusion minimal chordal graph $G'' = (V, E'')$, such that $E \subseteq E'' \subseteq E'$. We show that this can be done in $O(nm)$ time. This extends the results of [2]. The algorithm is also simpler and is based only on well known results on chordal graphs.

1 Introduction

One of the major problems in computational linear algebra is that of sparse Gauss elimination. The problem is to find a pivoting, such that the number of zero entries of the original matrix that become non zero entries in the elimination process is minimized. In case of symmetric matrices, we would like to restrict pivoting along the diagonal. The problem translates to the following graph theory problem [12].

Minimum Elimination Ordering: For an ordering $<$ on the vertices, we consider the fill-in graph $G'_< = (V, E')$ of $G = (V, E)$. $G'_<$ contains first the edges in E and secondly two vertices x and y form an edge in $G'_<$ if they have a common smaller neighbor in $G'_<$. *The problem of Minimum Elimination ordering is, given a graph $G = (V, E)$, find an ordering $<$, such that $G'_<$ has a minimum number of fill-in edges.* Note that this problem is NP-complete [17].
For this reason, we relativize the problem.

Minimal Elimination Ordering: *Given a graph G, find an ordering $<$, such that the edge set of $G'_<$ is minimal with respect to inclusion.* This problem can be solved in $O(nm)$ time [13].

* e-mail: dahlhaus@cs.uni-bonn.de

** e-mail: dahlhaus@informatik.uni-koeln.de

In case that $G = G'_<$ ($<$ has no fill-in edges) $<$ is a perfect elimination ordering, and graphs having a perfect elimination ordering are exactly the *chordal graph*, i.e. graphs with the property that every cycle of length greater three has an edge that joins two non consecutive vertices of the cycle.

There are two practical polynomial time heuristics to get "good" elimination orderings, the minimum degree heuristics (see for example [10]) and nested dissection heuristics (see for example [1] or [10]).

We are interested in the problem to combine one of the heuristics as mentioned above with minimal fill in, i.e. we first apply one of the heuristics and afterwards we further thin out the resulting chordal graph $G'_<$, such that we get a minimal fill in ordering $<'$ with $G'_{<'} \subseteq G'_<$.

For this purpose, we consider the following problem.

Relative Minimal Elimination Ordering: Given a graph $G = (V, E)$ and an ordering $<$, find another ordering $<'$, such that with $G'_< = (V, E')$ and $G'_{<'} = (V, E'')$,
1. $E \subseteq E'' \subseteq E'$ and
2. $<'$ is a minimal elimination ordering.

Blair, Heggernes, and Telle [2] were the first dealing with this problem and the run time of their algorithm is $O(f(m+f))$, where m is the number of original edges and f is the number of fill-in edges, i.e. additional edges of $G_< \setminus G$.

Here we present an algorithm with a time bound of $O(nm)$. The algorithm is also simpler than the algorithm of [2]. The algorithm is based on the fact that chordal graphs, i.e. graphs with a perfect elimination ordering are exactly the intersection graphs of subtrees of a tree [3, 9]. First we can compute, given a graph G and an ordering $<$, such tree representation of $G'_<$ in $O(n+m)\alpha(n,m)$ time. Here α is the "inverse" Ackermann function [15]. Although the basic idea of the algorithm has been presented in [6] we will repeat it here. Only the complexity analysis has been improved.

In section 2, we introduce the notation and basic results that are necessary for the paper. Section 3 describes an efficient algorithm to compute a tree representation for the fill-in graph $G'_<$ (see also [6]. In section 4, we discuss the problem to compute a relative minimal elimination ordering.

2 Notation

A *graph* $G = (V, E)$ consists of a *vertex set* V and an *edge set* E. Multiple edges and loops are not allowed. The edge joining x and y is denoted by xy.

We say that x is a *neighbor* of y iff $xy \in E$. The set of neighbors of x is denoted by $N(x)$ and is called the *neighborhood*. The set of neighbors of x and x is denoted by $N[x]$ and is called the *closed neighborhood* of x.

Trees are always directed to the root. The notion of the *parent*, *child*, *ancestor*, and *descendent* are defined as usual.

A *subgraph* of (V, E) is a graph (V', E') such that $V' \subseteq V$, $E' \subseteq E$.

We denote by n the number of vertices and by m the number of edges of G.

A graph is called *chordal* iff each cycle of length greater than three has a chord, i.e. an edge that joins two nonconsecutive vertices of the cycle. Note that chordal graphs are exactly those graphs having a *perfect elimination ordering* $<$, i.e. for each vertex v the neighbors $w > v$ induce a complete subgraph, i.e. they are pairwise joined by an edge [8].

Moreover, chordal graphs $G = (V, E)$ are exactly the intersection graphs of subtrees of a tree [9, 3], i.e. there is a tree T and a collection of subtrees T_v, $v \in V$, such that $vw \in E$ if and only if T_v and T_w share a node. We call $(T, T_v)_{v \in V}$ also a *tree representation of* G. Whenever the trees T_v are represented by their leaves and their roots, we call $(T, T_v)_{v \in V}$ a *compact tree representation* of G. Note that it is sufficient to mention the leaves and the root of T_v, because with s and t in T_v, all vertices of T on the unique path from s to t in T are also in T_v.

Note that in any chordal graph, the number of maximal cliques is bounded by n and the number of pairs (x, c) such that x is in the clique c is bounded by m.

$(G, <) = (V, E, <)$ is called the *compact representation* of a chordal graph $G' = (V, E')$ if E' is the smallest edge set containing E, such that $<$ is a perfect elimination ordering.

The problems we discuss in this paper are the following.

Tree Representation [6]: Given a compact representation $(G, <)$ of the chordal graph G', find a (compact) tree representation for G'.

Relative Minimal Elimination Ordering: Given $(G, <) = (V, E, <)$ as the compact representation of $G' = (V, E')$, find an inclusion minimal chordal graph $G'' = (V, E'')$ with $E \subseteq E'' \subseteq E'$ together with a perfect elimination ordering $<''$.

3 Compactly Represented Chordal Graphs and Basic Algorithms

Theorem 1. *For any compactly represented chordal graph, a compact tree representation can be computed in $O((n+m)\alpha(n, m))$ time, where α is the inverse Ackermann function [15].*

Proof. Let $(G, <) = (V, E, <)$ be a compact representation of the chordal graph $G' = (V, E')$. We compute for each vertex v, the next greater neighbor $P(v)$ of v in G'. Note that the tree T_P with the parent function P together with the subtrees $T_v = \{y \leq v | yv \in E' \textit{ or } y = v\}$ is a tree representation of G' (see for example [16]). Note that all leaves of T_v are also neighbors of v in G. Therefore we can get a compact tree representation in linear time. We represent T_v by the neighbors of $v \leq v$ and erase those neighbors of v that are descendents of other neighbors of v and not identical to v.

Let $G_v := G[\{y | y \leq v\}]$ and $G'_v = G'[\{y | y \leq v\}]$. Note that G_v and G'_v have the same connected components. Moreover, if $v < w$ and if there is an edge $v'w$, such that v and v' belong to the same connected component of G_v, then

$vw \in E'$, i.e. is an edge of G'. Therefore $P(v)$ is the smallest $w > v$ that is a neighbor of the connected component of G_v v belongs to.

To get P in linear time, we assume that $V = \{v_1, \ldots, v_n\}$ and $v_i < v_j$, for $i < j$. For $i = 1, \ldots, n$, we maintain the set C_i of connected components of G_{v_i} and for each component $c \in C_i$, we maintain the maximum element $max(c)$.

For $i = 1, \ldots, n$, for each $v < v_i$ with $vv_i \in E$,

1. **let $c(v)$ be the component $c \in C_{i-1}$ with $v \in c$;**
2. **$P(max(c(v))) := v_i$;**
3. **add $c(v)$ to D_i if it is not in D_i;**
4. **$C_i := (C_{i-1} \setminus D_i) \cup \{\bigcup_{c \in D_i} c\}$.**

This procedure computes the parent $P(w)$, for each vertex w, correctly and can be implemented in $O((n+m)\alpha(n,m))$ time, using union-find [15]. □

Corollary 2. *The size of a maximum clique of a compactly represented chordal graph can be determined in $O((n+m)\alpha(n,m))$ time. A minimum coloring of a compactly represented chordal graph can be found in the same time bound.*

Theorem 3. *Given a compact representation $(G, <)$ of a chordal graph G', then the edge set of G' and a (non compact) tree representation of G' can be computed in $O(n^2)$ time.*

Sketch of Proof: In principle, we proceed as in the parallel algorithm in [11]. Note that $P(v)$ is the smallest greater neighbor of v in G'. To get the edges of G', we start with the edges of G and the edges $vP(v)$, and for each edge xy in G' with $x < y$, we add the edge $P(x)y$ if $P(x) \neq y$. Since $\alpha(n, n^2)$ is bounded by a constant, the time is bounded by $O(n^2)$.

To get the full tree representation, let $T_v := \{x < v | xv \in E \,or\, x = v\}$ (compare also [16]).

This completes the proof. □

Remark. The procedure to find a tree representation, a minimum coloring and the size of a maximum clique in a compactly represented chordal graph can be parallelized. To find a the tree representation of a compactly represented chordal graph, we weight each edge uv with the number of the greater vertex with respect to the given perfect elimination ordering and apply the single linkage method [4]. To parallelize the computation of a maximum clique and a minimum coloring, we use the fact that we have a compact tree representation, i.e. we know only the leaves and the root of the subtrees T_v. The rest can be done as in [5]

4 Relatively Minimal Elimination Orderings

4.1 Basic Strategy

Let $(G,<) = (V,E,<)$ be a compact representation of the chordal graph $G' = (V,E')$ and $(T,T_v)_{v\in V}$ a tree representation of G'. We first compute a tree representation $(T_1,T_v^1)_{v\in V}$ of a chordal graph $G_1 = (V,E_1)$ with

1. $E \subset E_1 \subset E'$ and
2. all edges uv of E_1 such that the corresponding subtrees T_u and T_v have more than one node of T_1 in common and therefore an edge in common are edges in any minimal chordal extension E'' of E with $E'' \subseteq E_1$. This is the case if the following condition is satisfied.
   Let p_f be the set of vertices v, such that T_v passes the edge f of the tree T_1 of the tree representation of G_1. Then, for each edge f of T_1, there are two connected components C_1 and C_2 of $G[V\setminus p_f]$, such that all vertices in p_f are in the neighborhood of C_1 and of C_2. Moreover, the nodes of T appearing in $\{T_v|v \in C_1\}$ and the nodes appearing in $\{T_v|v \in C_2\}$ are separated by f. The graph G_1 together with its tree representation $(T_1,T_v)_{v\in V}$ with this property is also called *quasi minimal.* $(T_1,T_v)_{v\in V}$ is also called a *quasi minimal* tree representation.

Theorem 4. *Suppose $(T_1,T_v)_{v\in V}$ is a quasi minimal tree representation of $G_1 = (V,E_1)$. Then all edges uv, such that T_u and T_v share an edge of T_1, appear in each $G'_{<'} = (V,E'')$, such that $<'$ is a minimal elimination ordering and $E'' \subseteq E_1$.*

Proof. Suppose T_u and T_v share an edge f of T_1. Let C_1 and C_2 be the connected components of $G[V\setminus p_f]$ associated with f. Since the subtrees $T_x, x \in C_1$ and $T_y, y \in C_2$ are separated by the edge f of T_1, there is no edge $xy \in E_1$ and therefore no edge $xy \in E''$, such that $x \in C_1$ and $y \in C_2$. Consider any path p_1 from u to v with inner vertices in C_1 and any path p_2 from v to u with inner vertices in C_2 in the original graph G. The concatenation of p_1 and p_2 forms a cycle in $G'_{<'}$ of length ≥ 4. Assume there is no edge $uv \in E''$. Then consider any chordless path p'_1 and p'_2 in $G'_{<'}$ from u to v and v to u respectively, such that their vertices are in p_1 and p_2 respectively. Then the concatenation of p'_1 and p'_2 forms a cycle in $G'_{<'}$ of length at least four. Therefore in $G'_{<'}$, it must contain a chord in E''. Since p'_1 and p'_2 are chordless, one incident vertex must be in p'_1 and therefore in C_1, and the other incident vertex must be in p'_2 and therefore in C_2. This is a contradiction to the fact that there is no edge $xy \in E''$ with $x \in C_1$ and $y \in C_2$. □

The step to produce a quasi minimal tree representation is also called the *tree splitting procedure.*

Knowing the tree representation of G_1, we apply a variation of the extended lexical breadth-first search procedure of [13] to eliminate edges that appear in only one maximal clique.

4.2 Tree Splitting

Let $(T_0, T_v^0)_{v\in V}$ be the initial tree representation, i.e. a tree representation of G'. For each t, let G_t be $G[\{v|root(T_v^0)$ is a descendent of $t\}]$.

Remark. For each node t of T_0, G_t is connected and the set of vertices v with T_v passing the edge $tP(t)$ joining t with its parent is the set of neighbors of G_t that are not in G_t.

This follows immediately from the construction of T_0 by the parent function P as mentioned above.

We compute a sequence $(T_i, T_v^i)_{v\in V}$ of tree representations, such that $(T_{i+1}, T_v^{i+1})_{v\in V}$ represents a chordal subgraph G^{i+1} of the chordal graph G^i represented by $(T_i, T_v^i)_{v\in V}$. The final tree representation $(T_k, T_v^k)_{v\in V}$ represents a chordal graph with the properties of the chordal graph G_1 as mentioned above.

Let $e_1, \ldots, e_k$ be an enumeration of the edges of T_0, such that if e_i is an ancestor edge of e_j then $i < j$. We call such an enumeration a *top down enumeration.* Let $e_i = s_i t_i$ where t_i is the parent of s_i. During the algorithm, for each edge $f = st$, let $c_{(s,t)}$ be a connected subset of G, such that all T_u with $u \in c_{(s,t)}$ appear on the s-side of st in T and all vertices w, such that T_w pass st, are in the neighborhood of $c_{(s,t)}$. Note that an edge satisfies the condition of quasi minimality if $c_{(s,t)}$ and $c_{(t,s)}$ are defined. Initially, let $c_{(s_i,t_i)}$ be the set of vertices u such that T_u appears only at the s_i-side of e_i. By construction of $(T_0, T_v^0)_{v\in V}$, all these sets are connected in G.

Algorithmically we proceed as follows.

For $i = 1, \ldots, k$,
compute T_i **from** T_{i-1}, **i.e.**

1. **compute the set** C_i **of connected components of**

   $$G[\{v|T_v \textbf{ appears only on the } t_i\textbf{-side of } T_0\}];$$

2. **for each** $c \in C_i$, **mark** c **as** *good* **if there is a** $v \in c$, **such that** $t_i \in T_v^{i-1}$;
3. **for each good connected component** $c \in C_i$, **create a tree node** t_c **and a tree edge** $s_i t_c$;
   $c_{(s_i,t_c)} := c_{(s_i,t_i)}$; $c_{(t_c,s_i)} := c$;
4. **construct** T_i **from** T_{i-1} **as follows: for each edge** $t_i u$ **of** T_{i-1}, **let** d_u **be the component** $c \in C_i$ **that contains** $c_{(u,t_i)}$;
   if d_u **is a good component then**
   begin replace $t_i u$ **by** $t_{d_u} u$;
   $c_{(t_{d_u},u)} := c_{(t_i,u)}$ **if defined;** $c_{(u,t_{d_u})} := c_{(u,t_i)}$; **if** $t_i u$ **was an** e_j, $j > i$ **then** e_j **is updated by** $t_{d_u} u$, **i.e.** $s_j := u$ **and** $t_j := t_{d_u}$;
   end
   else
   begin

replace $t_i u$ by $s_i u$; $c_{(s_i,u)} := c_{(t_i,u)}$ if defined; $c_{(u,s_i)} := c_{(u,t_i)}$; if $ut_i = e_j$, for some $j > i$, then e_j is updated to us_i ($s_j = u$; $t_j = s_i$);
end;
erase t_i;

5. **(updating T_v) for v with $t_i \in T_v^{i-1}$, construct $T_v = T_v^i$ from $T_v = T_v^{i-1}$ as follows: for any good component $c \in C_i$, add t_c to T_v if and only if $v \in c$ or v is a neighbor of some vertex in c;**

To prove the correctness, we have to show that the tree representation $(T_k, T_v^k)_{v \in V}$ is quasi minimal and that the edge set E^k of G^k contains E and is contained in E^1.

We say that an edge f of T_j *arises from* e_i if either

1. $j = i$ and f is an edge $s_i t_c$, for some good component c of C_i or
2. $j > i$ and (f is also an edge of T_{j-1} and arises from e_i or there is an edge f' of T_{j-1} that arises from e_i and is replaced by f in T_j).

Note that in each T_i, every edge is either some e_j, $j > i$ or arises from some e_j, $j \leq i$. To show that $(T_k, T_v^k)_{v \in V}$ is quasi minimal, we show that for each $j \leq i$ and each edge $f = st$ that arises from j, $c_{(s,t)}$ and $c_{(t,s)}$ are both defined, each $c_{(s,t)}$ defines, for each i, an in G connected subset of V that is adjacent to all vertices of $p_f = \{v | T_v^i \text{ passes } f\}$.

By induction on i, we show

Lemma 5. *For each i:*

1. *In $(T_i, T_v^i)_{v \in V}$, for all edges $f = st$ arising from som e_j, $j \leq i$, $c_{(s,t)}$ and $c_{(t,s)}$ are defined.*
2. *For all edges ut of T_i with $t = s_j$, $j \leq i$ or $t = t_c$, $c \in C_j$, $j \leq i$, $c_{(u,t)}$ is defined.*
3. *For each edge ut_{i+1} of T_i, $c_{(u,t_{i+1})}$ is defined.*
4. *If $c_{(s,t)}$ is defined in $(T_i, T_v^i)_{v \in V}$ then $c_{(s,t)}$ is an in G connected subset of V and*
5. *for all T_v^i passing st, vw is an edge in G, for some $w \in c_{(s,t)}$*
6. *T_v^i is a tree, i.e. defines a connected subset of T_i.*
7. *For $j > i$, if us_j is an edge of T_i and $u \neq t_j$ then $c_{(u,s_j)}$ is defined.*

Proof. We simultaneously prove all the statements by induction.

For $i = 0$, statements 1 and 2 are trivially true, because e_j, $j \leq i$ do not exist. Statements 4, 6, and 7 are true, by construction of $(T_0, T_v^0)_{v \in V}$. Note that t_1 is the root of T^0, and therefore also statement 3 is true, for $i = 0$. , since statement 7 is true.

To show the inductive step, observe that whenever $ut = ut_i$ is replaced by ut', $c_{(u,t)}$ is always defined, since statement 3 is true for $i - 1$, $c_{(u,t')} = c_{(u,t)}$, and $c_{(t,u)} = c_{(t',u)}$. Moreover, observe that a new $c_{(t,u)}$ is created if and only if $u = s_i$

and $t = t_c$, for some good component c of C_i. $c_{(t_c,s_i)} = c$ is an in G connected subset of V and for all T_v^i passing t_cs_i, v is adjacent to some vertex in c in G. Therefore statement 1 and statement 4 are true, for all i.

Statement 2 is true for $t = s_j$, $j < i$ and $t = t_c$ with $c \in C_j$, $j < i$, because it is true in $(T_{i-1}, T_v^{i-1})_{v \in V}$ and t does not get new incident edges in T_i. If $t = s_i$ then all edges incident with s_i in T_i are either edges incident with s_i in T_{i-1} or edges $us_i = t_cs_i$ or replaced edges us_i (i.e. ut_i was an edge in T_{i-1}). In either cases $c_{(u,s_i)}$ is defined and therefore statement 2 is true for s_i. Suppose now that $t = t_c$ and c is a good component of C_i. Incident edges in T_i are s_it_c and edges ut_c with ut_i in T_{i-1}. Statement 3 follows for $u = s_i$ from statement 1. For the remaining u, statement 2 follows from the observation that whenever ut_i is replaced by some ut' then $c_{(u,t')}$ remains defined.

Statement 7 is always preserved, because any s_j, $j > i$ is not a t_l, $l \leq i$, because $e_1, \ldots, e_k$ is a top down enumeration of the edges of T_0, and therefore s_j is not of the form t_c and not an s_l, $l \leq j$ and therefore up to replacements, s_j has the same incident edges in T_i as in T_{i-1}.

To show statement 3, observe that t_{i+1} is an s_j, $j \leq i$ or a t_c, $c \in C_j$, $j \leq i$, because $e_1, \ldots, e_k$ is a top down enumeration of the edges of T_0. Therefore statement 3 follows from statement 2.

Next observe that, when we create $(T_i, T_v^i)_{v \in V}$ and replace any edge ut_i by ut then either $t = t_c$, for some good component c or $t = s_i$. In the first case, exactly for those T_v^{i-1} passing ut_i, T_v^i contains u and t_c. In the second case, no good component contains $c_{(u,t)}$ and therefore no vertex of $c_{(u,t)}$ is adjacent to some vertex in a good component and therefore for no T_v^{i-1} containing t_i but not s_i, v is adjacent to some vertex in $c_{(u,t)}$. Since for all $T^{i-1}v$ passing ut_i, v is adjacent to some vertex of $c_{(u,t_i)}$, all these T_v^{i-1} pass s_it_i. In either cases T_v^i passes ut if and only if T^{i-1} passes ut_i. Therefore statement 5 is preserved by edge replacements. Moreover observe that if T_v^i passes s_it_c then T_v^{i-1} passes s_it_i and therefore in the neighborhood of $c_{(s_i,tc)} = c_{(s_i,t_i)}$ and statement 5 is preserved in any way.

It remains to show statement 6. Replacing ut_i by us_i takes place only in the case that the subtrees passing ut_i form a subset of the subtrees passing s_it_i, and all subtrees T_v^{i-1} remain subtrees. Note that all subtrees $T^{i-1}v$ containing t_i but not s_i are in some good component c. Since when ut_i is replaced by ut_c the set $c_{(u,t_i)}$ is defined, for all T_v^{i-1} passing ut_i, $v \in c$ or T_v passes s_it_i. The only isolated vertex of T_v^{i-1} that might arise from such a replacement is t_i. But t_i will be deleted, and T_v^i is a tree again. □

It remains to show

Lemma 6. *Let E^i be the set of edges of G^i, i.e. $vw \in E^i$ if and only if T_v^i and T_w^i share a node of T. Then*

1. *$E^{i+1} \subseteq E^i$, for $i = 0, \ldots, k-1$ and*
2. *$E \subseteq E_i$, for $i = 0, \ldots, k$.*

Proof. Note that t_{i+1} is the only node that is in T_i, but not in T_{i+1} and the nodes t_c arising from t_{i+1} are those that appear in T_{i+1} but not in T_i. The first statement follows immediately, because in case that $vw \in E^{i+1}$ then either T_v and T_w share in T_{i+1} a node that is also in T_i or they share a node t_c and therefore the node t_i of T_i.

The second statement can be proved by induction on i. For $i = 0$, the statement is true, by construction. Now suppose $vw \in E$ and T_v and T_w share a node in T_i. If they share a node $\neq t_{i+1}$ then also in T_{i+1}, T_v and T_w share a node. If T_v and T_w share only t_{i+1} then at least one of T_v and T_w does not contain s_{i+1}. If they both do not contain s_{i+1} then v and w are in the same good component c and therefore T_v and T_v share t_c in T_{i+1}. If, for example T_v contains s_{i+1} and T_w does not contain s_{i+1} then w belongs to a good component of C_{i+1} and v is adjacent to some vertex (this is w) of the good component c, w belongs to. Therefore also in this case, T_v and T_w share t_c in T_{i+1}. □

The complexity of this algorithm can be checked as follows. We show that the algorithm works, for each i, in $O(n + m)$ time and therefore the overall time bound is $O(nm)$.

The set C_i can be computed in $O(n+m)$ time, because connected components can be computed in the same time bound.

The good components can be computed in $O(n)$ time. We have a list L_i of those vertices v, such that $t_i \in T_v^{i-1}$. For all these vertices v, we mark the $c \in C_i$ it belongs to as good if v belongs to such a c.

The creation of t_c, for each c, can be done in $O(n)$ time.

The connected component $c \in C_i$ that contains $c_{(u,t_i)}$ can be computed, for all u in $O(n)$ time by picking a vertex $x \in c_{(u,t_i)}$ and determining the $c \in C_i$ x belongs to. The edge replacements can be done in the same time bound.

The update procedure for the T_v's can be done in $O(n + m)$ time. First one has to compute in $O(n)$ time the set of all $Ti - 1_v$ passing s_it_i, by initially labelling all vertices v with 0, then labelling all vertices v with $t_i \in T_v^{i-1}$ by 1 and then labelling all 1-labelled vertices v with $s_i \in T_v^{i-1}$ with 2. If $T_v = T_v^{i-1}$ passes s_it_i and $vw \in E$ then one has to check whether w is in a good component (in one step), and if it is in a good component c then one has to add c to T_v. If T_v does not contain s_i but contains t_i (i.e. is 1-labelled) then one has to determine the good component c its belongs to and to add t_c to T_v.

4.3 The Improved RTL-Algorithm

It remains to eliminate superfluous edges that appear in only one maximal clique, i.e. edges uv, such that T_u and T_v share only a node, but not an edge of $T = T_k$. Here we apply a variation of the algorithm of Rose, Tarjan, and Lueker [13], also called the RTL-algorithm.

The RTL-algorithm works as follows.

Initialize: We start with one list $L_1 := V$;

For $i = n, \ldots, 1$**:** 1. Select a vertex v_i from the nonempty list L_j of the largest index and remove v_i from L_j;

2. for each j and each $y \in L_j$, let v_iy be an edge in E' iff $v_iy \in E$ or y and v_i are neighbors of a connected component C of $G[\bigcup_{\mu<j} L_\mu]$;

3. split each L_j into a list of smaller index containing the non neighbors of v_i with respect to E' and a list of larger index containing the neighbors of v_i with respect to E'; renumber the new lists L_i.

Note that in the last section, we have computed a tree representation, such that all vw such that T_v and T_v have at least two nodes in common then they appear in any chordal extension G'' of G that is a subgraph of G_1.

We select a root r of the tree T representing G_1. Let $t_1, \ldots, t_k$ be an enumeration of the nodes of T, such that if t_j is the parent of t_i then $i < j$. Such an ordering is called a *bottom up ordering.* Such an ordering can be computed in linear time, for example by postorder.

Let L_i be the list of vertices with $root(T_v) = t_i$. The only difference between the original RTL-algorithm is that we do not start with one list $L_1 = V$, but with the lists $L_i = \{v | root(T_v) = t_i\}$. This produces a minimal elimination ordering. To verify this, one shows that the algorithm does the same as if we would apply the original RTL-algorithm to each graph $G_t = (V_t, E_t)$ where V_t consists of those v with $t \in T_v$ and $vw \in E_t$ if v and w are in V_t and $vw \in E$ or $T_v \cap T_w$ contains at least two nodes of T, i.e. there is an edge of T incident with t that is passed by T_v and T_w.

Lemma 7. *Suppose v is numbered, i.e. v becomes v_i in the improved RTL-algorithm, $w \in L_j$ is not yet numbered, and $v, w \in V_t$. Then vw becomes an edge in E' (i.e. v and w are adjacent in G or are both adjacent to a common connected component of $G[\bigcup_{j'<j} L_{j'}]$) if and only if vw is an edge in E_t or v and w are adjacent to a common connected component of $G_t[\bigcup_{j'<j} L_{j'}]$.*

Proof. Since w is not numbered, $root(T_v)$ is an ancestor of $root(T_w)$ (this includes also equality). If $t \neq root(T_w)$ then $vw \in E_t$, because T_v and T_w share t and $root(T_w)$. Therefore T_v and T_w pass the edge $t\, parent(t)$ of T, and since $(T, T_v)_{v \in V}$ is a quasi minimal tree representation, v and w are adjacent to a connected subset of vertices u, such that all root of T_u are descendents of t, and therefore all these u are in $L_{j'}$, $j' < j$. therefore vw becomes an edge in E'.

Now assume that $t = root(T_w)$. First suppose there is a path p from v to w in G, such that all inner vertices u are in $L_{j'}$, $j' < j$. Note that if $u_1u_2 \in E$ then T_{u_1} and T_{u_2} share a node of T. Therefore the roots $root(T_u)$ of all inner vertices u of p are descendents of t (equality is possible. let p' be a subpath of p, such that, for the end vertices v', w', $root(T_{v'}) = root(T_{w'}) = t$, and for the inner vertices u, $root(T_u)$ is a proper ancestor of t. Then there is a child t' of t, such that for all these u, T_u is an ancestor of t' (equality is included). Therefore $T_{v'}$ and $T_{w'}$ share the nodes t and t' and therefore $v'w' \in E_t$. Replacing all these subpaths p' by edges in E_t, we get a path q from v to w in E_t with all inner vertices in $L_{j'}$, $j' < j$.

No we assume there is a path q from v to w in E_t, such that all inner vertices u are in $L_{j'}$, $j' < j$. note that all these vertices u are in V_t, and for all these u, $root(T_u) = t$ (not a proper ancestor of t). Suppose v' and w' are consecutive vertices of q. Since $v'w' \in E_t$ either $v'w' \in E$ or there is another node t' that is contained in $T_{v'}$ and $T_{w'}$. t' must be a descendent of t and can be chosen as a child of t. Since $(T, T_v)_{v \in V}$ is quasi minimal, there is a connected subset of u with $root(T_u)$ descendent of t' that is adjacent to v' and w'. Therefore there is a path from v' to w' in G, say p' with inner vertices in $L_{j'}$, $j' < j$. Concatenating all these paths p', we get a path from v to w in G with inner vertices in $L_{j'}$, $j' < j$. □

The complexity of the original RTL-algorithm and the improved RTL-algorithm are the same. Therefore we get the following final result.

Theorem 8. *Relative Minimal Elimination Ordering can be solved in $O(nm)$ time.*

5 Conclusions

We developed a sequential algorithm to compute a minimal elimination ordering, such that the fill-in graph is inside a given greater chordal graph. The time bound is $O(nm)$. A better time bound is not to expect, because the minimal elimination ordering problem without the restriction of a larger chordal graph has a time bound of $O(nm)$. Using union find as in finding compact tree representations, the tree splitting procedure might be speeded up a little bit. This is more a practical aspect. One does not get a lower time bound in the order. Another aspect that might be discussed is the parallelization. The components of the tree split procedure are $O(n)$ computations of connected components and reorganization of the tree. First can be parallelized very easily [14]. The parallelization of the second component of the tree split procedure might be a topic for a masters or honors thesis. The improved RTL-algorithm might be replaced by a variation of the algorithm of [7].

References

1. A. Agrawal, P. Klein, R. Ravi, Cutting Down on Fill-in Using Nested Dissection, in *Sparse Matrix Computations: Graph Theory Issues and Algorithms*, A. George, J. Gilbert, J.W.-H. Liu ed., IMA Volumes in Mathematics and its Applications, Vol. 56, Springer Verlag, 1993, pp. 31-55.
2. J. Blair, P. Heggernes, J.A. Telle, *Making an Arbitrary Filled Graph Minimal by Removing Fill Edges*, Algorithm Theory-SWAT96, R. Karlsson, A. Lingas ed., LLNCS 1097, pp. 173-184.
3. P. Bunemann, *A Characterization of Rigid Circuit Graphs*, Discrete Mathematics 9 (1974), pp. 205-212.
4. E. Dahlhaus, *Fast parallel algorithm for the single link heuristics of hierarchical clustering*, Proceedings of the fourth IEEE Symposium on Parallel and Distributed Processing (1992), pp. 184-186.
5. E. Dahlhaus, *Efficient Parallel Algorithms on Chordal Graphs with a Sparse Tree Representation*, Proceedings of the 27-th Annual Hawaii International Conference on System Sciences, Vol. II (1994), pp. 150-158.

6. Elias Dahlhaus, *Sequential and Parallel Algorithms on Compactly Represented Chordal and Strongly Chordal Graphs*, STACS 97, R. Reischuk, M. Morvan ed., LLNCS 1200 (1997), pp. 487-498.
7. Elias Dahlhaus, Marek Karpinski, *An Efficient Parallel Algorithm for the Minimal Elimination Ordering (MEO) of an Arbitrary Graph*, Theoretical Computer Science 134 (1994), pp. 493-528.
8. M. Farber, *Characterizations of Strongly Chordal Graphs*, Discrete Mathematics 43 (1983), pp. 173-189.
9. F. Gavril, *The Intersection Graphs of Subtrees in Trees Are Exactly the Chordal Graphs*, Journal of Combinatorial Theory Series B, vol. 16(1974), pp. 47-56.
10. A. George, J.W.-H. Liu, *Computer Solution of Large Sparse Positive Definite Systems*, Prentice Hall Inc., Englewood Cliffs, NJ, 1981.
11. J. Gilbert, H. Hafsteinsson, *Parallel Solution of Sparse Linear Systems*, SWAT 88 (1988), LNCS 318, pp. 145-153.
12. D. Rose, *Triangulated Graphs and the Elimination Process*, Journal of Mathematical Analysis and Applications 32 (1970), pp. 597-609.
13. D. Rose, R. Tarjan, G. Lueker, *Algorithmic Aspects on Vertex Elimination on Graphs*, SIAM Journal on Computing 5 (1976), pp. 266-283.
14. Y. Shiloach, U. Vishkin, *An O(log n) Parallel Connectivity Algorithm*, Journal of Algorithms 3 (1982), pp. 57-67.
15. R. Tarjan, *Efficiency of a Good but not Linear Set Union Algorithm*, Journal of the ACM 22 (1975), pp. 215-225.
16. R. Tarjan, M. Yannakakis, *Simple Linear Time Algorithms to Test Chordality of Graphs, Test Acyclicity of Hypergraphs, and Selectively Reduce Acyclic Hypergraphs*, SIAM Journal on Computing 13 (1984), pp. 566-579. Addendum: SIAM Journal on Computing 14 (1985), pp. 254-255.
17. M. Yannakakis, Computing the Minimum Fill-in is NP-complete, *SIAM Journal on Algebraic and Discrete Methods* 2 (1981), pp. 77-79.

On-Line Algorithms for Networks of Temporal Constraints*

Fabrizio d'Amore and Fabio Iacobini

Università di Roma "La Sapienza", Dipartimento di Informatica e Sistemistica, Via Salaria 113, I-00198 Roma, Italy

Abstract. We consider a semi-dynamic setting for the Temporal Constraint Satisfaction Problem, where we are requested to maintain the path-consistency of a network under a sequence of insertions of new (further) constraints between pairs of variables. We show how to maintain path-consistent a network in the defined setting in $O(nR^3)$ amortized time on a sequence of $\Theta(n^2)$ insertions, where n is the number of vertices of the network and R is its range, defined as the maximum size of the minimum interval containing all the intervals of a single constraint. Furthermore we extend our algorithms to deal with more general temporal networks where variables can be points and/or intervals and constraints can be also defined on pairs of variables of different kind. For such cases our algorithms maintain their performance. Finally we adapt our algorithms for maintaining also the arc-consistency of such general networks, which is a particular kind of path-consistency limited to paths of length 1. The property is maintained in $O(R)$ amortized time for $\Theta(n^2)$ insertions. In case of constraints consisting of simple intervals the algorithm also gives a solution to the satisfaction problem.

1 The problem

Given a totally ordered universe U, a *network of temporal constraints* [6] can be defined as an edge-labeled graph $(G, \mathcal{L})$, where $G = (V, E)$ is a graph and $\mathcal{L}$ is a function associating edges with sets of (closed) intervals of U. We assume an (arbitrary) order among vertices and we rename them according to this order so that $V = \{v_1, \ldots, v_n\}$. Each vertex is a *variable* varying in U and each edge e is a *constraint* on the difference between the incident variables, namely, if $\mathcal{L}(e) = \{I_1, \ldots, I_k\}$, I_j being an interval of U and $e = (v_i, v_j)$, then such difference must be in $\bigcup_{I \in \mathcal{L}(e)} I$. For clarifying the order of the variables in the difference we refer to the difference $v_{\max\{i,j\}} - v_{\min\{i,j\}}$; we do not loose generality because a constraint $u - v \in [a, b]$ is equivalent to $v - u \in [-b, -a]$. A typical choice for the universe U is the set $\mathbb{Z}$ of signed integers.

Given a network of temporal constraints $((V, E), \mathcal{L})$ the *Temporal Constraint Satisfaction Problem* (TCSP) is the problem of finding an n-ple $(z_1, \ldots, z_n) \in U^n$

* Work supported by the EU ESPRIT LTR Project "ALCOM-IT" under contract n. 20244 and by the Italian MURST National Project "Efficienza di Algoritmi e Progetto di Strutture Informative".

which satisfies the constraints, where z_i is the value to be assigned to v_i; namely, for each $(v_i, v_j) \in E$, it results $z_j - z_i \in \bigcup_{I \in \mathcal{L}(v_i, v_j)} I$ (assuming $j > i$).

Problems involving temporal constraints arise in several areas of computer science such as scheduling [2,9], program verification [11,3,12], real time systems [17], temporal databases [18] and artificial intelligence (see [6] for an extensive bibliography).

In [5] it is proved that the problem of deciding whether a given network admits a solution is NP-hard, also if we restrict each constraint to consist of no more than two intervals. Conversely, the particular problem occurring when each constraint is defined through a simple interval, known as the *Simple Temporal Problem* (STP), is polynomial. Typically the TCSP is solved through the use of backtracking and this causes exponential running time. In order to improve performance of these algorithms a preliminary step is normally carried out, consisting of making the network arc- or path-consistent. In fact, in a network it frequently happens that the constraint defined on a pair of variables can be made more restrictive without altering the set of the solutions because of the constraints between pairs of vertices in a path linking the initial pair. An example is illustrated in Fig. 1 where the interval $[6, 20]$ can be replaced by $[7, 11]$ without altering the solutions of the network. Informally, the process of

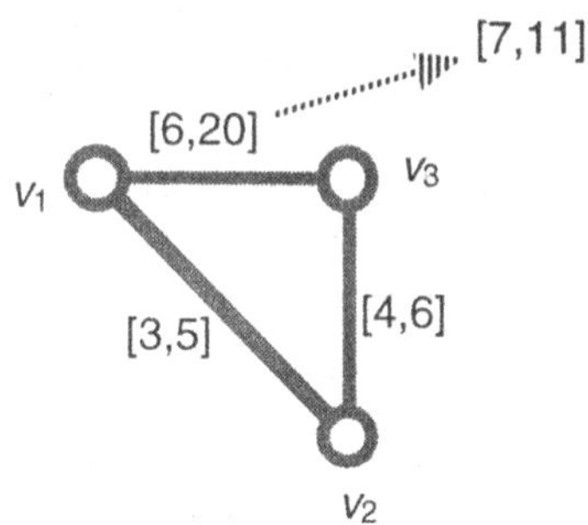

Fig. 1. Constraint $[6, 20]$ can by replaced by $[7, 11]$.

making path-consistent a given network consists of restricting its constraints to keep into account the effects of the propagation of the other constraints. This pre-processing in the general case is very useful for lowering the running time of the successive backtracking phase and in some cases it is sufficient for deciding the whole problem. In fact, in the case of STP, the arc-consistency is sufficient to guarantee the existence of a solution [4], while a solution can be built in $O(n)$ time as long as a path-consistent network is available [6]. Other particular cases where the path-consistency practically provides a solution include networks with qualitative constraints (see Sect. 4.1).

However, it should be clear that in general neither the the path-consistency of a network nor its arc-consistency provide us with significant information about the existence of a solution: a network path-consistent (or arc-consistent) can admit 0 or more solutions. It is true that the process of restricting constraints could lead to empty ones (i.e., $\mathcal{L}(e) = \emptyset$): this is obviously a sufficient (but not necessary) condition for the network to admit no solutions.

In this paper we consider a semi-dynamic setting for the TCSP, where we are requested to maintain the path-consistency of a network under a sequence of insertions of a new (further) constraint between pairs of variables; in terms of graphs this corresponds to restricting, for a given edge e, the set of intervals $\mathcal{L}(e)$. The operation we consider is very general, allowing also to model the case where we add a constraint between two non-adjacent variables: this is achieved by creating an edge e between the pair of variables and defining $\mathcal{L}(e)$ to be equal to the inserting constraint. We show how to maintain path-consistent a network in the defined setting in $O(nR^3)$ amortized time on a sequence of $\Theta(n^2)$ insertions, where n is the number of vertices and R is the range of the network, defined as the maximum size of the minimum interval containing all the intervals of a single constraint. To the best of our knowledge, ours is the first algorithm dealing with the problem in a non-static setting. The best off-line algorithm known for the problem is PC-2 [6] which runs in $O(n^3R^3)$ worst case time, allowing to maintain the path-consistency on $\Theta(n^2)$ constraint insertions in $O(n^5R^3)$. It is worthwhile to point out that (off-line) algorithms exist in the literature for the non-temporal case, where constraints are described by explicitly listing pairs of allowed values thus giving raise to quadratic space per constraint. For this case the best algorithm is PC-4 [8] which runs in $O(n^3a^3)$ where a denotes the size of the largest involved domain.

Furthermore we extend our algorithms to deal with more general temporal networks [13], where variables can be points and/or intervals and constraints can be defined on pairs of variables of different kind [1,21]. For such cases our algorithms maintain their performance. Finally we show how to adapt our algorithms for maintaining also the arc-consistency of such general networks, which is a particular kind of path-consistency limited to paths of length 1. The property is maintained in $O(R)$ amortized time for $\Theta(n^2)$ insertions. This result should be compared to that in [4], where a fully dynamic algorithm is presented for the STP only, capable of performing insertions/deletions of constraints in $O(mR)$ worst case time per operation, where m is the number of edges (so in the worst case the bound is $O(n^2R)$).

2 Preliminaries

Networks of temporal constraints are defined in Sect. 1 by introducing only binary constraints. In general it is convenient to consider also unary constraints. The purpose of this is to manage cases where each variable v_i is constrained to vary in a *domain* $D_i \subseteq U$, thus giving raise to a constraint which can be thought "unary". Even if unary constraints could be managed by admitting self-loops in the graph representing the network, we observe that it is easy to transform a unary constraint into a binary one: it suffices to introduce a special vertex (variable) v_0, constrained[1] to assume a fixed value (e.g., 0), and an edge (v_0, v_i) whose label is a set of intervals of U whose union yields D_i, so from a network $\mathcal{N}$ with unary and binary constraints we can derive a network $\mathcal{N}'$ with only binary ones. Premising that two networks $\mathcal{M}$ and $\mathcal{M}'$ with the same set of solutions are said to be *equivalent* ($\mathcal{M} \equiv \mathcal{M}'$), it is clear that $\mathcal{N}$ and $\mathcal{N}'$ are not equivalent, but a solution for $\mathcal{N}'$ "automatically" provides a solution for $\mathcal{N}$ iff it is an $(n+1)$-ple of the kind $(0, z_1, \ldots, z_n)$.

Also, it is worth noting that we can assume to deal with complete graphs. In fact we can equivalently image the existence of all the complementary edges with label $\{U\}$ (or U), if U is (is not) an interval; we call this ***trivial constraint***. It is clear that this does not require extra storage because we do not need to explicitly represent complementary edges. The labeling function can be consequently extended to the complete graph.

A few words on the cardinality of $\mathcal{L}(e)$ are due. We do not bound a priori the number of intervals which constitute a constraint, however we should note that they can be assumed disjoint. We call *extension* of a constraint the difference between the maximum and the minimum values it allows. The *range* R of a network is defined as the maximum extension of its non-trivial constraints.

A network is said to be *consistent* if the corresponding TCSP admits a solution. An edge (v_i, v_j) is *arc-consistent* [10] iff for each $x_i \in D_i$ there is $x_j \in D_j$ such that the pair (x_i, x_j) satisfies the constraint $\mathcal{L}(v_i, v_j)$.

A path $(v_{i_1}, \ldots, v_{i_p})$ is *path-consistent* [14] iff for each $x_{i_1} \in D_{i_1}$ and $x_{i_p} \in D_{i_p}$ satisfying $\mathcal{L}(v_{i_1}, v_{i_p})$ there are $p-2$ values x_{i_j} such that $(x_{i_j}, x_{i_{j+1}})$ satisfies $\mathcal{L}(v_{i_j}, v_{i_{j+1}})$, for $j = 2, .., p-1$.

A network is said to be *arc-consistent* iff all its edges are arc-consistent; analogously, it is said to be *path-consistent* iff all its paths are path-consistent. The set of networks that are path-consistent is denoted by $\mathcal{PC}$. Moreover a network of n vertices is *i-consistent* [7], with $1 < i \leq n$, iff for any subset $V' \subseteq V$ such that $|V'| = i-1$ there exists a $(i-1)$-ple $\in U^{i-1}$ satisfying the constraints defined by the edges of the subgraph induced by V' and for each variable $v \in V \setminus V'$ the $(i-1)$-ple can be extended into a i-ple $\in U^i$ satisfying all constraints defined by the edges of the subgraph induced by $V' \cup \{v\}$.

We define the following operations on constraints. Let $A = \{I_1, \ldots, I_{|A|}\}$ and $B = \{J_1, \ldots, J_{|B|}\}$ be two constraints.

- $A \oplus B$
  Intersection: values are admitted that are allowed by both the constraints: $A \oplus B = \{K_i \mid K_i = I_j \cap J_k, \text{ for some } j \text{ and } k\}$. The operation is commutative and associative.
- $A \otimes B$
  Composition: values r are admitted such that $r = s + t$, where s is allowed by A and t is allowed by B: $A \otimes B = \{K_i \mid K_i = [a+c, b+d] \text{ for some } I_j = [a,b] \text{ and some } J_k = [c,d]\}$. The operation is commutative and associative.

[1] This is not an explicit constraint in the network.

Also, we introduce an order in the set of all constraints $\mathcal{C}$. Given two constraints $C_1, C_2 \in \mathcal{C}$, we write $C_1 \sqsubseteq C_2$ (C_1 is not less *restrictive* than C_2) if each pair allowed by C_1 is also allowed by C_2.

We conclude the section by giving a few preliminary results.

Lemma 1. *A network is path-consistent iff for each ordered m-ple* $(v_{i_1}, \ldots, v_{i_m})$ *it results* $\mathcal{L}(v_{i_1}, v_{i_m}) \sqsubseteq \mathcal{L}(v_{i_1}, v_{i_2}) \otimes \cdots \otimes \mathcal{L}(v_{i_{m-1}}, v_{i_m})$.

Proof. It easily follows from the definition of operation '$\otimes$'.

Lemma 2. *A network is arc-consistent iff it is 2-consistent.*

Proof. It immediately follows from the definition of i-consistency for $i = 2$.

Lemma 3. *A network is path-consistent iff it is 3-consistent.*

Proof. $\Longrightarrow$ Immediate.

$\Longleftarrow$ By induction on the length of the path. By hypothesis the network is 3-consistent and this implies that each path of length 2 is path-consistent. Suppose now that each path of length m is path-consistent and consider a path $(v_{i_1}, \ldots, v_{i_{m+2}})$ of length $m+1$. By inductive hypothesis the path $(v_{i_1}, \ldots, v_{i_{m+1}})$ is path-consistent and by Lemma 1 it holds $\mathcal{L}(v_{i_1}, v_{i_{m+1}}) \sqsubseteq \mathcal{L}(v_{i_1}, v_{i_2}) \otimes \cdots \otimes \mathcal{L}(v_{i_m}, v_{i_{m+1}})$. On the other hand by the base of the induction we know that $\mathcal{L}(v_{i_1}, v_{i_{m+2}}) \sqsubseteq \mathcal{L}(v_{i_1}, v_{i_{m+1}}) \otimes \mathcal{L}(v_{i_{m+1}}, v_{i_{m+2}})$. So we obtain

$$\mathcal{L}(v_{i_1}, v_{i_{m+2}}) \sqsubseteq \mathcal{L}(v_{i_1}, v_{i_2}) \otimes \cdots \otimes \mathcal{L}(v_{i_{m+1}}, v_{i_{m+2}})$$

which by Lemma 1 proves the thesis.

An immediate consequence of the above lemma is the following.

Lemma 4. *A network is path-consistent iff for each triple* (v_i, v_j, v_k), *with* $i \neq j$, $i \neq k$ *and* $j \neq k$, *it results* $\mathcal{L}(v_i, v_k) \sqsubseteq \mathcal{L}(v_i, v_j) \otimes \mathcal{L}(v_j, v_k)$.

The following results characterizes networks with unary and binary constraints which have been modified by inserting the special vertex v_0, according to what previously described.

Lemma 5. *A network* $\mathcal{N}$ *of* n *vertices in which all unary constraints are managed through the insertion of vertex* v_0 *and of* n *edges making* v_0 *adjacent to all the vertices of* $\mathcal{N}$ *in such a way that* $\mathcal{L}(v_i, v_0)$ *is a set of intervals of* U *whose union yields* D_i *is arc-consistent iff all paths of length 2 originating from* v_0 *are path-consistent.*

Proof. $\Longrightarrow$ As the network is arc-consistent any edge (v_i, v_j) is such that for each $z_i \in D_i$ there is $z_j \in D_j$ such that (z_i, z_j) is allowed by $\mathcal{L}(v_i, v_j)$. After the insertion of v_0 the above property can be expressed as $\mathcal{L}(v_0, v_i) \sqsubseteq \mathcal{L}(v_0, v_j) \otimes \mathcal{L}(v_j, v_i)$.

$\Longleftarrow$ Symmetric reasoning.

3 Incremental path-consistency

We use the notation C_{ij}, for denoting the representation of the constraint between variables v_i and v_j of the temporal network made path-consistent. Note that $C_{ij} \sqsubseteq \mathcal{L}(v_i, v_j)$. We store C_{ij} assuming wlog $i < j$, however, if during the computation we explicitly need C_{ji} we can easily obtain it from C_{ij} in $O(R)$ time.

Algorithm IPC (Incremental Path-Consistency) is illustrated below: it inserts a new constraint A on edge (v_i, v_j) and maintains the path-consistency of the network.

```
1.   Algorithm IPC(i, j, A)
2.   begin
3.       Z := A ∩ C_ij
4.       if Z ≠ C_ij then
5.           C_ij := Z
6.           Q := Q ∪ {(i, j)}
7.           while Q ≠ ∅ do
8.               (k, l) := an element of Q
9.               Q := Q \ {(k, l)}
10.              LocalPathConsistency(k, l)
11.          endwhile
12.      endif
13.  end
```

The effect of the new constraint is computed and stored in Z (cf. line 3) and if A changes (restricts) C_{ij} (line 4) then we store into a set Q (line 6) the pairs whose constraints have been restricted and we have to restore the local consistency of the $2(n-2)$ triples (v_k, v_l, v_m) and (v_l, v_k, v_m), for $v_m \in V \setminus \{v_k, v_l\}$ (inside the cycle (line 7) C_{kl} is a constraint that has been restricted): this is achieved through the procedure LocalPathConsistency called at line 10.

```
procedure LocalPathConsistency(i, j)
begin
    for k ∈ {1, ..., n} \ {i, j} do
        if Revise(i, j, k) then
            Q := Q ∪ {(i, k)}
        endif
        if Revise(j, i, k) then
            Q := Q ∪ {(j, k)}
        endif
    endfor
end
```

Procedure LocalPathConsistency has the task of propagating the effect of the constraint restriction to the other variables. In doing this it can insert other pairs in the set Q, causing the cycle of algorithm IPC to be prolonged. Procedure LocalPathConsistency uses a function Revise, inspired to an analogous function in [10], which takes in input three indices i, j and k and makes path-consistent the path (v_i, v_j, v_k).

```
1.  function Revise(i, j, k)
2.  begin
3.      Z := C_ik ⊕ (C_ij ⊗ C_jk)
4.      if Z = C_ik then
5.          return false
6.      else
7.          C_ik := Z
8.          return true
9.      endif
10. end
```

Remind that `Revise`(i, j, k) is called because C_{ij} has been restricted: at line 3 the function computes the effect on C_{ik} of such restriction, possibly restricting it. At the end of the function it surely holds $C_{ik} \sqsubseteq C_{ij} \otimes C_{jk}$. The function returns `true` iff C_{ik} has been restricted: in this case set Q is augmented (lines 5 and 8 of procedure `LocalPathConsistency`).

3.1 Analysis

We first prove that algorithm IPC terminates its computation.

Theorem 6. *Given a network $\mathcal{N}$ with n vertices, algorithm IPC runs in time $O(n^3R^3)$, where R is the range of $\mathcal{N}$ augmented by the new constraint.*

Proof. We first note that function `Revise` runs in $O(R^2)$ time. In fact the operations carried out at line 3 may take $O(R \log R)$ time for the intersection and $O(R^2)$ time for the composition.

Now observe that the cycle at line 7 of algorithm IPC is executed as many times as the number of pairs inserted in Q. Each insertion corresponds to a constraint which is restricting: since changes on the constraint are monotonic each constraint can be restricted at most $O(R)$ times. As there are at most $O(n^2)$ constraints, it follows that the cycle is executed at most $O(n^2R)$ times.

Each cycle executes a call to procedure `LocalPathConsistency` which, on its turn, executes $\Theta(n)$ calls to `Revise`.

The total running time is therefore $O(n^2R) \cdot \Theta(n) \cdot O(R^2) = O(n^3R^3)$.

Let $\mathcal{N}$ and $\mathcal{N}'$ be two networks such that $\mathcal{N} \equiv \mathcal{N}'$ and $\mathcal{N}' \in \mathcal{PC}$. Let us denote by $\mathcal{M} + A_{ij}$ the network obtained through the addition of the constraint A_{ij} between v_i and v_j in $\mathcal{M}$ and by $\text{IPC}\,(\mathcal{M}, i, j, A_{ij})$ the network built by applying IPC(i, j, A_{ij}) to $\mathcal{M}$. In order to prove the correctness of algorithm IPC we first provide the following initial result.

Theorem 7. *For any constraint A_{ij},* $\text{IPC}\,(\mathcal{N}', i, j, A_{ij}) \in \mathcal{PC}$.

Proof. Suppose by contradiction that $\text{IPC}\,(\mathcal{N}', i, j, A_{ij}) \notin \mathcal{PC}$. Then, by Lemma 4, it must exist a triple (v_r, v_s, v_t) such that $C_{rt} \not\sqsubseteq C_{rs} \otimes C_{st}$. This means that at least one among C_{rs} and C_{st} has been restricted (remind that $\mathcal{N}' \in \mathcal{PC}$ by hypothesis). Suppose without loss of generality C_{rs} has been restricted. As a consequence (r, s) has been inserted in Q (cf. line 6 of algorithm IPC and lines 5 and 8 of procedure `LocalPathConsistency`) and later on procedure `LocalPathConsistency` with parameters (r, s) has been called. This on its turn has called **Revise**(r, s, t) which has necessarily made true that $C_{rt} \sqsubseteq C_{rs} \otimes C_{st}$, which contradicts the above assumption.

Now it remains to prove that $\mathrm{IPC}(\mathcal{N}', i, j, A_{ij}) \equiv \mathcal{N} + A_{ij}$. To this purpose we first show that $\mathrm{IPC}(\mathcal{N}', i, j, A_{ij}) \equiv \mathcal{N}' + A_{ij}$.

Theorem 8. *For any constraint* A_{ij}, $\mathrm{IPC}(\mathcal{N}', i, j, A_{ij}) \equiv \mathcal{N}' + A_{ij}$.

Proof. We denote by $\mathcal{N}^{(l)}$, $l = 1, \ldots, r$, for some r, the network whose constraints are defined by the values of the C_{hk}'s after the l-th restriction carried out by **Revise**. Moreover we denote by $C_{hk}^{(l)}$ such values, and by $C_{hk}^{(0)}$ the initial ones. First of all note that if $C_{ij}^{(0)} = C_{ij}^{(0)} \oplus A_{ij}$ the theorem holds (the test at line 4 fails). We prove by induction on the number of restrictions carried out by **Revise** that $\mathcal{N}^{(l)} \equiv \mathcal{N}' + A_{ij}$ for $l = 1, \ldots, r$. First we prove that $\mathcal{N}^{(1)} \equiv \mathcal{N}' + A_{ij}$. To this purpose consider the first restriction and suppose wlog that it has been made on C_{ip}, for some p. This occurred because $C_{ip}^{(0)} \oplus ((C_{ij}^{(0)} \oplus A_{ij}) \otimes C_{jp}^{(0)}) \neq C_{ip}^{(0)}$. Obviously it results $C_{ip}^{(1)} = C_{ip}^{(0)} \oplus ((C_{ij}^{(0)} \oplus A_{ij}) \otimes C_{jp}^{(0)})$. Now consider a solution $(z_1, \ldots, z_n)$ of $\mathcal{N}' + A_{ij}$. By construction $z_p - z_i$ satisfies $C_{ip}^{(0)}$; but it must also satisfy the constraint obtained composing those on $z_j - z_i$ and on $z_p - z_j$, namely, $(C_{ij}^{(0)} \oplus A_{ij}) \otimes C_{jp}^{(0)}$. In other words, it must satisfy $C_{ip}^{(1)}$.

Now suppose $\mathcal{N}^{(l)} \equiv \mathcal{N}' + A_{ij}$ for $1 \leq l \leq h - 1$: we prove that $\mathcal{N}^{(h)} \equiv \mathcal{N}' + A_{ij}$. For simplicity of notation we re-define the value of $C_{ij}^{(0)}$ as follows: $C_{ij}^{(0)} := C_{ij}^{(0)} \oplus A_{ij}$. Suppose that the h-th constraint that is restricted is C_{st}, for some s and t, and that this occurred because $C_{st}^{(h-1)} \oplus (C_{su}^{(h-1)} \otimes C_{ut}^{(h-1)}) \neq C_{st}^{(h-1)}$, for some u. Again, consider a solution $(z_1, \ldots, z_n)$ of $\mathcal{N}' + A_{ij}$. By construction $z_t - z_s$ satisfies $C_{st}^{(0)}$ and $C_{su}^{(0)} \otimes C_{ut}^{(0)}$ and by inductive hypothesis it also satisfies $C_{st}^{(h-1)}$ and $C_{su}^{(h-1)} \otimes C_{ut}^{(h-1)}$, that is $C_{st}^{(h-1)} \oplus (C_{su}^{(h-1)} \otimes C_{ut}^{(h-1)})$. But this is exactly $C_{st}^{(h)}$.

We are now ready to prove the following important result.

Theorem 9. *Algorithm* IPC *is correct, namely, for any constraint* A_{ij} *it results* $\mathrm{IPC}(\mathcal{N}', i, j, A_{ij}) \equiv \mathcal{N} + A_{ij}$.

Proof. By Theorem 8 we know that $\mathrm{IPC}(\mathcal{N}', i, j, A_{ij}) \equiv \mathcal{N}' + A_{ij}$. For proving the theorem it suffices to show that $\mathcal{N}' + A_{ij} \equiv \mathcal{N} + A_{ij}$. This is immediate. In fact, let S be the set of solutions of $\mathcal{N}$ (and of $\mathcal{N}'$, which, by hypothesis, is equivalent to $\mathcal{N}$). If all n-ples in S satisfy the new constraint A_{ij} the theorem holds. Otherwise, let $S' \subset S$ be the set of n-ples in S satisfying A_{ij}. Note that the insertion of a new constraint cannot introduce new solutions in a network, hence the sets of solutions of $\mathcal{N}' + A_{ij}$ and of $\mathcal{N} + A_{ij}$ are both contained in S. But, being S' the set of solutions satisfying all the constraints, it follows the it is the set of solutions of $\mathcal{N}' + A_{ij}$ and of $\mathcal{N} + A_{ij}$, which are therefore equivalent.

Theorem 9 shows an important propriety of algorithm IPC: it does not need to be the only algorithm manipulating the network. In other words, while maintaining the path-consistency of a network, this can be processed even by other algorithms (e.g., for optimization purposes), as long as they do not change the set of solutions and do not make the network path-inconsistent.

After proving the correctness of IPC, we want deepen the study of its computational complexity. We have already proved (Theorem 6) that IPC $\in O(n^3 R^3)$. However this is a worst case time and this does not necessarily mean that repeated applications of IPC can incur the worst case. In fact we will show that $\Theta(n^2)$ executions of IPC globally take $O(n^3 R^3)$ time. To this purpose we make use of *amortized analysis* [19]. This technique of analysis considers a ***potential*** function Φ which maps configurations of the data structures in use to $\mathbb{R}$. The *amortized time* of an operation is defined as $a = t + \Phi_1 - \Phi_0$, where t is the actual cost of the operation and Φ_0 and Φ_1 are the values of the potential respectively before and after the operation. For a sequence of m operations we have $\sum_{i=1}^{m} t_i = \sum_{i=1}^{m} a_i + \Phi_0 - \Phi_m$, where t_i is the actual time of the i-th operation, a_i is its amortized time, Φ_i is the potential at its termination and Φ_0 is the potential before carrying out the m operations. If we choose Φ so that $\Phi_0 = 0$ and $\Phi_m \geq 0$ it follows that the time cost of the whole sequence is upper bounded by $\sum_{i=1}^{m} a_i$. For other concepts and notations we refer to [19].

For our amortized analysis, let us denote by q_i the number of pairs inserted in Q during the i-th execution of IPC and by $\overline{q}_i = \sum_{j=1}^{i} q_i$ the total number of pairs inserted in Q during the first i executions of IPC. We choose the following potential:

$$\Phi_i = 2(n-2)\,(Ri - \overline{q}_i)$$

Note that it results $\Phi_0 = 0$; moreover $\Phi_{n^2} = 2(n-2)\left(Rn^2 - \overline{q}_{n^2}\right) > 0$ because $\overline{q}_{n^2} < Rn^2$.

Theorem 10. *Any sequence of $\Theta(n^2)$ executions of IPC runs in $O(n^3 R^3)$ worst case time.*

Proof. Let us first compute the (relative) cost of the sequence in terms of number of executions of Revise. The amortized relative complexity of the i-th execution is $a'_i = t'_i + \Phi_i - \Phi_{i-1} = t'_i + 2(n-2)R - 2(n-2)q_i$. Now note that t'_i is equal to $2(n-2)q_i$. From this it follows that the amortized relative cost is $a'_i = 2(n-2)R \in O(nR)$. Thus a sequence of $\Theta(n^2)$ executions has amortized relative complexity $O(n^3 R)$. The thesis follows by recalling that Revise runs in $O(R^2)$ worst case time (see proof of Theorem 6).

We conclude the section mentioning two further properties of IPC, which will be deepened in the full version of the paper. First, IPC does not care about unconstrained variables. It follows that we can insert a new (unconstrained) variable into a path-consistent network in $O(1)$ time. Second, IPC always computes an equivalent path-consistent network by restricting the constraints as little as possible. This shows that the performance of IPC cannot be improved by reducing the number of restrictions on the constraints.

4 Extensions

In this section we briefly show how algorithm IPC can be easily adapted for maintaining the path-consistency of networks of (general) temporal constraints, and how it can be specialized for maintaining the arc-consistency of such general networks.

4.1 Networks of general temporal constraints

In the general case of networks of temporal constraints each variable can independently be a point or an interval. For point variables we have already seen what kind of unary and binary (quantitative) constraints can be considered. For interval variables the situation is different; both in the case of constraints between point and interval and in that of constraints between intervals we deal with *qualitative* constraints (e.g., an interval can before another interval, or a point can be to the right of an interval, etc.). We do not want here to go into details but we note that 5 kinds of constraints can be defined between a point and an interval [13], and 13 kinds of constraints can be defined between two intervals [1]. Also, large and significant classes of networks with qualitative constraints can be directly solved through the path-consistency (see [22,20] for the point algebra and [15] for a significant subclass of the interval algebra).

The operations between constraints introduced in Sect. 2 can be extended to the general case: in particular, the extension of the intersection between constraints is immediate (only constraints of the same kind can be intersected), while the extension of the composition operation is more involved, needing to use the so-called "composition tables" [21,1].

Algorithm IPC maintains the path-consistency of networks of general temporal constraints without needing to be adapted: the only operation which in practice changes is that involving intersection and composition of constraints (see line 3 of function Revise), to reflect that fact that in the general case the definition of such operations can be different. Thus all results for the TCSP directly extend to networks of general temporal constraints without changes. However, in the particular case of qualitative constraints only, the time complexity of Revise becomes constant because qualitative constraints have fixed size. Thus in the case of only qualitative constraints, it is easy to verify that Theorems 6 and 10 can be re-formulated as follows.

Theorem 11. *Given a network with n vertices having only qualitative constraints, IPC runs in time $O(n^3)$.*

Theorem 12. *Any sequence of $\Theta(n^2)$ executions of IPC on a network with n vertices having only qualitative constraints runs in $O(n^3)$ worst case time.*

These results should be compared to those in [16] where an incremental setting is considered for the problem of inserting a new vertex (variable) into the network together with all the qualitative constraints with the other variables.

For this problem the authors provide an algorithm capable of maintaining the path-consistency of the network in $O(n^3)$ worst case time. Evidently, on the base of what seen in this section and at the end of Sect. 3.1, we can support the same operation in the same $O(n^3)$ worst case time and in $O(n^2)$ time if amortized over a sequence of operations.

4.2 Incremental arc-consistency

The arc-consistency of a network $\mathcal{N}$ is a property guaranteeing the satisfaction of all the unary constraints of $\mathcal{N}$ (as long as no edge has the empty constraint). In the particular case of the STP it is sufficient to guarantee also the consistency of the network; however we note that in this case there is no automatic provision of a solution.

Another interesting case occurs when the graph of the network is a tree: in this case the arc-consistency guarantees that a solution can be found in polynomial time by a backtrack-free algorithm [7].

Now we show how algorithm `IPC` can be adapted for maintaining the arc-consistency of a network of general temporal constraints. The idea consists of exploiting the result of Lemma 5, namely of maintaining the path-consistency only of the paths originating from v_0. We name the new algorithm `IAC`: it is obtained from `IPC` by replacing procedure `LocalPathConsistency` by procedure `LocalArcConsistency`, described below.

```
1.  procedure LocalArcConsistency(i, j)
2.  begin
3.      if i ≠ 0 then
4.          if Revise(0, i, j) then
5.              Q := Q ∪ {(0, j)}
6.          endif
7.          if Revise(0, j, i) then
8.              Q := Q ∪ {(0, i)}
9.          endif
10.     else
11.         for k ∈ {0, ..., n} \ {i, j} do
12.             if Revise(0, j, k) then
13.                 Q := Q ∪ {(0, k)}
14.             endif
15.         endfor
16.     endif
17. end
```

First we note that the procedure only tests index i (cf. line 3): index j is not tested because it is surely greater than 0 (in fact, in Q it is never inserted a pair of the kind $(h, 0)$). Second, if $i \neq 0$ then C_{ij} has been restricted, therefore, if needed, we restrict the domains of v_j (line 4) and of v_i (line 7). Otherwise (line 10) the domain of v_j has been restricted and we have to maintain the arc-consistency of the $n - 1$ edges (v_k, v_j) (line 12).

All the analyses carried out on algorithm IPC can be repeated for IAC. Since the techniques of analyses are conceptually the same as those seen in Sect. 3.1 and 4.1 we here only summarize the results achieved by IAC in a unique theorem.

Theorem 13. *Given a temporal network $\mathcal{N}$ with n vertices, algorithm IAC maintain its arc-consistency running in $O(R^3)$ amortized time and in $O(n^2R^3)$ worst case time. If $\mathcal{N}$ only contains qualitative binary constraints then the above running times respectively lower to $O(1)$ and to $O(n^2)$.*

5 Conclusions

We have presented two algorithms, IPC and IAC, which maintain path-consistent and arc-consistent a network of temporal constraints under a sequence of constraint insertions (restrictions). While processing the sequence the algorithms incur their worst case at most a constant number of times, thus guaranteeing amortized performance. Fig. 2 summarizes the running times achieved by the algorithms. It is assumed wlog the presence of qualitative constraints; the third column reports running worst case running times while the fourth column reports running times amortized over a sequence of $\Theta(n^2)$ insertions.

algorithm	quantitative constraints	worst case	amortized
path-consistency (IPC)	yes	$O(n^3R^3)$	$O(nR^3)$
path-consistency (IPC)	no	$O(n^3)$	$O(n)$
arc-consistency (IAC)	yes	$O(n^2R^3)$	$O(R^3)$
arc-consistency (IAC)	no	$O(n^2)$	$O(1)$

Fig. 2. Performance of IPC (incremental path-consistency) and of IAC (incremental arc-consistency).

The case of constraint deletions appears to be computationally more expensive. This issue deserves further investigation.

References

1. J.F. Allen. Maintaining knowledge about temporal intervals. *Communication of the ACM*, 26(11), 1983.
2. F.D. Anger and R.V. Rodriguez. Effective scheduling of tasks under weak temporal interval constraints. *Lecture Notes in Computer Science*, 945, 1995.
3. J. Carmo and A. Sernadas. A temporal logic framework for a layered approach to systems specification and verification. In *Proceedings of the Conference on Temporal Aspects in Information Systems*, pages 31–47, France, May 1987. AFCET.
4. R. Cervoni, A. Cesta, and A. Oddi. Managing dynamic temporal constraint networks. In *Proceedings of AIPS '94*, 1994.
5. E. Davis, 1989. Private communication reported in [6].

6. R. Dechter, I. Meiri, and J. Pearl. Temporal constraint networks. *Artificial Intelligence*, 49, 1991.
7. E.C. Freuder. A sufficient condition for backtrack-free search. *Journal of the ACM*, 29, 1982.
8. C.-C. Han and C.H. Lee. Comments on Mohr and Hendersons path consistency algorithms. *Artificial Intelligence*, 36, 1988.
9. C.-C. Han, K.-J. Lin, and J.W.-S. Liu. Scheduling jobs with temporal distance constraints. *SIAM Journal on Computing*, 24(5):1104–1121, 1995.
10. A.K. Mackworth. Consistency in networks of relations. *Artificial Intelligence*, 8, 1977.
11. Z. Manna and A. Pnueli. Verification of concurrent programs: Temporal proof principle. In D. Kozen, editor, *Logics of Programs (Proceedings 1981)*, LNCS 131, pages 200–252. Springer-Verlag, 1981.
12. Z. Manna and A. Pnueli. Verification of concurrent programs: the temporal framework. In R.S. Boyer and J.S. Moore, editors, *The Correctness Problem in Computer Science*, pages 215–273. Academic Press, 1981.
13. I. Meiri. Combining qualitative and quantitative constraints in temporal reasoning. In *Proc. of the 10th National Conference of the American Association for Artificial Intelligence (AAAI '91)*, 1991.
14. U. Montanari. Networks of constraints: Fundamental properties and applications to picture processing. *Information Sciences*, 7, 1974.
15. B. Nebel and H.J. Bürckert. Reasoning about temporal relations: a maximal tractable subclass of Allen's interval algebra. *Journal of the ACM*, 42, 1995.
16. H. Noltemeier and G. Schmitt. Incremental temporal constraint propagation. In *Proc. of the 9th Florida Artificial Intelligence Research Symp. (FLAIRS '96)*, pages 25–29, 1996.
17. J.S. Ostroff. Temporal Logic of Real-Time Systems. *Research Studies Press*, 1990.
18. A. Tansel, J. Clifford, S. Gadia, S. Jajodia, A. Segev, and R. Snodgrass, editors. *Temporal Databases: Theory, Design, and Implementation.* Database Systems and Applications Series. Benjamin/Cummings, Redwood City, CA, 1993.
19. R.E. Tarjan. Amortized computational complexity. *SIAM J. Algebraic Discrete Methods*, 6(2):306–318, 1985.
20. P. van Beek. Reasoning about qualitative temporal information. *Artificial intelligence*, 58, 1992.
21. M. Vilain and H.A. Kautz. Constraint propagation algorithms for temporal reasoning. In *Proc. of the 5th National Conference of the American Association for Artificial Intelligence (AAAI '86)*, 1986.
22. M. Vilain, H.A. Kautz, and P. van Beek. Constraint propagation algorithms for temporal reasoning: a revised report. In D.S. Weld and J. de Kleer, editors, *Readings in qualitative reasoning about physical systems.* Morgan Kaufman, 1989.

Parallel Algorithms for Treewidth Two*

Babette de Fluiter[1] and Hans L. Bodlaender[2]

[1] Centre for Quantitative Methods
P.O. Box 414, 5600 AK, Eindhoven, the Netherlands
e-mail: deFluiter@cqm.nl

[2] Department of Computer Science, Utrecht University
P.O. Box 80.089, 3508 TB Utrecht, the Netherlands
e-mail: hansb@cs.ruu.nl

Abstract. In this paper we present a parallel algorithm that decides whether a graph G has treewidth at most two, and if so, constructs a tree decomposition or path decomposition of minimum width of G. The algorithm uses $O(n)$ operations and $O(\log n \log^* n)$ time on an EREW PRAM, or $O(\log n)$ time on a CRCW PRAM. The algorithm makes use of the resemblance between series-parallel graphs and partial two-trees. It is a (non-trivial) extension of the parallel algorithm for series-parallel graphs that is presented in [6].

1 Introduction

In this paper we consider the problem of finding a tree decomposition of width at most two of a graph, if one exists.

Many important graph classes have bounded treewidth. Given a tree decomposition of bounded width of a graph, many (even NP-hard) problems can be solved sequentially in linear time, and in parallel in $O(\log n)$ time with $O(n)$ operations on an EREW PRAM, where n denotes the number of vertices of the graph (see e.g. [4, 8]) (the number of operations that an algorithm uses is the product of the number of processors and the time it uses). Therefore, the problem of finding a tree decomposition of bounded width of a graph is well studied.

Sequentially, there exist linear time algorithms for each fixed k that, when given a graph G, decide whether the treewidth of G is at most k, and if so, build a tree decomposition of minimum width for G. Practical algorithms exist for $k = 1$, 2, 3, and 4 [3, 16, 18]; in [5], linear time algorithms are given for each fixed k.

* This research was carried out while the first author was working at the Department of Computer Science at Utrecht University, with support by the Foundation for Computer Science (S.I.O.N) of the Netherlands Organization for Scientific Research (N.W.O.). This research was partially supported by ESPRIT Long Term Research Project 20244 (project ALCOM IT: *Algorithms and Complexity in Information Technology*).

The best known parallel algorithm for *recognizing* graphs of treewidth at most k was found by Bodlaender and Hagerup [8]. It uses $O(n)$ operations, with $O(\log n)$ time on a CRCW PRAM or $O(\log n \log^* n)$ time on an EREW PRAM. They also gave a parallel algorithm for *building* a tree decomposition of width at most k, which uses $O(n)$ operations and $O(\log^2 n)$ time on a CRCW or EREW PRAM. Related, earlier results can be found e.g. in [14, 15].

For treewidth one there is a more efficient algorithm than the one of [8]. A connected simple graph has treewidth one if and only if it is a tree, and a tree can be recognized by using a tree contraction algorithm. This takes $O(\log n)$ time with $O(n)$ operations on an EREW PRAM [1]. One can easily construct a tree decomposition of a tree in $O(1)$ time with $O(n)$ operations on an EREW PRAM. The algorithm can be modified such that it can be used on input graphs which are not necessarily connected (see also Section 4).

In this paper, we improve on the algorithm of [8] for treewidth two. Our algorithm constructs a tree decomposition of width at most two of a graph, if the graph has treewidth at most two. It uses $O(n)$ operations, $O(\log n)$ time on a CRCW PRAM and $O(\log n \log^* n)$ time on an EREW PRAM. We also obtain an algorithm solving the problem on multigraphs, which uses $O(n + m)$ operations with $O(\log(n+m) \log^*(n+m))$ time on an EREW PRAM and $O(\log(n+m))$ time on a CRCW PRAM. From these results and a result from [8] we immediately obtain parallel algorithms for the problem of finding a path decomposition of width at most two of a graph, if it has pathwidth at most two, both for the case of simple graphs and multigraphs. These algorithms run in the same time and resource bounds as the algorithms for treewidth two.

A central technique in this paper is *graph reduction*, introduced in [2]. In [7] and [8] it is shown how the technique can be used to obtain parallel algorithms for graphs of bounded treewidth. In [6] this technique is used for checking whether a given graph is series-parallel, and if so, finding a decomposition of the graph in series and parallel compositions. Our algorithm for treewidth two uses the resemblance between series-parallel graphs and graphs of treewidth two: we show that a graph has treewidth at most two if and only if its biconnected components are series-parallel. We modify the reduction algorithm that is presented in [6] for recognizing series-parallel graphs in order to obtain an algorithm for graphs of treewidth at most two: we add extra reduction rules and show how a tree decomposition of width at most two is constructed. (Also, the counting arguments needed for showing the time bounds for this algorithm are different from those of the algorithm in [6].)

This paper is organized as follows. In Sect. 2 we start with definitions and preliminary results. In Sect. 3 we give a reduction algorithm which checks whether a given graph has treewidth at most two, and if so, finds a tree decomposition of width at most two of the graph. We give this algorithm for a special type of input graph, namely a *connected* B*-labeled multigraph.* Finally in Sect. 4, we show how this algorithm can be used for general multigraphs and simple graphs, and we give some additional results.

More details can be found in [11, 12].

2 Preliminaries

The graphs we consider are undirected and contain no self-loops, but they may have parallel edges, unless stated otherwise.

A *tree decomposition* of a graph $G = (V, E)$ is a pair $(T, \mathcal{X})$ with $T = (I, F)$ a tree, and $\mathcal{X} = \{X_i \mid i \in I\}$ a family of subsets of V, one for each node of T, such that

- $\bigcup_{i \in I} X_i = V$,
- for all edges $\{v, w\} \in E$ there exists an $i \in I$ with $v \in X_i$ and $w \in X_i$, and
- for all $i, j, k \in I$: if j is on the path from i to k in T, then $X_i \cap X_k \subseteq X_j$.

The *width* of a tree decomposition $((I, F), \{X_i \mid i \in I\})$ is defined as $\max_{i \in I} |X_i| - 1$. The *treewidth* of a graph G is the minimum width over all tree decompositions of G [17].

A *path decomposition* of a graph G is a tree decomposition $(\mathcal{X}, T)$ of G in which the tree T is a path, i.e. each node in T has at most degree two. The *pathwidth* of a graph G is the minimum width over all path decompositions of G.

A *series-parallel graph* is a triple (G, s, t), with $G = (V, E)$ a graph and s and t vertices in V called *source* and *sink*, respectively, such that one of the following cases holds.

- G has two vertices, s and t, and one edge between s and t.
- There are two series-parallel graphs (G_1, s_1, t_1) and (G_2, s_2, t_2), such that (G, s, t) is obtained by a *parallel composition* of these graphs, i.e. G is obtained by first taking the disjoint union of G_1 and G_2, and then identifying vertices s_1 and s_2 to the new source s, and vertices t_1 and t_2 to the new sink t.
- There are two series-parallel graphs (G_1, s_1, t_1) and (G_2, s_2, t_2), such that (G, s, t) is obtained by a *series composition* of these graphs, i.e. G is obtained by first taking the disjoint union of G_1 and G_2, then identifying vertex t_1 with vertex s_2, and letting $s = s_1$ and $t = t_2$.

If (G, s, t) is a series-parallel graph, we also say that G is a series-parallel graph.

Theorem 1. *A graph G has treewidth at most two if and only if each biconnected component of G is a series-parallel graph.*

Proof. As the treewidth of a graph equals the maximum treewidth of its biconnected components, it suffices to show that a biconnected graph has treewidth at most two if and only if it is series parallel.

The 'if' part was shown in [6]: each series-parallel graph has treewidth at most two (with induction, show that each series-parallel graph with source s and sink t has a tree decomposition $(T, \mathcal{X})$, with $s, t \in X_i$ for some $i \in I$; both for the series and for the parallel composition one can also compose the corresponding tree decompositions).

For the 'only if' part, suppose G has treewidth at most two. It is shown in [13] that a graph is series-parallel if and only if it can be reduced to the graph consisting of one edge by a sequence of series and parallel reductions. (A series reduction is the removal of a vertex v of degree two, and the addition of an edge between the neighbors of v. A parallel reduction is the removal of an edge which has a parallel edge.) By induction we can show that each biconnected component of G can be reduced to a single edge by a sequence of series and parallel reductions: every simple graph of treewidth two has a vertex of degree at most two. Take a vertex of degree at most two in the simple graph underlying multigraph G: simple case analysis shows that either a series or parallel reduction is possible, or G is a single edge. (We use that series and parallel reductions keep biconnectivity intact as long as the number of edges of G is at least three.) □

A bridge in a graph G is an edge $e \in E(G)$ for which the graph $(V, E - \{e\})$ has more components than G. Our main algorithm is given first for connected multigraphs in which some edges have label B. Such a multigraph is called a *B-labeled multigraph.* A B-labeled multigraph is said to have *treewidth at most two* if the underlying multigraph has treewidth at most two and each edge with label B is a bridge. A *tree decomposition of width at most two* of a B-labeled graph G is a tree decomposition $(T, \mathcal{X})$ of width at most two of the underlying graph, with $T = (I, F)$ and $\mathcal{X} = \{X_i \mid i \in I\}$, such that for each edge $e = \{u, v\}$ with label B, there is a node $i \in I$ with $X_i = \{u, v\}$ and no component in $T[I - \{i\}]$ contains vertices of two components of $(V, E - \{e\})$. It is easy to prove (by induction) that a B-labeled graph has treewidth at most two if and only if it has a tree decomposition of width at most two.

A *terminal graph* G is a triple (V, E, X) with (V, E) a B-labeled multigraph, and $X \subseteq V$ an ordered subset of the vertices ($|X| \geq 0$), denoted by $\langle x_1, \ldots, x_l \rangle$, and called the set of *terminals.* For each i, x_i is called the ith terminal of G. Vertices in $V - X$ are called *inner vertices.* A terminal graph (V, E, X) is called an l-terminal graph if $|X| = l$.

The operation $\oplus$ maps two terminal graphs G and H with the same number l of terminals to a graph $G \oplus H$, by taking the disjoint union of G and H, and then for $i = 1, \ldots, l$, identifying the ith terminal of G with the ith terminal of H.

Two terminal graphs $(V_1, E_1, \langle x_1, \cdots, x_k \rangle)$ and $(V_2, E_2, \langle y_1, \cdots, y_l \rangle)$ are said to be *isomorphic*, if $k = l$ and there exist bijective functions $f : V_1 \to V_2$ and $g : E_1 \to E_2$ with for all $v, w \in V_1$ and $e \in E_2$, $e = \{v, w\} \Leftrightarrow g(e) = \{f(v), f(w)\}$ and for all i, $1 \leq i \leq k$, $f(x_i) = y_i$.

A *reduction rule* r is an ordered pair (H_1, H_2), with H_1 and H_2 l-terminal graphs for some $l \geq 0$, and $|E(H_2)| < |E(H_1)|$. An *application* of reduction rule $r = (H_1, H_2)$ is the operation that replaces a graph G of the form $G_1 \oplus G_3$, with G_1 isomorphic to H_1, by a graph $G_2 \oplus G_3$, with G_2 isomorphic to H_2. We also say the subgraph G_1 is rewritten (to the subgraph G_2).

More background information about (parallel) graph reduction for graphs of bounded treewidth and series-parallel graph can be found in [2, 6, 7, 8, 9, 11].

3 A Constructive Reduction Algorithm

In this section we give an algorithm for finding a tree decomposition of width at most two of a connected B-labeled multigraph, if one exists. The algorithm is a *constructive reduction algorithm.* It consists of two phases: the first phase is the reduction phase, the second one the construction phase. The algorithm uses a set $\mathcal{R}$ of reduction rules, which we define later. Basically, the two phases work as follows, given a connected B-labeled multigraph G.

Phase one. The first phase consists of a number of reduction rounds which are executed subsequently. In each reduction round, a number of applications of rules from $\mathcal{R}$ is carried out simultaneously: if the graph has treewidth at most two, this number is $\Omega(|E(G)|)$. In this phase, the input graph is reduced to a single vertex if and only if it has treewidth at most two. If G has treewidth more than two, i.e., we do not have a single vertex after the first phase, then the algorithms stops. Otherwise, we proceed with the second phase.

In order to make the first phase work correctly, the set $\mathcal{R}$ of reduction rules must be *safe*, i.e. for each $r \in \mathcal{R}$ if a graph G' can be obtained from a graph G by applying r, then G has treewidth at most two if and only if G' has treewidth at most two.

In each reduction round, all reductions that are carried out must be *non-interfering*: no inner vertex of a subgraph that is rewritten may occur in another subgraph that is rewritten (so the subgraphs that are rewritten may share terminals). This is to assure that the graph that results after applying all reductions of one round simultaneously is the same graph as the graph that would result if the reductions were applied subsequently in any order. Hence to be sure that in each reduction round, $\Omega(|E(G)|)$ reductions can be carried out, if the graph has treewidth at most two, it must be the case that in each connected B-labeled multigraph G of treewidth at most two with at least two vertices, there is a set of $\Omega(|E(G)|)$ possible non-interfering reductions.

Phase two. In the second phase, all reductions are undone in reversed order, in a number of *construction rounds.* The number of construction rounds equals the number of reduction rounds. In the first construction round, the reductions of the last reduction round of phase one are undone, in the second construction round, the reductions of the one-but-last reduction round are undone, etc., until all reductions are undone and the input graph is obtained. During the undoing of the reductions, a tree decomposition of width at most two of the current graph is maintained. Each time a reduction is undone, the current tree decomposition is 'locally' modified in such a way that it becomes a tree decomposition of width at most two for the new current graph. When the last construction round is finished, we obtain a tree decomposition of width at most two of the input graph. All undo-actions and local modifications in one construction round are carried out simultaneously.

In the remainder of this section, we describe all ingredients of the algorithm in more detail. In Sect. 3.1 we define the set $\mathcal{R}$ of reduction rules. We show that this set is safe, and that in each connected B-labeled multigraph G of treewidth at most two which has two or more vertices, there is a set of $\Omega(|E(G)|)$ possible reductions. In Sect. 3.2 we show how, in phase two of the algorithm, the reconstruction of the tree decomposition is done for each undoing of a reduction. Finally in Sect. 3.3, we describe the complete algorithm in more detail.

3.1 A Set of Reduction Rules

The set $\mathcal{R}$ of reduction rules for treewidth at most two is depicted in Fig. 1. It is an extension of the set of reduction rules for series-parallel graphs that is presented in [6]: rules 1 – 18 form the set of rules for series-parallel graphs, except for the labeling of the edges in rule 1. Rule 1 consists of two parts, which distinguish between the case in which none of the edges that are involved in the reduction have label B (rule 1a), and the case in which at least one of the edges that are involved in the reduction has label B (rule 1b).

The sequential algorithm for recognizing graphs of treewidth two from [3] uses only rules 1a, 2, and 20 without degree restriction (and a rule to remove isolated vertices). For a parallel algorithm, these rules are not sufficient. Rules 3 – 18 for instance are necessary to reduce subgraphs that form a chain of triangles and squares fast enough in parallel (see e.g. Figure 2). Without these rules, only two concurrent reductions are possible on such a sequence. Rules 21 – 23 are necessary to reduce long sequences of 'small' biconnected components as shown in Figure 3 quickly enough: in such a sequence, only two concurrent reductions are possible without rules 21 – 23.

In rule 20, we pose a degree constraint of eight on the terminal vertex. This means that if we rewrite a terminal subgraph G_1 in G which is isomorphic to the left-hand side of rule 20, then the terminal vertex of G_1 has degree at most eight in G. This degree constraint is added to avoid problems with writing conflicts in the parallel algorithm. It also makes the presence of rule 19 necessary: without rule 19, a large star-like graph can not be reduced.

In rules 3 – 18, we pose a degree constraint of seven on all edges between two terminal vertices. This means that if we rewrite a terminal subgraph G_1 in G which is isomorphic to the left-hand side of one of the rules 3 – 18, then for each edge $e = \{u, v\}$ between two terminals u and v in G_1 either u or v has degree at most seven in G. These degree constraints are used to be able to quickly find applications of these rules.

The B-labeled edges are introduced to decrease the number of reduction rules that is needed: without the B-labelings, it would not be able to rewrite the left-hand sides of rules 21 – 23 to small graphs. The intuition behind B-labeled edges is that these must be a bridge in the graph, or the original graph had treewidth more than three. (Note that originally, no edges are labeled, so bridges can be unlabeled edges, but not vice versa.) For instance, consider rule 21. If the B-labeled edge would not be a bridge in the graph resulting after the reduction, then there is an additional path between the terminals of the occurrence of the left-hand side of rule 21 in the original graph — but then the original graph contains a K_4 as a minor and thus has treewidth at least three.

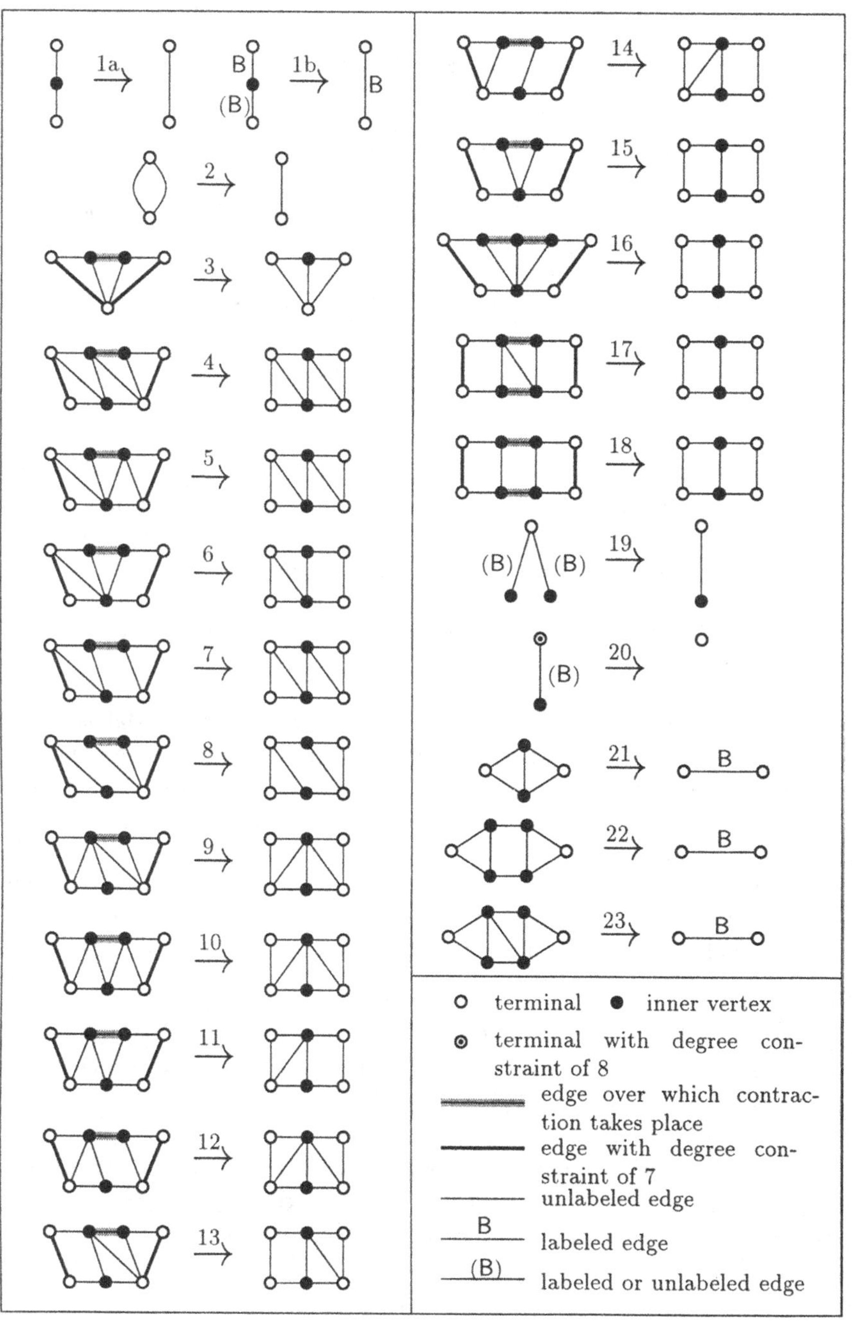

Fig. 1. The set of reduction rules for treewidth at most two on connected B-labeled multigraphs.

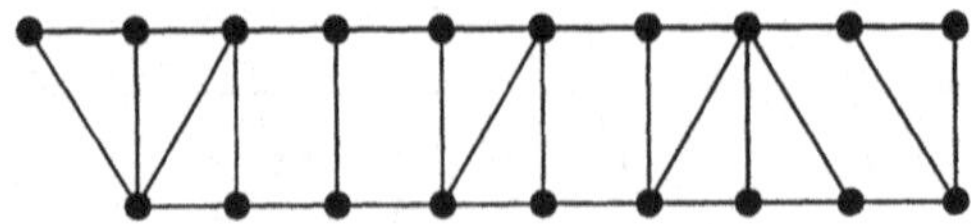

Fig. 2. A chain of triangles and squares

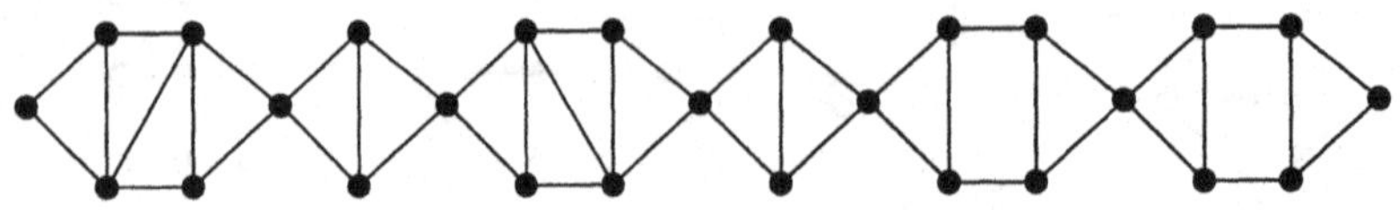

Fig. 3. A chain of 'small' biconnected components

Consider also rule 1b. When we remove a vertex v of degree two and make its neighbors adjacent, and one of the edges adjacent to v was a bridge, then the new edge between the neighbors of v must also be a bridge, hence there is a B-label on the edge in the right-hand side of rule 1b.

Lemma 2. *The set $\mathcal{R}$ is safe for treewidth at most two on connected* B*-labeled multigraphs.*

Proof. We only sketch the proof. Let G and G' be connected B-labeled multigraphs. We have to show that if G' is obtained from G by applying one of the rules in $\mathcal{R}$, then G has treewidth at most two if and only if G' has treewidth at most two. From [6] we know that rules 1 – 18 (without the edge labelings in rule 1) are safe for series-parallel graphs. This can be used to show that these rules are also safe for treewidth at most two. For rules 19 – 23, we can easily show that they are safe for treewidth at most two: a tree decomposition of width at most two of G can be transformed into a tree decomposition of width at most two of G', and vice versa. □

Lemma 3. *There is a constant $k > 0$ such that in each connected* B*-labeled multigraph G of treewidth at most two with $|V(G)| > 1$ a set of at least $k|E(G)|$ non-interfering reductions can be applied.*

This lemma is proven with a detailed case analysis, using the result from [6] that there is a constant $k' > 0$ such that in each series-parallel graph G with at least two edges there is a set of at least $k'|E(G)|$ non-interfering applications of rules 1 – 18 in G.

3.2 Constructing a Tree Decomposition

In this section we show how the construction of a tree decomposition of width at most two is done in phase two of the constructive reduction algorithm. The construction is such that after each construction round, the constructed tree decomposition is a *special* tree decomposition of the current graph.

Definition 4. Let $G = (V, E)$ be a connected B-labeled multigraph with treewidth at most two. Let $TD = (T, \mathcal{X})$ be a tree decomposition of width two of G with $T = (I, F)$ and $\mathcal{X} = \{X_i \mid i \in I\}$. Then TD is a *special tree decomposition* of G if it satisfies the following conditions,

1. For each vertex $u \in V$ there is a unique node i with $X_i = \{u\}$, called the node associated with u.
2. Each edge $e \in E$ with end points u and v has a node i with $X_i = \{u, v\}$ associated with it. Distinct edges have distinct associated nodes.
3. Let u be a cut vertex of G, let i denote the node associated with u. Then each component of $T[I - \{i\}]$ contains vertices of at most one component of $G[V - \{u\}]$.
4. Let e be a bridge of G with end points u and v and let i be the node associated with e. Then each component of $T[I - \{i\}]$ contains vertices of exactly one component of $(V, E - \{e\})$.
5. Let $u, v \in V$. If there is an edge between u and v, and $\{u, v\}$ is a minimal x, y-separator for some vertices x and y, then there is a node i associated with some edge between u and v such that x and y occur in different components of $T[I - \{i\}]$. ($\{u, v\}$ is a minimal x, y-separator if x and y are in different components of $G[V - \{u, v\}]$ but in the same component of $G[V - \{u\}]$ and of $G[V - \{v\}]$.)
6. For each two adjacent nodes $i, j \in I$, $||X_i| - |X_j|| = 1$, unless if $X_i = X_j = \{u, v\}$ and i and j are nodes associated with different edges between u and v.
7. For each $u, v \in V$, the nodes associated with edges between u and v induce a subtree of T.

Suppose we start phase two of the algorithm with the graph G consisting of one vertex v. Then the algorithm first constructs a tree decomposition $TD = (T, \mathcal{X})$ of G which consists of one node i with $X_i = \{v\}$. Node i is the node associated with vertex v. Note that TD is a special tree decomposition of G.

Each time a reduction is undone in phase two, the special tree decomposition of the graph is locally modified. This is done as follows. Suppose the current graph is the graph G, and a terminal graph G_2 in G is replaced by terminal graph G_1. Let $r = (H_1, H_2)$ be the rule that is undone (note that G_i is isomorphic to H_i, $i = 1, 2$). Then the algorithm takes an edge $e = \{u, v\} \in E(G_2)$, and looks up the node i in the tree decomposition that is associated with e (hence $X_i = \{u, v\}$). Then it 'locally' finds the part of the tree decomposition that contains all vertices and edges of G_1 (except possibly some nodes that are associated with edges between two terminals). A subpart of this part is replaced by a new subpart such that the vertices and edges of G_2 occur in this new subpart, and the resulting tree decomposition is a special tree decomposition of the new graph.

Figure 4 shows how this replacement can be done for the case that r is rule 1b. Part 1 of this figure shows graphs G_2 and G_1. The local structure of the tree decomposition will look like the lefthand side of part II of the figure. The righthand side of the figure shows the local structure of the new tree decomposition. The light-gray parts in the tree decompositions are the parts that are involved in the modification. It is easy to see that the new tree decomposition remains special.

Figure 5 shows the reconstruction for rule 3. For the nodes associated with vertices a, b, c and d in the special tree decomposition of G there are a number of possible adjacencies. We denote this in the figure by dotted lines: for each of these nodes, only one of the dotted lines corresponds to an actual adjacency.

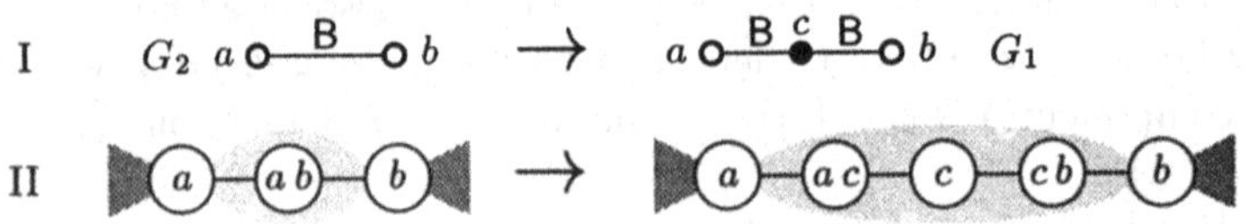

Fig. 4. The reconstruction of the tree decomposition if rule 1b is undone.

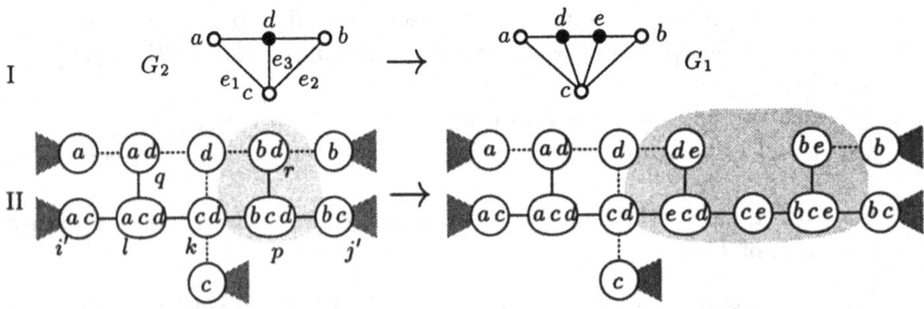

Fig. 5. The reconstruction of the tree decomposition if rule 3 is undone.

For the other reduction rules, the reconstruction can be done in a similar way. By the properties of a special tree decomposition it can be seen that, for each rule $r \in \mathcal{R}$, there are only few different possibilities for the local structure of the tree decomposition that is involved in the undo-action of r. The replacement of this local part of the tree decomposition can be done in constant time. Furthermore, by a careful analysis it can be concluded that if two non-interfering applications of reduction rules are undone simultaneously, then the modifications of the tree decomposition can also be done simultaneously, without interference.

3.3 The Complete Algorithm

In this section, we describe the constructive reduction algorithm in more detail. As said before, the first phase consists of a number of reduction rounds. In each reduction round, $\Omega(|E(G)|)$ non-interfering reductions are applied, if the graph has treewidth at most two. This is done as follows. First, every edge of the current graph 'looks around' to see whether it can take part in a reduction. The set of reductions is not necessarily non-interfering, and hence a subset of non-interfering reductions is selected next. Finally, the reductions of this subset are carried out simultaneously— some bookkeeping is done such that later the reductions can be undone.

The first step, in which each edge looks around, is done as follows. An edge e can easily check whether it can occur in an application of one of the rules 1, 3 – 18 or 20 – 23: follow all paths of length at most eight from e which visit only vertices of degree at most eight (except for the last vertex of a path). This can be done in $O(1)$ time per edge. In this way, all possible choices for applications of these rules are found. However, for rules 2 and 19, probably not all possible applications can be found in this way. Instead, for rule 2, every edge $e = \{u, v\}$ searches in the adjacency lists of u and v for all edges that have distance at most ten to e in this list. Edge e proposes an application of rule 2 if one of the edges it found also has end points u and v. For rule 19, every edge $e = \{u, v\}$ of which end point u has degree one searches in the adjacency list of v for all edges which have distance at most ten to e in this list. The edge e proposes an application of rule 19 if it finds an edge $e' = \{v, w\}$ for which w has degree one. Thus, these rule applications can also be found in $O(1)$ time. (Adjacency lists are assumed to be cyclic.)

Clearly, when looking only at distances at most ten in adjacency lists, we will not find *all* pairs of parallel edges. However, a detailed counting argument based upon properties of graphs of bounded treewidth, using a technique first introduced in [8], shows that we find *sufficiently many* pairs.

Each reduction found in the way described above is said to be *enabled*. It can be shown that, if the graph has treewidth at most two, then there are at least $k|E|$ enabled reductions for some constant k, in a similar but more detailed way as in [6].

Next, a subset of non-interfering reductions of all enabled reductions must be found. This set must be large, i.e. it must have size at least $k'|E|$ for some $k' > 0$. This is solved in the same way as in [8]: a 'conflict graph' is built; one can note that this conflict graph has bounded degree, and a large independent set in the conflict graph is then found (see [8] for more details).

Finally, the set of selected reductions is carried out. Each reduction can be carried out in $O(1)$ time on one processor.

As each reduction round reduces the number of edges with a constant fraction when the input has treewidth at most two, after $O(\log m)$ reduction rounds we can conclude whether the input graph has treewidth at most two or not, depending on whether we end up with a single vertex or not. By using the same approach as in [8], we can carry out all reductions in $O(\log m \cdot \log^* m)$ time with $O(m)$ operations and $O(m)$ space on an EREW PRAM, and with $O(\log m)$ time and $O(m)$ operations and $O(m)$ space on a CRCW PRAM (see also [6]).

The second phase builds the (special) tree decomposition of width at most two, in case G has treewidth at most two. We start with the simple tree decomposition with one node, containing the only vertex of the current graph. Then, we undo each reduction round. Given a special tree decomposition for the reduced graph, we build a special tree decomposition for the graph as it was just before this reduction round was carried out in the first phase, using the constructions as described in Sect. 3.2. The processor that carried out the reduction in the first round will be the same processor that carries out the undoing of the reduction and the local modification of the tree decomposition for that reduction. Note that each undoing of a single reduction, with reconstruction of the tree decomposition, can be done in $O(1)$ time on a single processor without concurrent reading or writing. Hence phase two takes $O(\log m)$ time with $O(m)$ operations.

The presented algorithm uses techniques based on work reported in [6, 7, 8], where also more details can be found.

Theorem 5. *The following problem can be solved with $O(m)$ operations in $O(\log m \log^* m)$ time on a EREW PRAM, and in $O(\log m)$ time on a CRCW PRAM: given a connected (*B*-labeled) multigraph $G = (V, E)$, determine whether G has treewidth at most two, and if so, find a tree decomposition of width at most two of G.*

4 Additional Results

We can use the algorithm presented in Sect. 3 for the same problem, but without requiring that the input graph is connected. In that case, we use the same technique as is used in [8]. This does not increase the amount of resources.

Theorem 6. *There is a parallel algorithm which checks whether a given (*B*-labeled) multigraph G has treewidth at most two, and if so, returns a tree decomposition of width at most two of G. The algorithm uses $O(n+m)$ operations and space, and $O(\log(n+m))$ time on a CRCW PRAM, or $O(\log(n+m)\log^*(n+m))$ time on an EREW PRAM.*

Using that a simple graph $G = (V, E)$ of treewidth at most two has $|E| \leq 2|V|$, a variant of Theorem 6 can be obtained for simple graphs, where the running times are $O(\log n)$, respectively $O(\log n \log^* n)$ and the number of operations is $O(n)$.

Many problems can be solved in $O(\log n)$ time, and $O(n)$ operations and space, when the input graph is simple, and is given together with a tree-decomposition of bounded treewidth. These include all problems that can be formulated in monadic second order logic and its extensions, all problems that are 'finite state', etc. [10]. A large number of interesting and important graph problems can be dealt in this way, including CHROMATIC NUMBER, MAXIMUM CLIQUE, MAXIMUM INDEPENDENT SET, HAMILTONIAN CIRCUIT, STEINER TREE, LONGEST PATH, etc. [8]. As a consequence, a very large class of graph problems can be solved on graphs of treewidth at most two in $O(\log n \log^* n)$

time with $O(n)$ operations on an EREW PRAM, or in $O(\log n)$ time with $O(n)$ operations on an EREW PRAM.

One of the problems which can be solved if a tree decomposition of bounded width of the input graph is given, is the pathwidth problem: given a graph G and an integer constant k, check whether G has pathwidth at most k, and if so, find a path decomposition of width at most k of the graph [8]. Hence we have the following result.

Theorem 7. *Let $k \geq 1$ be an integer constant. There is a parallel algorithm which checks whether a given graph G has treewidth at most two and pathwidth at most k, and if so, returns a path decomposition of width at most k of G. The algorithm uses $O(n)$ operations and space, and $O(\log n)$ time on a CRCW PRAM, or $O(\log n \log^* n)$ time on an EREW PRAM.*

Note that the theorem also holds for multigraphs, if we replace n by $n + m$ in the time and operations bounds. As graphs of pathwidth at most two also have treewidth at most two, the theorem implies that we can find a path decomposition of width at most two of a graph, if one exists, within the same resource bounds.

Acknowledgement. We thank Torben Hagerup for helpful discussions and information.

References

1. K. R. Abrahamson, N. Dadoun, D. G. Kirkpatrick, and T. Przytycka. A simple parallel tree contraction algorithm. *J. Algorithms*, 10:287–302, 1989.
2. S. Arnborg, B. Courcelle, A. Proskurowski, and D. Seese. An algebraic theory of graph reduction. *J. ACM*, 40:1134–1164, 1993.
3. S. Arnborg and A. Proskurowski. Characterization and recognition of partial 3-trees. *SIAM J. Alg. Disc. Meth.*, 7:305–314, 1986.
4. H. L. Bodlaender. NC-algorithms for graphs with small treewidth. In J. van Leeuwen, editor, *Proceedings 14th International Workshop on Graph-Theoretic Concepts in Computer Science WG'88*, pages 1–10. Springer Verlag, Lecture Notes in Computer Science, vol. 344, 1988.
5. H. L. Bodlaender. A linear time algorithm for finding tree-decompositions of small treewidth. *SIAM J. Comput.*, 25:1305–1317, 1996.
6. H. L. Bodlaender and B. de Fluiter. Parallel algorithms for series parallel graphs. In J. Diaz and M. Serna, editors, *Proceedings 4st Annual European Symposium on Algorithms ESA'96, Lecture Notes on Computer Science, vol. 1136*, pages 277–289, Berlin, 1996. Springer Verlag.
7. H. L. Bodlaender and B. de Fluiter. Reduction algorithms for constructing solutions in graphs with small treewidth. In J.-Y. Cai and C. K. Wong, editors, *Proceedings 2nd Annual International Conference on Computing and Combinatorics, COCOON'96*, pages 199–208. Springer Verlag, Lecture Notes in Computer Science, vol. 1090, 1996.

8. H. L. Bodlaender and T. Hagerup. Parallel algorithms with optimal speedup for bounded treewidth. In Z. Fülöp and F. Gécseg, editors, *Proceedings 22nd International Colloquium on Automata, Languages and Programming*, pages 268–279, Berlin, 1995. Springer-Verlag, Lecture Notes in Computer Science 944. To appear in SIAM J. Computing, 1997.
9. B. Courcelle. Graph rewriting: an algebraic and logical approach. In J. van Leeuwen, editor, *Handbook of Theoretical Computer Science, volume B*, pages 192–242, Amsterdam, 1990. North Holland Publ. Comp.
10. B. Courcelle. The monadic second-order logic of graphs I: Recognizable sets of finite graphs. *Information and Computation*, 85:12–75, 1990.
11. B. de Fluiter. *Algorithms for Graphs of Small Treewidth.* PhD thesis, Utrecht University, 1997.
12. B. de Fluiter and H. L. Bodlaender. Parallel algorithms for graphs of treewidth two. Technical Report UU-CS-1997-23, Dept. of Computer Science, Utrecht University, Utrecht, the Netherlands, 1997.
13. R. J. Duffin. Topology of series-parallel graphs. *J. Math. Anal. Appl.*, 10:303–318, 1965.
14. D. Granot and D. Skorin-Kapov. NC algorithms for recognizing partial 2-trees and 3-trees. *SIAM J. Disc. Meth.*, 4(3):342–354, 1991.
15. J. Lagergren. Efficient parallel algorithms for graphs of bounded tree-width. *J. Algorithms*, 20:20–44, 1996.
16. J. Matoušek and R. Thomas. Algorithms finding tree-decompositions of graphs. *J. Algorithms*, 12:1–22, 1991.
17. N. Robertson and P. D. Seymour. Graph minors. II. Algorithmic aspects of tree-width. *J. Algorithms*, 7:309–322, 1986.
18. D. P. Sanders. On linear recognition of tree-width at most four. *SIAM J. Disc. Meth.*, 9(1):101–117, 1996.

On Optimal Graphs Embedded into Paths and Rings, with Analysis Using l_1-Spheres

Yefim Dinitz   Marcelo Feighelstein   Shmuel Zaks
(email: {dinitz,marcelof,zaks}@cs.technion.ac.il)

Department of Computer Science
Technion, Haifa 32000, Israel

Abstract. In this paper we study path layouts in communication networks. Stated in graph-theoretic terms, these layouts are translated into embeddings (or linear arrangements) of the vertices of a graph with N nodes onto the points $1, 2, \cdots, N$ of the x-axis. We look for a graph with minimum diameter $D_c^L(N)$, for which such an embedding is possible, given a bound c on the cutwidth of the embedding. We develop a technique to embed the nodes of such graphs into the integral lattice points in the c-dimensional l_1-sphere. Using this technique, we show that the minimum diameter $D_c^L(N)$ satisfies $\mathcal{R}_c(N) \leq D_c^L(N) \leq 2\mathcal{R}_c(N)$, where $\mathcal{R}_c(N)$ is the minimum radius of a c-dimensional l_1-sphere that contains N points. Extensions of the results to augmented paths and ring networks are also presented. Using geometric arguments, we derive analytical bounds for $\mathcal{R}_c(N)$, which result in substantial improvements on some known lower and upper bounds.

1 Introduction

1.1 Background

In this paper we study path layouts in several models of communication networks. We study communication networks in which pairs of nodes exchange messages along pre-defined paths in the network, termed ***virtual paths***. Given a physical network, the problem is to design these paths optimally. Each such design forms a layout of paths in the network, and each connection between two nodes must consist of a concatenation of such virtual paths. The smallest number of these paths between two nodes is termed the ***hop count*** for these nodes, and the ***congestion*** of a layout is the maximum number of virtual paths that go through any (physical) communication line. The two principal parameters that determine the optimality of the layout - and thus its quality of service - are the maximum congestion of any communication line and the maximum hop count between any two nodes. The hop count corresponds to the time to set up a connection between the two nodes, and the congestion measures the load of the routing tables at the nodes and is also important for fault-tolerance considerations.

When the physical communication network is a path of N nodes, we can state the "all-to-all Chain VP Layout Problem" (see, *e.g.*, [CGZ94]) in graph-theoretic terms as follows. A layout can be viewed as an embedding (or linear arrangement) of the vertices of a graph with N nodes onto the points $1, 2, \ldots, N$ of the x-axis. We look for a graph with minimum diameter $D_c^L(N)$, for which such an embedding is possible, given a bound c on the cutwidth of the embedding (where the diameter $D_c^L(N)$ and the cutwidth c correspond to the hop count and the congestion of the layout, respectively). Determining bounds for $D_c^L(N)$ in several topologies is the subject of few recent studies [KKP95, SV96, ABCRS97], in which there is more than a constant gap between the upper and the lower bounds (more details on these results are mentioned in Section 6).

1.2 Results

We first show a technique to embed any N-node tree layout with cutwidth c and radius ρ into Z^c, preserving the distance of each node from the center of the tree. In Z^c, the distance is measured using the l_1-metric. Therefore, the tree is embedded into the c-dimensional l_1-sphere of radius ρ.

We present a family of trees that completely fill their corresponding spheres. Such a tree gives an exact optimal layout construction for the case where the maximum hop count is between one specified node and the rest of the network. Using this technique, we prove that the minimum diameter $D_c^L(N)$ satisfies $\mathcal{R}_c(N) \leq D_c^L(N) \leq 2\mathcal{R}_c(N)$, where $\mathcal{R}_c(N)$ is the minimum radius of a c-dimensional l_1-sphere that contains N points. Extensions of the results to augmented paths and ring networks are also presented. Using geometric arguments, we derive analytical bounds for $\mathcal{R}_c(N)$, which result in substantial improvements on some known lower and upper bounds.

Basic notions, including the geometric ones, are presented in Section 2. In Section 3 our basic embedding algorithm of a graph into an l_1-sphere is discussed. In Section 4 we define the trees that completely fill these spheres; these trees form the main building blocks for our results concerning the general layouts. In Section 5 we derive analytical bounds for the radii of these spheres using geometric considerations. Our basic theorem (8) together with the embedding algorithm of Section 3 imply our tight bounds presented in Section 6 for general layouts and planar layouts on paths and rings; this results in exact easily-computable bounds (to within a factor of at most 2) for $D_c^L(N)$. The discussion in Section 5 enables us to make a comparison of our results with existing analytical ones in Section 6. Summary and open problems are discussed in Section 7. Most proofs are omitted or only briefly sketched in this Extended Abstract.

2 Preliminaries

Given an undirected graph $G = (V, E)$, where V is the set of nodes and E the set of edges, the *distance* $d_G(u, v)$ (or $d(u, v)$ if there is no ambiguity) between two nodes u and v is the minimum number of edges in a path joining u and v, and ∞ if the two nodes are not connected by any path. The *eccentricity* $ecc_G(v)$ of a node v is the maximum value of $d_G(v,u)$, taken over all vertices u. The *diameter* $D(G)$ of G, is the maximum value of $ecc_G(v)$ taken over all vertices v, and the *radius* $R(G)$ of G is the minimum value of $ecc_G(v)$, taken over all vertices v. A *center* of a graph G is any node v satisfying $ecc_G(v) = R(G)$.

A *(linear) layout* (or *linear arrangement*) of a graph $G = (V, E)$ is a one-to-one mapping $\mathcal{L}$ of the nodes of the graph to the points $1, \ldots, |V|$ on the x-axis. A *tree layout* is a linear arrangement of a tree. If p is a non-integer point of the x-axis, then the *cut* of the layout $\mathcal{L}$ at p, denoted $cut_{\mathcal{L}}(p)$, is the number of edges that cross over p, i.e. the number of edges $(u, v) \in G$ with $\mathcal{L}(u) < p < \mathcal{L}(v)$. The *cutwidth* of a layout $\mathcal{L}$, denoted $cut(\mathcal{L})$, is the maximum cut of $\mathcal{L}$ over all possible values; namely, $cut(\mathcal{L}) = max_{1<p<|V|}cut_{\mathcal{L}}(p)$. The *cutwidth of a graph* G, denoted $cut(G)$, is the minimum cutwidth of any linear arrangement of G. A layout is termed *planar* (or *non-crossing*) if no edge crossing is allowed if all the edges are drawn above the x-axis; namely, there are no edges $(a, b), (c, d) \in E$, such that $\mathcal{L}(a) < \mathcal{L}(c) < \mathcal{L}(b) < \mathcal{L}(d)$. The node of a layout $\mathcal{L}$ mapped to 1 will be termed $first(\mathcal{L})$ and the node mapped to $|V|$ will be termed $last(\mathcal{L})$.

For our embeddings we will make use of the c-dimensional space R^c and its integral points Z^c. The l_1-metric (or *Manhattan* metric) associates with each point $\mathbf{x} = (x_1, \ldots, x_c) \in R^c$ a value $\|\mathbf{x}\| = \|(x_1, \ldots, x_c)\| = \sum_1^c |x_i|$. For every two points $\mathbf{x}, \mathbf{y} \in R^c$, the *$l_1$-distance* between them is defined as $d(\mathbf{x}, \mathbf{y}) = \|\mathbf{x} - \mathbf{y}\|$. A *$c$-dimensional l_1-Sphere of radius ρ* ($Sphere(c, r)$ for short) is the set of points in Z^c that can be reached from the origin $\mathbf{0} = (0, \ldots, 0)$ by at most ρ lattice steps, i.e.: $Sphere(c, \rho) = \{\mathbf{x} \in Z^c \mid \|\mathbf{x}\| \leq \rho\}$ (see Fig. 6). The *c-dimensional l_1-Radius of N*, denoted $\mathcal{R}_c(N)$, is defined as the radius of the smallest c-dimensional l_1-Sphere that contains at least N lattice points; *i.e.*, $|Sphere(c, \mathcal{R}_c(N) - 1)| < N \leq |Sphere(c, \mathcal{R}_c(N))|$.

3 Embedding Tree Layouts

In this section we present our key technique for embedding the nodes of any tree layout into high-dimensional spheres. In other words, we introduce an algorithm that assigns to each node v of a given tree layout T with cutwidth c and any specified node termed *root*, a c-dimensional vector $LABEL(v) = (x_1, \ldots, x_c) \in Z^c$, such that $d_T(root, v) = \|LABEL(v)\|$ for every node v (in particular, $LABEL(root) = \mathbf{0}$). For this purpose, we use the procedure EmbedTree (whose proof of correctness is omitted in this Extended Abstract). Given a tree layout T with root *root*, we first assign $LABEL(root) = \mathbf{0}$. We then consider the edges of the two paths joining $first(T)$ and $last(T)$ to *root*, P_1 and P_2. Each node on P_1 (P_2) will be assigned to a point whose 1st coordinate is equal to $d(v, root)$ $(-d(v, root))$. After this stage, we delete these edges of P_1 and P_2 from the tree layout, and we continue recursively to label the remaining non-trivial[1] tree layouts, whose roots are among the nodes of P_1 and P_2. Each such

[1] A trivial tree is one that has only one node.

subtree with root $root'$ and cutwidth $\leq c-1$ will be embedded into points whose 1st coordinate is equal to that of $LABEL(root')$. This recursive process continues until only trivial trees remain. The algorithm is executed by invoking the procedure *EmbedTree* with parameters T (the layout) , $root$ (its root), and $\xi = 1$, while initializing $LABEL(root)$ to $\mathbf{0}$.

Procedure *EmbedTree* $(T, root, \xi)$
Parameters:
T: tree layout; $root$: node of T; ξ : next coordinate to be assigned.
Variables:
For each node $v \in T$, $LABEL(v)$ is the vector to which this node is mapped. It is assumed that at the first function call, $LABEL(root)$ is initialized to $\mathbf{0}$.
Output:
An embedding of the tree layout T into a $cut(T)$-dimensional space, such that

1. $root$ is mapped to $\mathbf{0}$,
2. $d_T(root, v) = \|LABEL(v)\|$ for every node v in T, and
3. the first $\xi - 1$ coordinates of $LABEL(v)$ are equal to those of $LABEL(root)$, for every node v.

Algorithm:

1. For each vertex v in T: $LABEL(v) \leftarrow LABEL(root)$.
2. Call P_1 the path connecting $first(T)$ to $root$, and $V_1 = V(P_1)$.
3. Call P_2 the path connecting $last(T)$ to $root$, and $V_2 = V(P_2)$.
4. For each vertex v in P_1: $LABEL(v)[\xi]$[2] $\leftarrow dist_T(v, root)$.
5. For each vertex v in $P_2 - P_1$: $LABEL(v)[\xi] \leftarrow -dist_T(v, root)$.
6. Remove all edges of $P_1 \cup P_2$ from T.
7. For each non-trivial tree layout T' in T with root $root'$ in $V_1 \cup V_2$: do $EmbedTree(T', root', \xi + 1)$.

Note:

1. The algorithm does not assume that the input layout T is planar; indeed, it will embed any layout into high-dimensional spheres. The following exemplifies the algorithm for a planar layout.
2. The algorithm does not assume that the paths P_1 and P_2 are edge-disjoint. In this case, the algorithm will map each vertex only once into the sphere (note that in line 5 of the algorithm, only vertices in $P_2 - P_1$ are mapped).

Example 1. We illustrate our algorithm on the tree layout T shown in Figure 1(A). $first(T) = a$, $last(T) = d$, and $root = c$. The path P_1 is $first(T) = a - b - c = root$ and the path P_2 is $root = c - d = last(T)$. We thus map in the first stage ($\xi = 1$) the nodes a, b, c and d to the points (2,0,0), (1,0,0), (0,0,0) and (-1,0,0), respectively (see Figure 1(B)). We then delete these edges from T, and the remaining graph (forest) is shown in Figure 1(C). At this 2nd stage ($\xi = 2$),

[2] $\mathbf{x}[i]$ denotes the ith coordinate of $\mathbf{x}$.

the nodes b, c and d are roots of non-trivial layouts, and the algorithm maps the nodes e, f, and g to the points (1,-1,0), (0,-1,0) and (-1,1,0), respectively. Note that $LABEL(e)[1] = LABEL(b)[1] = 1$. The corresponding edges are then deleted from the layout, and we result in the graph depicted in Figure 1(D), which results in a similar mapping for nodes h and i.

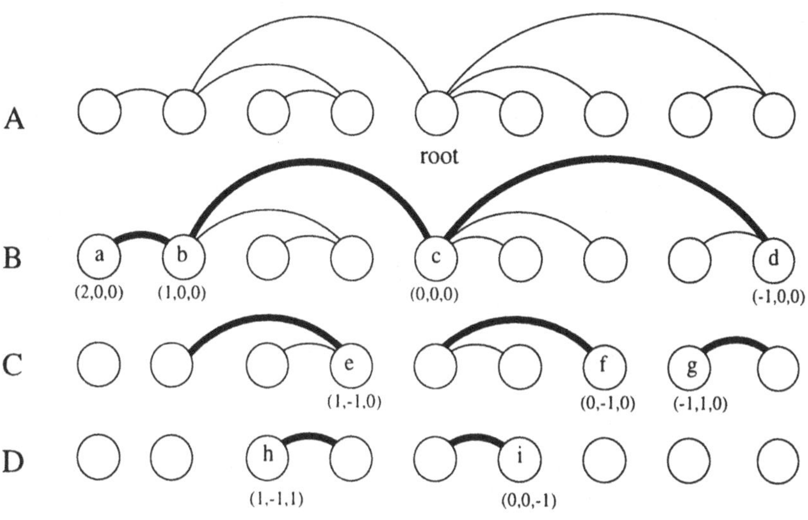

Fig. 1. Embedding of a tree layout of cutwidth 3

From our algorithm it follows immediately that:

Theorem 1. *Given a tree layout T with any node root and cutwidth c, it can be embedded into the c-dimensional l_1-sphere of radius $ecc_T(root)$. In particular, T can be embedded in the c-dimensional l_1-sphere of radius $R(T)$.*

4 Optimal Tree Layouts

In the previous section, we showed that every tree layout of cutwidth c and radius ρ can be embedded into a c-dimensional l_1-sphere of radius ρ. In this section, we present a family of tree layouts that completely fill these spheres. These trees will be later used (Section 6) to derive upper bounds.
We first introduce two operations on layouts. Given two layouts $\mathcal{L}_\infty$ and $\mathcal{L}_\in$ of the graphs $G_1 = (V_1, E_1)$ and $G_2 = (V_2, E_2)$ respectively, $V_1 \cap V_2 = \phi$, their *concatenation* is a new layout $\mathcal{L} = \mathcal{L}_1 \parallel \mathcal{L}_2$ of a graph $G = (V, E)$, such that: $V = V_1 \cup (V_2 - first(\mathcal{L}_2))\}$; $\mathcal{L}(v) = \mathcal{L}_1(v)$ for all $v \in V_1$, $\mathcal{L}(v) = \mathcal{L}_2(v) + |V_{\mathcal{L}_1}| -$

1 for all $v \in V_2 - first(\mathcal{L}_2)$, and if $(u,v) \in E_2$ where $u = first(\mathcal{L}_2)$, then $(\mathcal{L}(last(\mathcal{L}_1), \mathcal{L}(v)) \in E$, otherwise $(\mathcal{L}(u), \mathcal{L}(v)) \in E$ for all $(u,v) \in E_1 \cup E_2$. Note that the last node of $\mathcal{L}_1$ and the first node of $\mathcal{L}_2$ are identified with a node in $\mathcal{L}$ termed *glue-point* and denoted as v^* (see Figure 2(a,b), in which node 3 is the glue-point).
The *join* of two layouts $\mathcal{L}_\infty$ and $\mathcal{L}_\in$ on their two nodes $v_1 \in V_1$ and $v_2 \in V_2$ (denoted $\mathcal{L}_\infty(v_1) \odot \mathcal{L}_\in(v_2)$) of the graphs $G_1 = (V_1, E_1)$ and $G_2 = (V_2, E_2)$ respectively, $V_1 \cap V_2 = \phi$, is a new layout $\mathcal{L}$ of a graph $G = (V, E)$, such that: $V = V_1 \cup V_2$; $\mathcal{L}(v) = \mathcal{L}_1(v)$ for all $v \in V_1$, $\mathcal{L}(v) = \mathcal{L}_2(v) + |V_1|$ for all $v \in V_2$, and $(\mathcal{L}(u), \mathcal{L}(v)) \in E$ for all $(u,v) \in E_1 \cup E_2$. In addition, $(\mathcal{L}(v_1), \mathcal{L}(v_2)) \in E$ (see Figure 2(a,c)).

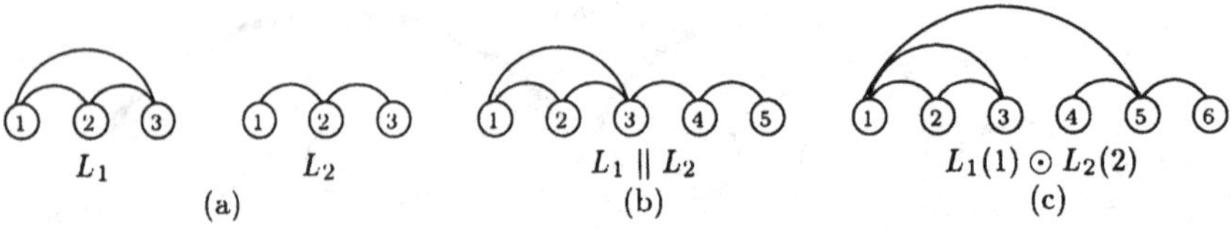

Fig. 2. Concatenation and Join Examples

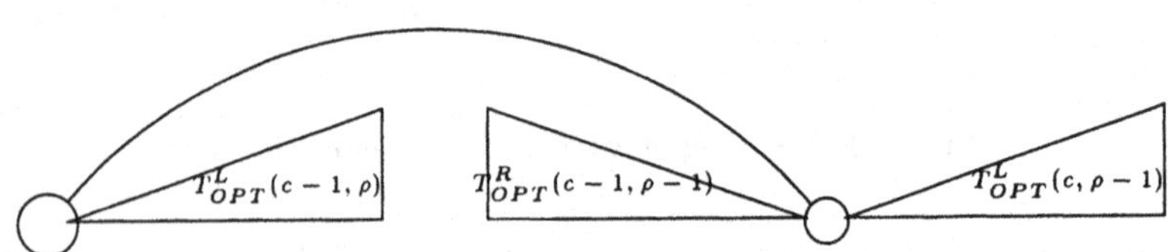

Fig. 3. The recursive definition of $T^L_{OPT}(c,\rho)$

Definition 2. The Optimal Left Tree Layout with cutwidth c and radius ρ ($T^L_{OPT}(c,\rho)$ for short) is the tree layout defined recursively as follows. $T^L_{OPT}(c,0)$ and $T^L_{OPT}(0,\rho)$ are tree layouts with a unique node. Otherwise (c,$\rho > 0$) $T^L_{OPT}(c,\rho) = T_1(first(T_1)) \odot (T_2 \parallel T_3)(v^*)$, where $T_1 = T^L_{OPT}(c-1,\rho)$, $T_2 = T^R_{OPT}(c-1,\rho-1)$, and $T_3 = T^L_{OPT}(c,\rho-1)$ (see Figures 3 and 4; note that this tree is termed "left" since its center is the left most node). Let Optimal Right Tree Layout with cutwidth c and radius ρ ($T^R_{OPT}(c,\rho)$ for short) be defined similarly as $T^L_{OPT}(c,\rho)$, viewed as a mirror image of $T^L_{OPT}(c,\rho)$.

Proposition 3. *Let* $c, \rho \in I\!N$. *Then*

1. $cut(T^R_{OPT}(c,\rho)) = cut(T^L_{OPT}(c,\rho)) = c$.

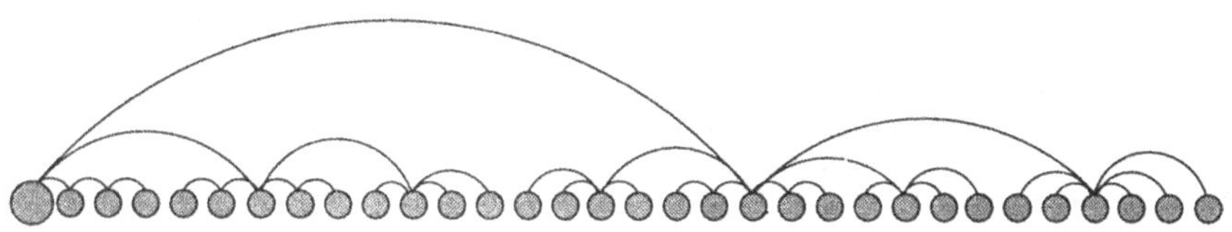

Fig. 4. $T^L_{OPT}(3,3)$

2. $R(T^L_{OPT}(c,\rho)) = \rho$ *and* $first(T^L_{OPT}(c,\rho))$ *is the center of* $T^L_{OPT}(c,\rho)$.
3. $R(T^R_{OPT}(c,\rho)) = \rho$ *and* $last(T^R_{OPT}(c,\rho))$ *is the center of* $T^R_{OPT}(c,\rho)$.
4. *The tree layouts* $T^L_{OPT}(c,\rho)$ *and* $T^R_{OPT}(c,\rho)$ *are planar.*

We make use of the following properties of $Sphere(c,\rho)$:

Lemma 4. [G70] *For all* $c,\rho \in I\!N$ *:*

$$|Sphere(c,0)| = |Sphere(0,\rho)| = 1, \quad \text{and}$$

for all $c,\rho \in I\!N^+$*:*

$$|Sphere(c,\rho)| = |Sphere(c-1,\rho)| + |Sphere(c-1,\rho-1)| + |Sphere(c,\rho-1)|.$$

Let $|T|$ denote the number of nodes in the tree T. We have:

Theorem 5. *For every* $c,\rho \in I\!N$ *:*

$$|T^L_{OPT}(c,\rho)| = |T^R_{OPT}(c,\rho)| = 1/2 \cdot |Sphere(c,\rho)| + 1/2.$$

Proof. The left equality is trivial. We prove the second equality by induction on c and ρ.
If $c = 0$, $|T^L_{OPT}(0,\rho)| = 1 = 1/2 \cdot 1 + 1/2 = 1/2 \cdot |Sphere(0,\rho)| + 1/2$ (and similarly for $\rho = 0$).
Assuming that the inequality holds for all $c' < c$ and $\rho' \leq \rho$ or for all $c' \leq c$ and $\rho' < \rho$, we prove it for c and ρ. From the definition of $T^L_{OPT}(c,\rho)$:
$|T^L_{OPT}(c,\rho)| = |T^L_{OPT}(c-1,\rho)| + |T^R_{OPT}(c-1,\rho-1)| + |T^L_{OPT}(c,\rho-1)| - 1.$
By the inductive hypothesis, we thus have: $|T^L_{OPT}(c,\rho)| = 1/2 \cdot (|Sphere(c-1,\rho)| + |Sphere(c-1,\rho-1)| + |Sphere(c,\rho-1)|) + 1/2$. Hence, by Lemma 4 the inequality holds.

Definition 6. The Optimal Tree Layout with cutwidth c and radius ρ, denoted $T_{OPT}(c,\rho)$, is defined as the concatenation of a $T^R_{OPT}(c,\rho)$ and a $T^L_{OPT}(c,\rho)$ with glue-point termed *center* (see Figure 5).

The key observation that leads to our results is the following:

Lemma 7. *For every* c,ρ *in* $I\!N$ *:*

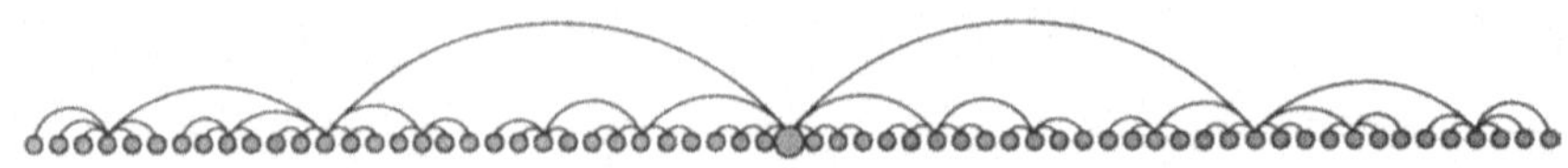

Fig. 5. $T_{OPT}(3,3)$

1. *The tree layout* $T_{OPT}(c, \rho)$ *is planar.*
2. $|T_{OPT}(c,\rho)| = |Sphere(c,\rho)|$.
3. $cut(T_{OPT}(c,\rho)) = c$ *and* $ecc_{T_{OPT}(c,\rho)}(center) = \rho$.

Theorem 8. $T_{OPT}(c,\rho)$ *is an optimal layout with cutwidth c and radius* ρ; i.e., *every layout with the same parameters has at most* $|T_{OPT}(c,\rho)|$ *nodes.*

Proof. Since by Theorem 1, any BFS tree T of any layout with cutwidth c and radius ρ can be embedded in the $Sphere(c,\rho)$, $|T| \leq |Sphere(c,\rho)|$. But, by Lemma 7, $|T_{OPT}(c,\rho)| = |Sphere(c,\rho)|$, and the theorem follows.

5 Geometric Bounds

In this section, we use a geometric technique to obtain analytically closed-form bounds for the radius $\mathcal{R}_c(N)$. Note that this results are not needed for our discussion - in particular, they are not needed in the proofs of our main theorems in Section 6 - except for the fact that they are used in Section 6 for a comparison with known results.

Definition 9. The *host* of every c-dimensional real point $\mathbf{x}$, denoted $host(\mathbf{x})$, is defined as $host(\mathbf{x}) = \{(y_1,\ldots,y_c) \mid y_i = \lceil x_i - 0.5 \rceil)\}$. Note that $host(\mathbf{x})$ always belongs to Z^c. Clearly, $host(\mathbf{x})$ is one of the points in Z^c which are closest (in l_1-metric) to $\mathbf{x}$, and $d(\mathbf{x}, host(\mathbf{x})) \leq c/2$.

Definition 10. The *c-dimensional expanded* l_1*-Sphere of radius* ρ, denoted $ESphere(c,\rho)$, is the set of all c-dimensional real points whose hosts belong to $Sphere(c,\rho)$; *i.e.*, $ESphere(c,\rho) = \{\mathbf{x} \in R^c \mid host(\mathbf{x}) \in Sphere(c,\rho)\}$. In other words, the $ESphere(c,\rho)$ can be viewed as the union of all the c-dimensional unit volume cells, whose hosts belong to $Sphere(c,\rho)$ (see Figure 6).

Denote the volume of a c-dimensional body A by $vol(A)$. We have:

Proposition 11. *For every* $c,\rho \in I\!N$: $|Sphere(c,\rho)| = vol(ESphere(c,\rho))$.

Proof. Each $ESphere(c,\rho)$ is the disjoint union of the c-dimensional unit volume cells surrounding the points in $Sphere(c,\rho)$.

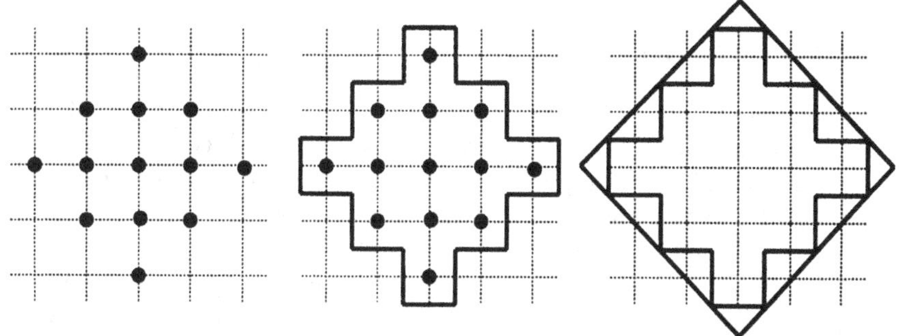

Fig. 6. l_1–*Sphere*$(2,2)$,*ESphere*$(2,2)$ and covering *Cross–Polytope*$(2,3)$

Definition 12. The c-dimensional l_1-*Cross–Polytope* of radius ρ, denoted *Cross–Polytope*(c,ρ), is the set of all c-dimensional real points whose l_1-distance from the origin is at most ρ; *i.e.*, *Cross–Polytope*$(c,\rho) = \{\mathbf{x} \in R^c \mid \|\mathbf{x}\| \leq \rho\}$. (It is the convex hull of *Sphere*(c,ρ), and can be viewed as the c-dimensional generalization of the 3-dimensional octahedron.)

Clearly, for all $c,\rho \in I\!N$: $vol(\textit{Cross–Polytope}(c,\rho)) = \frac{(2\rho)^c}{c!}$.

Definition 13. A c-dimensional l_1-*Cube* of radius ρ, denoted *Cube*(c,ρ), is the set of c-dimensional real points whose maximal coordinate is no greater than ρ; *i.e.*, *Cube*$(c,\rho) = \{\mathbf{x} = (x_1,\ldots,x_c) \in R^c \mid max\{|x_1|,\ldots,|x_c|\} \leq \rho\}$.

Clearly, for all $c,\rho \in I\!N$: $vol(Cube(c,\rho)) = (2\rho)^c$.

With the above definitions we prove the next lemma and its two corollaries. The proofs of this lemma and corollaries are omitted in this Extended Abstract.

Lemma 14. *For all* $c,\rho \in I\!N$:

1. *Cross–Polytope*$(c,\rho) \subseteq$ *ESphere*$(c,\rho) \subseteq$ *Cross–Polytope*$(c,\rho + c/2)$, *and*
2. *Cube*$(c,\rho/c) \subseteq$ *ESphere*$(c,\rho) \subseteq$ *Cube*$(c,\rho + 1/2)$.

From the definitions, the volume formulas above, Lemma ??, and the symmetry between c and ρ in $|Sphere(c,\rho)|$ (followed from Lemma 4 or Lemma 18 in the sequel), we derive the following corollary.

Corollary 15. *For all* $c,\rho \in I\!N$:

1. $max\{\frac{(2\rho)^c}{c!}, \frac{(2c)^\rho}{\rho!}\} \leq |Sphere(c,\rho)| \leq min\{\frac{(2\rho+c)^c}{c!}, \frac{(2c+\rho)^\rho}{\rho!}\}$.
2. $max\{(\frac{(2\rho)}{c})^c, (\frac{(2c)}{\rho})^\rho\} \leq |Sphere(c,\rho)| \leq min\{(2\rho+1)^c, (2c+1)^\rho\}$.

Corollary 16. *For all* $c,\rho \in I\!N$:

1. $1/2 \cdot (c!N)^{\frac{1}{c}} - c/2 \leq \mathcal{R}_c(N) \leq 1/2 \cdot (c!N)^{\frac{1}{c}} + 1$.
2. $1/2 \cdot N^{\frac{1}{c}} - 1/2 \leq \mathcal{R}_c(N) \leq c/2 \cdot N^{\frac{1}{c}} + 1$.
3. $\frac{\log N}{\log(2\cdot c+1)} \leq \mathcal{R}_c(N)$.

6 Applications

6.1 Chain ATM Networks

The *Asynchronous Transfer Mode* (*ATM* for short) is a new model of multi-purpose broadband high-speed networks thoroughly described in the literature (see *e.g.* [ITU90]). The mathematical model of this networks was introduced in [CGZ94] and further developed in [KKP95] and [SV96].

The *"all-to-all* VP *Layout problem"* in an N-node chain topology network (see, *e.g.*, [CGZ94]) can be reduced to the problem of finding a layout of an N-node graph with minimum diameter $D_c^L(N)$, given a bound c on the cutwidth of this layout.

So far, only analytical bounds have been presented for $D_c^L(N)$, with a larger than constant gap between the upper bound and the lower bound. In [KKP95] it its shown that $\frac{1}{2}N^{\frac{1}{c}} \leq D_c^L(N) \leq cN^{\frac{1}{c}}$. In [SV96] it is shown that $D_c^L(N) = \Theta(\frac{\log n}{\log c})$, for a partial range of possible values (specifically, for $c \geq \log^{1+\epsilon} n$, for any fixed $\epsilon > 0$). Our technique enables us to derive the following two consequences: first, we present tight bounds for $D_c^L(N)$. Moreover, using our geometric technique, we are able to further improve known analytical bounds.

Our tight bounds are presented in the following theorem.

Theorem 17. *Let* $N, c \in I\!N$. *Then:*

$$\mathcal{R}_c(N) \leq D_c^L(N) \leq 2 \cdot \mathcal{R}_c(N). \tag{1}$$

Sketch of Proof of Theorem 17: We first prove the upper bound. For all $c, N \in I\!N$, an N-node layout $\mathcal{L}$ with $cut(\mathcal{L}) \leq c$ can be constructed, where $\mathcal{L}$ has a node denoted as *root* and $ecc_{\mathcal{L}}(root) \leq \mathcal{R}_c(N)$. To do this, construct a tree layout $\mathcal{L} = T_{OPT}(c, \mathcal{R}_c(N))$, and denote its center as *root*. Repeatedly remove leafs until $|\mathcal{L}| = N$. By Lemma 7, $cut_T(\mathcal{L}) \leq c$ and $ecc_{\mathcal{L}}(root) \leq \mathcal{R}_c(N)$, as required. Since $D(\mathcal{L}) \leq 2 \cdot \mathcal{R}_c(N)$, then $D_c^L(N) \leq 2 \cdot \mathcal{R}_c(N)$.
As for the lower bound, consider the optimal N-node layout $\mathcal{L}_{OPT}$ with $cut(\mathcal{L}_{OPT}) \leq c$ and $D(\mathcal{L}_{OPT}) = D_c^L(N)$. Choose any node as *root* and build a BFS tree T from *root* over $\mathcal{L}_{OPT}$. Since $cut(T) \leq cut(\mathcal{L}_{OPT}) \leq c$ and $|T| = N$, T can be embedded in a c-dimensional l_1-Sphere of radius $ecc_T(root)$ with at least N internal lattice points. From the minimality of $\mathcal{R}_{cut(T)}(N)$, $\mathcal{R}_c(N) \leq \mathcal{R}_{cut(T)}(N) \leq ecc_T(root)$, but $ecc_T(root) \leq D(\mathcal{L}_{OPT}) = D_c^L(N)$, and the theorem follows. □

Note that the bounds in (1) are tight; their exact computation can be easily done by the following:

Lemma 18. [GW70] *For all* $c, \rho \in I\!N$:

$$|Sphere(c, \rho)| = \sum_{i=0}^{min(c,\rho)} 2^i \binom{c}{i} \binom{\rho}{i}.$$

6.2 Augmented Path and Ring Networks

We now extend the results to another family of topologies termed *augmented networks*. Let an *N-node augmented path AP_N* be an N-node layout $\mathcal{L}$, all of whose pairs of contiguous nodes are connected by an edge; *i.e.*, $(i, i+1) \in E_{\mathcal{L}}$ for all $1 \leq i \leq N-1$.

Lemma 19. *Every planar layout $\mathcal{L}$ with cutwidth c and diameter $D(\mathcal{L})$ can be transformed into an N-node augmented path $\mathcal{L}'$ of the same cutwidth and $D(\mathcal{L}') \leq D(\mathcal{L})$.*

Proof. Construct a new layout $\mathcal{L}'$ copying $\mathcal{L}$. For every pair of contiguous nodes i and $i+1$ in $\mathcal{L}'$, add an edge $(i, i+1)$ to $E_{\mathcal{L}'}$ if they are not connected by an edge. It can be easily proved that this operation does not increase the cutwidth of the layout and does not increase the diameter. Hence, $\mathcal{L}'$ is an N-node planar augmented path with the needed parameters and the lemma follows (see Figure 7).

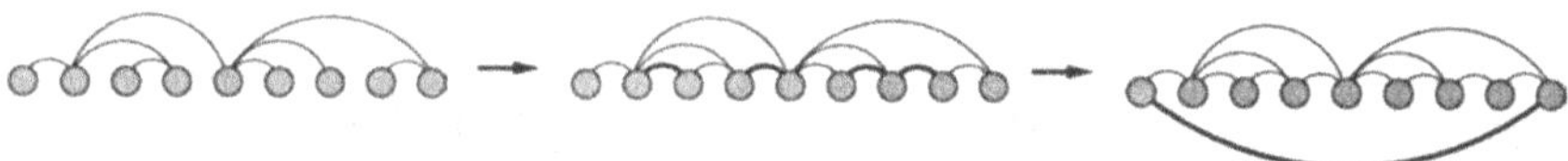

Fig. 7. Transforming a layout into an augmented path, and then into an augmented ring

Theorem 20. *Let $N, c \in \mathbb{N}$. Then:*

$$\mathcal{R}_c(N) \leq D_c^{AP}(N) \leq 2 \cdot \mathcal{R}_c(N),$$

when $D_c^{AP}(N)$ is the minimal diameter of an N-node augmented path with cutwidth c.

Sketch of Proof: For this proof, we use the same techniques developed in Theorem 17. For the upper bound we use Lemma 19, and for the lower bound we use BFS trees. □

Another interesting family of augmented networks is that of *augmented rings (AR)*. In [ABCRS97] three equivalent network models were presented for rings with some additional edges: *chordal rings* with minimal ring-cutwidth c, c-level *express-rings* and c-level *multi-rings*. Their definitions is as follows:

- A *chordal ring* is a ring network with non-crossing "shortcut" edges, which can be viewed as chords of the ring.

- An *express ring* is obtained from a chordal ring by orienting its chords and turning them into external arcs (we term it *augmented ring* in this paper; see Figure 7). The *ring-cutwidth*[3] of an express ring is the number of arcs that cross "above" any ring edge (counting the edge itself).
- A *multi-ring* is obtained from a ring network by appending subsidiary rings to edges of the ring and, recursively, to edges of subsidiary sub-rings. The *depth* of the multi-ring is the depth of the recursive appending of subsidiary rings.

This paper shows that these three models are different views of the same family of graph and investigates the minimum diameter of any ring-cutwidth c, N-node express ring.

It is shown that for all positive integers c and N: $\frac{1}{4e}cN^{\frac{1}{c}} - \frac{c}{2} \leq D_c^{AR}(N) \leq (\frac{1}{2})^{\frac{1}{c}}cN^c$, for the planar case (termed "non-crossing" in [ABCRS97]), and that for the general case this diameter can decrease by at most a factor of 2. Particularly interesting is their lower bound, proved by an embedding of the cycles in a c-dimensional mesh with dilation 2. We use our technique to present tight bounds for augmented rings as well.

Theorem 21. *Let $N, c \in \mathbb{N}$. Then:*

$$\mathcal{R}_c(N) \leq D_c^{AR}(N) \leq 2 \cdot \mathcal{R}_c(N). \tag{2}$$

Sketch of Proof: We use the same techniques developed in Theorem 17. For the upper bound we first show that every planar layout $\mathcal{L}$ with cutwidth c can be transformed into an N-node express ring of a ring-cutwidth c and no larger diameter, by transforming $\mathcal{L}$ into an augmented path then adding to it the edge $(first(\mathcal{L}), last(\mathcal{L}))$. For the lower bound we use BFS trees. □

7 Summary and Open Problems

In this paper we presented a geometric correspondence between virtual path layouts and embeddings in high dimensional spheres. These correspondences result in near-optimal constructions of layouts for all-to-all VP layout problem in ATM chain networks, for augmented paths, and for augmented rings.

Our combinatorial bounds for $D_c^L(N)$, $D_c^{AP}(N)$ and $D_c^{AR}(N)$ are exact to within a factor of 2, which by far improves the known results, presented in Section ??. In addition, using our geometric technique, we manage to derive analytical bounds, which also substantially improve upon existing results. From Theorem 17,20, 21 and Corollary 16 we get:

Corollary 22. *Let $N, c \in \mathbb{N}$. Then:*

$$max\{\frac{1}{2}[(c!N)^{1/c} - c], \frac{1}{2}[N^{1/c} - 1], \frac{\log N}{\log(2c+1)}\} \leq D_c^{L,AP,AR}(N) \leq (c!N)^{1/c} + 2. \tag{3}$$

[3] This notion is termed *cutwidth* in [ABCRS97]; we refer to it here as ring-cutwidth to avoid confusion.

Therefore, we conclude that:

- our analytical upper bound in (3) improves the one in [KKP95] by a factor of $2e$, and generalizes the one in [SV96],
- our analytical lower bounds in (3) coincide with those of [ABCRS97] and [KKP95],
- our analytical upper bound in (2) improves the one in [ABCRS97] by a factor of e, and
- our analytical lower bound in (2) significantly improves the one in [ABCRS97] for large values of c.

References

[ABCRS97] W. Aiello, S. Bhatt, F. Chung, A. Rosenberg, and R. Sitaraman, *Augmented Rings Networks*, to appear in *J. Math. Modelling and Scientific Computing*; see also, *11th Intl. Conf. on Math. and Computer Modelling and Scientific Computing (ICMCM & SC)* (1997).

[CGZ94] I. Cidon, O. Gerstel and S. Zaks, *A scalable approach to routing in ATM networks. 8th International Workshop on Distributed Algorithms, Lecture Notes in Computer Science 857* Springer Verlag, Berlin, 1994, pp.209-222.

[GWZ96] O. Gerstel, A. Wool and S. Zaks, *Optimal Layouts on a Chain ATM Network*, to appear in *Discrete Applied Mathematics*; also: *European Symp. on Algorithms. Lecture Notes in Computer Sciences 979 (P. Spirakis ed.) Springer-Verlag, Berlin,* pp. 508-522.

[GW70] S. W. Golomb and L. R. Welch, *Perfect Codes in the Lee Metric and the Packing of Polyominoes. SIAM Journal on Applied Math.*,vol.18,no.2, January, 1970, pp. 302-317.

[G70] S. W. Golomb, *Sphere Packing, Coding Metrics, and Chess Puzzles. Chapel Hill Conference on Combinatorial Mathematics and its Applications.* May 1970, pp.176-189.

[KKP95] E. Kranakis, D. Krizanc, and A. Pelc, *Hop-congestion tradeoffs for ATM networks. 7th IEEE Symp. on Parallel and Distributed Processing.* pp.662-668.

[ITU90] *ITU recommendations. I series (B-ISDN).* Blue Book, November 1990.

[SV96] L. Stacho and I. Vrt'o, *Virtual Path Layouts for Some Bounded Degree Networks.* 3rd International Colloquium on Structural Information and Communication Complexity (*SIROCCO'96*), Siena, Italy, June 1996.

On Greedy Matching Ordering and Greedy Matchable Graphs *

(Extended Abstract)

Feodor F. Dragan

Universität Rostock, Fachbereich Informatik,
Lehrstuhl für Theoretische Informatik, D–18051 Rostock, Germany
e-mail: *dragan@informatik.uni-rostock.de*

Abstract. In this note a greedy algorithm is considered that computes a matching for a graph with a given ordering of its vertices, and those graphs are studied for which a vertex ordering exists such that the greedy algorithm always yields maximum cardinality matchings for each induced subgraph. We show that these graphs, called greedy matchable graphs, are a subclass of weakly triangulated graphs and contain strongly chordal graphs and chordal bipartite graphs as proper subclasses. The question when can this ordering be produced efficiently is discussed too.

1 Introduction and The Greedy Algorithm

Throughout this note all graphs $G = (V, E)$ are finite, undirected and simple (i.e. without loops and multiple edges). The *(open) neighborhood* of a vertex v is the set $N(v) = \{u \in V : uv \in E\}$ and the *closed neighborhood* is $N[v] = N(v) \cup \{v\}$.

A *matching* of a graph G is a subset M of E such that no two edges share a vertex. A matching of maximal size is called a *maximum matching*. The problem of constructing maximum matchings is one of the most extensively studied problems of graph theory and this is mostly due to the wide variety of applications (see [17, 20]).

Assume that we are given a graph with an ordering on vertices $\sigma = (v_1, v_2, \ldots, v_n)$ (in practice, when we are given a graph, usually it is given with ordered vertex set) and we want to solve the maximum matching problem. One of the simplest way to construct a matching is to choose the smallest (with respect to σ) neighbor of the first vertex to match this vertex, and continue in a similar fashion stepping through the ordering: for each unmatched vertex, we encounter, choose its smallest unmatched neighbor if there is one. A formal description of this method is presented below.

Algorithm (greedy matching)

Input: A graph G with an ordering of vertices $\sigma = (v_1, v_2, \ldots, v_n)$.
Output: A matching M of G.

* Research supported by the VW, Project No. I/69041, and by the DFG.

Complexity: $O(|V| + |E|)$.
Method:

$V' := V$; $M := \emptyset$;
for $i := 1$ **to** n **do**
  if $v_i \in V'$ and $N(v_i) \cap V' \neq \emptyset$ **then**
    choose a neighbor $u \in V'$ of v_i which is smallest with respect to σ;
    add the edge $v_i u$ to M;
    delete the vertices v_i and u from V'

Such a strategy is used by QUEYRANNE ET AL. [21] to solve special transport problems and by DAHLHAUS and KARPINSKI [6] to compute a maximum matching for strongly chordal graphs. A graph G is called *strongly chordal* if it admits a *strong simplicial ordering* [8], i.e. an ordering σ of the vertices of G such that

1. if $a < \{b, c\}$ and $ab, ac \in E$ then $bc \in E$,
2. if $ab, ac, bd \in E$, $a < d$ and $b < c$ then $cd \in E$.

(We write $a < b$ whenever in a given ordering σ vertex a has a smaller number than vertex b.) DAHLHAUS and KARPINSKI have shown in [6] that along a strong simplicial ordering the greedy matching algorithm always yields a maximum matching. Moreover, this ordering allows to compute a maximum matching for every induced subgraph of a strongly chordal graph, i.e. to solve a kind of constrained matching problem, when we want to find a maximum matching M of a graph G such that both end vertices of each edge in M are from a given set $S \subseteq V$. Unfortunately, to date, there are no linear algorithms to give a strong simplicial ordering of a strongly chordal graph $G = (V, E)$: the fastest such algorithm takes $O(|E| \log |V|)$ [19] or $O(|V|^2)$ [23] time. When a strong simplicial ordering is given then the maximum matching problem is solved in linear time.

In this note we introduce and study the graphs for which a vertex ordering exists such that the greedy matching algorithm always gives maximum matchings for each induced subgraph.

Definition 1. An ordering $\sigma = (v_1, v_2, \ldots, v_n)$ of the vertex set of a graph $G = (V, E)$ is a *greedy matching ordering* if, for each induced subgraph F of G with ordering σ_F induced by σ, the matching computed by the above algorithm for F is a maximum matching.

Definition 2. A graph G is *greedy matchable* if it admits a greedy matching ordering.

We give a characterization of greedy matching orderings in terms of forbidden induced suborderings and present some properties of greedy matchable graphs. We show that these graphs are a subclass of weakly triangulated graphs and contain both strongly chordal graphs and chordal bipartite graphs as proper subclasses. Moreover, not only strongly chordal graphs and chordal bipartite graphs are greedy matchable but all graphs which admit an ordering satisfying the second condition of strong simplicial orderings. In the last section we characterize those greedy matchable graphs for which a greedy matching ordering can be produced efficiently by Lexicographic Breadth–First–Search or lexical ordering.

2 Greedy Matching Ordering

Here we characterize greedy matching orderings by forbidden induced suborders and show that greedy matching orderings can be recognized in $O(|V||E|)$ time.

In what follows the expression $\{a, b\} < \{c, d\}$ will mean that in a given ordering σ both vertices a and b have smaller numbers than vertices c and d.

Definition 3. *An ordering σ of the vertex set of a graph G is admissible if for every four vertices a, b, c, d of G such that $ab, ac, bd \in E$ the following holds.*

(i) *If $\{a, b\} < \{c, d\}$ then $cd \in E$.*
(ii) *If $a < d < b < c$ and $ad \notin E$ then $cd \in E$.*

In other words, σ is admissible for G if G ordered with respect to σ does not contain induced ordered subgraphs listed below.

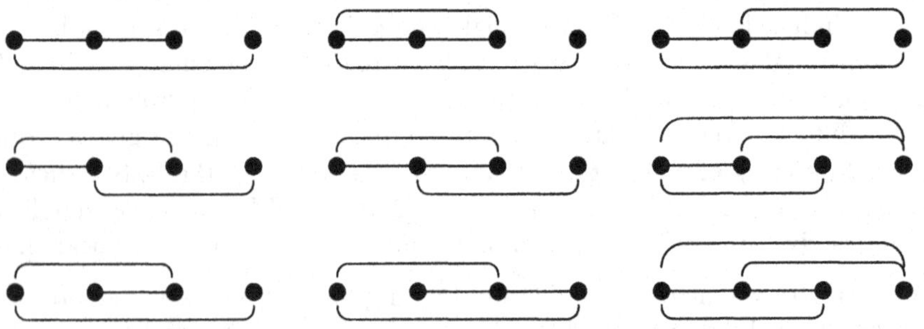

FIGURE 1. Forbidden ordered subgraphs.

Theorem 1. *An ordering $\sigma = (v_1, v_2, \dots, v_n)$ of the vertex set of a graph G is a greedy matching ordering if and only if it is admissible.*

Proof. It is easy to see that, for ordered graphs presented in Figure 1, the greedy matching algorithm will give a matching of cardinality one while all these graphs have matching consisting of two edges. Hence, any greedy matching ordering of G must be admissible.

The converse it is sufficient to prove only for graph G itself; note that an ordering σ, admissible for G, is admissible for every induced subgraph of G as well.

For a given matching M of G, we denote by $M_{\leq x}$ the matching restricted to the set $\{v \in V : v \leq x\}$. Let M^* be an arbitrary maximum matching of G and

M' be the matching of G computed by the above algorithm. We will show that it is possible to transform M^* to another maximum matching that coincides with M'. Consider the smallest in σ vertex x such that $M^*_{\leq x}$ and $M'_{\leq x}$ are different. Then $M^*_{<x}$ and $M'_{<x}$ coincide. We distinguish between three cases.

Case 1. There is an edge xt' in $M'_{\leq x}$ but no edge incident to x in $M^*_{\leq x}$.
If vertex t' is not incident to any edge from M^* then in M^* we can replace the edge incident to x by xt' (since M^* is a maximum matching such an edge xt with $t > x$ exists). Analogously, if vertex x is not incident to any edge from M^* then in M^* we can replace the edge incident to t' by xt'. Hence, we may assume that there are two edges in M^* of the form $t'y$ and xt. Moreover, since $M^*_{<x}$ and $M'_{<x}$ coincide and x has no incident edges in $M^*_{\leq x}$, we have $x < y$ and $x < t$.
Thus, the vertices t', x, y, t with $t'x$,$t'y$,$xt \in E$ fulfill the condition $t' < x < \{y, t\}$. So far as σ is an admissible ordering of G, the vertices y and t must be adjacent. But now we can replace in M^* xt and $t'y$ by $t'x$ and ty.

Case 2. There is an edge xt in $M^*_{\leq x}$ but no edge incident to x in $M'_{\leq x}$.
Since there is no edge in $M'_{\leq x}$ incident to x on step when the vertex t is considered by algorithm either t is not in V' or there is in V' a neighbor of t smaller than x. In both cases vertex t is incident to an edge from $M'_{<x}$. This is a contradiction to $M^*_{<x} = M'_{<x}$.

Case 3. There exists an edge xt in $M^*_{\leq x}$ and an edge xt' in $M'_{\leq x}$.
Assume that $t < t'$. Since $xt \notin M'$ and x belongs to V' on step when the vertex t is considered by algorithm there exists a vertex $y < x$ such that $ty \in M'$. But this contradicts $M^*_{<x} = M'_{<x}$. Therefore, $t' < t < x$.
If vertex t' is not incident to any edge from M^* then in M^* we can replace the edge xt by xt'. So, assume that there is an edge $t'y \in M^*$. From $M^*_{<x} = M'_{<x}$ we conclude that $y > x$.
Suppose now that the vertices t' and t are adjacent. Since $t < x$ and $t'x$ belongs to M', by algorithm the vertex t is not in V' on the step when the vertex t' is considered. This means that there exists a vertex $a < t'$ such that $at \in M'$. Again a contradiction to $M^*_{<x} = M'_{<x}$ arises (recall that $tx \in M^*$). Hence, the vertices t' and t are not adjacent. Additionally we had $t' < t < x < y$ and $t'x$,$t'y$,$xt \in E$. Since σ is an admissible ordering, the vertices y and t must be adjacent in G. As before we can replace in M^* xt and $t'y$ by $t'x$ and ty.

Thus, we have shown how to transform M^* to a new maximum matching that coincides with M' in $\{v \in V : v \leq x\}$. By induction we get a maximum matching that coincides with M' in V. □

From Theorem 1 we can derive that a graph G is greedy matchable if and only if there is an ordering of V which contains no suborderings listed in Figure 1.

Recall that an ordering σ is a strong simplicial ordering if the following two conditions hold.

1. If $a < \{b, c\}$ and $ab, ac \in E$ then $bc \in E$.
2. If $ab, ac, bd \in E$, $a < d$ and $b < c$ then $cd \in E$.

If an ordering σ satisfies only the first condition then σ is called *simplicial ordering*. It is well-known that a graph G has a simplicial ordering if and only if G is *chordal*, i.e. it has no induced cycles of length greater than three (cf. [9]). An ordering satisfying only the second condition we will call *strong ordering*.

Definition 4. *An ordering σ of the vertex set of a graph G is a strong ordering if for every four vertices a, b, c, d of G such that $ab, ac, bd \in E$, $a < d$ and $b < c$ we have $cd \in E$.*

An immediate consequence of the above theorem is the following.

Corollary 1. *Every strong ordering is a greedy matching ordering.*

Hence, all graphs which admit a strong ordering are greedy matchable graphs. Among these graphs are also the well-known class of *chordal bipartite graphs*, i.e. bipartite graphs having no induced cycles of length greater than four (cf. [9]). It follows from the results in [8, 1, 18] that chordal bipartite graphs are exactly those bipartite graphs that have a strong ordering.

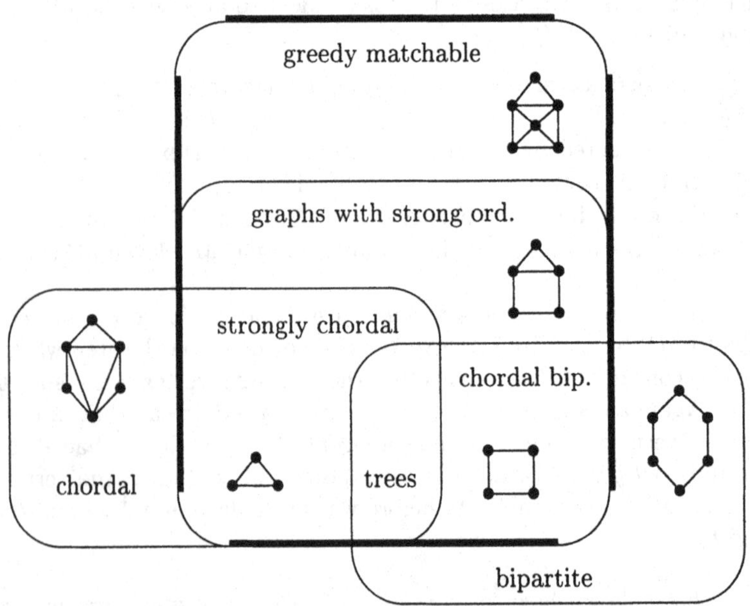

FIGURE 2. Some subclasses of greedy matchable graphs.

The next result indicates how we can check a given ordering for admissibility.

Theorem 2. *An ordering σ of the vertex set of a graph G is admissible if and only if the following holds.*

(*i*) *If* $ad \in E$, $a < d$ *and there exists a vertex* $b \in N(a) \cap N(d)$ *such that* $a < b < d$ *then* $cd \in E$ *for every* $c \in N(a)$, $c > d$.
(*ii*) *If* $ab \in E$, $a < b$ *and* c *is the smallest neighbor of* a *with respect to* σ *satisfying* $c > b$ *then* $cd \in E$ *for every vertex* $d \in N(b)$ *such that* $d < c$ *and* $ad \notin E$ *or* $d > c$.

Proof is omitted.

Since both these conditions can be verified in $O(|V||E|)$ time, it can be decided in $O(|V||E|)$ time whether a given ordering σ is a greedy matching ordering.

3 Greedy Matchable Graphs

By P_k and C_k we denote a path and a cycle on k vertices. An induced cycle C_k with $k \geq 5$ is called a *hole*. An *antihole* is the complement of a hole. We call a graph *nontrivial* if it has more than one vertex.

For a given ordering σ of the vertex set of a graph G, by $G_{\geq v}$ we denote a subgraph of G induced by the set $\{u \in V : u \geq v\}$. A vertex v of a graph G is called *simplicial* if every two neighbors of v are adjacent, i.e. the neighborhood $N(v)$ of v induces a complete subgraph of G. Using this notions one can give an alternative definition of simplicial orderings. An ordering σ is a ***simplicial ordering*** of G if vertex v is simplicial in $G_{\geq v}$ for all $v \in V$. In [10] the notion of simpliciality was adapted for bipartite graphs. An edge ab of a bipartite graph G is called *bisimplicial* if $N(a) \cup N(b)$ induces a complete bipartite subgraph of G. Analogously to this we define a simplicial edge of an arbitrary graph.

Definition 5. *An edge* ab *of a graph* G *is simplicial if every two vertices* $v \in N(a)$ *and* $u \in N(b)$ $(u \neq v)$ *are adjacent in* G.

Lemma 1. *Every connected nontrivial induced subgraph of a greedy matchable graph has a simplicial edge.*

Proof. Let σ be an admissible ordering of a graph G and $F = (V', E')$ be an arbitrary connected nontrivial induced subgraph of G. Consider leftmost vertices $a \in V'$ and $b \in N(a) \cap V'$ with respect to σ. From the choice of a and b we have $a < d$ and $a < b < c$ for all vertices $c \in N(a) \cap V'$ and $d \in N(b) \cap V'$. Moreover, if $d < b$ then $ad \notin E$. Now $cd \in E$ follows from $a < b < \{c, d\}$, when $b < d$, or from $a < d < b < c$ and $ad \notin E$, when $b > d$. □

In what follows we will need the following definition.

Definition 6. *A pseudo-sun* S_k *of size* k $(k \geq 3)$ *is a graph whose vertex set can be partitioned into two sets,* $U = \{u_1, \ldots, u_k\}$, $W = \{w_1, \ldots, w_k\}$, *so that* U *forms a cycle* $C_k = (u_1, \ldots, u_k, u_1)$, *the graph induced by* W *has no connected components with only two vertices, and* w_i *is adjacent to* u_j *if and only if* $i = j$ *or* $i = j + 1$ (*mod* k). *A sun is a pseudo-sun in which* U *is a clique (we call it inner clique) and* W *is an independent set.*

A pseudo–sun S_3 (hexahedron) and suns S_3 and S_4 are presented in Figure 3.

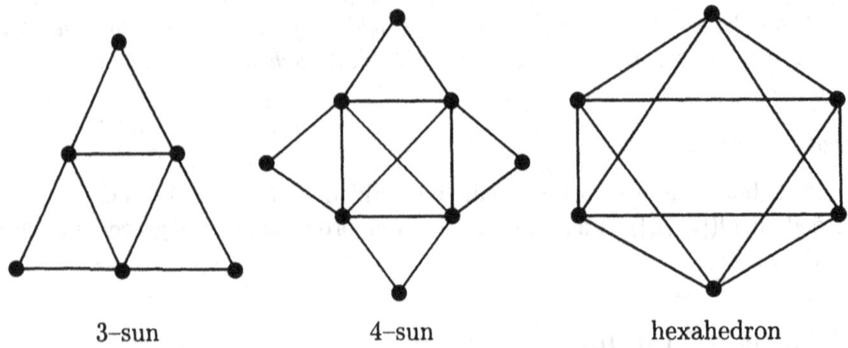

FIGURE 3. Some forbidden subgraphs.

In [5, 8] it was proven that a graph G is strongly chordal if and only if it is chordal and does not contain any sun S_k as an induced subgraph. A graph is called *weakly triangulated* [11] if it has no hole or antihole as an induced subgraph.

Corollary 2. *Every greedy matchable graph is a weakly triangulated graph that does not contain pseudo–suns and graphs from Figure 4 as induced subgraphs.*

Proof. Straightforward verification shows that no edge of hole or any graph from Figure 4 is simplicial. By Lemma 1, these graphs cannot be induced subgraphs of a greedy matchable graph. We will see also that no antihole and no pseudo–sun has a simplicial edge.

Consider an antihole $\bar{C}_k = (v_1, \ldots, v_k, v_1)$ with the edge set E and an arbitrary edge v_iv_j, $i \neq j \pm 1 \pmod{k}$, of it. Vertices v_i and v_j divide the cycle C_k into two induced paths. Let v_lv_{l+1} be a middle edge of a longest path, and assume that v_l is closer than v_{l+1} to v_j in that path. Since $k \geq 5$ in $\bar{C}_k$ we have $v_lv_i, v_iv_j, v_jv_{l+1} \in E$ but $v_lv_{l+1} \notin E$, that is the edge v_iv_j is not simplicial.

Now let S_k be a pseudo–sun with cycle $C_k = (u_1, \ldots, u_k, u_1)$ and vertices $\{w_1, \ldots, w_k\}$ of W. Consider an edge w_iw_j of S_k. By definition of pseudo–sun, vertex u_i cannot be adjacent to both w_{j-1} and w_{j+1}. Assume without loss of generality that $w_{j+1}u_i \notin E$. Then from $u_i \in N(w_i)$, $w_{j+1} \in N(w_j)$ and $w_{j+1}u_i \notin E$ it follows that the edge w_iw_j is not simplicial. Now consider the edges w_iu_i and w_iu_{i-1} of S_k. Since $u_iw_{i+1}, u_{i-1}w_{i-1} \in E$ and $u_iw_{i-1}, u_{i-1}w_{i+1} \notin E$ both these edges cannot be simplicial. Finally consider a possible edge u_iu_j of S_k. Since a subgraph of S_k induced by W has no connected components with only one edge, we can find a third vertex u_l in W adjacent to u_i or/and to u_j. Assume without loss of generality that $u_iu_l \in E$. As $l \neq j$, vertex u_l cannot be adjacent to both w_{j+1} and w_j. Let $w_ju_l \notin E$. Then again from $u_l \in N(u_i)$, $w_j \in N(u_j)$ and $w_ju_l \notin E$ we get that the edge u_iu_j is not simplicial. This completes the proof. □

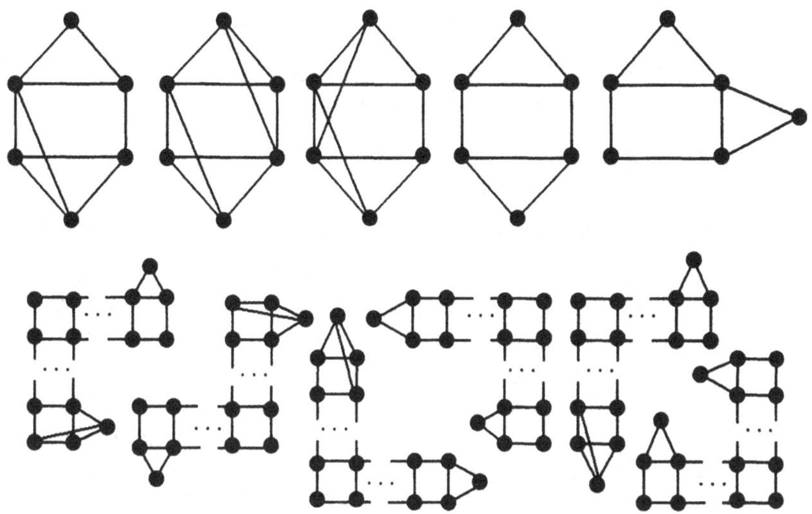

FIGURE 4. More forbidden subgraphs.

It seems, simplicial edges play an important role in the structure of greedy matchable graphs. We could not find any minimal forbidden subgraph of greedy matchable graphs which has a simplicial edge. Motivating by this we formulate the following conjecture.

Conjecture. *If every nontrivial induced subgraph of a graph G has a simplicial edge, then G is a greedy matchable graph.*

Below we will show that in two particular cases this conjecture is true.

Definition 7. *An ordering σ of the vertex set of a graph G is a lexicographic ordering if the following property holds.*

$(P\star)$ *For every three vertices a, b, c with $a < b$, $ac \in E$ and $bc \notin E$ there exists a vertex d with $d > c$, $db \in E$ and $da \notin E$.*

Lemma 2. *Any graph has a lexicographic ordering. Moreover, a lexicographic ordering of a graph $G = (V, E)$ can be produced in $O(|V||E|)$ time.*

Proof is omitted. In our proof we show that, for a given graph $G = (V, E)$, a graph theoretic variant of the algorithm, proposed by HOFFMAN, KOLEN and SAKAROVITCH in [12] for doubly lexical ordering of (0, 1)-matrices, gives a lexicographic ordering of G in $O(|V||E|)$ time.

A house is an induced C_4 with one additional vertex adjacent to exactly two adjacent vertices of C_4. A tent is an induced C_4 with an additional vertex adjacent to exactly three vertices of C_4 (see Figure 6).

Theorem 3. *Let G contain no induced subgraphs isomorphic to tent, house, sun, hexahedron or hole. Then every lexicographic ordering of G is a strong ordering.*

Proof is omitted.

Since suns, holes and hexahedron do not have any simplicial edges, house has only one and tent has two adjacent simplicial edges, we have

Theorem 4. *If every induced subgraph of a graph G, enjoying at least two nonadjacent edges, has two nonadjacent simplicial edges, then G is a greedy matchable graph and a greedy matching ordering of G can be found in $O(|V||E|)$ time.*

Strongly chordal graphs can be characterized also by another elimination scheme [8]. A vertex v of a graph G is called *simple* if for all $x, y \in N(v)$, $N[x] \subseteq N[y]$ or $N[y] \subseteq N[x]$ holds, i.e. $\{N[x] : x \in N[v]\}$ is linearly ordered by inclusion. A *simple elimination ordering* of a graph G is an ordering σ such that vertex v is simple in $G_{\geq v}$ for all $v \in V$. In [8] it is shown that G is strongly chordal if and only if G admits a simple elimination ordering. Note that a strong simplicial ordering is a simple elimination ordering σ such that for all $v \in V$ in $G_{\geq v}$ we have $N[x] \subseteq N[y]$ whenever $x < y$ in σ and $x, y \in N(v)$. Here we define a new ordering of the vertex set of a graph which has similar relation to strong orderings.

Definition 8. *A vertex v of a graph G is quasi–simple if for all $x, y \in N(v)$ the following holds:*
1. if $x, y \in E$ then $N[x] \subseteq N[y]$ or $N[y] \subseteq N[x]$,
2. if $x, y \notin E$ then $N(x) \subseteq N(y)$ or $N(y) \subseteq N(x)$.
A quasi–simple elimination ordering *of a graph G is an ordering σ such that vertex v is quasi–simple in $G_{\geq v}$ for all $v \in V$.*

We will need also the following notion of a *simplicial-edge-without-vertex elimination ordering*. It generalizes the known notion of a *bisimplicial-edge-without-vertex elimination ordering* (see [3] and [16]) and refers to an edge elimination ordering such that no vertices are deleted in the process.

Definition 9. *Let $(e_1, \ldots, e_m)$ be an ordering of the edges of $G = (V, E)$ and $G_i = (V, E_i)$ be a subgraph of G with vertex set V and edge set $E_i = \{e_j : j \geq i\}$. The ordering $(e_1, \ldots, e_m)$ is a simplicial-edge-without-vertex elimination ordering for G if each edge e_i is simplicial in G_i.*

It is known from [3] (see also [16]) that a graph G is chordal bipartite if and only if G has a bisimplicial-edge-without-vertex elimination ordering.

The next result shows that both dismantling schemes defined above characterize the graphs with strong orderings.

Theorem 5. *The following three conditions are equivalent for a graph G.*

(1) *G admits a strong ordering.*
(2) *G admits a quasi–simple elimination ordering.*
(3) *G admits a simplicial-edge-without-vertex elimination ordering.*

Proof is omitted. Our proof is constructive and leads to $O(|V|^3)$ time algorithm for computing a strong ordering of G.

Theorem 6. *Every graph G that has a simplicial-edge-without-vertex elimination ordering is a greedy matchable graph and a greedy matching ordering of G can be found in $O(|V|^3)$ time.*

A greedy matchable graph that does not have any simplicial-edge-without-vertex elimination ordering is given in Figure 5. It has only one simplicial edge. After removing this edge we get a sun S_3 which is not a greedy matchable graph by Corollary 2.

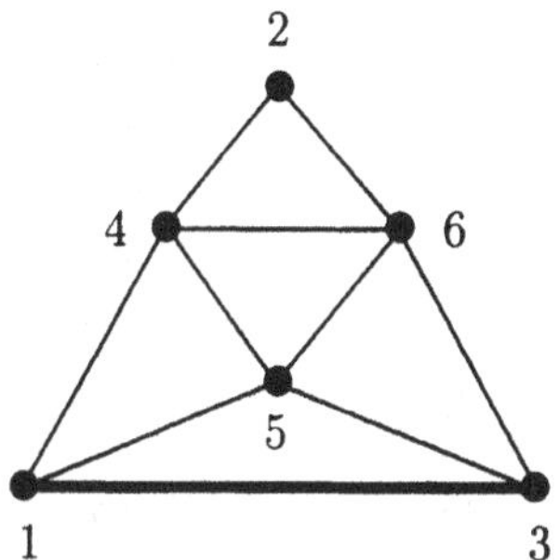

FIGURE 5. A greedy matchable graph that does not have any simplicial-edge-without-vertex elimination ordering.

4 More Subclasses

Here we describe two subclasses of greedy matchable graphs where greedy matching orderings can be found more efficiently. One of these subclasses contains both strongly chordal and chordal bipartite graphs. It is well-known [18, 23] that for these graphs any lexical ordering of LUBIW [18], defined below, is a strong ordering and hence strong orderings for these graphs can be produced in $O(min\{|V|^2, |E|\log|V|\})$ time [19, 23]. Note that strongly chordal graphs are exactly the class of chordal greedy matchable graphs, while chordal bipartite graphs are exactly bipartite greedy matchable graphs.

In what follows we will use properties :

($P0$) If $a < b$ and $ac \in E$ and $bc \notin E$ then there exists a vertex d such that $c < d$, $d \in N(b) \cup \{b\}$ and $da \notin E$. ($d = b$ is allowed.)

($P1$) If $a < b < c$ and $ac \in E$ and $bc \notin E$ then there exists a vertex d such that $c < d$, $db \in E$!and $da \notin E$.

Evidently, ($P1$) is a relaxation of ($P0$) and ($P0$) is a relaxation of ($P\star$). It is well-known that any ordering of a graph G produced by Lexicographic Breadth-First-Search (LexBFS [22]) has property ($P1$) (cf. [9]). Moreover, any ordering with property ($P1$) can be produced by LexBFS as shown in [4]. Recall that LexBFS orders vertices of a graph by assigning numbers from $n = |V|$ to 1 in the following way: assign the number k to a vertex v (as yet unnumbered) which has lexically largest vector $(s_i : i = n, n-1, \ldots, k+1)$, where $s_i = 1$ if v is adjacent to the vertex numbered i, and $s_i = 0$ otherwise.

Definition 10. *An ordering $\sigma = (v_1, v_2, \ldots, v_n)$ of the vertex set of a graph G is called a LexBFS–ordering if it satisfies the property $(P1)$. If σ satisfies the property $(P0)$ then it is called a lexical ordering of G.*

Note that for a given graph G, a LexBFS–ordering of G can be generated by LexBFS in linear time [9] while to date the fastest method – doubly lexical ordering of the (closed) neighborhood matrix of G [18] – producing a lexical ordering of G takes $O(|E|\log|V|)$ [19] or $O(|V|^2)$ [23] time. Notice that every lexical ordering of a graph is a LexBFS–ordering but not conversely.

Simplicial vertex can be defined also as a vertex which is not midpoint of an induced P_3. In [15] this notion was relaxed: A vertex is *semi–simplicial* if it is not a midpoint of an induced P_4. An ordering σ is a *semi–simplicial ordering* if vertex v is semi–simplicial in $G_{\geq v}$ for all $v \in V$. In [15] (see also [7]) the authors characterized the graphs for which every LexBFS–ordering is a semi–simplicial ordering as the HHD–free graphs, i.e. the graphs which contain no house, hole and domino as an induced subgraph (cf. Figure 6). We will use this fact in what follows.

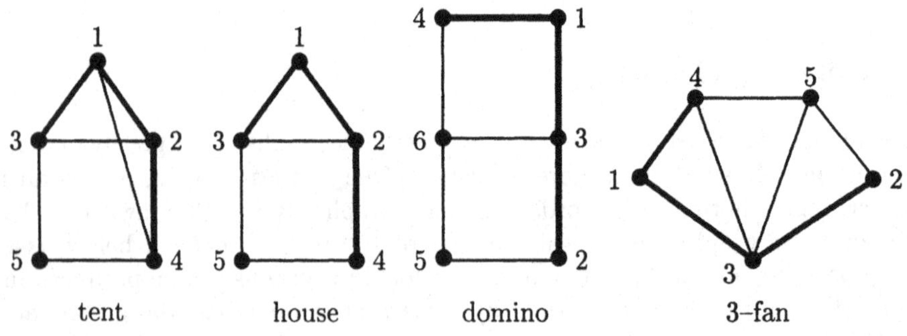

FIGURE 6. Orderings produced by LexBFS that are not admissible.

Lemma 3. *Let G contain no induced subgraphs isomorphic to tent, house, domino, hexahedron or hole, and let σ be a LexBFS–ordering (or a lexical ordering) of G. For every four vertices a, b, c, d of G such that $ab, ac, bd \in E$ and $\{a, b\} < \{c, d\}$, we have $cd \in E$.*

Proof. We prove the assertion for LexBFS–ordering only. This is enough because every lexical ordering of a graph G is a LexBFS–ordering of G. Assume that $cd \notin E$ for some vertices a, b, c, d with $ab, ac, bd \in E$ and $\{a, b\} < \{c, d\}$. We may suppose that $a < b$. Then $cb \in E$ or $ad \in E$ since a is a semi–simplicial vertex in $G_{\geq a}$.

If $cb \notin E$ then $ad \in E$, and applying $(P1)$ to $a < b < c$ yields a vertex $t > c$ adjacent to b but not to a. From semi-simpliciality of a we have $ct \in E$. But now the vertices a, b, c, d, t induce a house or a tent, a contradiction. Therefore, c and b are adjacent.

If $d < c$ then $(P1)$ applied to $b < d < c$ gives a vertex $t > c$ adjacent to d but not to b. Since the vertex b is minimal in the path formed by c, b, d, t it cannot be a middle vertex of an induced P_4. Hence, tc is an edge of G. To avoid a house and a tent we must have $at, ad \in E$. Now we can apply $(P1)$ to $a < d < c$ yielding a vertex $s > c$ adjacent to d but not to a. Note that $s \neq t$. Again, using semi-simpliciality of b in $G_{\geq b}$, we obtain $sc \in E$ or $sb \in E$. If $sc \in E$ then $sb \in E$, otherwise we get a tent. So, in every case the vertices s and b are adjacent. To avoid a house and a possible tent formed by c, b, d, t, s we must have both edges sc and st. Now the vertices c, b, d, t, s and a induce a hexahedron, a contradiction.

Thus, $d > c$. Applying $(P1)$ to $b < c < d$ we will get a vertex $t > d$ adjacent to c and not to b. From semi-simpliciality of b in $G_{\geq b}$ the vertices t and d must be adjacent. To avoid a house and a tent we must have $at, ad \in E$. Now we apply $(P1)$ to $a < b < t$ and find a vertex $s > t$ adjacent to b and not to a. From semi-simpliciality of a in $G_{\geq a}$, we have $ts \in E$. Furthermore, $cs \in E$ and $sd \in E$, otherwise we obtain a tent induced by $\{c, a, b, s, t\}$ or by $\{d, a, b, s, t\}$. But then all vertices together induce a hexahedron, a contradiction.

This settles the proof. □

The following theorem gives a characterization of those greedy matchable graphs for which every LexBFS-ordering is a greedy matching ordering.

Theorem 7. *Any LexBFS-ordering of a graph G is a greedy matching ordering if and only if G does not contain an induced subgraph isomorphic to tent, house, domino, 3-fan, hexahedron or hole.*

Proof. Suppose that G has no induced subgraphs isomorphic to tent, house, domino, 3–fan, hexahedron or hole, but however σ produced by LexBFS is not a greedy matching ordering. Then we can find four vertices a, b, c, d in G such that $ab, ac, bd \in E$, $a < d < b < c$, $ad \notin E$ and $cd \notin E$. Note that the case when $\{a, b\} < \{c, d\}$ is handled by Lemma 3. Again from semi-simpliciality of a in $G_{\geq a}$ we conclude that $cb \in E$. Property $(P1)$ applied to $a < d < c$ gives a vertex $t > c$ adjacent to d but not to a. Since the vertex d is minimal in the path formed by c, b, d, t this path cannot be induced, i.e. $tb \in E$ or $tc \in E$ holds in G. If $tb \notin E$ then $tc \in E$ and we obtain a house. Therefore, tb must be an edge, and the vertices a, b, c, t fulfill conditions $ab, ac, bt \in E$ and $\{a, b\} < \{c, t\}$. By Lemma 3, $ct \in E$. Hence, we have constructed in G an induced 3–fan that is impossible.

By Corollary 2, for the converse we need to show only that the graphs from Figure 6 must be forbidden. Assume that G contains a house or a tent as an induced subgraph. We start LexBFS with the vertex labeled by 5 in Figure 6 yielding number n. Let k, l, s and t be the numbers of vertices 4, 3, 2 and 1, respectively. Since vertices 2 and 1 are not adjacent to 5 but 4 and 3 are adjacent, we get $\{t,s\} < \{k,l\}$. From $kl \notin E$ we conclude that the obtained LexBFS–ordering is not a greedy matching ordering. Now assume that G contains a 3–fan. We start LexBFS with the vertex labeled by 5 yielding number n. Now we may number vertex 4 by $n-1$ and vertex 3 by $n-2$. Let k and l be the numbers of vertices 2 and 1, respectively. By the rules of LexBFS we get $k > l$. Since $l < k < n-2 < n-1$, $kl \notin E$ and the vertices $n-1$ and k are not adjacent, this LexBFS–ordering is not a greedy matching ordering. Finally assume that G contains a domino. We start LexBFS with the vertex labeled by 6 yielding number n. Then we may number vertex 5 by $n-1$. Let k, l, s and t be the numbers of vertices $4, 3, 2$ and 1, respectively. By the rules of LexBFS we get $t < s < \{k,l\}$. Suppose that we cannot number vertex k before vertex l in this LexBFS–ordering. Then $l > k$ and there exists a vertex v such that $v > l$, $vl \in E$ and $vk \notin E$. Since $v > l$ by the rules of LexBFS we must have $vn \in E$. But now the vertices n, k, l, t and v induce either a hose or a tent. This contradicts to the proof above. So, we may number the vertex k before l and get $t < s < l < k$. Since $st, sk \notin E$ again the obtained LexBFS–ordering is not a greedy matching ordering. □

The graphs that do not contain house, hole, domino and 3–fan as induced subgraphs are known as *distance–hereditary graphs* [13]. *Ptolemaic graphs* are exactly the chordal distance–hereditary graphs [14] while *bipartite (6,2)-chordal graphs* are the bipartite distance–hereditary graphs [2].

Corollary 3. *Let G be a ptolemaic graph, or a bipartite (6,2)-chordal graph, or a distance-hereditary graph without induced subgraphs isomorphic to tent and hexahedron. Then a maximum matching of G can be computed in linear time.*

Denote by S_k^- the graph isomorphic to sun S_k without one vertex from the independent set W (see Definition 6).

Theorem 8. *Any lexical ordering of a graph G is a greedy matching ordering if G does not contain an induced subgraph isomorphic to tent, house, domino, sun, hexahedron or hole.*

Proof. By contradiction, as in the proof of Theorem 7 we will find four vertices x_1, y_1, y_2, x_2 in G such that $x_1y_1, x_1y_2, y_1x_2 \in E$, $x_1 < x_2 < y_1 < y_2$, $x_1x_2 \notin E$ and $y_2x_2 \notin E$. We may choose these vertices x_1, y_1, y_2, x_2 with maximal sum $\Sigma = x_1 + y_1 + y_2 + x_2$ of their numbers in σ.

Repeating the arguments of the proof of Theorem 7 (we need only to replace $(P1)$ by $(P0)$ and a, b, c, d, t by x_1, y_1, y_2, x_2, y_3, respectively), we can construct in G an induced 3–fan (S_3^-) formed by x_1, y_1, y_2, x_2, y_3 with inner clique $\{y_1, y_2, y_3\}$, independent set $\{x_1, x_2\}$ and $x_1 < x_2 < y_1 < y_2 < y_3$. We may choose the vertex y_3 with $y_3 > y_2$, $y_3x_2 \in E$ and $y_3x_1 \notin E$ rightmost in σ.

In what follows we show how to extend this S_3^- to an induced S_4^-. Since $y_1 < y_2$ and $x_2y_1 \in E$ but $x_2y_2 \notin E$ we can apply $(P0)$ and get a vertex $x_3 > x_2$ adjacent to y_2 and not to y_1. Note that $x_3 \neq y_2$. The vertices x_3 and y_3 are not adjacent, for otherwise we would have a 3–sun or a house or a tent as an induced subgraph (it depends on adjacency between x_3 and x_1, x_2).

We claim that $x_3 < y_1$. Indeed, if $x_3 > y_1$ then either $y_1 < x_3 < y_2 < y_3$ or $y_1 < y_2 < \{x_3, y_3\}$. Since $y_1y_2, y_1y_3, y_2x_3 \in E$ and $x_3y_1 \notin E$ we obtain $x_3y_3 \in E$ from maximality of Σ (note that $x_1 + y_1 + y_2 + x_2 < y_1 + y_2 + y_3 + x_3$) or by Lemma 3, respectively. But this is impossible. Hence, $x_3 < y_1$.

Now we apply $(P0)$ to $x_2 < x_3 < y_3$ yielding a vertex $y_4 > y_3$ adjacent to x_3 and not to x_2. We choose the vertex y_4 rightmost in σ. Since $\{x_3, y_2\} < \{y_3, y_4\}$, by Lemma 3, y_3 and y_4 are adjacent. Moreover, to avoid a house or a tent formed by $\{y_1, y_2, y_3, y_4, x_3\}$ or a 3–sun formed by $\{y_1, y_2, y_3, y_4, x_1, x_2\}$, we must have $y_2y_4, y_1y_4 \in E$. The vertices y_4 and x_1 cannot be adjacent. For otherwise, from $x_1 < x_2 < y_1 < y_4$, $x_1x_2 \notin E$ and maximality of Σ we would get $y_4x_2 \in E$ that is impossible. Furthermore, $x_3x_1, x_3x_2 \notin E$ because induced house and tent are forbidden subgraphs for G.

Thus, we have constructed in G an induced S_4^- with inner clique $\{y_1, y_2, y_3, y_4\}$, independent set $\{x_1, x_2, x_3\}$ and $x_1 < x_2 < x_3 < y_1 < y_2 < y_3 < y_4$.

Next we show how to extend induced S_{k-1}^- to an induced S_k^- for an arbitrary $k \geq 5$ (this is done in the full version). Since we deal with finite graphs, a contradiction arises. □

Notice that the class of graphs described in Theorem 8 does not contain all chordal bipartite graphs. To get a greedy matching ordering for such a graph we have to apply the doubly lexical ordering method to the bipartite adjacency matrix instead to the neighborhood matrix, as in the case of general graphs[18, 16].

As far as the orderings of the first three graphs presented in Figure 6 are lexical but not admissible, we can conclude

For every induced subgraph F of a graph G any lexical ordering is a greedy matching ordering if and only if G does not contain an induced subgraph isomorphic to tent, house, domino, sun, hexahedron or hole.

Summarizing we have

Corollary 4. *For a graph G that does not contain an induced subgraph isomorphic to tent, house, domino, sun, hexahedron or hole, a maximum matching can be computed in time* $O(min\{|V|^2, |E| \log |V|\})$. *If G is given together with a lexical ordering then a maximum matching of G can be computed in linear time.*

Acknowledgement. The author is grateful to Andreas Brandstädt and Jeremy Spinrad for many useful discussions. Thanks also to anonymous referees for constructive comments.

References

1. R.P. Anstee and M. Farber, Characterizations of totally balanced matrices, *J. Algorithms*, 5 (1984), 215–230.
2. G. Ausiello, D. D'Atri and M. Moscarini, Chordality properties on graphs and minimal conceptual connections in semantic data models, *J. Computer and System Sciences*, 33 (1986), 179–202.

3. A. Brandstädt, Special graph classes - A survey, *Schriftenreihe des Fachbereichs Mathematik*, SM-DU-199, Universität Duisburg, 1991.
4. A. Brandstädt, F.F. Dragan and F. Nicolai, LexBFS-orderings and powers of chordal graphs, *Discrete Math.*, 171 (1997), 27-42.
5. G. J. Chang and G.L. Nemhauser, The k-domination and k-stability problems on sun-free chordal graphs, *SIAM J. Alg. Discrete Meth.*, 5 (1984), 332-345.
6. E. Dalhaus and M. Karpinski, On the computational complexity of matching and multidimensional matching in chordal and strongly chordal graphs, Manuscript.
7. F.F. Dragan, F. Nicolai and A. Brandstädt, Convexity and HHD-free graphs, *Technical Report* Gerhard-Mercator-Universität - Gesamthochschule Duisburg SM-DU-290, 1995.
8. M. Farber, Characterization of strongly chordal graphs, *Discrete Math.*, 43 (1983), 173-189.
9. M.C. Golumbic, Algorithmic Graph Theory and Perfect Graphs, *Academic Press, New York* 1980.
10. M.C. Golumbic and C.F. Goss, Perfect elimination and chordal bipartite graphs, *J. Graph Theory*, 2 (1978), 155-163.
11. R. Hayward, Weakly triangulated graphs, *Journal of Combin. Theory (B)* , 39 (1985), 200-209.
12. A.J. Hoffman, A.W.J. Kolen and M. Sakarovitch, Totally-balanced and greedy matrices, *SIAM J. Alg. Disc. Meth.*, 6 (1985), 721-730.
13. E. Howorka, A characterization of distance-hereditary graphs, *Quart. J. Math. Oxford* Ser. 2, 28 (1977), 417-420.
14. E. Howorka, A characterization of ptolemaic graphs, *J. Graph Theory*, 5 (1981), 323-331.
15. B. Jamison and S. Olariu, On the semi-perfect elimination, *Advances in Applied Math.* 9 (1988), 364-376.
16. T. Kloks and D. Kratsch, Computing a perfect edge without vertex elimination ordering of a chordal bipartite graph, *Information Processing Letters* 55 (1995) 11-16.
17. L. Lovász and M.D. Plummer, Matching Theory, *Ann. Discrete Math.* 29 (1986).
18. A. Lubiw, Doubly lexical orderings of matrices, *SIAM J. Comput.* 16 (1987), 854-879.
19. R. Paige and R.E. Tarjan, Three partition refinement algorithms, *SIAM J. Comput.*, 16 (1987), 973-989.
20. W.R. Pulleyblank, Matching and Extensions, in *Handbook of Combinatorics* (L.R. Graham, et al., eds.), Elsevier (Amsterdam) 1995, vol. 1, pp. 179-232.
21. M. Queyranne, F. Spieksma and F. Tardella, A general class of greedily solvable linear programs, *3rd IPCO Conference* (G. Rinaldi and L. Wolsey, eds.), (1993) 385-399.
22. D. Rose, R.E. Tarjan and G. Lueker, Algorithmic aspects on vertex elimination on graphs, *SIAM J. Comput.* 5 (1976), 266-283.
23. J.P. Spinrad, Doubly lexical ordering of dense 0-1- matrices, *Information Processing Letters* 45 (1993) 229-235.
24. R.E. Tarjan and M. Yannakakis, Simple linear time algorithms to test chordality of graphs, test acyclicity of hypergraphs, and selectively reduce acyclic hypergraphs, *SIAM J. Comput.* 13 (1984), 566-579.

Off-Line and On-Line Call-Scheduling in Stars and Trees*

Thomas Erlebach[1] and Klaus Jansen[2]

[1] Institut für Informatik, TU München, D–80290 München,
erlebach@informatik.tu-muenchen.de

[2] Fachbereich IV – Mathematik, Universität Trier, Postfach 3825, D–54286 Trier,
jansen@dm3.uni-trier.de

Abstract. Given a communication network and a set of call requests, the goal is to find a minimum makespan schedule for the calls such that the sum of the bandwidth requirements of simultaneously active calls using the same link does not exceed the capacity of that link. In this paper the call-scheduling problem is studied for star and tree networks. Lower and upper bounds on the worst-case performance of List-Scheduling (LS) and variants of it are obtained for call-scheduling with arbitrary bandwidth requirements and either unit call durations or arbitrary call durations. LS does not require advance knowledge of call durations and, hence, is an on-line algorithm. It has performance ratio (competitive ratio) at most 5 in star networks. A variant of LS for calls with unit durations is shown to have performance ratio at most $2\frac{2}{3}$. In tree networks with n nodes, a variant of LS for calls with unit durations has performance ratio at most 6, and a variant for calls with arbitrary durations has performance ratio at most $5 \log n$.

1 Introduction

Call-scheduling problems arise naturally in modern communication networks, e.g., ATM networks. ATM (*asynchronous transfer mode*) is a network protocol that allows high-bandwidth connections with a guaranteed quality of service [15]. A connection request (call) can specify a certain bandwidth requirement, and the network guarantees that, once the connection is established, this bandwidth is available to it as long as it remains active. Consequently, this bandwidth must be reserved on all links along a path that connects the endpoints of the call in the network. The high bandwidth and guaranteed quality of service in ATM networks are essential for upcoming applications like multimedia servers or real-time medical imaging.

Formally, the communication network is given by a connected, undirected graph $G = (V, E)$ such that each edge $e \in E$ has a certain capacity $c(e)$. We assume that all edges have the same capacity, and that this capacity is normalized to 1. A call request r is a tuple (u_r, v_r, b_r, d_r), where u_r and v_r are different nodes

* Partly supported by German Science Foundation (DFG), Contract: SFB 342 TP A7.

of G representing the endpoints of the connection, $b_r \in]0;1]$ is the requested bandwidth, and $d_r \in \mathbb{N}$ is the duration of the call. Given a graph G and a (multi-)set R of call requests, a feasible schedule S assigns to each request $r \in R$ a starting time $t_r \in \mathbb{N}_0$ and an undirected path P_r from u_r to v_r in G such that the sum of bandwidths of simultaneously active calls using the same edge does not exceed the capacity of that edge. Precisely speaking, call r is active during the time interval $[t_r; t_r + d_r[$, and it occupies bandwidth b_r on all edges of P_r during that time. Several active calls can share an edge if the sum of their bandwidth requirements is at most 1.

The length $|S|$ of a schedule S is the latest finishing time of all calls, i.e., $|S| = \max_{r \in R} t_r + d_r$. We denote by $OPT = OPT(R)$ the length of a shortest feasible schedule for R, and by $A(R)$ the length of the schedule produced by algorithm A. Since it is in general $\mathcal{NP}$-hard to compute a minimum makespan schedule [5], one is interested in polynomial-time approximation algorithms with provable performance guarantee. An algorithm A has *performance ratio* at most ρ if $A(R) \le \rho \cdot OPT(R)$ for all request sets R.

If an algorithm does not require advance knowledge of call durations, we refer to it as an *on-line* algorithm even if it requires that all call requests are given to the algorithm at once. Such *batch-style* on-line algorithms can easily be converted into *fully on-line* algorithms, i.e., algorithms that can deal with additional call requests that arrive on-line while other calls have already been scheduled, increasing the competitive ratio by no more than a factor 2 [14, 7]. An on-line algorithm has *competitive ratio* ρ if it always produces a schedule with makespan at most a factor ρ longer than the optimum (off-line) schedule.

If G is a tree, the path P_r is already completely determined by u_r and v_r. For an edge e of a tree network and a request set R we call $L(e) = \sum_{r \in R: e \in P_r} b_r \cdot d_r$ the *load* of edge e. Furthermore, L_{max} is the maximum of $L(e)$ over all $e \in E$. Obviously, L_{max} is a lower bound for the optimum schedule length. A special case of a tree is a graph that consists of a central node c and an arbitrary number of nodes $v_1, v_2, \ldots, v_k$ that are adjacent to c but not adjacent to each other. We refer to such graphs as *stars*. In the following, we will always assume that G is a star or a tree. Scheduling calls with unit bandwidth requirements in stars with unit edge capacities is equivalent to scheduling multiprocessor tasks with prespecified processor allocations if each task requests one or two processors [11].

One of the earliest heuristics for the solution of scheduling problems was List-Scheduling (LS), introduced by Graham [9]. In the call-scheduling context, the input to LS is a star or tree network G and a set R of call requests arranged in a list L. LS starts to schedule calls at time 0. If there is a call r in L such that bandwidth b_r is available on all edges along path P_r, LS schedules the first such call r in L and removes it from L; otherwise, it waits until one of the active calls finishes. This is repeated until all calls have been scheduled.

One important property of list-schedules is that, if a call request r is established at time t_r, it follows that at any time prior to t_r at least one of the edges on path P_r did not have bandwidth $\ge b_r$ available. We will use this property as a tool to prove performance guarantees for list-schedules. Note that this property holds only because there are no precedence constraints for the call requests.

Allowing arbitrary bandwidth requirements, we will show that $DBLS(L) \leq \frac{8}{3}OPT(L)$ for calls with unit durations in stars, that $LS(L) \leq 5 \cdot OPT(L)$ for calls with arbitrary durations in stars, that $LLS(L) \leq 6 \cdot OPT(L)$ for calls with unit durations in trees, and that $LSL(L) \leq 5 \log n \cdot OPT(L)$ for calls with arbitrary durations in trees with n nodes. DBLS, LLS, and LSL are variants of LS that will be defined later.

1.1 Related Work

Wavelength Allocation. The off-line call-scheduling problem with unit durations and unit bandwidth requirements is equivalent to the wavelength allocation problem in all-optical networks with wavelength-division multiplexing, where a minimum makespan schedule corresponds to a wavelength assignment with the minimum number of distinct wavelengths. Routing and wavelength allocation in all-optical networks have received considerable attention lately, see, e.g., [1] and the references contained in there.

A variation of the problem dealing with directed instead of bidirectional calls has also been studied. Here, calls using the same edge can receive the same wavelength if they use the edge in different directions. The best approximation algorithm known up to now requires $\frac{5}{3}L_{max}$ wavelengths in the worst case [12]. It is known that the bidirectional call-scheduling problem with unit durations and unit bandwidths is $\mathcal{NP}$-hard in trees of arbitrary degree, but solvable in polynomial time in trees whose degree is bounded by a constant [5, 6]. The directed version is $\mathcal{NP}$-hard already for binary trees [6].

In the on-line version of the wavelength allocation problem the algorithm is given requests one by one and must assign wavelengths immediately without knowledge about future requests. Bartal and Leonardi [3] obtain deterministic on-line algorithms with competitive ratio $O(\log n)$ for networks with n processors whose topology is that of a tree, a tree of rings, or a mesh. In addition, they present a matching lower bound of $\Omega(\log n)$ for all on-line algorithms for wavelength allocation in meshes, and a lower bound of $\Omega(\frac{\log n}{\log \log n})$ for trees. Note that the on-line version of the wavelength allocation problem corresponds to a call-scheduling problem where the algorithm must assign starting times to call requests one by one before the first call is established. Hence, the lower bounds in [3] do not apply to the call scheduling problem we study in this paper. Furthermore, their algorithms work for the call-scheduling problem only in the case of unit durations and unit bandwidths.

Scheduling File-Transfers. Coffman *et al.* study a file-transfer scheduling problem that corresponds to call-scheduling in a star with varying edge capacities and calls with unit bandwidth requirements and arbitrary durations [4]. They present complexity results for various restricted versions of the problem, approximation results, and distributed implementations. Many of their results for arbitrary edge capacities and unit bandwidth requirements do not apply to our call-scheduling problem with unit edge capacities and arbitrary bandwidth requirements, however.

Previous Work on On-Line Call-Scheduling. Feldmann *et al.* initiated research on on-line call-scheduling in [7] and [8]. They analyze the GREEDY algorithm (equivalent to LS) and show that running GREEDY once on the calls with bandwidth requirements $\leq \frac{1}{2}$ and once on the calls with bandwidth requirements $> \frac{1}{2}$ yields an on-line algorithm for call-scheduling in binary trees with competitive ratio $12 \log n$. In addition, they obtain results for linear array networks, meshes, complete graphs, and graphs with small separators.

2 Approximation Results for Stars

Stars are the subgraphs of trees that are induced by an arbitrary node of the tree and its neighbors. Hence, call-scheduling problems in stars are encountered as subproblems of call-scheduling in trees. Note that there are two kinds of calls in a star G. First, there are calls that connect the central node to one of the other nodes. Second, there are calls that connect two nodes that are both adjacent to the central node. We refer to these calls as 1-calls and 2-calls, respectively.

In the case of calls with unit durations ($d_r = 1$ for all $r \in R$) and unit bandwidths ($b_r = 1$ for all $r \in R$), call-scheduling in a star is equivalent to edge-coloring a multigraph and thus $\mathcal{NP}$-hard [5]. The algorithm from [13] colors any multigraph G with at most $\lfloor 1.1 \cdot OPT(G) + 0.8 \rfloor$ colors and can be used for the call-scheduling problem with the same performance guarantee, even in trees [5]. The equivalence between call-scheduling and edge-coloring is lost once we allow arbitrary bandwidth requirements or arbitrary call durations. It is known, however, that the performance ratio of LS for call-scheduling with unit bandwidth requirements and arbitrary durations in a star is 2 [4, Corollary 12.2].

2.1 Unit Durations and Arbitrary Bandwidth Requirements

In this section we assume that all call durations are 1, while bandwidth requirements can be arbitrary numbers in $]0;1]$. Note that call-scheduling with arbitrary bandwidth requirements is a generalization of bin-packing and hence $\mathcal{NP}$-complete in the strong sense, even if the network is a single link. Theorem 7, which will be proved in Sect. 2.2, implies that the worst-case performance of LS for calls with unit durations is at most 5. The following tighter result can be proved similar to Lemma 2 below.

Theorem 1. *LS has performance ratio at most* 4.875 *for call-scheduling with arbitrary bandwidth requirements and unit durations in stars.*

Given a schedule S computed by LS for a list L of call requests, it turns out that estimates on the performance ratio of LS on that particular instance L depend heavily on the smallest bandwidth requirement of a call that finishes last in S, i.e., at time $|S|$. The following lemmas make this relationship clearer.

Lemma 2. *Let S be a list-schedule for a list L of calls with arbitrary bandwidth requirements and unit durations. If there is a call r with bandwidth requirement $b_r \leq \frac{1}{2}$ that finishes last in S, then $|S| \leq \lceil 3.875 \cdot OPT(L) \rceil$.*

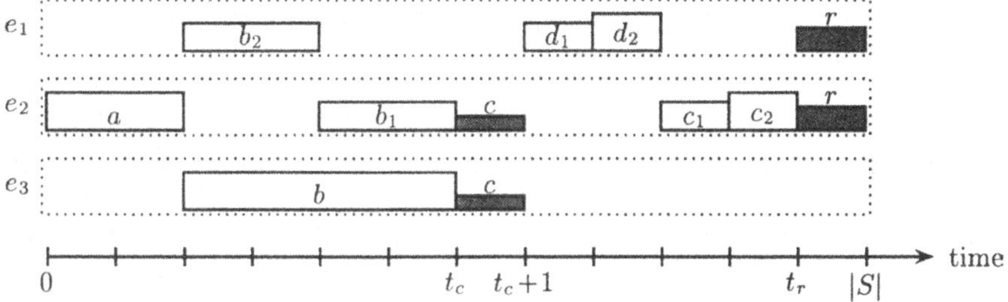

Fig. 1. List-Schedule S, $b_r \leq \frac{1}{2}$, $b_c \leq \frac{1}{3}$

Proof. Assume that r is a 2-call. (If r is a 1-call, the proof is much simpler.) If $b_r \leq \frac{1}{3}$, it follows that at least one of the two edges used by r has less than $\frac{1}{3}$ bandwidth available during at least $\lceil t_r/2 \rceil$ (not necessarily consecutive) time steps prior to t_r. Hence, $OPT > \lceil t_r/2 \rceil \cdot \frac{2}{3} \geq \frac{t_r}{3}$ and $|S| = t_r + 1 \leq 3 \cdot OPT$ in this case. Therefore, assume that $\frac{1}{3} < b_r \leq \frac{1}{2}$. Consider all calls with bandwidth requirement $\leq \frac{1}{3}$ that use at least one edge that is also used by r. Assume that there are such calls, and let c be a call with latest finishing time $t_c + 1$ among them. Furthermore, assume that c is a 2-call and that c uses only one edge that is also used by r. (The cases that no call c exists, that c is a 1-call, or that c is a 2-call using the same edges as r can be treated in a similar way.) Let the edges used by r be e_1 and e_2, and let the edges used by c be e_2 and e_3. Introduce the following variables (cf. Fig. 1): $a =$ (number of) time steps during which c is blocked on e_2; $b =$ time steps during which c is blocked on e_3, but not on e_2; $b_1 =$ time steps prior to t_c during which r is blocked on e_2, but not c; $b_2 =$ time steps prior to t_c during which r is blocked on e_1, but not on e_2; $c_1 =$ time steps after t_c during which r is blocked on e_2 by a single call; $c_2 =$ time steps after t_c during which r is blocked on e_2 by a combination of at least two calls; $d_1 =$ time steps after t_c during which r is blocked on e_1 by a single call, but not blocked on e_2; $d_2 =$ time steps after t_c during which r is blocked on e_1 by a combination of at least two calls, but not blocked on e_2. Note that the time steps accounted for by these variables need not be consecutive. Using these definitions, it is clear that $|S| = a + b_1 + b_2 + c_1 + c_2 + d_1 + d_2 + 2$. If $a + c_2 + d_2 \leq \frac{3}{8} OPT$, the easily observed inequalities $OPT > \frac{2}{3}(b_1 + b_2)$ (follows from $b \geq b_1 + b_2$ and the load on e_3), $OPT \geq c_1 + 1$, and $OPT \geq d_1 + 1$ imply $|S| \leq 3.875 \cdot OPT$. If $a + c_2 + d_2 > \frac{3}{8} OPT$, consider the sum of the loads on e_1 and e_2 (note that there is load $> \frac{1}{2}$ on e_1 or e_2 at time t_c):

$$L(e_1) + L(e_2) > \frac{2}{3}(a + c_2 + d_2) + \frac{1}{2}(b_1 + b_2 + c_1 + d_1 + 1) \qquad (1)$$

Since $L(e_1) + L(e_2) \leq 2 \cdot OPT$, we get $|S| - 1 < 4 \cdot OPT - \frac{1}{3}(a + c_2 + d_2) \leq 3.875 \cdot OPT$. Hence, $|S| \leq \lceil 3.875 \cdot OPT \rceil$. □

Lemma 3. *There are stars and lists of calls with arbitrary bandwidth requirements and unit durations such that the schedule computed by LS is longer than the optimum schedule by a factor arbitrarily close to 3.7. The call scheduled last by LS has bandwidth requirement $\frac{1}{2}$.*

Proof. We use a well-known worst-case input to first-fit bin-backing (cf. [10, pp. 211-213]) with ratio $\approx \frac{17}{10}$ to construct a call-scheduling input with ratio ≈ 3.7. For any positive integer ℓ divisible by 17, we obtain a list L of calls with optimum schedule length $10\ell/17 + 1$ and list-schedule length $37\ell/17 + 1$. The worst-case input to first-fit bin-packing consists of $30\ell/17$ items with sizes approximately $\frac{1}{6}$, $\frac{1}{3}$, and $\frac{1}{2}$. The optimum packing uses at most $10\ell/17 + 1$ bins for these items, while first-fit requires exactly ℓ bins. Furthermore, each bin is filled to at least $\frac{1}{2} + \delta$ in the packing produced by first-fit, where δ is a parameter that must be chosen sufficiently small.

A list of $30\ell/17$ calls with bandwidth requirements equal to the item sizes in such a bin-packing instance is called a 1.7-*list*. Let $\ell' = 10\ell/17$. The input list L for LS contains calls in a star with $2 + 2\ell' + 3\ell'(\ell' + 1)$ nodes adjacent to the central node c. These nodes are denoted $u, u_1, \ldots, u_{\ell'}, v_0, \ldots, v_{\ell'}$, and $w_{i,j}$ for $0 \le i \le \ell'$ and $1 \le j \le 3\ell'$. The list L contains the following calls ($\varepsilon < 1/(6\ell')$):

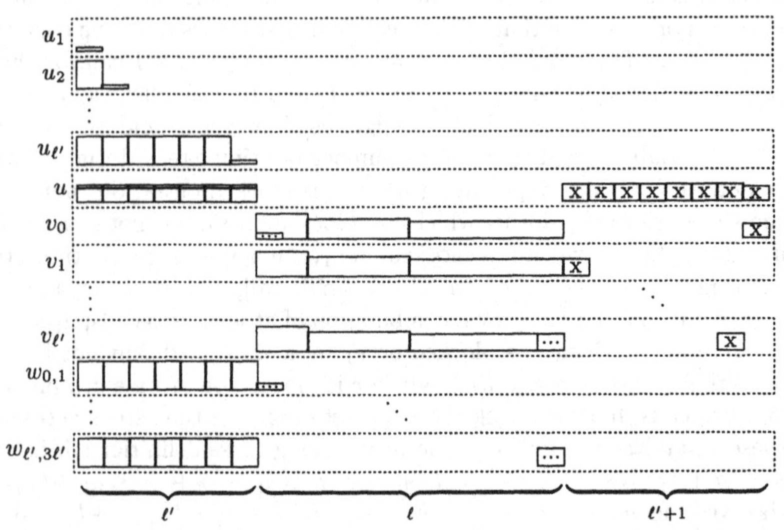

Fig. 2. Example with $LS(L)/OPT(L) \approx 3.7$

1. For $1 \le i \le \ell'$, $i - 1$ calls with bandwidth 1 connecting u_i and c.
2. For $1 \le i \le \ell'$, a call with bandwidth $\frac{1}{2} - \varepsilon$ connecting u and c and a call with bandwidth 3ε connecting u and u_i.

3. For $0 \le i \le \ell'$ and $1 \le j \le 3\ell'$, ℓ' calls with bandwidth 1 connecting $w_{i,j}$ and c.
4. For $0 \le i \le \ell'$, a 1.7-list of calls connecting v_i and some node $w_{i,j}$, such that no two calls connect v_i to the same $w_{i,j}$.
5. For $1 \le i \le \ell'$, a call with bandwidth $\frac{1}{2} + \varepsilon$ connecting u and v_i.
6. A call with bandwidth $\frac{1}{2}$ connecting u and v_0.

It is easy to verify that LS will produce the schedule sketched in Fig. 2. The edge $\{u, c\}$ is occupied by a call with bandwidth $\frac{1}{2} - \varepsilon$ and a call with bandwidth 3ε during each of the first ℓ' time steps. All 1.7-lists are scheduled in time steps ℓ' to $\ell' + \ell - 1$, because every call in a 1.7-list is blocked during the first ℓ' time steps on an edge $\{w_{i,j}, c\}$. The calls with bandwidth $\frac{1}{2} + \varepsilon$ connecting u and v_i are scheduled in time steps $\ell' + \ell$ to $2\ell' + \ell - 1$, because they are blocked on $\{u, c\}$ during the first ℓ' time steps and subsequently on $\{v_i, c\}$ during the next ℓ time steps. (Recall that the 1.7-lists occupy at least $\frac{1}{2} + \delta$ bandwidth in time steps ℓ' to $\ell' + \ell - 1$ on all edges $\{v_i, c\}$.) Finally, the call $(u, v_0, \frac{1}{2}, 1)$ is scheduled at time step $2\ell' + \ell$. Hence, $LS(L) = 37\ell/17 + 1$.

On the other hand, it is clear that L can be scheduled in $\ell' + 1$ time steps. In particular, on edge $\{u, c\}$ one can schedule one call with bandwidth $\frac{1}{2} - \varepsilon$ and one call with bandwidth $\frac{1}{2} + \varepsilon$ during each of the first ℓ' time steps. Since ε has been chosen small enough, all calls with bandwidth 3ε together with the call $(u, v_0, \frac{1}{2}, 1)$ can then be scheduled together at time ℓ'. The 1.7-lists can be scheduled in $\ell' + 1$ time steps, such that one of the time steps has bandwidth $\ge \frac{1}{2} + \varepsilon$ available. Hence, the schedule for the 1.7-list on v_i can be arranged such that the call connecting u and v_i is scheduled at that time step. Finally, the remaining 1-calls can be filled in without making the schedule longer. Therefore, $OPT(L) \le \ell' + 1$. The ratio between $LS(L)$ and $OPT(L)$ is at least $\frac{37\ell+17}{10\ell+17}$, which is arbitrarily close to 3.7 for large ℓ. □

Note that Lemma 2 and Lemma 3 show that the exact bound on the worst-case performance ratio of LS lies between 3.7 and 3.875 if a call with bandwidth requirement $\le \frac{1}{2}$ finishes last in the list-schedule. Next, we investigate the case that a call with bandwidth requirement $\le \frac{1}{k}$ for some $k \ge 3$ finishes last in a list-schedule.

Lemma 4. *Let S be a list-schedule for a list L of calls with arbitrary bandwidth requirements and unit durations. If there is a call r with bandwidth requirement $b_r \le \frac{1}{k}$ for some $k \ge 3$, $k \in \mathbb{N}$, that finishes last in S, then $|S| \le \left\lceil \frac{2k}{k-1} OPT \right\rceil$.*

Proof. Since r is blocked during the first t_r time steps, at least one of the edges used by r has less than $b_r \le \frac{1}{k}$ bandwidth available during at least $\lceil t_r/2 \rceil$ time steps. Hence, the load on that edge is greater than $\lceil t_r/2 \rceil \cdot \frac{k-1}{k} < OPT$, and we obtain $t_r < \frac{2k}{k-1} OPT$ and, consequently, $|S| = t_r + 1 \le \left\lceil \frac{2k}{k-1} OPT \right\rceil$. □

In [10, pp. 217–219], first-fit bin-packing is analyzed under the restriction that all items have size $\le \alpha$ for some $\alpha \le \frac{1}{2}$. With $k' = \lfloor 1/\alpha \rfloor$, it is shown that, for any list L of items with sizes $\leq \alpha$, $FF(L) \leq \frac{k'+1}{k'} OPT + 2$ and that there are examples with $FF(L) \geq \frac{k'+1}{k'} OPT - \frac{1}{k'}$. We adapt the construction of these examples to obtain call-scheduling inputs that show that the bound from Lemma 4 is tight.

Lemma 5. *For every $k \geq 3$, $k \in \mathbb{N}$, there are stars and lists L of calls with unit durations and bandwidth requirements $\leq \frac{1}{k-1}$ such that a call with bandwidth requirement $\leq \frac{1}{k}$ finishes last in the list-schedule for L and $\frac{LS(L)}{OPT(L)}$ is arbitrarily close to $\frac{2k}{k-1}$.*

Proof. Let $k' = k-1$. Let ℓ be a positive integer such that k' divides $\ell(k'+1)-1$. We construct a list L of calls with optimum schedule length $\ell + 1$ and list-schedule length $2\frac{\ell(k'+1)-1}{k'}$. Let $b_j^\delta = 1/(k'+1) - k'^{2j+1}\delta$ ($j = 1, 2, \ldots, \ell - 1$) and $a_{1j}^\delta = \cdots = a_{k'j}^\delta = 1/(k'+1) + k'^{2j}\delta$ ($j = 1, 2, \ldots, \ell$), where δ is chosen sufficiently small. A list of calls with exactly one call with bandwidth requirement b_j^δ for each $j = 1, 2, \ldots, \ell - 1$ and one call with bandwidth requirement a_{ij}^δ for each $i = 1, 2, \ldots, k'$ and $j = 1, 2, \ldots, \ell$ is called a *δ-list* if the calls are ordered as follows: the a_{ij}^δ-calls appear in order of non-increasing bandwidths, the b_j^δ-calls appear in order of strictly increasing bandwidths, there are k' a_{ij}^δ-calls between every pair of successive b_j^δ-calls, and the call with bandwidth requirement $b_{\ell-1}^\delta$ is the second call in the list. Note that a δ-list contains $l(k'+1) - 1$ calls.

Consider a δ-list L_δ such that all calls in L_δ are 1-calls using the same edge e. Since first-fit bin-packing is equivalent to LS for calls with unit durations on one edge, [10, pp. 217–219] implies $LS(L_\delta) = \frac{\ell(k'+1)-1}{k'}$ and $OPT(L_\delta) = \ell$. Furthermore, LS schedules exactly k' calls in every time step, and no time step has more than $1/(k'+1) - k'^3\delta$ bandwidth available on edge e in the resulting schedule. In addition, the call with bandwidth $b_1^\delta < \frac{1}{k'+1} = \frac{1}{k}$ is scheduled in the last time step.

We use $\ell(k'+1) - 1$ such δ-lists with 1-calls on separate edges (one edge for each δ-list). These δ-lists come first in the list L. At the end of L, we append one additional δ'-list $L_{\delta'}$, with δ' such that $k'^{2\ell-1}\delta' < k'^3\delta$. Let v be a node of the star that has not been used by any of the 1-calls. The calls in $L_{\delta'}$ all connect the node v to one of the nodes used by the $\ell(k'+1) - 1$ δ-lists, such that no two calls in $L_{\delta'}$ connect v to the same node v'. Obviously, LS will schedule the calls in $L_{\delta'}$ in $\frac{\ell(k'+1)-1}{k'}$ successive time steps starting from $\frac{\ell(k'+1)-1}{k'}$. Hence, the list-schedule has length $2\frac{\ell(k'+1)-1}{k'}$, whereas $OPT = \ell + 1$. Therefore, the performance ratio of LS is arbitrarily close to $\frac{2(k'+1)}{k'} = \frac{2k}{k-1}$. □

While our best general upper bound for the worst-case performance of LS for calls with unit durations and arbitrary bandwidth requirements in stars is 4.875, a slightly modified algorithm gives a much better performance guarantee. The algorithm Decreasing-Bandwidth List-Scheduling (DBLS) behaves just like standard List-Scheduling, but it sorts the given list of call requests according to non-increasing bandwidth requirements before it begins to schedule the calls.

Theorem 6. *DBLS has performance ratio at most $\frac{8}{3}$ for call-scheduling with arbitrary bandwidth requirements and unit durations in stars. There are instances for which the performance ratio of DBLS is arbitrarily close to $\frac{22}{9}$.*

Proof. First, we prove the upper bound. Given a set R of call requests, let $L = L(R)$ be the list of call requests obtained by sorting R in order of non-increasing bandwidth, and denote by S the schedule produced by DBLS. Note that a call c scheduled at time t_c in S is blocked during all time steps prior to t_c entirely by calls that precede c in L. (This holds only because we assume unit call durations.) For a call $c \in R$, denote by L_c the sublist of L that contains all requests from the beginning of the list up to and including c. Taking into account the above argument, it is clear that all calls in L_c are scheduled at the same time step in a list-schedule for L and in a list-schedule for L_c.

We claim that the finishing time $t_c + 1$ of any call $c \in R$ with bandwidth requirement $b_c > \frac{1}{3}$ satisfies $t_c + 1 \leq 2 \cdot OPT(R)$. If c has bandwidth requirement $b_c > \frac{1}{2}$, no two calls in L_c can be scheduled at the same time if they use the same edge. Therefore, scheduling L_c is just like scheduling calls with unit bandwidth requirements, and [4, Corollary 12.2] implies $t_c + 1 = LS(L_c) \leq 2 \cdot OPT(L_c) \leq 2 \cdot OPT(R)$. If c has bandwidth requirement b_c satisfying $\frac{1}{3} < b_c \leq \frac{1}{2}$, note that during all time steps prior to t_c more than $1 - b_c$ bandwidth was occupied by other calls from L_c on at least one of the edges used by c. Hence, an edge e was occupied to this extent during at least $\lceil t_c/2 \rceil$ time steps prior to t_c. During each such time step, that edge must have been used either by a single call occupying more than $1 - b_c$ bandwidth or by two calls occupying at least b_c bandwidth each. It is clear that even an optimum schedule requires $\lceil t_c/2 \rceil$ time steps for these calls and an additional time step for c, and thus $t_c + 1 \leq 2 \cdot OPT(L_c) \leq 2 \cdot OPT(R)$.

Now let r be a call with maximum bandwidth requirement among the calls that finish last in S. If $b_r > \frac{1}{3}$, the previous argument shows that $|S| = t_r + 1 \leq 2 \cdot OPT$. If $b_r \leq \frac{1}{4}$, note that an edge used by r has less than b_r bandwidth available during at least $\lceil t_r/2 \rceil$ time steps prior to t_r. Hence, $\lceil t_r/2 \rceil \cdot (1 - b_r) < OPT$, implying $t_r < \frac{2}{1-b_r} OPT$ and, therefore, $|S| = t_r + 1 \leq \left\lceil \frac{2}{1-b_r} OPT \right\rceil$. With $b_r \leq \frac{1}{4}$, this implies $|S| \leq \lceil \frac{8}{3} OPT \rceil$.

Finally, consider the case that $\frac{1}{4} < b_r \leq \frac{1}{3}$. If r is a 1-call, the edge used by r is occupied to more than $\frac{2}{3}$ during all time steps prior to t_r, and we have $\frac{2}{3} t_r < OPT$, implying $|S| \leq \lceil \frac{3}{2} OPT \rceil$. If r is a 2-call, denote by C the set of all calls with bandwidth requirement $> \frac{1}{3}$ that use at least one edge also used by r. If C is empty, r is blocked during the first t_r time steps entirely by calls d with bandwidth requirement b_d satisfying $b_r \leq b_d \leq \frac{1}{3}$. In addition, it is clear that two such calls are not enough to block r, because $2 \cdot \frac{1}{3} + b_r \leq 1$. Therefore, whenever r is blocked on an edge during one of the first t_r time steps, that edge is occupied to at least $3b_r$. Since r is blocked on an edge during at least $\lceil t_r/2 \rceil$ time steps, we have $\lceil t_r/2 \rceil \cdot 3b_r < OPT$. This implies $t_r + 1 \leq \left\lceil \frac{2}{3b_r} OPT \right\rceil \leq \lceil \frac{8}{3} OPT \rceil$, where the last inequality follows from $b_r > \frac{1}{4}$.

If C is not empty, let c be a call with the latest finishing time among all calls in C. Note that $t_c + 1 \leq 2 \cdot OPT$. Furthermore, note that starting from $t_c + 1$

call r is blocked entirely by calls d with bandwidth requirement b_d satisfying $b_r \leq b_d \leq \frac{1}{3}$, and that three such calls are necessary in each time step to block r. Hence, the sum of the loads on the two edges used by r is more than $(t_c+1)(1-b_r)+(t_r-t_c-1)3b_r+2b_r < 2 \cdot OPT$. This can simply be transformed into $3b_r(t_r-t_c-1+\frac{2}{3}+t_c+1) < 2\cdot OPT+(t_c+1)(4b_r-1)$. Using $t_c+1 \leq 2\cdot OPT$ and $4b_r-1 \geq 0$, we obtain $t_r+\frac{2}{3} < \frac{8}{3}\cdot OPT$ and, consequently, $t_r+1 \leq \frac{8}{3}\cdot OPT$. This concludes the proof of the upper bound.

Now we give the construction of the instances L that provide the lower bound, using a well-known family of worst-case instances I for first-fit-decreasing bin-packing with $FFD(I) = \frac{11}{9}OPT(I)$ [10, p. 220, Fig. 5.40]. The calls in L have bandwidth requirements $\alpha = \frac{1}{2}+\varepsilon$, $\beta = \frac{1}{4}+2\varepsilon$, $\gamma = \frac{1}{4}+\varepsilon$, and $\delta = \frac{1}{4}-2\varepsilon$. For a given $n \in \mathbb{N}$, let $N = 62208n^5+3888n^3+30n$ and $M = 5184n^4+252n^2+1$, and consider a star with $N+M$ edges $e_1,\ldots,e_N, f_1,\ldots,f_M$. L contains the following calls: (1) for $i = 1,\ldots,N$, we have $6n$ calls with bandwidth α using only edge e_i; (2) for $i = 1,\ldots,M$ and $j = 0,\ldots,6n-1$, we have one call with bandwidth α using edges f_i and $e_{N-(i-1)\cdot 6n-j}$; (1') for $i = 1,\ldots,N$, we have $6n$ calls with bandwidth β using only edge e_i; (2') for $i = 1,\ldots,M$ and $j = 0,\ldots,6n-1$, we have one call with bandwidth β using edges f_i and $e_{N-M\cdot 6n-(i-1)\cdot 6n-j}$; (3) for $i = 1,\ldots,N-M\cdot 12n = 864n^3+18n$ and $j = 0,\ldots,6n-1$, we have one call with bandwidth γ using edges e_i and $f_{M-(i-1)\cdot 6n-j}$; (4) for $i = 1,\ldots,M-(864n^3+18n)\cdot 6n = 144n^2+1$ and $j = 0,\ldots,6n-1$, we have one call with bandwidth γ using edges f_i and $e_{N-M\cdot 12n-(i-1)\cdot 6n-j}$; (5) for $i = 1,\ldots,12n$ and $j = 0,\ldots,12n-1$, we have one call with bandwidth δ using edges e_i and $f_{2+(i-1)12n+j}$; (6) for $j = 0,\ldots,12n-1$, we have one call with bandwidth δ using edges f_1 and e_{1+j}. It is not difficult to show that $DBLS(L) = 22n$ and $OPT(L) = 9n+1$. An optimum schedule can combine calls such that the full capacity of the edges is exploited most of the time. (Note that $\alpha+\gamma+\delta = 1$ and $2\beta+2\delta = 1$.) Details are omitted. □

2.2 Arbitrary Durations and Arbitrary Bandwidth Requirements

In this section, call durations can be arbitrary positive integers, and bandwidth requirements can be arbitrary numbers in $]0;1]$.

Theorem 7. *If S is the schedule computed by LS for a list L of call requests with arbitrary durations and bandwidth requirements in a star, then $LS(L) \leq 5\cdot OPT(L)$. If there is a call with bandwidth requirement $\leq \frac{1}{2}$ that finishes last in S, then $|S| \leq 4\cdot OPT(L)$. If there is a call with bandwidth requirement $\leq \frac{1}{3}$ that finishes last in S, then $|S| \leq 3\cdot OPT(L)$.*

Proof. Let r be a call with the smallest bandwidth requirement b_r among all calls that finish last in S, i.e., at time $|S|$. Since call r is blocked during all time steps prior to t_r, the load on at least one edge used by r is more than $\lceil t_r/2\rceil \cdot (1-b_r)+d_r b_r < OPT$. This implies $t_r \leq \frac{2}{1-b_r}OPT - \frac{2b_r}{1-b_r}d_r$, and we obtain $t_r+d_r \leq \frac{2}{1-b_r}OPT+\frac{1-3b_r}{1-b_r}d_r$. For $b_r \leq \frac{1}{3}$, we have $1-3b_r \geq 0$ and, using

$d_r \leq OPT$, obtain $|S| = t_r + d_r \leq \frac{2+(1-3b_r)}{1-b_r} OPT = 3 \cdot OPT$; for $\frac{1}{3} < b_r \leq \frac{1}{2}$, $\frac{2}{1-b_r}$ is at most 4, and with $1 - 3b_r < 0$ we obtain $|S| = t_r + d_r \leq 4 \cdot OPT$.

Assume now that $b_r > \frac{1}{2}$ and that r is a 2-call; if r is a 1-call, similar arguments can be applied. Consider all calls with bandwidth requirement $\leq \frac{1}{2}$

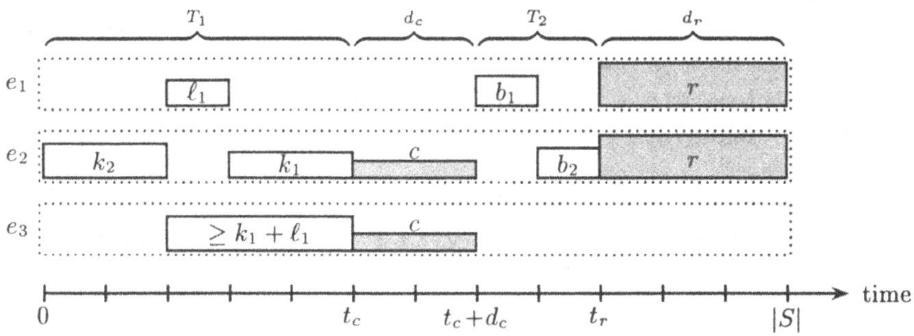

Fig. 3. List-Schedule S, $b_r > \frac{1}{2}$, $b_c \leq \frac{1}{2}$

that use at least one edge that is also used by r. If there is no such call, at least one of the edges used by r is blocked by a call with bandwidth $> \frac{1}{2}$ in at least $\lceil t_r/2 \rceil + d_r \leq OPT$ time steps and, consequently, $|S| = t_r + d_r \leq 2 \cdot OPT$. Otherwise, let c be a call with the latest finishing time $t_c + d_c$ among all such calls. Assume that c is a 2-call that uses only one edge that is also used by r. (The cases that c is a 1-call and that c is a 2-call that uses the same edges as r can be treated similarly.) Furthermore, assume that c finishes before r is established. (Otherwise, $|S| \leq 5 \cdot OPT$ follows directly from $t_c + d_c \leq 4 \cdot OPT$ and $d_r \leq OPT$.) Let the edges used by r be e_1 and e_2, and let the edges used by c be e_2 and e_3. The list-schedule is partitioned into the following disjoint time intervals: (A) T_1 time steps from the beginning of the schedule until t_c (the time when call c is scheduled), (B) d_c time steps during which call c is active, (C) T_2 time steps from the finishing time of c until t_r (the time when call r is scheduled), and (D) d_r time steps during which call r is active. Obviously, $|S| = T_1 + d_c + T_2 + d_r$. Introduce the following variables (cf. Fig. 3): b_1 = number of time steps in part (C) during which r is blocked on e_1, but not on e_2; b_2 = number of time steps in part (C) during which r is blocked on e_2; k_1 = number of time steps in part (A) during which r is blocked on e_2, but not c; k_2 = number of time steps in part (A) during which c (and r) is blocked on e_2; ℓ_1 = number of time steps in part (A) during which r is blocked on e_1, but not on e_2. Considering the load on edge e_3, we obtain $(k_1 + \ell_1)(1 - b_c) + d_c b_c \leq OPT$. This implies $k_1 + \ell_1 \leq \frac{1}{1-b_c} OPT - \frac{b_c}{1-b_c} d_c$. Adding d_c on both sides of this inequality, we get $k_1 + \ell_1 + d_c \leq \frac{1}{1-b_c} OPT + \frac{1-2b_c}{1-b_c} d_c$. Since $d_c \leq OPT$ and $1 - 2b_c \geq 0$, this implies $k_1 + \ell_1 + d_c \leq \frac{1+(1-2b_c)}{1-b_c} OPT = 2 \cdot OPT$. In addition, it is easy to observe that $k_2 + b_2 \le 2 \cdot OPT$ and $b_1 + d_r \le OPT$. Taking into account that $|S| = k_1 + k_2 + \ell_1 + d_c + b_1 + b_2 + d_r$, these inequalities can be combined to obtain $|S| \le 5 \cdot OPT$. □

For the case that a call with bandwidth requirement $\le \frac{1}{3}$ finishes last in a list-schedule, the following lemma shows that the upper bound 3 on the worst-case performance of LS is tight.

Lemma 8. *For arbitrary $k > 2$, $k \in \mathbb{N}$, there are stars and lists of call requests with arbitrary durations and bandwidth requirements $\le \frac{1}{k-1}$ such that a call with bandwidth requirement $\le \frac{1}{k}$ finishes last in the list-schedule and the performance ratio of LS is arbitrarily close to 3.*

Proof. Fix arbitrary integers $k > 2$ and $\ell > 1$. We construct a list L of call requests such that $LS(L) = 3\ell$, $OPT(L) = \ell + 1$, and the call that finishes last in the list-schedule for L has bandwidth requirement $\frac{1}{k}$.

The star used for the construction has $k\ell + 3$ nodes: the central node c and nodes u, v, $u_1, \ldots, u_\ell$, $v_1, \ldots, v_{(k-1)\ell}$ adjacent to c. The list L contains the following call requests ($\varepsilon \ll 1$):

(1) For $i = 1, \ldots, \ell$: $k-1$ calls $(u, c, \frac{1}{k}, 1)$, $k(i-1)$ calls $(u_i, c, \frac{1}{k}, 1)$, and one call $(u_i, u, \varepsilon, 1)$.
(2) For $i = 1, \ldots, (k-1)\ell$: $k\ell$ calls $(v_i, c, \frac{1}{k}, 1)$.
(3) For $i = 0, \ldots, \ell - 1$: for $j = 1, \ldots, k-1$: one call $(v, v_{i(k-1)+j}, \beta_{i,j}, 1)$.
(4) One call $z = (u, v, \frac{1}{k}, \ell)$.

The bandwidth requirements $\beta_{i,j}$ are defined by $\beta_{i,1} = \frac{1}{k} + k\delta_i$ and $\beta_{i,2} = \cdots = \beta_{i,k-1} = \frac{1}{k} - \delta_i$, where $\delta = \delta_0$ is chosen sufficiently small and $\delta_{i+1} = \frac{k-2}{k}\delta_i$.

What schedule is produced by LS for the list L? The calls (1) fill the edge $\{u, c\}$ to $\frac{k-1}{k} + \varepsilon$ during the first ℓ time steps. Each of the calls with bandwidth ε is blocked on one of the edges $\{u_i, c\}$ in all time steps before its starting time. The calls (2) fill the edges $\{v_i, c\}$ completely during the first ℓ time steps. The calls (3) are scheduled in time steps ℓ to $2\ell - 1$, because each call is blocked on a different edge $\{v_i, c\}$ during the first ℓ time steps and blocked on the edge $\{v, c\}$ from time step ℓ up to its starting time. Exactly $k-1$ calls (3) are scheduled in each time step, because their bandwidths add up to $\frac{k-1}{k} + 2\delta_i$ and, therefore, block all subsequent calls (3). Finally, call z is scheduled at time 2ℓ, because it is blocked on $\{u, c\}$ during the first ℓ time steps and on $\{v, c\}$ during the second ℓ time steps. Hence, $LS(L) = 3\ell$.

In an optimum schedule, call z is scheduled at time 0. In each of the first ℓ time steps, $k-1$ calls from (1) using edge $\{u, c\}$ and with bandwidth requirement $\frac{1}{k}$ can be scheduled together with z. All the calls from (1) with bandwidth ε are scheduled together at time ℓ. The remaining calls from (1) can easily be scheduled in free time slots during the first ℓ time steps.

Among the calls from (3), the $k-1$ calls with bandwidth requirements $\beta_{i,2}, \ldots, \beta_{i,k-1}, \beta_{i+1,1}$ are scheduled together in time step i, for $0 \le i \le \ell - 1$. (For $i = \ell - 1$, there is no call with bandwidth $\beta_{i+1,1}$, and only $k-2$ of the calls from (3) are scheduled in time step $\ell-1$.) The bandwidths of the calls from (3) scheduled during one of the time steps $0, \ldots, \ell-1$ add up to at most $\frac{k-1}{k}$. Hence, they can be scheduled concurrently with call z. The call with bandwidth $\beta_{0,1}$ is scheduled at time ℓ. The calls from (2) can easily be scheduled in the remaining free time slots during the first $\ell+1$ time steps. Therefore, $OPT = \ell + 1$. □

3 Approximation Results for Trees

3.1 Unit Durations and Arbitrary Bandwidth Requirements

It is known that the performance of LS can be arbitrarily bad in trees or even in chains if arbitrary bandwidth requirements are allowed. Feldmann *et al.* give a list of call requests with unit durations on a chain with $n+1$ nodes such that the performance ratio of LS is $\Omega(n)$ [7]. Therefore, we consider a variation of the basic List-Scheduling algorithm. Pick an arbitrary node of the tree network as the root and assign each node of the tree a *level* according to its distance from the root. (The root has level 0.) Let m_r be that node on P_r (the path corresponding to call r) whose level is minimum among all nodes on P_r. The level of a call r is defined to be equal to the level of the node m_r. We consider the Level-List-Scheduling algorithm (LLS), which is identical to List-Scheduling except that it sorts the list of calls according to non-decreasing levels before it starts to schedule the calls.

Theorem 9. *LLS has performance ratio at most* 6 *for call-scheduling with arbitrary bandwidth requirements and unit durations in trees.*

Proof. Let S be a schedule computed by LLS for a given set R of call requests. First, we show that any call r with bandwidth requirement $b_r \leq \frac{1}{2}$ finishes no later than at time $4 \cdot OPT$. To see this, consider the node m_r, and let e_1 and e_2 be the edges incident to m_r that are used by r. (If r uses only one edge incident to m_r, it can be proved by similar arguments that $t_r + 1 \leq 2 \cdot OPT$.) It is clear that call r is blocked either on edge e_1 or on edge e_2 by calls with equal or smaller level during all time steps prior to t_r. Hence, at least one of these edges has less than $\frac{1}{2}$ bandwidth available during at least $\lceil \frac{t_r}{2} \rceil$ time steps prior to t_r. Therefore, $OPT > \frac{1}{2} \cdot \lceil \frac{t_r}{2} \rceil$ and, consequently, $t_r + 1 \leq 4 \cdot OPT$.

Now, let r be a call with minimum bandwidth requirement b_r among all calls that finish last in S. If $b_r \leq \frac{1}{2}$, the argument above implies $|S| \leq 4 \cdot OPT$. Therefore, assume that $b_r > \frac{1}{2}$. Let e_1 and e_2 be the edges incident to m_r that are used by r. Again, it is clear that call r is blocked either on edge e_1 or on edge e_2 by calls with equal or smaller level during all time steps prior to t_r. Let c be a call with bandwidth requirement $b_c \leq \frac{1}{2}$ that has the latest finishing time among all such calls. (If no such call exists, call r is blocked only by calls with smaller or equal level and with bandwidth requirements $> \frac{1}{2}$, and $|S| \leq 2 \cdot OPT$.) The argument above implies $t_c + 1 \leq 4 \cdot OPT$, and $t_r - t_c \leq 2 \cdot OPT$ follows from the fact that call r is blocked by calls with bandwidth requirements $> \frac{1}{2}$ either on e_1 or on e_2 during all time steps from $t_c + 1$ to t_r. Combining these inequalities, we obtain $|S| = t_r + 1 \leq 6 \cdot OPT$. □

3.2 Arbitrary Durations and Arbitrary Bandwidth Requirements

Given a tree network T with n nodes, we use a well-known technique [2] based on a tree separator [16] to assign levels to the nodes of T as follows:

1. Choose a node v whose removal splits T into subtrees $T_1, T_2, \ldots, T_k$ with at most $n/2$ nodes each. Assign node v the level 0.
2. In each subtree T_i with n_i nodes, find a node v_i whose removal splits T_i into subtrees with at most $n_i/2$ nodes. Assign all such nodes v_i the level 1.
3. Continue recursively until every node of T is assigned a level.

This way every node of T is assigned a level ℓ, $0 \leq \ell \leq \log n$. For each call request $r = (u, v, b, d)$ in T, the level of r is defined to be the smallest level of all nodes on the path P_r from u to v. In addition, the *root* node of r is defined to be that node on P_r whose level is equal to the level of r. (Note that the root node is uniquely determined; if two nodes of equal level are on a path P, there must exist a node of smaller level on P.) Given a list L of call requests in T, let L_ℓ be the sublist of L that contains all call requests of level ℓ, $0 \leq \ell \leq \log n$. Note that scheduling a list L_ℓ is equivalent to scheduling calls in a number of disjoint stars: calls in L_ℓ with the same root node intersect if and only if they use the same edge incident to that root node; calls in L_ℓ with different root nodes never intersect. Therefore, $LS(L_\ell) \leq 5 \cdot OPT(L_\ell)$ as a consequence of Theorem 7. The algorithm List-Scheduling by Levels (LSL) simply uses List-Scheduling to schedule the lists L_ℓ, $0 \leq \ell < \log n$ one after another. ($L_{\log n}$ is empty, because the root node of a call can never have level $\log n$.) LSL begins to schedule $L_{\ell+1}$ only when all calls from L_ℓ have finished. Note that LSL is an on-line algorithm because it does not require advance knowledge of call durations. Hence, we obtain the following theorem:

Theorem 10. *LSL is an on-line algorithm for scheduling calls with arbitrary bandwidth requirements and arbitrary durations in trees. Its competitive ratio is at most* $5 \log n$.

4 Conclusion

We have analyzed List-Scheduling and variants of it for the call-scheduling problem in stars and trees. It was shown that variants of LS have good, constant performance ratio in all cases except for call-scheduling with arbitrary bandwidths and arbitrary durations in trees, where the ratio is $5 \log n$. Hence, List-Scheduling variants, which are easy to implement, can be applied in practice to schedule connections in networks with guaranteed quality of service.

Regarding possible directions for future research, it will be interesting to study call-scheduling algorithms for the cases that edge capacities may vary, that directed and undirected calls as well as calls with release times are allowed, and that the topology of the network is such that multiple paths between the endpoints of each connection exist.

References

1. Y. Aumann and Y. Rabani. Improved bounds for all optical routing. In *Proceedings of the 6th Annual ACM-SIAM Symposium on Discrete Algorithms SODA '95*, pages 567–576, 1995.
2. B. Awerbuch, Y. Bartal, A. Fiat, and A. Rosén. Competitive non-preemptive call control. In *Proceedings of the 5th Annual ACM–SIAM Symposium on Discrete Algorithms SODA '94*, pages 312–320, 1994.
3. Y. Bartal and S. Leonardi. On-line routing in all-optical networks. In *Proceedings of the 24th International Colloquium on Automata, Languages and Programming ICALP '97*, LNCS 1256, pages 516–526. Springer-Verlag, 1997.
4. E. Coffman, Jr., M. Garey, D. Johnson, and A. Lapaugh. Scheduling file transfers. *SIAM J. Comput.*, 14(3):744–780, August 1985.
5. T. Erlebach and K. Jansen. Scheduling of virtual connections in fast networks. In *Proceedings of the 4th Parallel Systems and Algorithms Workshop PASA '96*, pages 13–32. World Scientific Publishing, 1997.
6. T. Erlebach and K. Jansen. Call scheduling in trees, rings and meshes. In *Proceedings of the 30th Hawaii International Conference on System Sciences HICSS-30*, volume 1, pages 221–222. IEEE Computer Society Press, 1997.
7. A. Feldmann, B. Maggs, J. Sgall, D. D. Sleator, and A. Tomkins. Competitive analysis of call admission algorithms that allow delay. Technical Report CMU-CS-95-102, School of Computer Science, Carnegie Mellon University, Pittsburgh, PA, January 1995.
8. A. Feldmann. On-line call admission for high-speed networks (Ph.D. Thesis). Technical Report CMU-CS-95-201, School of Computer Science, Carnegie Mellon University, Pittsburgh, PA, October 1995.
9. R. Graham. Bounds on multiprocessing timing anomalies. *SIAM J. Appl. Math.*, 17(2):416–429, March 1969.
10. R. Graham. Bounds on the performance of scheduling algorithms. In E. G. Coffman, Jr., editor, *Computer and Job-Shop Scheduling Theory*, pages 165–227. John Wiley & Sons, Inc., New York, 1976.
11. J. Hoogeveen, S. van de Velde, and B. Veltman. Complexity of scheduling multiprocessor tasks with prespecified processor allocations. *Discrete Appl. Math.*, 55:259–272, 1994.
12. C. Kaklamanis, P. Persiano, T. Erlebach, and K. Jansen. Constrained bipartite edge coloring with applications to wavelength routing. In *Proceedings of the 24th International Colloquium on Automata, Languages and Programming ICALP '97*, LNCS 1256, pages 493–504. Springer-Verlag, 1997.
13. T. Nishizeki and K. Kashiwagi. On the 1.1 edge-coloring of multigraphs. *SIAM J. Disc. Math.*, 3(3):391–410, August 1990.
14. D. B. Shmoys, J. Wein, and D. P. Williamson. Scheduling parallel machines online. In *Proceedings of the 32nd Annual Symposium on Foundations of Computer Science FOCS '91*, pages 131–140, 1991.
15. The ATM Forum, Upper Saddle River, NJ. *ATM User-Network Interface (UNI) Specification Version 3.1.*, 1995.
16. J. van Leeuwen, editor. *Handbook of Theoretical Computer Science. Volume A: Algorithms and complexity.* Elsevier North-Holland, Amsterdam, 1990.

Computational Complexity of the Krausz Dimension of Graphs

Petr Hliněný and Jan Kratochvíl

Department of Applied Mathematics,
Faculty of Mathematics and Physics, Charles University, Czech Republic
{hlineny,honza}@kam.ms.mff.cuni.cz

Abstract. A Krausz partition of a graph G is a partition of the edges of G into complete subgraphs. The Krausz dimension of a graph G is the least number k such that G admits a Krausz partition in which each vertex belongs to at most k classes. The graphs with Krausz dimension at most 2 are exactly the line graphs, and graphs of the Krausz dimension at most k are intersection graphs of k-uniform linear hypergraphs.
This paper studies the computational complexity of the Krausz dimension problem. We show that deciding if Krausz dimension of a graph is at most 3 is *NP*-complete in general, but solvable in polynomial time for graphs of maximum degree 4. We pay closer attention to chordal graphs, showing that deciding if Krausz dimension is at most 6 is *NP*-complete for chordal graphs in general, while the Krausz dimension of a chordal graph with bounded clique size can be determined in polynomial time. We also show that for any fixed k, it can be decided in polynomial time if an interval graph has Krausz dimension at most k.

1 Introduction

A *Krausz partition* of a graph G is a partition of the edge set $E(G)$ into complete subgraphs (that are also called the *clusters* of the partition). The number of clusters containing a vertex v is called the *order* of v (in the Krausz partition). The order of the partition is the maximum order over all vertices of G. The *Krausz dimension* of G is defined as the minimum partition order over all Krausz partitions of G, and denoted by $dim(G)$. Note that if G is not connected, its dimension is the maximum dimension over all of its components; so we consider only connected graphs in our paper.

Every graph can be partitioned just by taking each edge alone as a cluster — thus for every graph, $dim(G) \leq \Delta(G)$. This bound is optimal for triangle-free graphs. On the other hand, complete graphs have dimension 1. Another important class of graphs, for which the Krausz dimension is known is described by Krausz's characterization of line graphs [6], which in fact inspired the definition of the Krausz dimension:

Theorem 1 (Krausz, 1943) *The Krausz dimension of a graph is at most* 2 *if and only if this graph is a line graph.*

The same issue was studied in several papers ([8],[4],[9]) under the notion of intersection graphs of uniform linear hypergraphs. A k-uniform linear hypergraph is a family of k-element sets where any two of them have at most 1 point in common; its intersection graph has the sets as vertices and the intersecting pairs of sets as edges. The following proposition can be found in [8]:

Proposition 1.1 *A graph has Krausz dimension k iff it is an intersection graph of some k-uniform linear hypergraph.*

Our notation of the Krausz dimension was inspired by [1], which deals mainly with the problem of dimension of a complete graph without one edge.

2 Computational complexity of the Krausz dimension

We summarize the main results of our paper in this section. The paper studies the computational complexity of the Krausz dimension for various classes of graphs.

The general problem of determining the dimension of a given graph is denoted by *KrauszDim*, the question whether the dimension is at most k is denoted by *KrauszDim*(k), and the same question restricted to graphs with maximum degrees at most d is denoted by *KrauszDim*(k,d). The decision forms of the Krausz dimension problem clearly belong to *NP*.

First we mention previous research concerning the computational complexity of the Krausz dimension. In [8], Naik, Rao, Shrikhande and Singhi found a finite list of forbidden subgraphs for graphs having the Krausz dimension at most 3 and minimum degree at least 69. That implies a polynomial algorithm for the *KrauszDim*(3) problem provided the minimum degree of the graph is at least 69. For Krausz dimension higher than 3, a finite list of forbidden subgraphs is found provided that each edge lies in sufficiently many triangles. In [4], the polynomial algorithm for (restricted) *KrauszDim*(3) is improved to lower bound on minimum degree equal to 19. Also, an *NP*-completeness result concerning a slightly different problem of dependency graphs of uniform hypergraphs is shown there.

Now we consider general graphs and the relation between the maximum degree of a graph and its Krausz dimension.

Theorem 2 **(a)** *The problem KrauszDim*$(3,4)$ *is solvable in polynomial time* $O(n^4)$.
(b) *The problem KrauszDim*$(D, D+2)$ *is NP-complete for all* $D \geq 3$, *even when restricted to planar graphs.*

We also show that the previous characterization of high-degree graphs of Krausz dimension at most 3 by a finite set of forbidden subgraphs probably cannot be extended to higher dimensions:

Corollary 2.1 *For any* δ, *the problem KrauszDim*(4) *is NP-complete for graphs of minimum degree* δ.

Next we pay closer attention to various classes of chordal graphs. It may seem somewhat surprising that Krausz dimension remains difficult even for chordal graphs, though here we are only able to prove the hardness result for larger dimension:

Theorem 3 *The problem KrauszDim(6) is NP–complete for chordal graphs.*

However, we have several results showing that special classes of chordal are easier:

Theorem 4 (a) *The problem KrauszDim is polynomial for graphs of bounded treewidth, and in particular for chordal graphs with bounded maximal clique size (or bounded maximum degree).*
(b) *For any fixed D, the problem KrauszDim(D) is polynomial for interval graphs.*

3 Graphs with maximum degree 4

In this section, we discuss the case of Krausz dimension being just by 1 smaller than the maximum degree. This question was suggested by K. Cechlárová [private communication]. Since $dim(\boldsymbol{G}) \leq \Delta(\boldsymbol{G})$ for any graph $\boldsymbol{G}$, this is the largest dimension (with respect to the maximum degree) for which nontrivial results may be expected.

Observation 3.1 The problem *KrauszDim*$(k, k+1)$ for a graph $\boldsymbol{G}$ is equivalent to the question whether there exists a collection of edge disjoint complete subgraphs of $\boldsymbol{G}$ of size at least 3 that cover each vertex of the maximum degree $k+1$.

In this sense we say that a graph is *CL*-coverable, if it contains a collection of edge disjoint complete subgraphs of size at least 3 covering all of its vertices. Then we can further reduce the considered problem:

Observation 3.2 Suppose that $\boldsymbol{G}$ contains a (not necessarily induced) subgraph $\boldsymbol{H}$ that is *CL*-coverable, and there is no triangle of $\boldsymbol{G}$ having just one or two edges in $\boldsymbol{H}$. Then, setting $\bar{\boldsymbol{G}} = (V(\boldsymbol{G}), E(\boldsymbol{G}) - E(\boldsymbol{H}))$, any Krausz order-$k$-partition of $\boldsymbol{G}$ is projected to an order-k-partition of $\bar{\boldsymbol{G}}$, and conversly, any order-k-partition of $\bar{\boldsymbol{G}}$ can be extended to whole $\boldsymbol{G}$ using the *CL*-cover of $\boldsymbol{H}$. Therefore $dim(\boldsymbol{G}) \leq k$ if and only if $dim(\bar{\boldsymbol{G}}) \leq k$.

Further we focus on the problem *KrauszDim*$(3,4)$.

Lemma 3.3 *To find a polynomial algorithm for KrauszDim$(3,4)$, it suffices to consider K_4-free graphs in which every triangle shares exactly one of its edges with other triangles.*

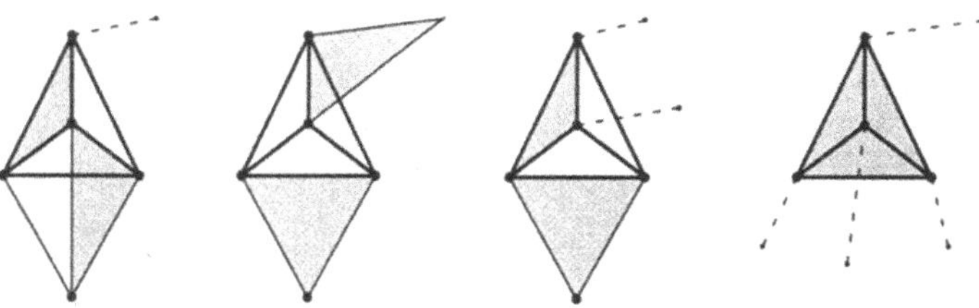

Fig. 1. Possible types of neighbourhood of a 4-clique

Proof: Let $\boldsymbol{G}$ be a graph satisfying the assumptions. If there is a 5-clique in $\boldsymbol{G}$, then $\boldsymbol{G} \cong \boldsymbol{K}_5$ (since the degrees are 4) and $dim(\boldsymbol{K}_5) = 1$.

A possible 4-clique $\boldsymbol{F}$ in $\boldsymbol{G}$ may be reduced as follows: Note that $\boldsymbol{F}$ has degrees 3, so each vertex of $\boldsymbol{F}$ is incident to at most one other edge. Discussing the positions of end vertices of these edges, there are only four possible configurations (except the 5-clique), depicted in Figure 1 (some of the dashed edges may be missing). Any of the cases leads to a subgraph in $\boldsymbol{G}$ that can be reduced using Observation 3.2—see the above picture where *CL*-covers are shaded.

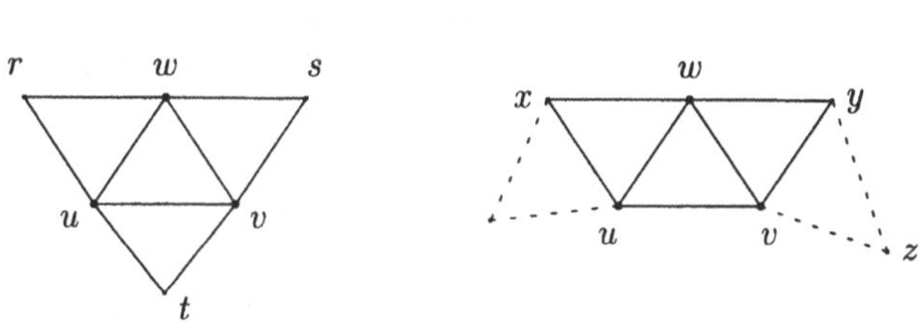

Fig. 2. Neighbourhood of a triangle uvw

So from now on, our graph $\boldsymbol{G}$ has maximum clique size 3. If some of its triangles is disjoint with all other triangles in $\boldsymbol{G}$, it can be simply reduced. Otherwise, let $T = uvw$ be a triangle in $\boldsymbol{G}$ that shares more than one edge with other triangles, as in Figure 2. The situation when T shares all its edges with other triangles is shown in the left ($r \neq s \neq t$ since there is no 4-clique). Respecting that the degrees of u, v, w are 4, there can be no other triangle sharing any of the edges in the picture. Thus we reduce by Observation 3.2.

A bit more complicated situation arises when precisely two edges of T are shared with other triangles, see Figure 2 right. In this situation, there may be other triangles using the edges xu or vy, such as the triangle vyz. Clearly $z \neq u, z \neq x$ hold and zu is not an edge, since that would lead to previous cases. Then we get the same situation for the triangle vyw, and we continue. Finally, we obtain either a "chain" or a "closed chain" of triangles, as shown in Figure 3. In

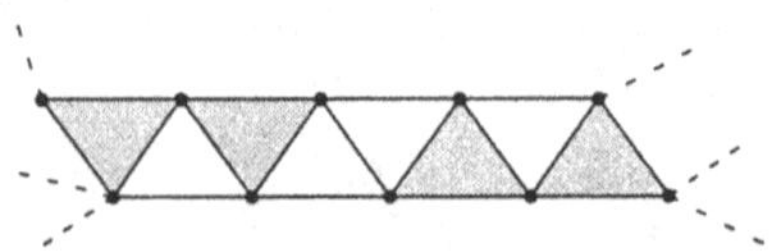

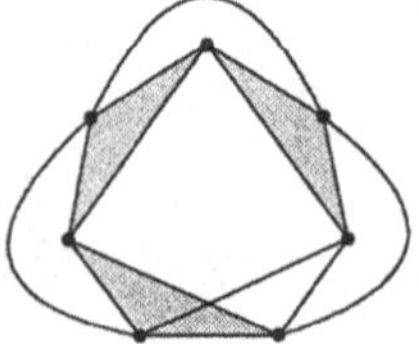

Fig. 3. A chain and a closed chain of triangles

both cases, for any length of the chain, we can choose a suitable collection of edge-disjoint triangles covering all vertices of the chain, and reduce via Observation 3.2 again.

□

Lemma 3.4 *The problem KrauszDim*(3, 4) *, for a graph satisfying the conditions of Lemma 3.3, can be reduced to finding maximum matching in a bipartite graph.*

Proof: Let $F \subseteq E(\boldsymbol{G})$ be the set of all edges of $\boldsymbol{G}$ that are contained in more than one triangle, and let $U \subseteq V(\boldsymbol{G})$ be the set of all vertices of $\boldsymbol{G}$ that have degree 4 and are not incident with any of the edges from F. The bipartite graph $\boldsymbol{B}$ is defined on the vertex set $F \cup U$, with edges of the form fu where $f = \{s,t\} \in F$, $u \in U$ and $\{u,s,t\}$ forming a triangle in $\boldsymbol{G}$.

We claim that $dim(\boldsymbol{G}) \leq 3$ if and only if $\boldsymbol{B}$ has a matching of size $|U|$: In one direction, having a Krausz order-3-partition of $\boldsymbol{G}$, each vertex $u \in U$ must be in some triangle T of the partition. Since every triangle of $\boldsymbol{G}$ shares one edge with other triangles, T contains one edge from F. Thus for each $u \in U$ there is an edge $\{f,u\} \in E(\boldsymbol{B})$, $f = \{s,t\} \in F$, corresponding to a triangle-cluster of the partition. These edges (one for each u) form a matching since the triangles in the partition are edge-disjoint.

For the opposite direction, suppose there exists a matching M in $\boldsymbol{B}$ covering each vertex of U, and let $f_u \in F$ be the edge matched to u, $u \in U$. Denote $\bar{F} = \{f_u | u \in U\}$ and $F' = F \setminus \bar{F}$. For every $f \in F'$, pick a triangle T_f containing the edge f. Note that since every triangle shares only one edge with other traingles, every triangle contains exactly one edge of F. Therefore the triangles $T_f, f \in F'$ and $\{u\} \cup f_u, u \in U$ are pairwise edge-disjoint and cover all vertices of degree 4. The desired Krausz partition of $\boldsymbol{G}$ consists of these triangles plus the remaining edges of $\boldsymbol{G}$ as single clusters.

□

Proof of Theorem 2 (a): A combination of the previous results gives a simple polynomial algorithm for testing *KrauszDim*(3, 4) , that uses a subroutine computing maximum matching in a bipartite graph (which is well known to be polynomial). To be accurate, we present a scheme of the whole algorithm here:

```
Algorithm 3.5 problem KrauszDim(3,4)
begin
   input a connected graph G = (V,E) with Δ(G) ≤ 4
   if G ≅ K_5 then output("dim(G) = 1") fi
   for every 4-tuple X ⊂ V(G) do
      if G|_X ≅ K_4 then
         examine the neighbours of X,
         find out which of the cases from Fig. 1 occurs,
         delete the edges found above from G
      fi
   done
   for every triple X ⊂ V(G) do
      if G|_X ≅ K_3 then
         if the situation from Fig. 2 left occurs then
            delete the edges of the subgraph from G
         else if the situation from Fig. 2 right occurs then
            go through the chain of edge-neighbouring triangles
            in both directions,
            delete the edges of the whole chain of triangles from G
         fi
      fi
   done
   construct the graph B on U ∪ F as defined in the proof of Lemma 3.4
   call MatchingInBipartiteGraph(B)
   if matching of size |U| exists then output("dim(G) ≤ 3")
   else output("dim(G) > 3")
end.
```

The running time of this algorithm is $O(n^4)$. This might be improved by more careful analysis of the reduction steps.

□

4 Graphs of bounded treewidth

We start with some essential definitions and considerations. A *tree decomposition* of a graph G is a pair $(\mathcal{S}, T)$ where $\mathcal{S} = \{X_i : i \in I\}$ is a collection of subsets of vertices of G, and T is a tree on the vertex set $V(T) = I$ (one node for each element of $\mathcal{S}$), satisfying the following conditions:

1. $\bigcup_{i \in I} X_i = V(G)$,
2. for every edge $\{u, v\} \in E(G)$ there is an $i \in I$ such that $u, v \in X_i$,
3. for each vertex $v \in V(G)$, the set of nodes $\{i : v \in X_i\}$ forms a subtree of T.

The *width* of the tree decomposition is defined as $\max_{i \in I}(|X_i| - 1)$, and the *treewidth of a graph* is the minimum width over all of its tree decompositions.

We may suppose that the tree T of the decomposition is a rooted binary tree—we just choose some root, duplicate nodes of high degrees (making a binary subtrees on them), and possibly add missing leaves. For a deeper discussion of the treewidth problematics, see the monograph [5].

In the next text the following notation is used: A tree decomposition is allways taken in the form presented above, with the same notation of the tree and the subsets. The set of all descendants of a node i is denoted by $\downarrow i$, and $\downarrow X_i$ stands for the union $\bigcup_{j \in \downarrow i} X_j$. A restriction of a Krausz partition $\mathcal{K}$ of a graph $\boldsymbol{G}$ onto a subset $X \subseteq V(\boldsymbol{G})$, i.e. a collection of clusters of $\mathcal{K}$ restricted to the vertex set X, is written as $\mathcal{K}|_X$.

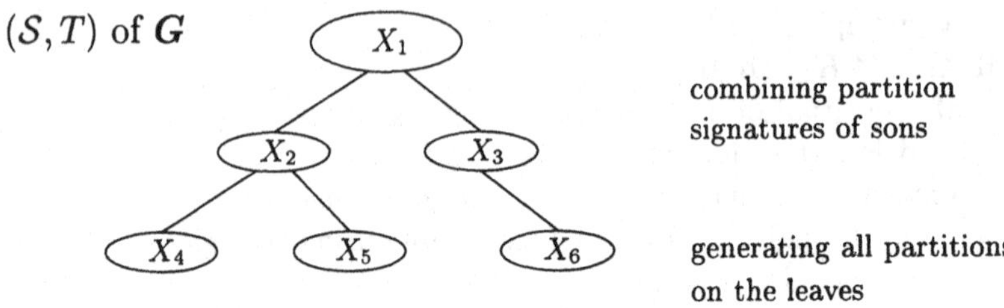

Fig. 4. A tree decomposition $(\mathcal{S}, T)$ of $\boldsymbol{G}$ and a scheme of the algorithm

The idea of our solution to *KrauszDim* problem for bounded treewidth graphs is rather simple (see also Figure 4)—we generate all possible Krausz partitions on all subgraphs of $\boldsymbol{G}$ corresponding to leaves of T in the tree decomposition, and then we process all their consistent combinations dynamically from leaves to the root of T, finally finding the lowest dimension used in the root node. Of course, it is impossible to maintain the whole partitioning during the dynamic process, so only a "partition signature" of a node (that describes everything relevant to the ancestors of this node) is carried. If the treewidth of $\boldsymbol{G}$ is bounded by some constant w, then it costs polynomial (though rather large in w) time to process every node with all possible signatures.

Proof of Theorem 4 a): Let $\boldsymbol{G}$ be a graph with treewidth bounded by w and let $(\mathcal{S}, T)$ be a corresponding tree decomposition (that can be found in linear time for a constant w, see [5]). Based on that, we present an algorithm that determines the Krausz dimension of $\boldsymbol{G}$ in time polynomial in the size of $\boldsymbol{G}$ for a constant w.

A *partition signature* in a node i of the tree decomposition $(\mathcal{S}, T)$ of $\boldsymbol{G}$ is defined as a 4-tuple $Sig(i) = (\mathcal{P}, q, o, m)$, where $\mathcal{P} = \{C_1, \ldots, C_p\}$ is a partition of the edges of $\boldsymbol{G}|_{X_i}$ into complete subgraphs, $q : \mathcal{P} \to \{0, 1\}$ are labels of these subgraphs, $o : X_i \to \{0, \ldots, \Delta\}$ are values of the vertices (here $\Delta = \Delta(\boldsymbol{G})$), and m is a number. For every node i, we construct the set $AllSig(i)$ of admissible partition signatures in i in the following sense: $(\mathcal{P}, q, o, m) \in AllSig(i)$ if and only if there exists a Krausz partition $\mathcal{K}$ of $\boldsymbol{G}|_{\downarrow X_i}$ of order m, such that $\mathcal{K}|_{X_i} = \mathcal{P}$, the order of each $v \in X_i$ is $o(v)$, and for every $C \in \mathcal{P}$, $q(C) = 1$ iff $C \in \mathcal{K}$. The aim of the attribute $q(C)$ is to indicate that the complete subgraph C is alone a cluster, and not a part of a larger cluster in $\mathcal{K}$, so it can be extended to a larger cluster in ancestor nodes.

To better understand this, realize that the properties of the tree decomposition imply $\downarrow X_r \cap \downarrow X_s \subseteq X_n$, $X_n \cap \downarrow X_r \subseteq X_r$, $X_n \cap \downarrow X_s \subseteq X_s$ for a node n and its two sons r, s of the decomposition, and $\boldsymbol{G}|_{\downarrow X_n} = \bigcup_{j \in \downarrow n} \boldsymbol{G}|_{X_j}$.

A simple computation shows that there are at most $2^{2^{w+1}} \cdot 2^{(w+1)^2} \cdot (\Delta + 1)^{w+1} \cdot (\Delta + 1)$ distinct signatures in one node, which is polynomial in Δ. So all of them can be efficiently generated in a leaf of T (in fact, in constant time for one leaf). One can also combine in polynomial time all pairs of signatures of the two sons of an inner node, thus generating all admissible signatures of that node (testing one triple of signatures is also in constant time), even if the computation is just "by brute force". Then dynamic processing from leaves to the root of the tree T gives all admissible signatures in the root node r. Among them, a signature with lowest m_r determines the Krausz dimension of $\boldsymbol{G}$.

Now it is enough to show how the signatures of sons are combined in a node of the decomposition. Let n be an inner node of T with two sons r, s, and $Sig(r) = (\mathcal{P}_r, q_r, o_r, m_r)$, $Sig(s) = (\mathcal{P}_s, q_s, o_s, m_s)$ be admissible (in the sense expressed above) signatures in the nodes r, s. If $Sig(n) = (\mathcal{P}_n, q_n, o_n, m_n)$ is *any* signature in the node n, it is an admissible combination of $Sig(r), Sig(s)$ if and only if the following conditions are satisfied:

- *(mutual consistency of $Sig(r), Sig(s)$)*
  If $C \in \mathcal{P}_r, C' \in \mathcal{P}_s$ and $|C \cap C'| > 1$, then either $C = C'$ or $C = C' \cap X_r$, $q_r(C) = 1$ or $C' = C \cap X_s$, $q_s(C') = 1$.
- *(correct clusters in node n)*
  For each $C \in \mathcal{P}_n$, one of $|C \cap X_r| \leq 1$ or $C \cap X_r = C_0 \in \mathcal{P}_r$, $q_r(C_0) = 1$ or $C = C_0 \cap X_n$, $C_0 \in \mathcal{P}_r$ should hold; and similarly for s.
- *(correct cluster labels in n)*
  For each $C \in \mathcal{P}_n$, the label is $q_n(C) = 0$ if and only if $C \subseteq X_r \cup X_s$, $\max\{|C \cap X_r|, |C \cap X_s|\} > 1$, and the following is satisfied: For $|C \cap X_r| > 1$, either $C \cap X_r \notin \mathcal{P}_r$ or $q_r(C \cap X_r) = 0$; and similarly for s.
- *(orders of vertices of X_n)*
  For each $v \in X_n$, the order of v is $o_n(v) = \left|\{C \in \mathcal{P}_n;\, v \in C\}\right|$ for $v \notin X_r \cup X_s$, $o_n(v) = o_r(v) + \left|\{C \in \mathcal{P}_n;\, v \in C, |C \cap X_r| \leq 1\}\right|$ for $v \in X_r - X_s$, $o_n(v) = o_s(v) + \left|\{C \in \mathcal{P}_n;\, v \in C, |C \cap X_s| \leq 1\}\right|$ for $v \in X_s - X_r$, and $o_n(v) = o_r(v) + o_s(v) - \left|\{C \in \mathcal{P}_r;\, v \in C, |C \cap X_r \cap X_s| > 1\}\right| + \left|\{C \in \mathcal{P}_n;\, v \in C, |C \cap (X_r \cup X_s)| \leq 1\}\right|$ for $v \in X_r \cap X_s$.
  The maximal order is $m_n = \max(o_n[X_n] \cup \{m_r, m_s\})$.

Finally, we summarize the above ideas in a scheme of the algorithm:

Algorithm 4.1 problem *KrauszDim* for graphs of constant treewidth
begin
  input graph $\boldsymbol{G} = (V, E)$
  suppose $\boldsymbol{G}$ connected, treewidth of $\boldsymbol{G}$ bounded by w

```
  call TreeDecomposition(G)  → (S,T)
  for l ∈ V(T), l leaf of T do
     AllSig(l) := a collection of all admissible signatures Sig(l)
           (derived from all Krausz partitions of X_l)
  done
  while exist n ∈ V(T) not processed yet do
     n := lowest node in V(T) not processed
     r, s := two sons of n (already processed)
     AllSig(n) := ∅
     for Sig(n) in all signatures possible on X_n do
        for [Sig(r), Sig(s)] ∈ AllSig(r) × AllSig(s) do
           check Sig(n) against Sig(r), Sig(s), as described above
           if Sig(n) consistent with Sig(r), Sig(s) then
              AllSig(n) := AllSig(n) ∪ {Sig(n)}
           fi
        done
     done
  done
  r = root of T
  D := min{m_r | Sig(r) ∈ AllSig(r)}
  output("dim(G) = D")
end.
```

□

Note that our result also applies to chordal graphs with bounded maximum clique size, since the treewidth of a chordal graph G is exactly $\omega(G) - 1$:

Corollary 4.2 *The problem KrauszDim is polynomially solvable for chordal graphs of bounded clique size.*

Since $\Delta(G) \geq \omega(G) - 1$, we have straightforwardly (compare to Theorem 2.(b)):

Corollary 4.3 *The problem KrauszDim is polynomially solvable for chordal graphs of bounded maximum degree.*

The following lemma, already presented in [8], is useful for dealing with several classes of graphs. We include our proof for the sake of completeness.

Lemma 4.4 *A clique of size at least $D^2 - D + 2$ must be a cluster in any Krausz order-D-partition of a graph.*

Proof: Let C be a clique of size ω in G that is not a cluster in a Krausz order-D-partition. If s is the size of the largest cluster used to cover C, then $\frac{\omega-1}{s-1} \leq D$ since there are at least $\frac{\omega-1}{s-1}$ clusters covering a fixed vertex $v \in C$ and its $\omega - 1$ neighbours. On the other hand, $s \leq D$ because for a cluster $A \subset C$, $|A| = s$ and a vertex $v \in C - A$, all the s edges $\{v, a\}$, $a \in A$ must be in different clusters.

Combining the previous inequalities, we get $\omega - 1 \leq Ds - D \leq D^2 - D$, thus the largest clique that need not be a cluster has size at most $D^2 - D + 1$. □

Split graphs are a special kind of chordal graphs, whose vertices can be split into a clique and an independent set.

Corollary 4.5 *For any fixed D, KrauszDim(D) is polynomial in the class of split graphs.*

Proof: Let G be a split graph and $\omega(G) = \omega$. If $\omega \geq D^2 - D + 2$, then this maximal clique must be a cluster of the partition, and the rest is a bipartite graph which has the only partition into single edges. Otherwise, the problem is reduced to the case of treewidth bounded by a constant $\omega - 1 \leq D^2 - D$. □

A similar result may be derived for complements of bipartite graphs. A more involved corollary of the previous results is a polynomial algorithm solving *KrauszDim(D)* for interval graphs and constant D (interval graphs are those that admit intersection representations by closed intervals on a line).

Proof of Theorem 4 (b): Suppose we are given an interval graph G decomposed into a sequence of cliques $Q_1, \ldots, Q_t$. Such a decomposition can be easily derived from an interval representation, or it can be viewed as a path decomposition of G (a tree decomposition where the tree is a path).

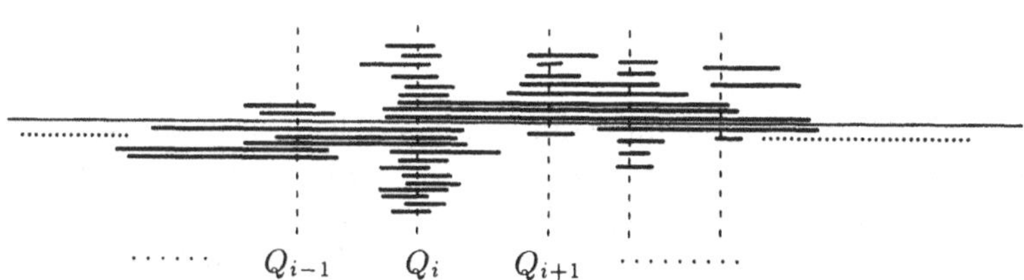

Fig. 5. A decomposition of an interval graph into a sequence of cliques

If there is an i such that $|Q_i \cap Q_{i+1}| \geq D^2 - D + 1$, then $dim(G) > D$ by Lemma 4.4 since $Q_i - Q_{i+1} \neq \emptyset$, $Q_{i+1} - Q_i \neq \emptyset$. Otherwise set $Q_i^0 = Q_i \cap \bigcup_{j \neq i} Q_j = Q_i \cap (Q_{i-1} \cup Q_{i+1})$ for every i. It follows that $|Q_i^0| \leq 2(D^2 - D)$. Thus if $|Q_k| > 2(D^2 - D)$ for some k, we can choose a subset $\bar{Q}_k$ of $2(D^2 - D)$ vertices such that $Q_k^0 \subseteq \bar{Q}_k \subset Q_k$.

In that case, the clique $\bar{Q}_k$ is a cluster in any order-D-partition of the graph $G' = \bar{Q}_k \cup \bigcup_{j \neq k} Q_j$; and since the vertices of $(Q_k - \bar{Q}_k)$ are disjoint with all other cliques in the graph, this partition can be extended using the cluster Q_k to an order-D-partition of G. That means $dim(G) \leq D$ iff $dim(G') \leq D$.

The above preprocessing either gives a negative answer, or produces an interval graph $\bar{G}$ of maximal clique size $2(D^2 - D)$, hence of treewidth at most $2(D^2 - D) - 1$. This graph is then passed to Algorithm 4.1.

Algorithm 4.6 problem *KrauszDim*(D) for interval graphs and fixed D

begin
  input a connected interval graph $\boldsymbol{G} = (V, E)$
  call *IntervalRepresentation*$(\boldsymbol{G})$
  got representation $\{L_i = \langle l_i, r_i \rangle;\ i \in V\}$, where all l_i, r_i are distinct
  define $q_1 = \min\{r_i\}$, $R_{k+1} = \{r_i \mid \exists j : r_i > l_j > q_k\}$
    (R_t last nonempty), $q_{k+1} = \min R_{k+1}$ for $k + 1 \leq t$
  $Q_k := \{i \mid q_k \in L_i\}$, $k = 1, \ldots, t$
  for $i :=$ 1 to $t - 1$ **do**
    if $|Q_i \cap Q_{i+1}| \geq D^2 - D + 1$ **then** **output**("*dim*$(\boldsymbol{G}) > D$")
  done
  $U := V$
  for $i :=$ 1 to t **do**
    if $|Q_i| > 2(D^2 - D)$ **then**
      choose $X \subseteq Q_i - (Q_{i-1} \cup Q_{i+1})$, $|X| = |Q_i| - 2(D^2 - D)$
      $U := U - X$
    fi
  done
  $\bar{\boldsymbol{G}} := \boldsymbol{G}|_U$
  output(*dim*$(\bar{\boldsymbol{G}})$ by Algorithm 4.1)
end.

□

5 NP reductions

For the *NP* reductions to the Krausz dimension problem, we use a special version of the well known satisfiability problem [2]. We consider a boolean formula Φ in the conjunctive normal form, with a set of clauses C over a set of variables V. By the *formula graph* we mean the bipartite graph $\boldsymbol{F}_\Phi$ on the vertex set $C \cup V$, and edges connecting each variable x to all clauses containing x or $\neg x$; formally $V(\boldsymbol{F}_\Phi) = C \cup V$, $E(\boldsymbol{F}_\Phi) = \{\{x, c\} \mid x \in V, c \in C,\ x \in c \vee \neg x \in c\}$. The *PLANAR 3-SAT* is defined as the satisfiability problem restricted to formulas with planar graphs of maximal degree 3.

The following lemma can be found in [7]:

Lemma 5.1 *The PLANAR 3-SAT problem is NP–complete.*

Now we show that the *PLANAR 3-SAT* can be reduced to the question, whether a given planar graph with degrees at most 5 has a Krausz dimension at most 3. We closely follow ideas used in [3] in the proof.

Proof of Theorem 2 (b): Given a formula Φ satisfying the conditions stated above, we construct a graph $\boldsymbol{R}_\Phi$ that has Krausz dimension at most 3 iff Φ is satisfiable. In the construction, every variable and every clause vertex of the graph $\boldsymbol{F_\Phi}$ is replaced by a special graph, see Figure 6. The variable graph $\mathcal{V}$ has two terminal vertices for its positive occurrences and two terminal vertices for negated occurrences, the clause graph $\mathcal{C}$ has three terminal vertices for its three literals. We may suppose that no variable has only positive (only negated) occurences; otherwise, we may set it true (false) and reduce the formula.

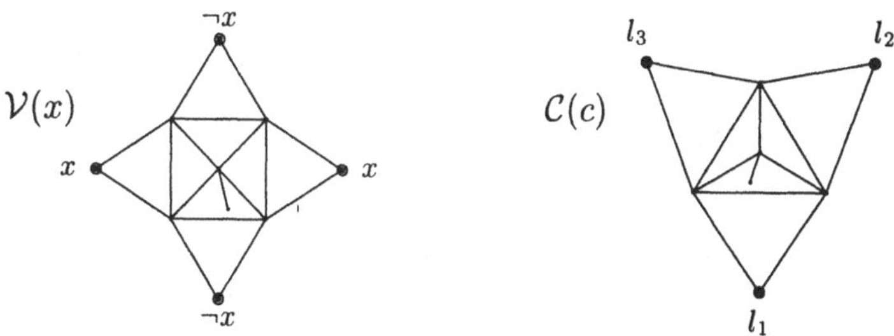

Fig. 6. The variable graph $\mathcal{V}(x)$ and the clause graph $\mathcal{C}(c)$

Clauses are connected with their variables by identifying the corresponding two terminal vertices. For clauses that contain less than 3 variables, a special false terminator is used on the remaining terminals. Clearly, if the formula graph is planar, so is the constructed graph $\boldsymbol{R_\Phi}$; and also $\Delta(\boldsymbol{R_\Phi}) = 5$ is fulfilled.

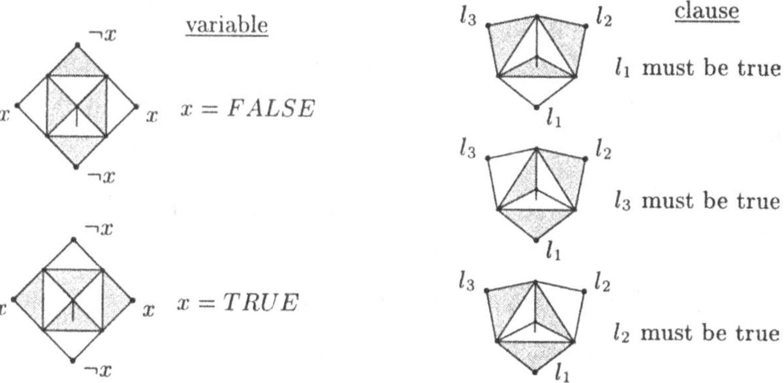

Fig. 7. Possible partitions of variable and clause graphs

To prove that the reduction is correct, see all the possible Krausz order-3-partitions depicted in Figure 7 (where clusters are the shaded triangles and the remaining edges). Note that the graphs do not contain 4-cliques, so a vertex of degree 5 must be in two triangles and a vertex of degree 4 in at least one triangle of the partition. Then focus on the central vertices in both graphs, their partitionings already determine the rest.

The two possible partitions of the variable graph $\mathcal{V}(x)$ encode the logical values *true* or *false* of the variable x, and the three partitions of the clause graph $\mathcal{C}(c)$ determine which of the three literals of c is chosen to satisfy the clause c. The false terminator is formed simply by adding two new vertices and two new edges connecting each of them to the terminal (i.e., adding two leaves).

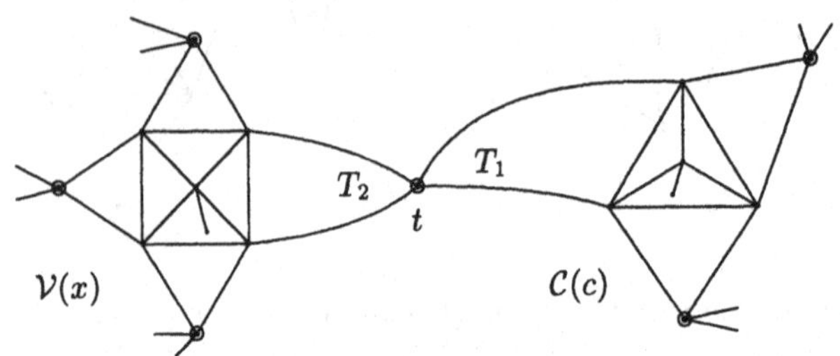

Fig. 8. A terminal connection between variable and clause graphs

Consider now a vertex t of the constructed graph $\boldsymbol{R}_{\Phi}$, that is a unified terminal vertex of $\mathcal{V}(x)$ and $\mathcal{C}(c)$ (Figure 8). In any Krausz order-3-partition, at least one of the triangles T_1, T_2 must be as a cluster. This implies, in the above presented interpretation, that either the literal of c represented by the terminal t is not the one that must be true in the clause c, or the occurence of the variable x in c is true (i.e., x or $\neg x$ depending on the terminal of $\mathcal{V}(x)$ used). So if $\boldsymbol{R}_{\Phi}$ has Krausz dimension 3, there exists an evaluation of variables in Φ such that every clause contains at least one true literal, and Φ is satisfiable. On the other hand, it is easy to construct, from a given satisfying assignment for Φ, a Krausz 3-partition of $\boldsymbol{R}_{\Phi}$.

□

Corollary 5.2 *For every $D \geq 3$, the problem KrauszDim$(D, D+2)$ is NP–complete, even when restricted to planar graphs.*

Proof: Pend $D-3$ vertices of degree one on each vertex of the input graph for *KrauszDim*$(3,5)$.

□

Proof Corollary 2.1: Let $\delta' = \max\{\delta, 13\}$. Pend a clique of size $\delta' + 1$ on each vertex of the input graph for *KrauszDim*$(3,5)$. Add a dummy $(\delta+1)$-clique if $\delta < 13$. Then the new graph has minimum degree δ; and, by Lemma 4.4, it has an order-4-partition iff the original graph has an order-3-partition and the added cliques are additional clusters.

□

If we consider chordal graphs, the Krausz dimension is much easier problem, as was shown in the previous section. Neverthless, the problem even then remains hard if the maximal clique size (treewidth) is not bounded. We show that the question whether a chordal graph $\boldsymbol{H}$ has Krausz dimension at most 6 is *NP*-complete.

Proof of Theorem 3: The proof is similar to that of Theorem 2 (b).

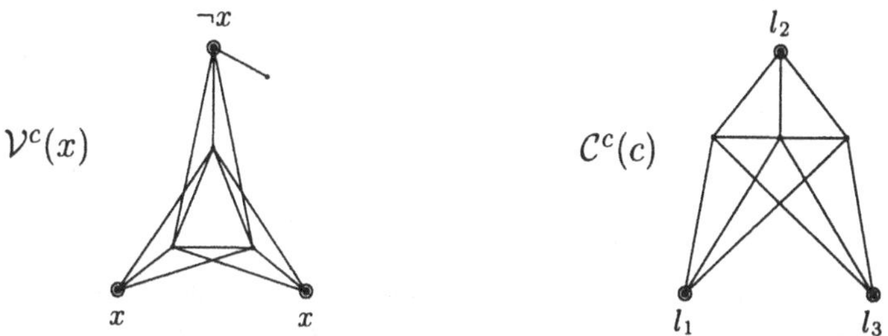

Fig. 9. The variable and clause graphs for a chordal reduction

The graph $\boldsymbol{R}_\Phi^c$ is constructed from the formula graph $\boldsymbol{F}_\Phi$ by replacing its vertices with variable and clause graphs $\mathcal{V}^c, \mathcal{C}^c$ from Figure 9, and by identifying corresponding terminals to represent its edges, similarly as above. Only variables with one negated and one or two positive occurences are considered (otherwise the variable is substituted by its negation). The false terminator is now formed by three leaves. Finally, a large clique Q containing all terminal vertices is added (to produce a chordal graph).

The key fact in the proof is that the clique Q must be a cluster in any order-6-partition of $\boldsymbol{R}_\Phi^c$ by Lemma 4.4, so we need not bother with it furthermore and the rest of the proof is similar to the previous one.

The reader can easily check the following facts about Krausz partitions of $\mathcal{V}^c$, $\mathcal{C}^c$:

- If a terminal vertex x of $\mathcal{V}^c(x)$ has order 2, then the terminal vertex $\neg x$ has order at least 3, and vice versa.
- At least one of the terminal vertices l_1, l_2, l_3 of $\mathcal{C}^c(c)$ has order more than 2.
- There exist partitions of $\mathcal{V}^c(x)$ such that the orders on terminals $x, \neg x, x$ are $2, 3, 2$ or $3, 2, 3$ respectively, and partitions of $\mathcal{C}^c(c)$ such that the orders on terminals l_1, l_2, l_3 are $2, 2, 3$ or $2, 3, 2$ or $3, 2, 2$ respectively.

These facts together imply: If Φ is satisfiable, then $dim(\boldsymbol{R}_\Phi^c) \leq 6$. Conversly, having a Krausz order-6-partition of $\boldsymbol{R}_\Phi^c$, each of its clause subgraphs has at least one terminal of order 3 (on the clause side), so the adjacent variable subgraph has order 2 on this terminal. That expresses the value true of this literal, and such evaluation is consistent over the whole formula.

□

6 Open problems

We have shown that for every fixed $D \geq 3$, deciding whether $dim(G) \leq D$ is NP–complete for graphs of maximum degree $D + 2$, while deciding $dim(G) \leq 3$ is polynomial for graphs of maximum degree 4. Thus we have partially answered a question first raised by Cechlárová [private communication]. The general question remains open:

Problem 1. Decide the complexity of $KrauszDim(D, D + 1)$ for $D > 3$.

We have shown that for any fixed D, the problem $KrauszDim(D)$ is solvable in polynomial time for split graphs and for complements of bipartite graphs. The following is, however, left open:

Problem 2. Decide the complexity of *KrauszDim* for split graphs and for complements of bipartite graphs.

Quite intriguing seems the question of Krausz dimension of general chordal graphs. We have proved that $KrauszDim(D)$ restricted to chordal graphs is NP–complete for $D \geq 6$, and of course this problem is polynomial for $D = 2$. The gap between 2 and 6 is open:

Problem 3. Decide the complexity of $KrauszDim(D)$ restricted to chordal graphs for $D = 3, 4, 5$.

References

1. L.W. Beineke, I. Broere, *The Krausz dimension of a graph*, preprint 1994.
2. M.R. Garey, D.S. Johnson, Computers and Intractability, W.H. Freeman and Company, New York 1979.
3. P. Hliněný, *Classes and recognition of curve contact graphs*, submitted to J. of Combinatorial Theory B, 1996.
4. M.S. Jacobson, A.E. Kézdy, J. Lehel, *Intersection graphs associated with uniform hypergraphs*, Congressus Numerantium 116 (1996), 173–192.
5. T. Kloks, Treewidth, computations and approximations, Lecture Notes in Computer Science 842, Springer-Verlag 1994.
6. J. Krausz, *Démonstration nouvelle d'un théorème de Whitney sur les résaux* (Hungarian with French summary), Mat. Fiz. Lapok 50 (1943), 75–85.
7. D. Lichtenstein, *Planar formulae and their uses*, SIAM J. of Computing 11 (1982), 329–343.
8. R.N. Naik, S.B. Rao, S.S. Shrikhande, N.M. Singhi, *Intersection graphs of k-uniform linear hypergraphs*, European J. of Combinatorics 3 (1982), 159–172.
9. E. Prisner, *Generalized octahedra and cliques in intersection graphs of uniform hypergraphs*, preprint 1996.

Asteroidal Sets in Graphs

Ton Kloks[1] Dieter Kratsch[2] Haiko Müller[2]

[1] Department of Applied Mathematics
University of Twente
P.O.Box 217
7500 AE Enschede, the Netherlands
A.J.J.Kloks@math.utwente.nl

[2] Friedrich-Schiller-Universität Jena
Fakultät für Mathematik und Informatik
07740 Jena, Germany
{kratsch,hm}@minet.uni-jena.de

Abstract. A set A of vertices of a graph $G=(V,E)$ is an *asteroidal set* if for each vertex $a \in A$, the set $A \setminus \{a\}$ is contained in one connected component of $G-N[a]$. The maximum cardinality of an asteroidal set of the graph G is said to be the *asteroidal number* of G. We show that there are efficient algorithms to compute the asteroidal number for claw-free graphs, HHD-free graphs, circular-arc graphs and circular permutation graphs, while the corresponding decision problem for graphs in general is NP-complete.

1 Introduction

An *asteroidal triple* of a graph $G=(V,E)$ is a set of three vertices, such that there exists a path between any two of them avoiding the neighborhood of the third. Asteroidal triples were introduced in [14] to characterize interval graphs as exactly those chordal graphs that do not have asteroidal triples.

Graphs without an asteroidal triple are called asteroidal triple-free graphs (short AT-free graphs) and attained much attention recently. Möhring has shown that every minimal triangulation of an AT-free graph is an interval graph which implies that for every AT-free graph the treewidth and the pathwidth of the graph are equal [16]. In fact, it was shown that every minimal triangulation of a graph is an interval graph if and only if the graph is AT-free [4, 17]. Furthermore a collection of interesting structural and algorithmic properties of AT-free graphs has been obtained by Corneil, Olariu and Stewart, among them an existence theorem for so-called dominating pairs in connected AT-free graphs and a linear time algorithm to compute a dominating pair for connected AT-free graphs (see [4, 5]).

Walter generalized the concept of asteroidal triples to so-called asteroidal sets [25]. He and later Prisner used asteroidal sets to characterize certain subclasses of the chordal graphs [18, 25]. A set of vertices A of a graph G is called an asteroidal set if for every vertex $a \in A$ the set $A \setminus \{a\}$ is contained in one connected component of $G - N[a]$. The asteroidal number of G denoted by $\mathsf{an}(G)$ is defined as the maximum cardinality of an asteroidal set in G.

The asteroidal number of a graph can be large, e.g. $\mathsf{an}(C_{2n}) = n$, for every $n \geq 2$. Clearly the AT-free graphs are exactly the graphs with asteroidal number at most two and the complete graphs are exactly the graphs with asteroidal number one. The asteroidal number is bounded above by the independence number for every graph, since every asteroidal set is an independent set.

The abovementioned theorem of [4, 17] generalizes in a natural way: A graph G has asteroidal number at most k if and only if $\mathsf{an}(H) \leq k$ for every minimal triangulation H of G [12]. For some problems algorithms for AT-free graphs naturally extend to graphs of bounded asteroidal number. The treewidth and the minimum fill-in are computable in polynomial time for graphs with bounded asteroidal number, assuming a polynomially bounded number of minimal separators [12]. A maximum independent set, a minimum independent dominating set and a minimum independent perfect dominating set can be computed by a polynomial time algorithm for graphs of bounded asteroidal number [3].

In this paper we show that the asteroidal number can be computed by efficient algorithms for a variety of well-studied graph classes such as claw-free graphs, HHD-free graphs and intersection graphs such as circular-arc graphs and circular permutation graphs.

2 Preliminaries

We denote the number of vertices of a graph $G = (V, E)$ by n and the number of edges by m. Let A be a set. Then we write $A + a$ instead of $A \cup \{a\}$. If $a \in A$ then we write $A - a$ instead of $A \setminus \{a\}$. For $W \subseteq V$ we write $G[W]$ for the subgraph of $G = (V, E)$ induced by the vertices of W. For a vertex a we write $G - a$ instead of $G[V - a]$, and for $X \subseteq V$ we write $G - X$ instead of $G[V \setminus X]$. We consider (connected) components of a graph as maximal connected subgraphs and as vertex subsets. For a vertex $x \in V$, $N(x)$ is the neighborhood of x and $N[x] = \{x\} \cup N(x)$ is the closed neighborhood of x.

2.1 Preliminaries on asteroidal sets

Definition 1. A set of vertices $A \subseteq V$ of a graph $G = (V, E)$ is an *asteroidal set* if for each $a \in A$, the set $A - a$ is contained in one component of $G - N[a]$.

Therefore each asteroidal set is an independent set. An asteroidal set with three elements is an asteroidal triple (short AT) that is usually defined as a triple of vertices such that between any two of the vertices there is a path avoiding the neighborhood of the third. We denote the set of all asteroidal triples of a graph G by $\mathsf{AT}(G)$.

Definition 2. The *asteroidal number* of a graph G denoted by $\mathsf{an}(G)$ is the maximum cardinality of an asteroidal set in G.

Graphs with asteroidal number at most two are commonly known as AT-free graphs. The class of AT-free graphs contains well-known graph classes such as interval, permutation and cocomparability graphs.

First we consider some elementary properties of the asteroidal number.

Lemma 3. *Let $G = (V, E)$ be a disconnected graph and let $G_1, G_2, \ldots, G_r$ be the components of G. Then $\mathsf{an}(G) = \max\left(2, \max_{i=1,2,\ldots,r} \mathsf{an}(G_i)\right)$.*

Proof. If G is the disjoint union of at least two cliques, then the asteroidal number is 2. If $\mathsf{an}(G) > 2$ then any asteroidal set must be contained in one component G_i of G. □

Lemma 4. *Let W be a subset of vertices of a graph $G = (V, E)$ and let $H = G[W]$. Then every asteroidal set in H is also an asteroidal set in G and $\mathsf{an}(H) \leq \mathsf{an}(G)$.*

Proof. Let A be an asteroidal set in H. Take a vertex $a \in A$. Let C be the component of $H - N_H[a]$ containing $A - a$. Then C induces a connected subgraph in G without any neighbors of a. Hence C is contained in a component of $G - N_G[a]$. This proves the lemma. □

Definition 5. A set S of vertices is a *separator* of G if $G - S$ is disconnected. A separator S is *minimal* if there are at least two components in $G - S$ such that every vertex of S has a neighbor in both.

Lemma 6. *Let A be an asteroidal set and let $a \in A$. Let S be a separator and assume there are two components C_1 and C_2 in $G - S$ such that a is in C_1 and $A - a$ is contained in C_2. Let y be any other vertex of C_1. Then $A - a + y$ is also an asteroidal set.*

Proof. Since $N[y]$ is contained in $C_1 \cup S$, it follows that C_2 is a connected subgraph in $G - N[y]$ containing $A - a$. Let $z \in A - a$. Since A is an asteroidal set, $A - z$ is contained in a component of $G - N[z]$. Since $N[z] \subseteq C_2 \cup S$, it follows that C_1 is a connected subgraph in $G - N[z]$. Hence a and y are in the same component of $G - N[z]$. □

Asteroidal sets of cardinality at most two are trivial. Any set consisting of one vertex is asteroidal. A set of two vertices is asteroidal if and only if these vertices are nonadjacent.

Lemma 7. *A set A with $|A| \geq 3$ is an asteroidal set if and only if every triple of A is an AT.*

Proof. If A is an asteroidal set with $|A| \geq 3$ then for every vertex $a \in A$, the set $A - a$ is also an asteroidal set. This shows that for every asteroidal set, all triples contained in it are AT's.

Now assume that every triple in a set A is an AT. Since $|A| \geq 3$, A must be an independent set. Let $a \in A$, and consider the components of $G - N[a]$. Let $y \in A - a$. Then every other vertex $z \in A - a$ is in the same component of $G - N[a]$ as y. Hence $A - a$ is contained in a component of $G - N[a]$. □

Definition 8. Two adjacent vertices x and y are *true twins* if $N[x] = N[y]$. Non adjacent vertices x and y are *false twins* if $N(x) = N(y)$.

Lemma 9. *Let p and q be (true or false) twins. If A is an asteroidal set with $p \in A$, then $A - p + q$ is also an asteroidal set. A is an asteroidal set* not *containing p if and only if A is an asteroidal set in $G - p$.*

Proof. Consider a path P. The path avoids $N[p]$ if and only if it avoids $N[q]$. Since p and q are true twins, all occurrences of p in P can be replaced by q. If P avoids some closed neighborhood, then so does the new path.

Now let A be an asteroidal set containing p. Since p and q are adjacent $q \notin A$. The above arguments show that $A - p + q$ is an asteroidal set. They also show that if A is an asteroidal set not containing p, then A is an asteroidal set in $G - p$.

By Lemma 4 every asteroidal set in $G - p$ is also an asteroidal set in G. □

A similar lemma of course also holds for false twins. Only in this case one other possibility occurs, namely there is an asteroidal set consisting exactly of the false twins.

2.2 The asteroidal graph

Definition 10. Let $G = (V, E)$ be a graph and let a be a vertex of G. The *asteroidal graph* $\mathsf{A}(G, a)$ is defined as follows. The graph $\mathsf{A}(G, a)$ has the vertex set $V - N[a]$ and two vertices p and q are adjacent in $\mathsf{A}(G, a)$ if $\{p, q, a\} \in \mathsf{AT}(G)$.

Theorem 11. *Let $G = (V, E)$ be a graph, $A \subset V$ and $a \in V \setminus A$. Then $A + a$ is an asteroidal set in G if and only if A induces a clique in $\mathsf{A}(G, a)$.*

Proof. The assertion of the theorem is trivial for $|A| \leq 1$ thus w.l.o.g. we may assume $|A| \geq 2$.

First let $A + a$ be an asteroidal set in G and let $p, q \in A$. Then p and q are vertices of $\mathsf{A}(G, a)$ since they are non adjacent to a. By Lemma 7 $\{p, q, a\} \in \mathsf{AT}(G)$. Hence p and q are adjacent in $\mathsf{A}(G, a)$.

Now we assume that A is a clique in $\mathsf{A}(G, a)$, $p, q \in A$ and $p \neq q$. We show that every triple in $A + a$ is an AT. The set $\{p, q, a\}$ is an AT of G since $\{p, q\}$ is an edge in $\mathsf{A}(G, a)$. Consider a third vertex $r \in A \setminus \{p, q\}$ if such a vertex r exists. By the above, $\{p, q, a\} \in \mathsf{AT}(G)$ and hence there exists a p, a-path avoiding $N[q]$. Similarly, there exists an r, a-path avoiding $N[q]$ since $\{q, r, a\} \in \mathsf{AT}(G)$. Concatenating these two paths we obtain a p, r-path without a vertex in $N[q]$. By symmetry this shows that $\{p, q, r\} \in \mathsf{AT}(G)$. By Lemma 7 this shows that $A + a$ is an asteroidal set. □

Lemma 12. *The following statements are equivalent.*

1. *For all a, $G - N[a]$ is connected.*

2. Every independent set is an asteroidal set.
3. Every triple of pairwise non adjacent vertices is an AT.
4. For every vertex a, $\mathsf{A}(G,a) = \overline{G - N[a]}$.

Proof. The equivalence of the first three statements follows from the definition of an asteroidal set and from Lemma 7.

The equivalence with 4 can be seen as follows. Assume that $\mathsf{A}(G,a) = \overline{G - N[a]}$ for all a. Consider a triple p, q, r of pairwise non adjacent vertices in G. Then q and r are adjacent in $\mathsf{A}(G,p)$. It follows by definition that $\{p,q,r\} \in \mathsf{AT}(G)$. Now assume that every independent triple is asteroidal, and consider a vertex p. Two vertices q and r of $G - N[p]$ are non adjacent if and only if $\{p,q,r\} \in \mathsf{AT}(G)$. Hence $\mathsf{A}(G,p) = \overline{G - N[p]}$. □

3 Hardness results

Let ASTEROIDAL NUMBER be the decision problem 'Given a graph G and an integer k, does G contain an asteroidal set with k vertices?'

Theorem 13. *The problem* ASTEROIDAL NUMBER *is NP-complete, even when restricted to triangle-free 3-connected 3-regular planar graphs.*

Proof. It is easy to see that ASTEROIDAL NUMBER is in NP. To prove NP-completeness for triangle-free, 3-connected, 3-regular and planar graphs we show the equivalence with INDEPENDENT SET restricted to this class, which is known to be NP-complete [23].

Every asteroidal set of a graph G is also an independent set of G. We use Lemma 12 to show that for any graph $G = (V,E)$ which is 3-connected, 3-regular and planar, each independent set $A \subset V$ is also an asteroidal set in G. Assume, by way of contradiction, that there is a vertex $a \in A$ such that $C_1, C_2 \subset V$ are different components of $G - N[a]$. Since G is 3-connected and 3-regular, $N(a)$ is a minimal separator in $G - a$, i.e., each vertex in $N(a)$ has a neighbour in C_1 and a neighbour in C_2. By contracting C_1 and C_2 to single vertices c_1 and c_2, respectively, we obtain a $K_{3,3}$ on the vertex set $N(a) \cup \{a, c_1, c_2\}$ as a minor of G. This is a contradiction since no planar graph has a $K_{3,3}$ as a minor [24]. Consequently $G - N[a]$ is connected for every vertex $a \in A$, and by Lemma 12, A is an asteroidal set of G. □

We repeat a construction due to Spinrad. Let $G = (V,E)$ be any graph. We define $H = (V \times \{1,2\}, E_1 \cup E_2 \cup E_3)$ by

$$E_1 = \{\{(v,1),(w,1)\} : \{v,w\} \in E\}$$
$$E_2 = \{\{(v,1),(w,2)\} : \{v,w\} \in E \text{ or } v = w\}$$
$$E_3 = \{\{(v,2),(w,2)\} : v,w \in V \text{ and } v \neq w\}.$$

It is easy to see that for every independent set $\{x,y,z\}$ of G the vertices $(x,1)$, $(y,1)$ and $(z,1)$ form an AT of H. Moreover, for each AT $\{(x,i),(y,j),(z,k)\}$

of H, $i, j, k \in \{1, 2\}$, the set $\{x, y, z\}$ is independent in G. Now, by Lemma 7 we have $\alpha(G) = \mathsf{an}(H)$. Since INDEPENDENT SET is complete for $W[1]$ (see [6]), the ASTEROIDAL NUMBER is also complete for $W[1]$. Similar, ASTEROIDAL NUMBER is as hard to approximate as INDEPENDENT SET, see [10].

4 The asteroidal number of circular-arc graphs

In this section we show that the asteroidal number of a circular-arc graph can be computed in $O(n^3)$ time.

Definition 14. A *circular-arc graph* is a graph for which one can associate with each vertex an arc on a circle such that two vertices are adjacent if and only if the corresponding arcs have a nonempty intersection.

There is an $O(n^2)$ time recognition algorithm for circular-arc graphs that also computes a circular-arc model within the same timebound if the input graph is indeed a circular-arc graph [7]. We may assume that no two arcs of the model have an endpoint in common.

Our algorithm is based on Theorem 11 and it computes for every vertex a the maximum cardinality of an asteroidal set of the graph G that contains a by computing a maximum cardinality clique of the graph $\mathsf{A}(G, a)$. The key observation is that for every vertex a of G, the asteroidal graph $\mathsf{A}(G, a)$ is a comparability graph.

Definition 15. An undirected graph $G = (V, E)$ is a *comparability graph* if there is a transitive orientation F of G obtained by taking for every edge $\{u, v\} \in E$ exactly one of the oriented edges (u, v) or (v, u) into F. Thereby the orientation F is transitive if $(x, y) \in F$ and $(y, z) \in F$ implies $(x, z) \in F$ for all $x, y, z \in V$.

Lemma 16. *Let G be a circular-arc graph and let a be a vertex of G. Then the asteroidal graph $\mathsf{A}(G, a)$ is a comparability graph.*

Proof. Let $\mathcal{D}(G)$ be a circular-arc model for G. We define an orientation on $\mathsf{A}(G, a)$. Consider two adjacent vertices p and q of $\mathsf{A}(G, a)$. Direct the edge from p to q if the arc corresponding to p appears before the arc corresponding to q, when going around the circle in clockwise order, starting at the arc a.

We show that this orientation is transitive. Let (p, q) and (q, r) be oriented edges. Clearly the arc corresponding to p appears before the arc corresponding to r, when going around in clockwise order starting at a. It remains to show that p and r are adjacent in $\mathsf{A}(G, a)$, which is equivalent to $\{p, r, a\} \in \mathsf{AT}(G)$.

We have a p, r-path avoiding $N[a]$ since there are a p, q-path avoiding $N[a]$ and a q, r-path avoiding $N[a]$. We show that there exists a p, a-path avoiding $N[r]$. Notice that the arcs corresponding to a, p, q, r appear in this order clockwise in $\mathcal{D}(G)$, hence every q, a-path must have an internal vertex of $N[p]$ or $N[r]$. There exists a q, a-path avoiding $N[r]$. Hence this path contains a vertex of $N[p]$. This shows the existence of a p, a-path avoiding $N[r]$. By symmetry, this shows that $\{p, r, a\} \in \mathsf{AT}(G)$. $\square$

Theorem 17. *There exists an $O(n^3)$ algorithm to compute the asteroidal number of a circular-arc graph.*

Proof. All asteroidal triples of a given graph can be computed in time $O(n^3)$ by computing the components of the graph $G - N[a]$ for every vertex a [4]. Next $\mathsf{A}(G,a)$ and a transitive orientation of $\mathsf{A}(G,a)$ can be computed in time $O(n^2)$ for every vertex a. Finally there is a linear time algorithm to compute a maximum cardinality clique in a comparability graphs if a transitive orientation is given [8]. □

This method also applies to circular permutation graphs and gives an $O(n^3)$ algorithm to compute the asteroidal number of a circular permutation graph. (For information on this graph class the reader is referred to [19].)

5 The asteroidal number of claw-free graphs

In this section we present an $O(n^3)$ algorithm to compute the asteroidal number in a claw-free graph.

Definition 18. A *claw-free graph* is a graph without a $K_{1,3}$ (i.e., a claw) as an induced subgraph.

The class of claw-free graphs contains the class of line graphs as a proper subclass. Furthermore the importance of claw-free graphs follows from matching and hamiltonicity properties [9]. Using standard techniques it is easy to see that connected claw-free graphs can be recognized in $O(n^{\alpha-1}m)$ time (where $O(n^\alpha)$ is the time required to multiply two $n \times n$ matrices; currently $\alpha < 2.376$.) Recently it was shown that there is an alternative recognition algorithm with running time $O(m^{\frac{\alpha+1}{2}}) = O(m^{1.69})$ [13].

Not many problems are known to be solvable in polynomial time when restricted to claw-free graphs. One exception is the INDEPENDENT SET problem, which was shown to be polynomial time solvable by Minty and Sbihi [15, 20]. We exploit this result to design an algorithm that computes the asteroidal number for claw-free graphs.

First we show that, in the case of connected claw-free graphs, we can restrict the search for asteroidal sets to some subset of vertices, namely those vertices x for which $G - N[x]$ is connected.

Definition 19. A vertex x is *extremal* if $G - N[x]$ is connected. A set A of vertices of G is *extremal* if every vertex in it is extremal.

Theorem 20. *Let G be a connected claw-free graph. There exists an extremal asteroidal set A with $|A| = \mathsf{an}(G)$.*

Proof. Let A be an asteroidal set in G such that $|A| = \mathsf{an}(G)$. Let $a \in A$ and let C be the component of $G - N[a]$ containing $A - a$. Assume a is not extremal.

Since a is not extremal, there exists a component C' of $G - N[a]$ different from C. Let y be any vertex of C'. Since G is connected, there exists a vertex $p \in N(a)$ that has a neighbor in C'. Notice that p cannot have a neighbor in C, otherwise there would be a claw.

We first show that $A' = A - a + y$ is an asteroidal set. Clearly, the vertices of C are contained in some component of $G - N[y]$, since y has no neighbors in C. Let $z \in A - a$. Then $A - z$ is in one component of $G - N[z]$. Since $p \in N(a) \setminus N[z]$, also p, and hence also all vertices of C' are in this component.

There exists a vertex $q \in N(a)$ which has a neighbor in C, since G is connected. Clearly $q \notin N[y]$ since the graph is claw-free. Now notice that the component of $G - N[y]$ containing $A - a$ is larger than C, since it contains a and q in addition to C. By induction, this proves that there exists a vertex $y \in V - A$ such that $A - a + y$ is an asteroidal set and y extremal. Hence there exists an extremal asteroidal set. □

Corollary 21. *Let $G = (V, E)$ be a connected claw-free graph and X the set of extremal vertices of G. Then every independent set of $G[X]$ is also a asteroidal set of G. Hence* $\mathsf{an}(G) = \alpha(G[X])$.

Proof. For every $a \in A$, $A - a$ is contained in one component of $G - N[a]$, since a is extremal and A is independent. □

Theorem 22. *There is a $O(n^3)$ time algorithm to compute the asteroidal number on claw-free graphs.*

Proof. By Lemma 3 it is sufficient to compute the asteroidal number for any component of the given claw-free graph. For a connected graph G, our algorithm computes a maximum independent set in $G[X]$ using an $O(n^3)$ algorithm given in [20], where X the set of extremal vertices of G. By Corollary 21 this is a maximum asteroidal set of G. □

6 The asteroidal number of HHD-free graphs

In this section we present an $O(n^3 + n^{3/2}m)$ time algorithm to compute the asteroidal number of an HHD-free graph.

A *house* is the complement of the path on five vertices. A *hole* is a chordless cycle of length at least five. A *domino* is a bipartite graph obtained from a chordless cycle of length six by adding exactly one chord.

Definition 23. A graph is *HHD-free* if it does not contain a hole, a house or a domino as an induced subgraph.

The class of HHD-free graphs contains such well-known graph classes as chordal graphs and distance-hereditary graphs [2]. Furthermore there is an $O(m^2)$ recognition algorithm for HHD-free graphs [21].

By Lemma 3 we may restrict to connected graphs. First we compute for all vertices x of the given HHD-free graph $G = (V, E)$ the components of $G - N[x]$.

These components are stored in two different data structures. The first one consists of a list of vertices for each component, and for every vertex x we have a list of such lists. The second one is a 3-dimensional boolean array B such that $B[x,y,z]$ is true if and only if y and z belong to one component of $G - N[x]$. Clearly this preprocessing can be done in time $O(n^3)$.

Next our algorithm computes for every vertex $a \in V$ an asteroidal set A containing a with maximum cardinality. In the following we describe the work of the algorithm for a fixed vertex a of G.

Let $G = (V, E)$ be a graph, $a \in V$ and $x \in V \setminus N[a]$. A minimal separator of G is *close to* x if it contains only neighbors of x. Let C be the component of $G - N[x]$ containing a, and let $S_x \subseteq N(x)$ be the set of neighbors of x with a neighbor in C.

Lemma 24. *For each $x \in V \setminus N[a]$, S_x is a minimal separator close to x.*

Proof. Let $x \in V \setminus N[a]$. Then x and a are in different components of $G - S_x$. Every vertex of S_x is adjacent to x and has a neighbor in the component of $G - S_x$ containing a. □

For a fixed vertex $x \in V \setminus N[a]$, let C_x be the component of $G - S_x$ containing x.

Lemma 25. *Let A be an asteroidal set of G containing x and a. Let y be any vertex of $C_x - x$. Then $A - x + y$ is also an asteroidal set of G.*

Proof. Since $a \in C$, this component of $G - N[x]$ contains the whole set $A - x$. Notice that, by definition of S_x, C is also a component of $G - S_x$. Now the lemma is a direct consequence of Lemma 6. □

In the **first phase** the algorithm reduces the number of candidates (red vertices) for an asteroidal set A containing a. All vertices in $N(a)$ are colored white and all other vertices of G are colored red. As long as there exists a red vertex $x \in V \setminus N[a]$ with another red vertex y in C_x, the vertex x is recolored white. As an invariant by Lemma 25 there is always an asteroidal set of maximum cardinality among all asteroidal sets containing a that contains red vertices only.

The first phase is implemented as follows. We look to the components of $G - N[x]$. At the beginning all components are unmarked. We loop through all vertices $z \in N(x)$. If $N(z)$ contains a vertex in C, then we do nothing, otherwise we mark all components containing vertices of $N(z)$. After this loop a vertex belongs to $C_x \setminus N[x]$ if and only if it is in marked components. Obviously this can be done in time $O(n^2)$ per vertex a.

Let $\mathcal{R}$ be the set of red vertices after this first step.

Theorem 26. *If G is HHD-free, then $\mathcal{R}$ is an independent set in G.*

Proof. Clearly a has no red neighbors.

Let x_1 and y_1 be two adjacent vertices of $\mathcal{R}-a$. Since y_1 is a red neighbor of x_1, there exists a y_1,a-path P_y such that y_1 is the only neighbor of x_1 on P_y (otherwise x_1 would have been colored white, during the first phase of the algorithm). Let y_2 be the neighbor of y_1 on P_y. Then $y_2 \neq a$ (otherwise y_1 would be white) and x_1 is not adjacent to y_2. Similarly, there exists an x_1,a-path P_x such that x_1 is the only neighbor of y_1 on P_x. Let x_2 be the neighbor of x_1 on P_x. Hence, $x_2 \neq a$ and y_1 is not adjacent to x_2.

If $\{x_2,y_2\}$ is not an edge of G, then the path (x_2,x_1,y_1,y_2) can be extended by some vertices from P_x or P_y to a hole in G. Since G is HHD-free, $\{x_2,y_2\}$ is an edge of G. If a common neighbor z of x_2 and y_2 exists, then x_1,x_2,y_1,y_2 and z induce a house in G. Since G is HHD-free such a common neighbor z does not exists. Consequently, there exist neighbor a x_3 of x_2 on P_x and a neighbor y_3 of y_2 on P_y such that x_1,x_2,x_3,y_1,y_2,y_3 are different vertices of G such that neither $\{x_2,y_3\}$ nor $\{y_2,x_3\}$ are edges in G. Again, If $\{x_2,y_2\}$ is not an edge of G, then the path (x_3,x_2,y_2,y_3) can be extended by some vertices from P_x or P_y to a hole in G. Otherwise $\{x_2,y_2\}$ is an edge of G, and x_1,x_2,x_3,y_1,y_2,y_3 induce a domino in G. But this implies the existence of an induced house, hole or domino. □

We now define a relation $\rightarrow$ on $\mathcal{R}$.

Definition 27. Let p and q be vertices of $\mathcal{R}-a$. We write $p \rightarrow q$ if $N[p]$ separates q and a in G, i.e., if q and a are in different components of $G-N[p]$.

Lemma 28. *Assume $p,q \in \mathcal{R}-a$, $p \rightarrow q$ and $q \rightarrow p$. An asteroidal set A, containing a, contains at most one of p and q. Furthermore, if A contains p then $A-p+q$ is also an asteroidal set.*

Proof. Assume A is an asteroidal set containing a and p. Then $q \notin A$, since $N[p]$ separates q and a.

Assume a is in a component induced by C_1 of $G-N[p]$ and q is in another component induced by C_2. Then $N[q] \subseteq N(p) \cup C_2$, hence C_1 induces a component of $G-N[q]$ also. This implies $S_p = S_q$ since $p \rightarrow q$ and $q \rightarrow p$.

Let $A' = C_1 \cup S_p \cup \{p,q\}$. Then $A \subseteq A'$ and p and q are twins in $G[A']$. By Lemma 9 applied to $G[A']$ the set $A-p+q$ is an asteroidal set in $G[A']$ and by Lemma 4 also in G. □

The **second phase** of the algorithm reduces the set of red vertices as follows. As long as there exists a pair of red vertices p and q in $\mathcal{R}-a$ with $p \rightarrow q$ and $q \rightarrow p$, then we recolor *one* of p and q white.

To implement the second phase we use the boolean array B. We have $p \rightarrow q$ if and only if $B[p,a,q]$ is false. This enables us to do the second phase in time $O(n^2)$ for every vertex a.

Let $\mathcal{R}^*$ be the new set of red vertices after this second phase. By Lemma 28, there exists a maximum asteroidal set A of G with $A \subseteq \mathcal{R}^*$.

Theorem 29. *Let $p, q, r \in \mathcal{R}^* - a$, $p \to q$ and $q \to r$. Then $p \to r$.*

Proof. Let C_1 be the component of $G - N[p]$ containing a and let C_2 be the component of $G - N[p]$ containing q.

Since p and q are both red, $q \not\to p$, hence $r \neq p$. If $r \notin C_1 \cup N(p)$ then clearly $p \to r$. Assume, by means of contradiction, that $r \in C_1 \cup N(p)$. Clearly $r \notin C_1$ since $q \to r$ and $N[q] \subseteq C_2 \cup N(p)$. Since $q \to r$, r and q are not adjacent. Hence $r \in N(p) \setminus N[q]$.

There cannot exist a vertex $\delta \in N(p) \setminus N[q]$ with a neighbor in C_1, otherwise there would be a r, a-path, via p and δ, avoiding $N[q]$, contradicting $q \to r$. This implies $q \to p$, which is a contradiction. □

Let H be the graph with vertex set $\mathcal{R}^*$ for which two vertices p and q of $\mathcal{R}^* - a$ are adjacent if $p \to q$ or $q \to p$. Then Theorem 29 shows that H is a comparability graph.

Theorem 30. *Let A be a set of vertices of $\mathcal{R}^*$ containing a. Then A is an independent set in H if and only if A is an asteroidal set in G.*

Proof. Let A be an independent set in H containing a. Since $A \subseteq \mathcal{R}^* \subseteq \mathcal{R}$, A is an independent set in G by Theorem 26.

Now $p, q \in A - a$ and $p \not\to q$ implies that q and a are in the same component of $G - N[p]$. This shows that for every pair $p, q \in A - a$, $\{p, q, a\} \in \mathsf{AT}(G)$. But this implies that $A + a$ is an asteroidal set in G (see Theorem 11).

Now let A be an asteroidal set in G with $a \in A$ and $A \subseteq \mathcal{R}^*$. Let $p, q \in A - a$. Then q and a are in the same component of $G - N[p]$ since $p, q, a \in \mathsf{AT}(G)$. Hence $p \not\to q$ and, by symmetry, also $q \not\to p$. Hence p and q are not adjacent in H. By definition, a is an isolated vertex in H. Hence $A + a$ is an independent set in H. □

In the **third phase** the algorithm computes a maximum independent set in H. Since every maximal independent set in H contains a, applying the $O(\sqrt{n}m)$ time algorithm to compute a maximum independent set in a comparability graph (see [22]) to H, the algorithm finds a maximum asteroidal set A of G containing the fixed vertex a.

Theorem 31. *There is an $O(n^3 + n^{3/2}m)$ time algorithm to compute the asteroidal number of an HHD-free graph.*

Proof. We summarize the work of our algorithm. First we compute the components of the input graph in time $O(n+m)$. All the other work is done separately for each component. Finally we apply Lemma 3 to compute the asteroidal number of the input graph. For simplicity we assume that the input graph was connected, and therefore $m \geq n - 1$. Next we do the preprocessing in time $O(n^3)$. For a fixed vertex a the running times of the three phases sum up to $O(n^2 + \sqrt{n}m)$. Hence the total running time of our algorithm is $O(n^3 + n^{3/2}m)$. □

Notice that in many special cases the algorithm can be simplified. Assume for example the graph is a k-*tree* without false twins. (If a graph contains twins then it can be reduced in linear time by Lemma 9.) In that case it is fairly easy to show, that a set A containing those simplicial vertices x such that $G-N[x]$ is connected, is a maximum asteroidal set. Hence for k-trees the asteroidal number is computable in time $O(nm)$ independent from k.

Consider the case where G is a *split graph.* Consider the partial order relation on the vertices of the independent set defined by neighborhood inclusion. It is easy to see that finding an asteroidal set in G with at least three vertices is equivalent with finding a maximum independent set in the comparability graph H given by the partial order. It takes $O(nm)$ time to construct H and $O(\sqrt{n}m)$ time to compute a maximum independent set in H, see [22].

7 Conclusions

In this paper we have shown that the asteroidal number is computable in polynomial time for claw-free graphs, HHD-free graphs and circular-arc graphs, but the corresponding decision problem remains NP-complete for planar graphs. Our method using the asteroidal graph extends to other graph classes such as circular permutation graphs.

For all $\ell \geq 0$ the predicate $\mathsf{an}(G) \geq \ell$ can be expressed in monadic second order logic (MSOL). This implies that for partial k-trees the asteroidal number is computable in time $O(f(k,\ell) \cdot n)$ for a suitable function f.

The algorithmic complexities of the problems ASTEROIDAL NUMBER and INDEPENDENT SET coincides on all graph classes for which it is known up to now. In this respect it would be interesting to know whether there is an efficient algorithm to compute the asteroidal number of bipartite graphs, since for these graphs the independence number is computable in time $O(n^{5/2})$ using a maximum matching algorithm.

References

1. Bodlaender, H., T. Kloks and D. Kratsch, Treewidth and pathwidth of permutation graphs, *SIAM Journal on Discrete Mathematics* **8** (1995), pp. 606–616.
2. Brandstädt, A., Special graph classes–A survey, Schriftenreihe des Fachbereichs Mathematik, SM-DU-199, 1991, Universität Duisburg Gesamthochschule.
3. Broersma, H., T. Kloks, D. Kratsch and H. Müller, Independent sets in asteroidal triple-free graphs, Memorandum No. 1359, University of Twente, Enschede, The Netherlands, 1996.
4. Corneil, D.G., S. Olariu and L. Stewart, The linear structure of graphs: Asteroidal triple-free graphs. *Proceedings of WG'93*, Springer-Verlag, LNCS 790, 1994, pp. 211–224; full version to appear in *SIAM Journal on Discrete Mathematics.*
5. Corneil, D.G., S. Olariu and L. Stewart, A linear time algorithm to compute dominating pairs in asteroidal triple-free graphs, *Proceedings of ICALP'95*, Springer-Verlag, LNCS 944, 1995, pp. 292–302.

6. Downey, R.G. and M.R. Fellows, Fixed-parameter tractability and completeness II: On completeness for $W[1]$, *Theoretical Computer Science* **141** (1995), pp. 109–131.
7. Eschen, E.M., J.P. Spinrad, An $O(n^2)$ algorithm for circular-arc graph recognition, *Proceedings of SODA '93*, 1993, pp. 128–137.
8. Golumbic, M.C., *Algorithmic graph theory and perfect graphs*, Academic Press, New York, 1980.
9. Faudree, R., E. Flandrin and Z. Ryjáček, Claw-free graphs - a survey, *Discrete Mathematics* **164** (1997), pp. 87–147.
10. Håstad, Clique is hard to approximate within $n^{1-\epsilon}$, in Proceedings of 37th Ann. IEEE Symp. on Foundations of Comput. Sci., IEEE Computer Society (1996), pp. 627–636.
11. Kloks, T., *Treewidth - Computations and Approximations*, Springer-Verlag, LNCS 842, 1994.
12. Kloks, T., D. Kratsch and H. Müller, A generalization of AT-free graphs and some algorithmic results, manuscript, 1996.
13. Kloks, T., D. Kratsch and H. Müller, Finding and counting small induced subgraphs efficiently, *Proceedings of WG'95*, Springer-Verlag, LNCS 1017, 1995, pp. 14–23.
14. Lekkerkerker, C.G. and J.Ch. Boland, Representation of a finite graph by a set of intervals on the real line, *Fundamenta Mathematicae* **51** (1962), pp. 45–64.
15. Minty, G.J., On maximal independent sets of vertices in claw-free graphs, *Journal on Combinatorial Theory Series B* **28** (1980), pp. 284–304.
16. Möhring, R.H., Triangulating graphs without asteroidal triples, *Discrete Applied Mathematics* **64** (1996), pp. 281–287.
17. Parra, A., Structural and algorithmic aspects of chordal graph embeddings, PhD. thesis, Technische Universität Berlin, 1996.
18. Prisner, E., Representing triangulated graphs in stars, *Abhandlungen des Mathematischen Seminars der Universität Hamburg* **62** (1992), pp. 29–41.
19. Rotem, D. and J. Urrutia, Circular permutations graphs, *Networks* **12** (1982), pp. 429–437.
20. Sbihi, N., Algorithme de recherche d'un stable de cardinalite maximum dans un graphe sans etoile, *Discrete Mathematics* **29** (1980), pp. 53–76.
21. Schäffer, A.A., Recognizing brittle graphs: remarks on a paper of Hoàng and Khouzam, *Discrete Applied Mathematics* **31** (1991), pp. 29–35.
22. Simon, K., *Effiziente Algorithmen für perfekte Graphen*, B.G. Teubner, Stuttgart, 1992.
23. Uehara, R., NP-complete problems on a 3-connected cubic planar graph and their applications, Technical Report TWCU-M-0004, Tokyo Woman's Christian Univ., 1996.
24. Wagner, K., Über eine Eigenschaft der ebenen Complexe, *Mathematische Annalen* **14** (1937), pp. 570–590.
25. Walter, J.R., Representations of chordal graphs as subtrees of a tree, *Journal of Graph Theory* **2** (1978), pp. 265–267.

Complexity of Colored Graph Covers I. Colored Directed Multigraphs

Jan Kratochvíl[1] * and Andrzej Proskurowski[2] and Jan Arne Telle[3] **

[1] Charles University, Prague, Czech Republic
[2] University of Oregon, Eugene, Oregon
[3] University of Bergen, Bergen, Norway

Abstract. A covering projection from a graph G onto a graph H is a "local isomorphism": a mapping from the vertex set of G onto the vertex set of H such that, for every $v \in V(G)$, the neighborhood of v is mapped bijectively onto the neighborhood (in H) of the image of v. We continue the investigation of the computational complexity of the H-cover problem – deciding if a given graph G covers H. We introduce a more general notion of covers of directed colored multigraphs (cdm-graphs) and show that a complete characterization of the complexity of covering of simple undirected graphs would necessarily resolve the complexity of covering of cdm-graphs as well. On the other hand, we introduce reductions that will enable to consider only multigraphs with minimum degree ≥ 3. We illustrate the methodology by presenting a complete characterization of the complexity of covering problems for two-vertex cdm-graphs.

1 Motivation and Overview

For a fixed graph H, the H-cover problem admits a graph G as input and asks about the existence of a "local isomorphism": a labeling of the vertices of G by vertices of H so that the label set of the neighborhood of every $v \in V(G)$ is equal to the neighborhood (in H) of the label of v and each neighbor of v is labeled by a different neighbor of the label of v. Such a labeling is referred to as a *covering projection* from G onto H. We trace this concept to Biggs' construction of highly symmetric graphs in [4], and to Angluin's discussion of "local knowledge" in distributed computing environment in [2]. More recently, Abello *et al.* [1] raised the question of computational complexity of H-cover problems, noting that there are both polynomial-time solvable (*easy*) and NP-complete (*difficult*) versions of this problem depending on the parameter graph H. We have studied the question of complexity of graph covering problems in [8], where one of our main results was a complete catalogue of the complexity of this problem for simple graphs on at most 6 vertices. In [9], we proved that the

* Research partially supported by Czech Research grants GAUK 194/1996 and GAČR 0194/1996.

** Third author supported by a fellowship from the Norwegian Research Council.

H-cover problem is NP-complete for any k-regular graph H, provided $k > 2$ and H is k-edge colorable or $(1 + \lfloor \frac{k}{2} \rfloor)$-edge connected. This is a significant headway towards the more general conjecture stating that the H-cover problem is NP-complete for every k-regular graph, $k > 2$. In particular, it follows that there are infnitely many rigid graphs for which the covering problem is NP-complete.

In this paper we introduce covers of colored directed multigraphs (cdm-graphs). Though this notion may seem too general at first sight, it is readily seen that covers of colored directed multigraphs can be encoded in terms of covers of simple graphs. On the other hand the language of cdm-graphs enables more compact description of the results. We will show that it suffices to consider covers of cdm-graphs with minimum degree ≥ 3.

2 Colored Directed Multigraphs

In this paper, we consider *colored directed multigraphs*, shortly *cdm-graphs.* A directed multigraph with vertex set V has edge set $E = D \cup F \cup L$, where D is the set of directed edges (including directed loops), F is the set of undirected edges and L is the set of undirected loops. A function $\mu : E \to (V \times V) \cup \binom{V}{2} \cup V$ describes the incidencies, i.e., for an undirected edge $e \in F$, $\mu(e) \in \binom{V}{2}$ is the pair of vertices connected by e, for a loop $e \in L$, $\mu(e) \in V$ is the vertex hosting the loop e and for a directed edge $e \in D$, $\mu(e) \in V \times V$ is the ordered pair of vertices connected by e. Vertices and edges are colored by a coloring $C : V \cup E \to C(V \cup E)$ (this coloring need not be proper in the sense that adjacent vertices and/or edges may receive the same color). Since vertices, directed and undirected edges may be distinguished regardless the color, we may assume that $C(V)$, $C(D)$ and $C(F \cup L)$ are disjoint. A cdm-graph is uncolored if $C(V)$, $C(D)$ and $C(F \cup L)$ are one-element set each.

For an edge-color c, the *c-colored degree* of a vertex $x \in V(G)$ is defined as

$$deg_G^c(x) = |\{e : x \in \mu(e), e \in F(G)\}| + 2|\{e : x = \mu(e), e \in L(G), C(e) = c\}|$$

for $c \in C(F \cup L)$, and

$$deg_G^c(x) = (deg_G^{c-}(x), deg_G^{c+}(x))$$

where

$$deg_G^{c-}(x) = |\{e : \mu(e) = (x, u) \text{ for some } u, e \in D(G), C(e) = c\}|$$

and

$$deg_G^{c+}(x) = |\{e : \mu(e) = (u, x) \text{ for some } u, e \in D(G), C(e) = c\}|.$$

The *total degree* of a vertex u is

$$deg_G u = \sum_{c \in C(F \cup L)} deg^c(u) + \sum_{c \in C(D)} (deg^{c+}(u) + deg^{c-}(u)).$$

Definition 1. A *covering projection of a cdm-graph G onto a cdm-graph H* is a mapping $f : V(G) \cup E(G) \to V(H) \cup E(H)$ such that

(1) $f(u) \in V(H)$ and $C(f(u)) = C(u)$ for every $u \in V(G)$,

(2) $f(e) \in D(H)$, $C(f(e)) = C(e)$ and $\mu(f(e)) = (f(u), f(v))$ for every $e \in D(G)$ such that $\mu(e) = (u, v)$,

(3) $f(e) \in F(H) \cup L(H)$, $C(f(e)) = C(e)$ and $\mu(f(e)) = \{f(u), f(v)\}$ for every $e \in F(G) \cup L(G)$ such that $\mu(e) = \{u, v\}$,

(4) for every $u \in V(G)$ and every $e \in D(H)$ such that $\mu(e) = (f(u), w)$ ($\mu(e) = (w, f(u))$) for some w, there is exactly one arc $e' \in D(G)$ such that $\mu(e') = (u, w')$ ($\mu(e') = (w', u)$, respectively) for some w' and $f(e') = e$,

(5) for every $u \in V(G)$ and every $e \in F(H)$ such that $f(u) \in \mu(e)$, there is exactly one edge $e' \in F(G)$ such that $u \in \mu(e')$ and $f(e') = e$,

(6) for every $u \in V(G)$ and every $e \in L(H)$ such that $\mu(e) = f(u)$, there is either exactly one loop $e' \in L(G)$ such that $\mu(e') = u$ and $f(e') = e$, or there are exactly two edges $e', e'' \in F(G)$ such that $u \in \mu(e'), u \in \mu(e'')$ and $f(e') = f(e'') = e$.

The above definition follows the usual definition of topological covering spaces. It is somewhat lengthy because of the presence of edges of different types. Note that (3) implies that every undirected loop of G is mapped (in a covering projection) again onto a loop in H, however the preimage of a loop need not be a loop itself. In general, the preimage of a loop in H is a disjoint union of cycles in G. In the case of simple undirected graphs a covering projection is obviously uniquely determined by its restriction to the vertex set. Theorem 3 shows that as far as the existence of a covering projection is concerned, this is true also for cdm-graphs.

Definition 2. A *vertex-covering projection* is a mapping $g : V(G) \to V(H)$ such that

(1') $C(g(u)) = C(u)$ for every $u \in V(G)$,

(4') for every $u \in V(G)$, $w \in V(H)$ and every edge color $c \in C(D)$,

$$|\{e \in D(G) : \mu(e) = (u, x), g(x) = w, C(e) = c\}| =$$

$$|\{e' \in D(H) : \mu(e') = (g(u), w), C(e') = c\}|$$

and

$$|\{e \in D(G) : \mu(e) = (x, u), g(x) = w, C(e) = c\}| =$$

$$|\{e' \in D(H) : \mu(e') = (w, g(u)), C(e') = c\}|$$

(5') for every $u \in V(G)$, $w \in V(H)$ such that $w \neq g(u)$, and every $c \in C(F \cup L)$,

$$|\{e \in F(G) : \mu(e) = \{x, u\}, g(x) = w, C(e) = c\}| =$$

$$|\{e' \in F(H) : \mu(e') = \{w, g(u)\}, C(e') = c\}|$$

(6') for every $u \in V(G)$ and every $c \in C(F \cup L)$,

$$|\{e \in F(G) : \mu(e) = \{x, u\}, g(x) = g(u), C(e) = c\}| +$$
$$2|\{e \in L(G) : \mu(e) = u, C(e) = c\}| =$$
$$2|\{e' \in L(H) : \mu(e') = g(u), C(e') = c\}|.$$

Theorem 3. *A cdm-graph G covers a cdm-graph H if and only if there exists a vertex-covering projection of G onto H.*

Proof. If f is a covering projection of G onto H then the restriction of f to the vertex set of G is a color-preserving projection satisfying (4'-6').

Suppose on the other hand that $g : V(G) \to V(H)$ is a color preserving mapping satisfying (4'-6'). In particular, (1) is fulfilled for g. We will show how to extend g to a mapping f defined also on the edges of G so that (2-6) are fulfilled as well.

Fix an edge-color, say c, and consider only the edges of color c (in G and in H as well). Suppose first that the edges of color c are undirected. Consider a vertex $u \in V(H)$ and let there be k loops $l_1, l_2, \ldots, l_k$ of color c around u in H. Let G^u be the subgraph of G induced by the vertices mapped onto u and edges among them of color c. (I.e., $V(G^u) = g^{-1}(u)$ and $F(G^u) \cup L(G^u) = \{e \in F(G) \cup L(G) : C(e) = c, \mu(e) \subset g^{-1}(u)\}$.) It follows from (6') that G^u is a $2k$-regular multigraph, and by Petersen theorem, G^u is 2-factorable. Let $E_1, \ldots, E_k$ be a collection of 2-factors that partitions $F(G^u) \cup L(G^u)$. We define

$$f(e) = l_i \text{ iff } e \in E_i.$$

Straightforwardly, (6) is satisfied for f.

Next consider vertices $u \neq v \in V(H)$ and let there be k edges $e_1, e_2, \ldots, e_k$ of color c in H such that $\mu(e_i) = \{u, v\}$. Let G^{uv} be the subgraph of G induced by the vertices mapped onto u or v and edges among them of color c. (I.e., $V(G^{uv}) = g^{-1}(\{u, v\})$ and $F(G^{uv}) = \{e \in F(G) : C(e) = c, \mu(e) = \{x, y\}$ for some $x \in g^{-1}(u), y \in g^{-1}(v)\}$.) It follows from (5') that G^{uv} is a k-regular multigraph, and since it is also bipartite, G^{uv} is 1-factorable (König-Hall theorem). Let $E_1, \ldots, E_k$ be a collection of perfect matchings that partitions $F(G^{uv})$. We define

$$f(e) = e_i \text{ iff } e \in E_i.$$

Straightforwardly, (5) is satisfied for f.

Now suppose that the edges of color c are directed. Though we formally do not distinguish directed edges and directed loops, we need to make the distinction for this proof. Consider a vertex $u \in V(H)$ and let there be k directed loops $l_1, l_2, \ldots, l_k$ of color c around u in H. Let G^u be the subgraph of G induced by the vertices mapped onto u and edges among them of color c. (I.e., $V(G^u) = g^{-1}(u)$ and $D(G^u) = \{e \in D(G) : C(e) = c, \mu(e) \in g^{-1}(u) \times g^{-1}(u)\}$.) It follows from (4') that G^u has all indegrees and outdegrees k. Similarly as in the undirected case, the edge set of G^u can be partitioned into sets $E_1, \ldots, E_k$, each of which is a disjoint union of directed cycles. We then set

$$f(e) = l_i \text{ iff } e \in E_i.$$

For non-loop directed edges, consider vertices $u \neq v \in V(H)$ and let there be k directed edges $e_1, e_2, \ldots, e_k$ of color c in H such that $\mu(e_i) = (u, v)$. Let G^{uv} be the subgraph of G induced by the vertices mapped onto u or v and edges $D(G^u) = \{e \in D(G) : C(e) = c, \mu(e) \in g^{-1}(u) \times g^{-1}(v)\}$.) It follows from (4') that each vertex of G^{uv} is either a sink of indegree k or a source of outdegree k. Thus G^{uv} is bipartite and, by König-Hall theorem, G^{uv} is 1-factorable. Let $E_1, \ldots, E_k$ be a collection of perfect matchings that partitions $F(G^{uv})$. We set

$$f(e) = e_i \text{ iff } e \in E_i.$$

Straightforwardly, (4) is satisfied for f.

It follows from the construction that f preserves also the edge-colors and that the mapping of the edges is compatible with the mapping of their endpoints, i.e., f satisfies (2-3) as well.

Let us note that the existence of a vertex-covering projection is an obvious necessary condition for a cdm-graph G to cover a cdm-graph H. Thus Theorem 3 describes an "oncas" situation ("obvious necessary conditions are sufficient").

It is clear that a nonconnected cdm-graph G covers H if and only if every connected component of G covers H, and a connected G covers a nonconnected H if and only if G covers at least one connected component of H. Therefore we assume in the rest of the paper that both G and H are connected. It is then easy to see that the preimages of the vertices of H have the same size, and every cdm-graph G that covers H is an h-fold cover for some h ($h = |\{x \in V(G) : g(x) = u\}|$ for any $u \in V(H)$).

Definition 4. The *degree partition* of a cdm-graph G is the coarsest partition of $V(G)$ into monochromatic equivalence classes $B_1, \ldots, B_k$ such that there exist numbers $r^c_{ij}, d^{c+}_{ij}, d^{c-}_{ij}$ ($i, j = 1, 2, \ldots, k$, $c \in C(D) \cup C(F) \cup C(L)$) such that

(i) for every i, j and every $u \in B_i$,

$$|\{e \in D(G) : \mu(e) \in \{u\} \times B_j, C(e) = c\}| = d^{c+}_{ij},$$

(ii) for every i, j and every $u \in B_i$,

$$|\{e \in D(G) : \mu(e) \in \{u\} \times B_j, C(e) = c\}| = d^{c-}_{ij},$$

(iii) for every $i \neq j$ and every $u \in B_i$,

$$|\{e \in F(G) : u \in \mu(e), \mu(e) \setminus \{u\} \in B_j, C(e) = c\}| = r^c_{ij},$$

(iv) for every i and every $u \in B_i$,

$$|\{e \in F(G) : u \in \mu(e), \mu(e) \setminus \{u\} \in B_i, C(e) = c\}| +$$
$$2|\{e \in L(G) : \mu(e) = u, C(e) = c\}| = r^c_{ii}.$$

The collection $r^c_{ij}, d^{c+}_{ij}, d^{c-}_{ij}$ $(i, j = 1, 2, \ldots, k,\ c \in C(D) \cup C(F) \cup C(L))$ is then called the *degree refinement* of G.

As in the case of simple undirected graphs, also for cdm-graphs, the degree partition is unique and can be determined in polynomial time (starting with the partition into vertex-color classes and refining this partition iteratively). The following is a direct corollary of Theorem 3:

Corollary 5. *If a cdm-graph G covers a cdm-graph H then G and H have the same degree refinements. Moreover, if $B_1, \ldots, B_k$ is the degree partition of G and $B'_1, \ldots, B'_k$ the degree partition of H (indexed so that $r^c_{ij} = r'^c_{ij}$, $d^{c+}_{ij} = d'^{c+}_{ij}$ and $d^{c-}_{ij} = d'^{c-}_{ij}$ for all $i, j = 1, 2, \ldots, k$ and $c \in C(D) \cup C(F) \cup C(L)$) then $f(B_i) = B'_i$ for every i and every covering projection $f : G \to H$.*

Proof. If $g : V(G) \to V(H)$ is a vertex-covering projection and $B'_i, i = 1, 2, \ldots, k$ the degree partition of H, define $B_i = \{u \in V(G) : g(u) \in B'_i\}, i = 1, 2, \ldots, k$. It follows from Theorem 3 that $B_i, i = 1, 2, \ldots, k$ is the degree partition of G and has the same degree refinement. The uniqueness of the degree partition of G implies that this partition is the same for every covering projection g.

3 Degree Reductions

In this section we show two reductions that can be performed on both the covering graph and covered graph and for which the existence of a covering projection is an invariant. These reductions enable us to consider graphs without small degrees. Recall that the degree of a vertex in a mixed graph is the number of undirected edges containing that vertex plus twice the number of undirected loops around that vertex plus the number of directed edges leaving and entering that vertex.

A cycle in a mixed graph is a sequence $u_1, e_1, u_2, e_2, \ldots, u_k, e_k$ such that

$$\mu(e_i) = \{u_i, u_{i+1}\} \text{ or } \mu(e_i) = (u_i, u_{i+1}) \text{ or } \mu(e_i) = (u_{i+1}, u_i)$$

for every $i = 1, 2, \ldots, k$ $(u_{k+1} = u_1)$.

3.1 Reduction I - Tree Liquidation

Definition 6. Given a cdm-graph G, denote by $Z(G)$ the maximal subgraph with all degrees greater than 1. Then $Z'(G) = (V(G), E(Z'(G)) = E(G) \setminus E(Z(G)))$ is acyclic, i.e., a disjoint union of trees. Each of these trees intersects the vertex set of $Z(G)$ in exactly one vertex. For every $u \in V(Z(G))$, denote by

T_u the connected component of $Z'(G)$ containing u. Let $\tau(u)$ be the isomorphism type of T_u as a colored tree rooted in u (i.e., $\tau(u) = \tau(v)$ iff there exists a color preserving isomorphism of T_u and T_v mapping u onto v). Redefine the coloring of the vertices of $Z(G)$ by setting $C(u) = \tau(u)$. The graph $Z(G)$ with this new coloring will be called the *dearborization of* G and denoted by $T(G)$ (edges of $T(G)$ are colored as in G).

Given a cdm-graph G, its dearborization can be found in polynomial time. One may want to see an argument why $Z'(G)$ is acyclic: Since G is connected, any cycle Q in $Z'(G)$ would be connected to $Z(G)$ by some path, say P. Then $Z(G) \cup Q \cup P$ would have all degrees ≥ 2 and would be larger than $Z(G)$. Similarly, one may argue that each component of $Z'(G)$ intersects $Z(G)$ in exactly one vertex: Since G is connected, every component does intersect $Z(G)$. On the other hand, if a component Q intersected $Z(G)$ in two vertices, say u, v, then Q would contain a path connecting u and v, say P. Then $Z(G) \cup P$ would have all degrees ≥ 2 and would be larger than $Z(G)$.

Theorem 7. *For any two cdm-graphs G and H, G covers H if and only if $T(G)$ covers $T(H)$.*

Proof. Let $f : G \to H$ be a covering projection of G onto H. Since a connected graph covers a tree only if it is isomorphic to the tree itself, for every vertex $u \in Z(H)$ and every $v \in V(G)$ such that $f(v) = u$, we observe that $v \in Z(G)$ and $T_u \cong T_v$. Thus $\tau(u) = \tau(v)$ and the restriction of f to $Z(G)$ is a covering projection of $T(G)$ onto $T(H)$.

On the other hand, suppose that $f : T(G) \to T(H)$ is a covering projection. Then $T_u \cong T_{f(u)}$ for every $u \in T(G)$, and let $\phi_u : T_u \to T_{f(u)}$ be an isomorphism. Then $g : G \to H$ defined by

$$g(x) = \phi_u(x) \text{ for every } x \in T_u \text{ and every } u \in T(G)$$

is a vertex-covering projection of G onto H. By Theorem 3, G covers H.

3.2 Reduction II - Dumping Small Degrees

A path in a mixed graph is a sequence $u_1, e_1, u_2, e_2, \ldots, u_k$ such that

$$\mu(e_i) = \{u_i, u_{i+1}\} \text{ or } \mu(e_i) = (u_i, u_{i+1}) \text{ or } \mu(e_i) = (u_{i+1}, u_i)$$

for every $i = 1, 2, \ldots, k-1$.

Definition 8. Given a cdm-graph G of minimum degree > 1, denote by $W(G)$ the subgraph induced by the vertices of degrees greater than 2. Let $W'(G)$ be the subgraph induced by $V(G) \backslash W(G)$. Then $W'(G)$ has all degrees ≤ 2, i.e., it is a disjoint union of paths. Since G is connected, the end vertices of each of these paths are connected to $W(G)$ by one edge each. Replace each such extended path P leading from a vertex u to a vertex v (it may be $u = v$) by and edge e_P with $\mu(e_P) = \{u, v\}$ (if P is symetric), or with $\mu(e_P) = (u, v)$ if P is not symetric (in the latter case we decide ad hoc a generic orientation of the edge obtained from a non-symetric path). We denote by $\pi(e_P)$ the isomorphism type of P. Denote the resulting cdm-graph by $S(G)$, again we assume that the newly added edges are colored via π, the original edges of G that remain as edges of $S(G)$ retain their original colors. Note that $S(G)$ can be found in polynomial time.

Theorem 9. *For any two cdm-graphs G and H, G covers H if and only if $S(G)$ covers $S(H)$.*

Proof. Let $f : G \to H$ be a covering projection of G onto H. Since a connected graph covers a path if and only if it is isomorphic to the path itself, every path P connecting vertices u and v in G maps onto a path P' connecting vertices $f(u)$ and $f(v)$ in H, and P and P' are isomorphic. Thus $\pi(e_P) = \pi(e_{P'})$ and the restriction of f to $W(G)$ is a vertex-covering projection of $S(G)$ onto $S(H)$.

On the other hand, suppose that $f : S(G) \to S(H)$ is a covering projection. Consider an edge e_P for a path $P = u, \ldots, v$ of G. Let P' be a path in H such that $f(e_P) = e_{P'}$ ($f(e_P) \notin E(W(H))$ because of its color). Then $e_{P'}$ is connecting vertices $f(u)$ and $f(v)$, and since $\pi(e_P) = \pi(e_{P'})$, $P \cong P'$. Let ϕ_P be an isomorphism of P and P'. Then $g : V(G) \to V(H)$ defined by

$$g(x) = \phi_P(x) \text{ iff } x \in P$$

is a vertex-covering projection of G onto H and G covers H.

3.3 Simple Graphs Versus cdm-graphs

It follows from Theorems 7 and 9 that in order to give a complete characterization of the computational complexity of the H-cover problem for cdm-graphs, it suffices to consider graphs H of minimum degree ≥ 3. On the other hand, if we are given a cdm-graph G as an input graph for a question if G covers a fixed cdm-graph H, we may encode the colors and directions of egdes by simple subgraphs: We assign a different tree T_c to every vertex color c, and both in G and H pend an isomorphic copy of T_c to each vertex u such that $C(u) = c$. Then we assign a different number $n_c > 1$ to every edge color c, and we replace every undirected edge of color c by a path of length n_c. Colors coresponding to directed edges will have assigned $n_c > 2$, and we pend a new tree on the second vertex of each path of length n_c (second from the tail determined by the orientation of the edge). In this way we obtain simple undirected graphs $U(G)$ and $U(H)$ such that G covers H if and only if $U(G) \cong U(H)$. See Figure 1 for an example. Thus we may conclude:

Corollary 10. *To achieve a complete characterization of the complexity of graph covering problems for simple undirected graphs, it is necessary and sufficient to give a complete characterization of the H-cover problem for colored directed multigraphs H of minimum degree ≥ 3.*

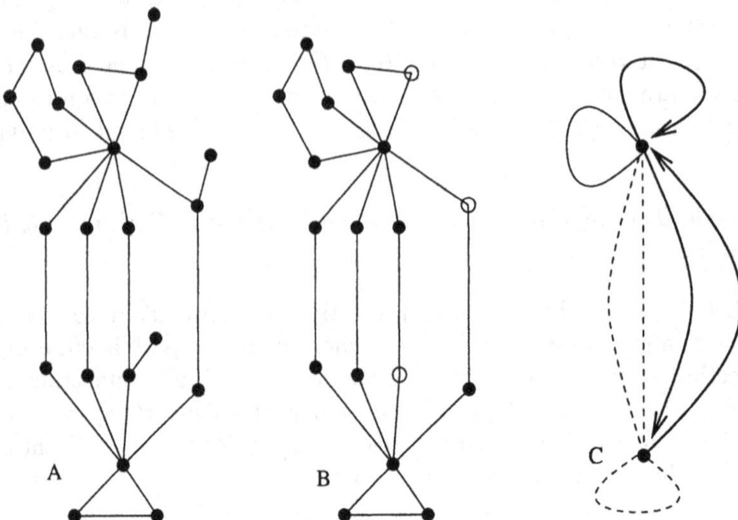

Fig. 1. A 21-vertex simple graph A encoded as a 2-vertex cdm-graph C, and vice-versa. Pending trees in graph A correspond to colored vertices in graph B. Paths of degree two in graph B correspond to colored directed edges and loops in graph C.

4 Two-vertex Graphs

To illustrate the methodology, we will now give a complete characterization of the complexity of the H-cover problem for cdm-graphs H with two vertices. (The case of H having one vertex only is straightforward - in such a case G covers H if and only if it has the same degree refinement as H.) Suppose H has two vertices, say L and R. For any edge color c, let l^c (r^c) be the number of loops of color c around the vertex L (R, respectively), let m^c be the number of edges of color c between L and R when edges of color c are undirected and let m^c_l (m^c_r) be the number of directed edges of color c starting in L and ending in R (starting in R and ending in L) when edges of color c are directed. (Thus $deg^c L = 2l^c + m^c$, $deg^c R = 2r^c + m^c$ in case of edges of color c being undirected, and $deg^{c-} L = l^c + m^c_l$, $deg^{c-} R = r^c + m^c_r$, $deg^{c+} L = l^c + m^c_r$, $deg^{c+} R = r^c + m^c_l$ in case of edges of color c being directed.) We denote by H^c the subgraph induced by the edges and loops of color c.

Theorem 11. *If*

(a) $C(L) \neq C(R)$, *or*

(b) $l^c \neq r^c$ *or* $m^c_l \neq m^c_r$ *for some color* c, *or*

(c) $m^c = 0$ ($m^c_l = m^c_r = 0$) *or* $l^c = r^c = 0$ *or* $l^c = r^c = m^c_l = m^c_r = 1$ *for every color* c

then the H*-cover problem is polynomially solvable. It is NP-complete in all other cases.*

In other words, assuming $P \neq NP$, the H-cover problem is polynomial iff H^c is non-regular for some color c or each H_c is either bipartite, or disconnected, or is regular of indegree 2 and outdegree 2. Rewording once more, the H-cover problem is NP-complete iff H^c is regular for every c and there is a color c for which H^c is connected nonbipartite and of degree at least 3 (resp. both indegree and outdegree at least 3). We will prove this result in several Lemmas. Note that it also follows that in the case of two-vertex cdm-graphs H, the H-cover problem is NP-complete if and only if H^c-cover is NP-complete for at least one edge-color c.

4.1 The Polynomial Cases

Lemma 12. *If*
(a) $C(L) \neq C(R)$, *or*
(b) $l^c \neq r^c$ *or* $m_l^c \neq m_r^c$ *for some color* c,
then the H*-cover problem is polynomially solvable.*

Proof. In both cases the vertices L and R are distinguishable in the degree partition of H. (This is trivial in case (a) as then they are distinguished by their colors. For case (b), note that $deg^c L = 2l^c + m^c = 2r^c + m^c = deg^c R$ implies $l^c = r^c$ if c is a color of undirected edges, while $deg^{c-} L = l^c + m_l^c = r^c + m_r^c = deg^{c-} R$ and $deg^{c+} L = l^c + m_r^c = r^c + m_l^c = deg^{c+} R$ imply $m_l^c = m_r^c$ and consequently $l^c = r^c$ if c is a color of directed edges.) It follows that G covers H if and only if G has the same degree refinement as H, and this can be decided in polynomial time.

The following lemma is a special case of a more general theorem [10]. We include a brief sketch of the proof for the sake of completeness.

Lemma 13. *If*
(i) $C(L) = C(R)$, *and*
for every color c,
(ii) $l^c = r^c$ *and* $m_l^c = m_r^c$, *and*
(iii) $m^c = 0$ $(m_l^c = m_r^c = 0)$ *or* $l^c = r^c = 0$ *or* $l^c = r^c = m_l^c = m_r^c = 1$
then the H*-cover problem is polynomially solvable.*

Proof. In this case H is symmetric and the degree partitions of G and H have each only one block. In particular, for any covering projection of G onto H, the mapping that interchanges the target vertices L and R is again a covering projection.

We will show how to reduce the H-cover problem to 2-SAT (which is well known to be solvable in polynomial time). Given a graph G which has the same degree refinement as H, we introduce a variable x_u for every vertex $u \in V(G)$. We then construct a formula $\Phi(G)$ over these variables so that G covers H if and only if $\Phi(G)$ is satisfiable, and in particular a covering projection $f : V(G) \to V(H)$ would correspond to a satisfying truth assignment so that x_u is true iff $f(u) = L$.

Our $\Phi(G)$ will be a conjuction of subformulas $\Phi(G) = \Phi_1 \wedge \Phi_2 \wedge \Phi_3$ defined as follows:

For every edge color c such that $m^c = 0$ (or $m_l^c = m_r^c = 0$ in case of directed color), Φ_1 will contain clauses

$$(x_u \vee \neg x_v) \wedge (\neg x_u \vee x_v)$$

for any pair of vertices $u, v \in V(G)$ connected by an edge $e \in F(G) \cup L(G)$ (or $e \in D(G)$) of color c (i.e., $\mu(e) = \{u, v\}$ resp. $\mu(e) = (u, v)$). Indeed, these two clauses guarantee that f maps u and v onto the same vertex of H.

For every edge color c such that $l^c = r^c = 0$, Φ_2 will contain clauses

$$(x_u \vee x_v) \wedge (\neg x_u \vee \neg x_v)$$

for any pair of vertices $u, v \in V(G)$ connected by an edge $e \in F(G) \cup L(G)$ (or $e \in D(G)$) of color c (i.e., $\mu(e) = \{u, v\}$ resp. $\mu(e) = (u, v)$). These clauses guarantee that f maps u and v onto distinct vertices of H.

Finally, for every directed edge color c such that $l^c = r^c = m_l^c = m_r^c = 1$, Φ_2 will contain clauses

$$(x_u \vee x_v) \wedge (\neg x_u \vee \neg x_v)$$

for any pair of vertices $u, v \in V(G)$ connected by directed edges $e, e' \in D(G)$ of color c to a common neighbor z so that $\mu(e) = (u, z)$ and $\mu(e') = (v, z)$ (or $\mu(e) = (z, u)$ and $\mu(e') = (z, v)$). These clauses guarantee that f maps u and v onto distinct vertices of H.

If G covers H then Φ is obviously satisfiable. On the other hand, a satisfying truth assignment yields a vertex-covering projection of G onto H. It follows from Theorem 3 that G covers H in such a case.

4.2 The NP-complete Cases

For the NP-complete cases, we assume that H is symmetric, i.e., $C(L) = C(R)$, $l^c = r^c$ and $m_l^c = m_r^c$ for every edge color c (the last equality is required for directed edge colors only). The impact of the first lemma is that we may study color-induced subgraphs separately.

Lemma 14. *The problem H-cover for two-vertex cdm-graphs is NP-complete provided H^c-cover is NP-complete for some edge color c.*

Proof. Given a graph G subject to the question of the existence of a covering projection from G to H^c, we construct $\widetilde{G}$ from two copies of G (say G_1 and G_2 with vertices named u_1 resp. u_2 for $u \in V(G)$). For every edge color $\varepsilon \neq c$, we add $l^\varepsilon = r^\varepsilon$ loops of color ε to every vertex of $\widetilde{G}$, and for every vertex $u \in V(G)$, we add m^ε undirected edges of color ε joining u_1 and u_2 (in case of an undirected edge color ε) and we add $m_l^\varepsilon = m_r^\varepsilon$ directed edges of color ε from u_1 to u_2 and the same number from u_2 to u_1 (in case of a directed edge color ε). Since H is symmetric, $f_2 : V(G_2) \to V(H)$ defined by $f_2(u_2) \neq f_1(u_1)$ is a covering projection of G_2 onto H^c whenever $f_1 : V(G_1) \to V(H)$ is a covering projection from G_1. It follows that $\widetilde{G}$ covers H if and only if G covers H^c.

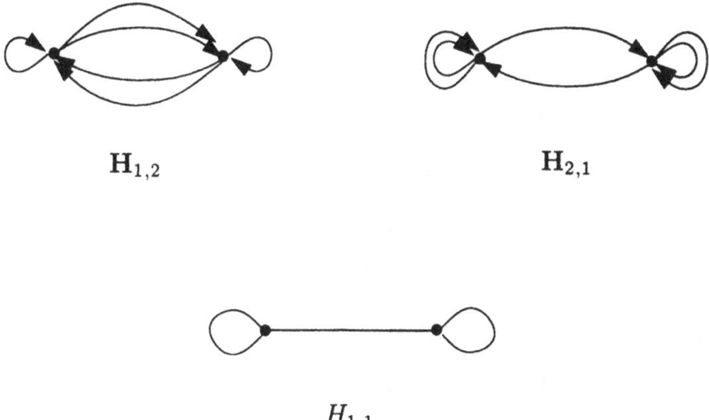

Fig. 2. Base NP-complete cdm-graphs.

In view of Lemmas 12, 13 and 14, it suffices to show that H-cover is NP-complete for monochromatic H such that

(1) $l = r \geq 1$ and $m \geq 1$ (in case of undirected graph H), or

(2) $l = r \geq 1$ and $m_l = m_r \geq 1$ and $l + m_l \geq 3$ (in case of directed H). We will show the NP-completeness by induction, starting with the graphs depicted in Fig. 2. For the sake of simplicity, we introduce notation $H(l, m)$ for an undirected graph with l loops around L and R and m edges joining L and R, and $\mathbf{H}(l, m)$ for a directed graph with l directed loops around L and R and m directed edges from L to R and m directed edges from R to L.

Proposition 15. [1] *The $H(1,1)$-cover problem is NP-complete.*

Though the NP-completeness result shown in [1] concerns multigraphs on the input, it is not difficult to show that the problem is NP-complete even if the input graph is simple, i.e., it is NP-complete to decide if the vertices of a simple cubic graph may be colored by two colors so that every vertex has two neighbors of the same color and one neighbor of the opposite color. This modification will be used in the sequel.

Lemma 16. *The $H(1,2)$-cover problem is NP-complete.*

Proof. Let G be a cubic graph subject to the question if G covers $H(1,1)$. Take two copies of G, say G_1 and G_2 (with vertices u_1 resp. u_2 for every $u \in V(G)$) and construct $\widetilde{G}$ from $G_1 \cup G_2$ by connecting each pair u_1, u_2 by a copy of the connector graph depicted in Fig. 3. This figure also shows a coloring of the connector graph such that every vertex has two black and two white neighbors. If $f : G \to H(1,1)$ is a covering projection, define $\widetilde{f} : \widetilde{G} \to H(1,2)$ so that $\widetilde{f}(u_1) = \widetilde{f}(u_2) = f(u)$ for $u \in V(G)$ and $\widetilde{f}(x) = f(u)$ if x is an inner vertex of the connector connecting u_1 and u_2 marked black in Fig. 3. This $\widetilde{f}$ is then a covering projection of $\widetilde{G}$ onto $H(1,2)$.

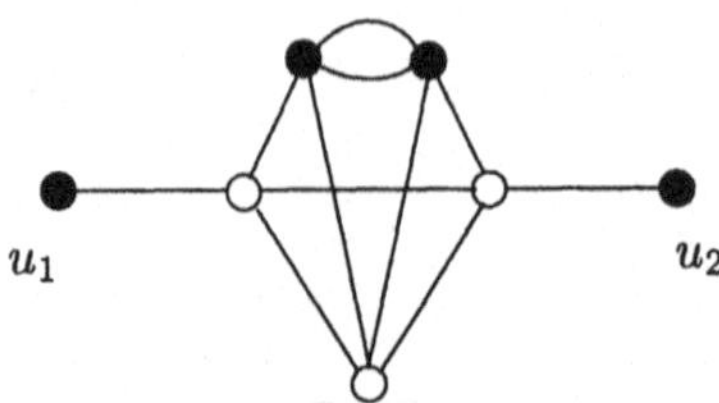

Fig. 3. Connector gadget for $H(1,2)$-cover.

Suppose on the other hand that $\widetilde{f} : \widetilde{G} \to H(1,2)$ is a covering projection. One can easily check that (upto the color reversal) the coloring depicted in Fig. 3 is the only coloring such that every inner vertex has two black and two white neighbors. Hence every vertex u_1 has two neighbors of its own color and one neighbor of the opposite color in G_1, and $f : G \to H(1,1)$ defined by $f(u) = \widetilde{f}(u_1)$ is a covering projection of G onto $H(1,1)$.

Lemma 17. *The $H(1,m)$-cover problem is NP-complete for every $m \geq 3$.*

Proof. Let G be a cubic graph subject to the question if G covers $H(1,1)$. Take two copies of G, say G_1 and G_2 (with vertices u_1 resp. u_2 for every $u \in V(G)$) and construct $\widetilde{G}$ from $G_1 \cup G_2$ by connecting each pair u_1, u_2 by $m-1$ parallel edges.

If $f : G \to H(1,1)$ is a covering projection, define $\widetilde{f} : \widetilde{G} \to H(1,m)$ so that $\widetilde{f}(u_1) = f(u)$, $\widetilde{f}(u_2) \neq f(u)$ for $u \in V(G)$. Since interchanging the values in a covering projection onto $H(1,1)$ results again in a covering projection, the restrictions of $\widetilde{f}$ to G_1 and G_2 are both covering projections onto $H(1,1)$. Each u_1 (u_2) has other $m-1$ neighbors of the opposite color in G_2 (resp. G_1). Hence this $\widetilde{f}$ is a covering projection of $\widetilde{G}$ onto $H(1,m)$.

Suppose on the other hand that $\widetilde{f} : \widetilde{G} \to H(1,m)$ is a covering projection. The pairs u_1, u_2 are the only pairs of vertices joined by parallel edges, and so $\widetilde{f}(u_1) \neq \widetilde{f}(u_2)$ for every $u \in V(G)$. Hence every vertex u_1 has two neighbors of its own color and one neighbor of the opposite color in G_1, and $f : G \to H(1,1)$ defined by $f(u) = \widetilde{f}(u_1)$ is a covering projection of G onto $H(1,1)$.

Proposition 18. *The $H(l,m)$-cover problem is NP-complete for every $l \geq 1, m \geq 1$.*

Proof. Let G be a multigraph subject to the question if G covers $H(1,m)$. Construct $\widetilde{G}$ from G by adding $l-1$ loops to each vertex of G. Any $f : V(G) \to \{L, R\}$

is a covering projection of G onto $H(1, m)$ if and only if it is also a covering projection of $\widetilde{G}$ onto $H(l, m)$.

This concludes the case of undirected graphs. For the rest of the section, we assume that H is directed.

Proposition 19. [7] *It is NP-complete to decide if the vertices of a simple cubic graph may be colored by two colors so that every vertex has exactly one neighbor of its own color.*

Lemma 20. *The* $\mathbf{H}(1,2)$*-cover problem is NP-complete.*

Proof. Let G be a simple cubic graph and let $\widetilde{G}$ be the symmetric orientation of G (i.e., every edge e of G is replaced by two directed edges joining the endpoints of e in opposite directions). Obviously, $\widetilde{G}$ covers $\mathbf{H}(1,2)$ if and only if G allows a coloring described in Proposition 19.

Lemma 21. *The* $\mathbf{H}(1,3)$*-cover problem is NP-complete.*

Proof. Take 6 copies of a cubic graph G, say $G_i, i = 1, 2, \ldots, 6$ (with vertices named $u_i \in V(G_i), i = 1, 2, \ldots, 6$ for $u \in V(G)$). Construct a 4-regular graph $\widetilde{G}$ from their disjoint union by adding connector graphs depicted in Fig. 4. We claim that the vertices of $\widetilde{G}$ can be colored by two colors so that each vertex has exactly one neighbor of its own color and three neighbors of the opposite color if and only if the vertices of G can be colored by two colors so that each vertex has exactly one neighbor of its own color.

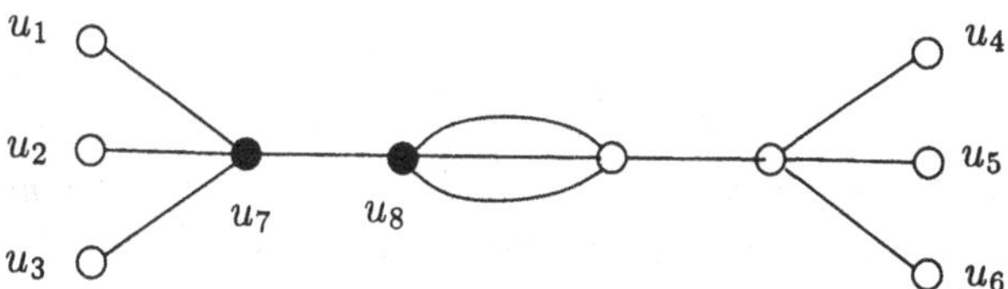

Fig. 4. Connector gadget for $\mathbf{H}(1,3)$-cover.

Indeed, suppose the vertices of $\widetilde{G}$ are colored so that each vertex has 3 neighbors of the opposite color. Then the middle two vertices of the connector graph have to get different colors. It follows that u_7 has the same color as u_8, and hence u_7 gets different color then u_1. Therefore u_1 has exactly one neighbor of its own color and 2 neighbors of the opposite color in G_1. Thus the restriction of this coloring to G_1 yields a coloring of G that satisfies Proposition 19.

On the other hand, if G admits such a coloring, we color the vertices of G_1, G_2, G_3 accordingly and the vertices of G_4, G_5 and G_6 with the colors interchanged. It is seen from the coloring depicted in Fig. 4 that this coloring of $\bigcup_{i=1}^{6} G_i$ can be extended to a coloring of $\widetilde{G}$ in which every vertex has one neighbor of its own color and 3 neighbors of the opposite color.

Finally, we let G' be the symmetric orientation of $\widetilde{G}$. It follows that G' covers $\mathbf{H}(1,3)$ if and only if G allows coloring satisfying Proposition 19, and it follows from this proposition that $\mathbf{H}(1,3)$-cover is NP-complete.

Lemma 22. *The $\mathbf{H}(1,m)$-cover problem is NP-complete for every $m \geq 4$.*

Proof. Let G be a directed graph subject to the question if G covers $\mathbf{H}(1,2)$. Take two copies of G, say G_1 and G_2 (with vertices u_1 resp. u_2 for every $u \in V(G)$) and construct $\widetilde{G}$ from $G_1 \cup G_2$ by connecting each pair u_1, u_2 by $m-2$ parallel edges directed from u_1 to u_2 and $m-2$ edges directed from u_2 to u_1.

If $f : G \to \mathbf{H}(1,2)$ is a covering projection, define $\widetilde{f} : \widetilde{G} \to \mathbf{H}(1,m)$ so that $\widetilde{f}(u_1) = f(u)$, $\widetilde{f}(u_2) \neq f(u)$ for $u \in V(G)$. Since interchanging the values in a covering projection onto $\mathbf{H}(1,2)$ results again in a covering projection, the restrictions of $\widetilde{f}$ to G_1 and G_2 are both covering projections onto $\mathbf{H}(1,2)$. Each u_1 (u_2) has other $m-2$ neighbors of the opposite color in G_2 (resp. G_1). Hence this $\widetilde{f}$ is a covering projection of $\widetilde{G}$ onto $\mathbf{H}(1,m)$.

Suppose on the other hand that $\widetilde{f} : \widetilde{G} \to \mathbf{H}(1,m)$ is a covering projection. The pairs u_1, u_2 are the only pairs of vertices joined by parallel edges, and so $\widetilde{f}(u_1) \neq \widetilde{f}(u_2)$ for every $u \in V(G)$. Hence every vertex u_1 has one neighbor of its own color and two neighbors of the opposite color in G_1, and $f : G \to \mathbf{H}(1,2)$ defined by $f(u) = \widetilde{f}(u_1)$ is a covering projection of G onto $\mathbf{H}(1,2)$.

Lemma 23. *The $\mathbf{H}(2,1)$-cover problem is NP-complete.*

Proof. Let G be a simple cubic graph and let $\widetilde{G}$ be the symmetric orientation of G. Then $\widetilde{G}$ covers $\mathbf{H}(2,1)$ if and only if G allows a coloring such that every vertex has two neighbors of its own color and one neighbor of the opposite color, i.e., if G covers $H(1,1)$. Thus the statement follows from Propostion 15.

Proposition 24. *The $\mathbf{H}(l,m)$-cover problem is NP-complete for every $l \geq 1, m \geq 1$ such that $l + m \geq 3$.*

Proof. Consider first the case $m > 1$. Let G be a directed multigraph subject to the question if G covers $\mathbf{H}(1,m)$. Construct $\widetilde{G}$ from G by adding $l-1$ loops to each vertex of G. Any $f : V(G) \to \{L,R\}$ is a covering projection of G onto $\mathbf{H}(1,m)$ if and only if it is also a covering projection of $\widetilde{G}$ onto $\mathbf{H}(l,m)$. The statement then follows from Lemmas 20, 21 and 22.

If $m = 1$, we have $l \geq 2$. Again, let $\widetilde{G}$ be obtained from G by adding $l-2$ loops to each vertex of G. Any $f : V(G) \to \{L,R\}$ is then a covering projection of G onto $\mathbf{H}(2,1)$ if and only if it is also a covering projection of $\widetilde{G}$ onto $\mathbf{H}(l,1)$. The statement then follows from Lemma 23.

5 Conclusion and Further Research

The primary purpose of this paper was to introduce covers of colored directed multigraphs, and to justify their introduction by showing that the discussion of their complexity is necessary for a complete solution of the complexity of covers of simple undirected graphs. We then illustrated our methodology by showing a complete discussion of the complexity of cdm-graph covers for two-vertex graphs (note that this classification contains as a proper substatement the classification of all simple graphs with exactly two vertices of degree higher than 2). Several of the lemmas used in the proof of this classification theorem can be actually stated in more general form, we have decided to state the simplified versions in order to keep the length of the paper reasonable. A full version of Lemma 13 will appear in [10]. It is rather a coincidence that all NP-complete 2-vertex graphs are symmetric. This is not the case for 3-vertex graphs. There the classification is more involved, as shown in [11]. In particular, it is no more true for 3-vertex cdm-graphs that H-cover is NP-complete if and only if H^c-cover is NP-complete for some edge color c.

References

1. J. Abello, M.R. Fellows and J.C. Stillwell, On the complexity and combinatorics of covering finite complexes, *Australasian Journal of Combinatorics 4* (1991), 103-112;
2. D. Angluin, Local and global properties in networks of processors, in *Proceedings of the 12th ACM Symposium on Theory of Computing* (1980), 82-93;
3. D. Angluin and A. Gardner, Finite common coverings of pairs of regular graphs, *Journal of Combinatorial Theory B 30* (1981), 184-187;
4. N. Biggs, *Algebraic Graph Theory*, Cambridge University Press, 1974;
5. M.R. Garey and D.S. Johnson, *Computers and Intractability*, W.H.Freeman and Co., 1978;
6. I. Holyer, The $\mathcal{NP}$-completeness of edge-coloring, *SIAM J. Computing 4* (1981), 718-720;
7. J. Kratochvíl: Regular codes in regular graphs are difficult, Discrete Math. 133 (1994), 191-205
8. J. Kratochvíl, A. Proskurowski, J.A. Telle: On the complexity of graph covering problems, In: Graph-Theoretic Concepts in Computer Science, Proceedings of 20th International Workshop WG'94, Munchen, Germany, 1994, Lecture Notes in Computer Science 903, Springer Verlag, Berlin Heidelberg, 1995, pp. 93-105.
9. J. Kratochvíl, A. Proskurowski, J.A. Telle: Covering regular graphs, to appear in Journal of Combin. Theory Ser. B
10. J. Kratochvíl, A. Proskurowski, J.A. Telle: Complexity of colored graph covers II. When 2-SAT helps. (in preparation)
11. J. Kratochvíl, A. Proskurowski, J.A. Telle: Complexity of colored graph covers III. Three-vertex graphs. (in preparation)

A Syntactic Approach to Random Walks on Graphs

M. MOSBAH N. SAHEB
LaBRI*, Université Bordeaux-I
351, cours de la Libération, 33405 Talence, France.
E-mail: {mosbah,saheb}@labri.u-bordeaux.fr

Abstract. We use formal language theory to study syntactic behaviour of random walks on graphs. The set of walks, viewed as sets of words, is a recognizable language. As a consequence, a set of random walks can be formally described by a rational fraction or equivalently by an automaton. Applying these techniques, we compute in a unified way various statistical parameters related to random walks, such as mean cover time, and the mean hitting time.

1 Introduction

Let G be a connected weighted undirected graph with n vertices. We consider random walks on G, where at each step the random walk moves from the current vertex to a neighbour chosen with a probability proportional to the weight of the traversed edge. Random walks have been studied extensively, and have numerous applications, including generation of random spanning trees [1, 6, 16, 18], online-algorithms[8], space-efficient algorithms for undirected connectivity [7], approximation algorithms [9, 14], assigning processes to nodes in networks [4], and token management schemes and self-stabilizing in distributed computing [13, 17].

The cover time of a random walk on a graph is the number of steps required for all vertices to be visited. The hitting time between two vertices is the number of steps from the first vertex to reach the second. Computing the expected cover time of a graph is a difficult problem in general, for which no deterministic polynomial time algorithm is known yet. It is known that for a general graph, the mean cover time can be computed in $O(n^3)$ expected time[2]. This complexity can be reduced to $O(n^2)$ if the graph is regular. The cover time is easy to study for particular classes of symmetric graphs, such as complete graphs or stars for which the above complexity is $O(n \ln n)$. The lack of simple characterizations of the expected cover time makes it difficult to exhibit efficient solutions. The goal of our work is to contribute to a syntactic characterization of the cover time, by giving a precise description of the walk behaviour before terminating. To do so, we introduce and study tools borrowed from formal languages and automata theory. This way, we can evaluate parameters of interests related to random walks

---

* Laboratoire associé au CNRS

on a large class of graphs. The key-point of this study is that these parameters may be reformulated in terms of statistics over regular languages, which in turn can be investigated, directly whenever they have a simple expression, or else indirectly using finite automata.

In this paper, we identify sets of random walks with automata. For a fixed graph and a given vertex, we construct an automaton that describes the behaviour of a random walk starting at this vertex. The aim of the automaton is to distinguish between the first visit and next visits of a vertex. Its states, as we will show, are subsets of vertices constructed gradually by considering new visited vertices. Although the size of the automaton is huge, it is possible to exploit it to compute the exact value of the mean cover time starting at a vertex of a given graph. In particular, we calculate the mean cover time starting at a leaf of a tree. Furthermore, it is known that an automaton recognizes a regular expression over an alphabet, in this case, the labels of the edges.

The regular expression characterizing a random walk on a graph can be written as a formal fraction using the similar notations as [10, 12, 3]. This fraction provides explicit formulae of the hitting time and the cover time, in a straightforward and elegant manner. It unifies the computation of many interesting statistical parameters related to random walks. More precisely, to compute a given parameter, such as the average number of visits to a vertex, we associate an appropriate generating function to the formal fraction. We apply this method to compute the cover time for internal vertices of a tree. We have used this technique in a previous paper[16] to prove that the probability of generating a random spanning tree is proportional to its weight. This provides a simple algorithm for generating a random spanning tree on a weighted graph. Moreover, we show that this formal fraction for the cover time, viewed as a syntactic structure, is a compound fraction, however it is very simple for the hitting time.

The method does not provide a simple explicit expression for the mean cover time as a function of the graph and the starting vertex in major cases. In fact, the study displays the exponential size of the recognizing automaton in terms of the size of the graph, for the language identifying the set of walks visiting all vertices and terminating once all vertices are visited. This implies that this set must be of exponential syntactic structure and, therefore, the computation of the cover time in general must be difficult as well. The same technique applied to the hitting time yields a rational language of linear complexity and provides simple expressions. Thus another interest of this work is that it introduces an explicit distinction between "hard" parameters, being of exponential syntactic complexity, and "easy" ones, of polynomial syntactic complexity.

Although the exhibited applications are often carried out on trees, our method seems sufficiently powerful to take into account larger classes of graphs. As far as the cover time is concerned, it is known that it is difficult to compute it even for trees[2].

The paper is organized as follows. In Section 2, we introduce a few notations and definitions related to random walks on graphs. The construction of the automata modelling random walks is explained in Section 3. In Section 4, we define formal fractions associated to sets of random walks on trees, and give useful properties. These techniques are applied to compute cover time and hitting time in Section 5. Applications to classes of graphs are given in Section 6.

2 Notations and definitions

Throughout this paper, $G = (V, E)$ denotes a fixed simple connected undirected graph. We consider the alphabet A whose letters are labels of edges of E directed in both senses. A walk over G is a finite path in G, which can be identified by a word over A. Using the same notations as Eilenberg[10], we can model interesting sets of finite walks by recognizable subsets of A^* which may be written as regular expressions.

For a vertex $i \in V$, a *cover tour* starting at i is a walk beginning at i, visiting all vertices of V and terminating once all vertices have been met. The language $C_i \in A^*$ containing all cover tours starting at i will be referred to as the *cover language* from i. Another useful notion is that of the hitting path. For two given vertices i and j, a walk w is a *hitting path* from i to j, if it is a path starting at i and ending as soon as it reaches j for the first time. The hitting language H_{ij} from i to j is the set of all such paths.

A random walk on a graph can also be simulated with a particle, which is initially located at the starting vertex, and moves from a vertex to one of its neighbours with a probability proportional to the weight of the traversed edge. It stops once all vertices have been visited.

3 Recognizing finite automata

It is easy to see that C_i and H_{ij} are recognizable languages. The recognizing automata, as we shall see, allow to compute interesting statistics relative to random walks over an undirected graph. The construction of an automaton for H_{ij} is quite obvious. In the sequel, we give an explicit construction, since it provides a powerful tool to compute the mean cover time (i.e. the expected length of a cover tour) for any given graph.

Let $G = (V, E)$ be a graph as above, and let i be a vertex of G, taken as the starting vertex. For a nonempty subset X of V, let $G_X = (X, E_X)$ denote the subgraph of G induced by X, i.e. $E_X = \{(x, y)/x \in X, y \in X\}$. The finite deterministic automaton $\mathcal{A}_i$ is introduced as follows. The set Q of states is the set of all ordered pairs (X, j), with $j \in X$, where X is any set of vertices containing i such that the induced subgraph G_X is connected. Roughly speaking, in a walk over G, the state (X, j) corresponds to the fact that j is the currently visited vertex and that X is the set of already visited vertices. The initial state of $\mathcal{A}_i$ will be $(\{i\}, i)$. The terminal states of $\mathcal{A}_i$ are those of the form (V, j), $\forall j \in V$.

These states correspond to the event that all vertices of G have been visited, and thus as we shall see, no transitions are defined for them. We now define the transition (partial) function θ over Q as follows.

1. For any pair of states $r = (X, j)$ and $s = (X, k)$, with $X \neq V$, such that j and k are neighbours in G, we let $\theta(r, a) = s$ (resp. $\theta(s, b) = r$), where $a \in A$ (resp. $b \in A$) is the letter labelling the oriented edge $(j, k) \in E_x$ (resp. (k, j)).
2. For a pair of states $r = (X, j)$ and $s = (X \cup \{k\}, k)$ with $k \notin X$, we let $\theta(r, a) = s$, where $a \in A$ is the label of the directed edge (j, k).

   Let us notice that it is possible to reduce the set of terminal states into a unique finite state.

Example 1. Let G be the following tree graph. Consider the alphabet $A =$

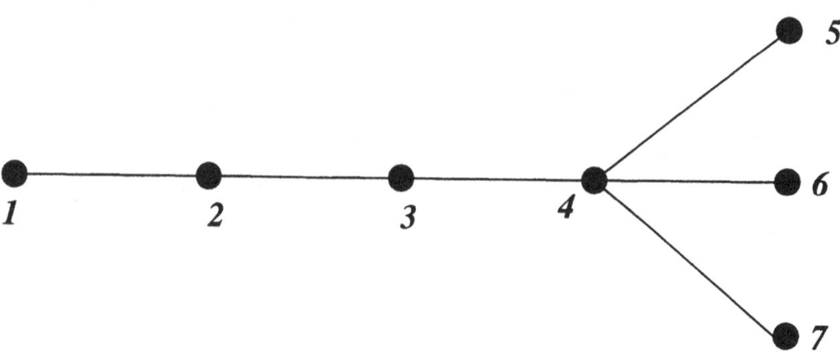

Fig. 1. A tree

$\{a, b, c, d, e, f, g, a', b', c', d', e', f', g'\}$ where $a, b, \ldots, f$ are labels of $(1,2), (2,3), \ldots, (4,7)$ and $a', b', \ldots, f'$ are those of $(2,1), (3,1), \ldots, (7,4)$. The automaton of Figure 2 recognizes the cover language C_1 starting at vertex 1. The first transition is to vertex 2, as it is the only possibility. But when the particle is at vertex 2, it can either go back to vertex 1 which is the state $(\{1,2\}, 1)$, or by reading b go to vertex 3 and so on. Observe that there are many levels in the automaton, a transition to a level below corresponds to a new visited vertex, however transitions on the same level correspond to already visited vertices. Note that in the figure, the first transition from vertex 4 to 5, is similar to that from 4 to 6 or from 4 to 7, since vertices 5,6 and 7 are symmetric w.r.t 4. In order to avoid heaviness of the figure, we represent only states from 4 to 5, the others (dashed transitions) are similar.

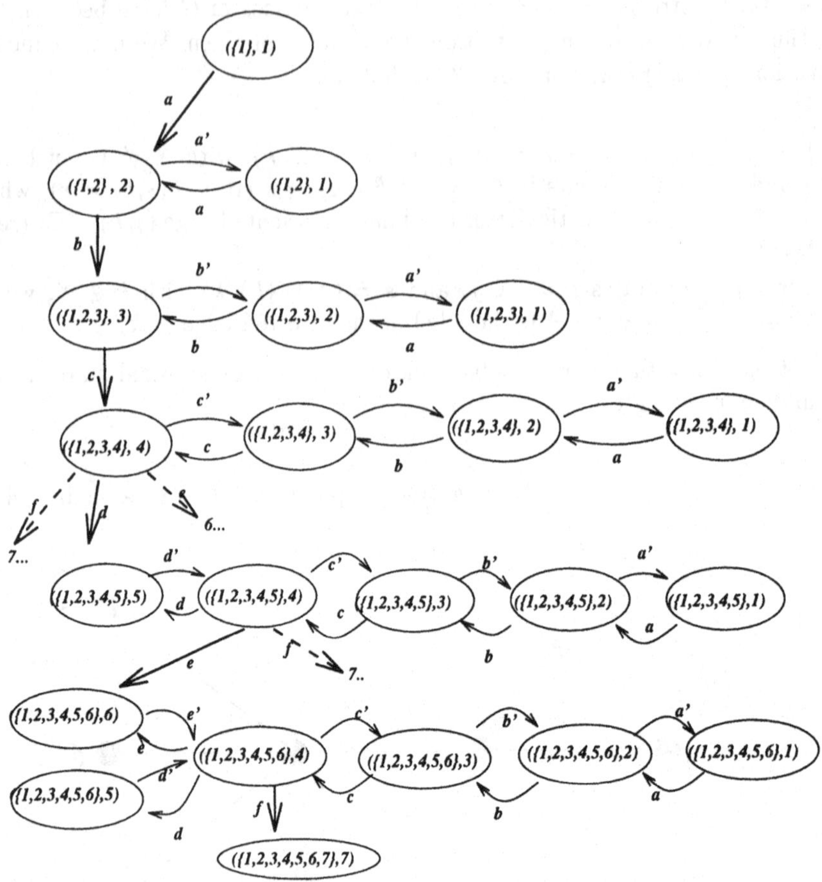

Fig. 2. Automaton recognizing C_1.

4 Trees

If $G = (V, E)$ is a tree, then there is a unique path between any pair of vertices. This characteristic property allows to derive the following relationships between hitting languages on the one hand and cover languages and hitting languages on the other hand.

Throughout this section $G = (V, E)$ is a tree. The hitting language $H_{i,j}$, for any pair of vertices $i, j \in E$, can be expressed by the hitting languages for neighbour vertices, as follows:

Proposition 4.1 *Let* $i, j \in V$. *If* $i_0 = i, i_1, \ldots, i_k = j$ *is the sequence of vertices on the path from* i *to* j, *then:*

$$H_{i,j} = H_{i_0,i_1}.H_{i_1,i_2}.\ldots.H_{i_{k-1},i_k}.$$

Proof. It follows from the fact that there is a unique path from i to j.

It is possible to express the cover language from i as a sum of products of hitting languages. For a given set X of vertices inducing a factor tree of G (i.e. X is such that $G_X = (X, E_X)$ is connected), let $H_{i,j}(X)$ denote the set of hitting paths from i to j in G_X. Now, let F be the set of leaves and I the set of internal vertices; hence $V = I \cup F$. Assume, without loss of generality, that the leaves are numbered $1, 2, \ldots, k$, where k is the size of F. Let π denote the set of permutations of F. Hence, an element σ of π will be written using the standard notation $\sigma(1), \ldots, \sigma(k)$. For a leaf i in F, we denote by π_i the set of permutations of F whose first element is i, i.e. the set permutations σ such that $\sigma(1) = i$.

The following proposition gives the cover language C_i, for an internal vertex i, in terms of cover languages C_j starting at leaves j.

Proposition 4.2 *For an internal vertex i of the tree G, we have*

$$C_i = \sum_{j \in F} H_{ij}(I \cup \{j\}) C_j.$$

Proof. If i is an internal vertex, then there is a unique path to each leaf j in the tree. Hence, a cover tour from i, first hits a leaf, say j, and then is a cover tour from j. This corresponds to the language $H_{ij}(I \cup \{j\})C_j$. By summing over all leaves, we obtain the cover language from i.

Remark. It is possible to extend the above proposition as follows. Let i be an internal vertex and let U be a subset of vertices, such that for every leaf $l \in F$, there exists exactly one vertex j in U on the unique path from i to l. We have thus for such a subset U of vertices

$$C_i = \sum_{j \in U} H_{ij}(U' \cup \{j\}) C_j$$

where U' is the set of all vertices, not in U, belonging to the path from i to a vertex of U. In other words, each term of the above sum corresponds to a cover language beginning at i, hitting j as the first element of U, and then beginning a cover language from j.

In particular, if $N(i)$ is the set of neighbours of i, then

$$C_i = \sum_{j \in N(i)} a_{ij} C_j$$

where a_{ij} is the label of the directed edge (i, j).

Proposition 4.3 *Let i be a vertex of the tree G.*
If $i \in F$, then

$$C_i = \sum_{\sigma \in \pi_i} H_{\sigma(1),\sigma(2)}(I \cup \{\sigma(1), \sigma(2)\}) H_{\sigma(2),\sigma(3)}(I \cup \{\sigma(1), \sigma(2), \sigma(3)\}) \ldots H_{\sigma(k-1),\sigma(k)}(V),$$

otherwise, if $i \in I$, then

$$C_i = \sum_{\sigma \in \pi} H_{i,\sigma(1)}(I \cup \{\sigma(1)\}) H_{\sigma(1),\sigma(2)}(I \cup \{\sigma(1), \sigma(2)\}) \dots H_{\sigma(k-1),\sigma(k)}(V).$$

Proof. Clearly, if the graph G is a tree, all vertices are visited iff all leaves have been visited. Thus, for a fixed σ, any term of the above sums corresponds to a set of cover paths whenever an ordering of leaf-visiting is imposed. The proposition follows.

It is also possible to give a simple regular expression under a formal rational fraction for a hitting language, and therefore a sum of formal rational fractions for a cover language. This representation has been explicitly introduced by Flajolet[3, 12]. It is based on the fact that, for $L \in A^\star$, $L^\star$ can be written formally as

$$L^\star = \sum_{n=0}^{\infty} L^n = \frac{1}{1-L}.$$

For more details see [3, 10, 12]. It should be noted here that, since a recognizable language is rational, there exist formal expressions under the form of a sum of formal rational fractions for the involved languages, whether the graph is a tree or not. But the following proposition allows to construct inductively the above stated formal fraction avoiding solving a system of formal linear equations in the case of trees.

For a pair of vertices i and j and a factor tree G' containing i and j, let $W_{i,j}(G')$ denote the set of walks starting at i and ending with j traversing only vertices in G'.

Proposition 4.4 *$W_{i,i}(G')$ can be written inductively as a formal rational fraction as follows.*

- *If G' is a single vertex i, then $W_{i,i}(G') = 1$.*
- *Otherwise, let $j_1, \dots, j_k$, $k \geq 1$, be the neighbouring vertices of i in G'. Suppose that the directed edges $(i, j_1), (i, j_2), \dots, (i, j_k)$ are labelled $a_1, a_2, \dots, a_k$ respectively; and are labelled $a_1', a_2', \dots, a_k'$ in the converse direction. We have then*

$$W_{i,i}(G') = \frac{1}{1 - \sum_{h=1}^{k} a_h W_{h,h}(G_h) a_h'}$$

where G_h is the subtree of G' containing j_h after removing the edge (i, j_h)

Proof. The first assertion corresponds to the language reduced the empty word. The second expression is the fractional form of the following identity

$$W_{i,i} = (\sum_{h=1}^{k} a_h W_{h,h}(G_h) a_h')^\star.$$

We deduce the following

Lemma 4.1 *Let i and j be two neighbour vertices in G, and let G_i denote the subtree of G containing i after removing the edge (i,j). If a is the label of the directed edge (i,j), then*

$$H_{i,j} = W_{i,i}(G_i)a.$$

Proof. To reach the neighbour j, a particle starting at i can first explore the tree rooted at i, return to i, and then go to j.

As a consequence of the above lemma and Proposition 4.4, the hitting language can be characterized by the inductive equation in the following proposition.

Proposition 4.5 *Let i and j be two neighbours, and let $j_1, j_2, \ldots, j_k$ be the neighbours of i other than j. If a is the label of the directed edge (i,j), and if $a_1, a_2, \ldots, a_k$ are the labels of the directed edges $(i,j_1), (i,j_2), \ldots, (i,j_k)$, then*

$$H_{ij} = \frac{1}{1 - \sum_{h=1}^{k} a_h H_{j_h i}} a.$$

Proof. It is obtained by combining Proposition 4.4 and Lemma 4.1.

Note that for hitting languages, the recognizing automata are very simple, and allow to calculate easily expressions for interesting statistics. This will be studied in the next section. Unfortunately, in the case of tour languages, the corresponding automata have an exponential number of states in the size of the graph. Even for trees, the automata must memorize the visited leaves, and hence the number of states remains exponential in the number of leaves. As a matter of fact, this implies that the persisting difficulty of computing some parameters related to random walks on graphs, such as mean cover time, is due to the syntactic complexity of the tour languages.

5 Hitting time and cover time

Many authors [1, 6, 5] have studied uniform random walks, by considering the hypothesis that the walk moves from a vertex v to one of its neighbours with the same probability $\frac{1}{d(v)}$, where $d(v)$ is the degree of v. It seems, however, that this assumption does not simplify the computation of interesting parameters and that the complexity of the computation is linked to the syntactic structure of the languages representing sets of walks. Here, we consider a positive real valued mapping w called weight over E. The probability of moving from i to a neighbour j will be $p_{ij} = \frac{w(i,j)}{w(i)}$, where $w(i) = \sum_{j \in N(i)} w(i,j)$. Thus p_{ij} is proportional to the weight of the edge (i,j). The probability $p_i(w)$ of a walk starting at i is the product of the probabilities of its directed edges.

C_i and H_{ij} are probabilistic languages, in the sense that the mapping p from 2^{C_i} into $[0,1]$ defined by $p(L) = \sum_{w \in L} p_i(w)$, $\forall L \in 2^{C_i}$, is σ-additive and $p(C_i) = 1$. This is due to the fact that C_i is a prefix language, i.e. whenever

$w \in C_i$, then for no $w' \neq 1$, $ww' \in C_i$. The equation $p(C_i) = 1$ means that the particle moves from a vertex i to a neighbour j with probability $\frac{w(i,j)}{w(i)}$, and will visit all vertices with probability 1 (see [16] for more details). The same assertions are true for H_{ij}. The regular expressions of these languages should allow to compute statistics such as hitting times or mean cover times. The *mean hitting time* h_{ij} or the *mean cover time* c_i is defined as the expected length of a random word in H_{ij} or C_i. That is,

$$h_{ij} = \sum_{w \in H_{ij}} p_i(w)|w|$$

and

$$c_i = \sum_{w \in C_i} p_i(w)|w|$$

where $|w|$ is the length of w.

To compute statistical parameters related to random walks, it suffices to substitute in the formal expression of the considered language L (cover language hitting language), each label a with $z\frac{w(a)}{w(I(a))}$, where $I(a)$ is the initial vertex of the edge labeled a. This yields the generating function $g(z)$ of the length of the words of L, interpreted as time. More precisely, if a word $w \in L$ is chosen with probability $p(w)$ from L, then its length is a random variable whose generating function is $g(z)$. Thus the random variable length is fully characterized by its generating function which is obtained by the above substitution.

Since the hitting languages have simpler syntactic expressions, we first consider the statistics of *hitting time* which is the length of words of H_{ij}. The following proposition generalizes the result of [5] to weighted trees,

Proposition 5.1 *Let i and j be two neighbour vertices of the tree T. Then the mean hitting time h_{ij} is*

$$h_{ij} = \frac{2\sum_{e \in T_i} w(e)}{w(i,j)} + 1$$

where T_i is the connected subtree of T containing i, and obtained from T by removing the edge (i,j).

Proof. Let $h_{ij}(z)$ be the generating function associated with H_{ij}, then by substituting $\frac{w(i,j_l)}{w(i)}z$ for a_l and a for $\frac{w(i,j)}{w(i)}z$ in the equation of Proposition 4.5, we get

$$h_{ij}(z) = \frac{1}{1 - \sum_{l=1}^{k} \frac{w(i,j_l)}{w(i)} z h_{j_l i}(z)} \frac{w(i,j)}{w(i)} z.$$

Differentiating $h_{ij}(z)$ and taking $z = 1$, we get

$$h'_{ij}(1) = \frac{w(i)}{w(i,j)}(1 + \sum_{l=1}^{k} \frac{w(i,j_l)}{w(i)} h'_{j_l i}(1))$$

which is similar to

$$w(i,j)h'_{ij}(1) = w(i) + \sum_{l=1}^{k} w(i,j_l)h'_{j_l i}(1).$$

By a structural induction on T_i, we finally get

$$h_{ij} = h'_{ij}(1) = \frac{w(i,j) + 2\sum_{e \in T_i} w(e)}{w(i,j)}.$$

The previous proposition can be extended to non neighbouring vertices i and j.

Corollary 5.1 *Let i and j be two arbitrary distinct vertices of a tree T. Let $i_0 = i, i_1, i_2, \ldots, i_k = j$ be the unique path from i to j. Define the tree T_{i_h}, $h = 0, \ldots, k-1$ as in Proposition 5.1, we have then*

$$h_{ij} = k + 2\sum_{h=0}^{k-1} \frac{\sum_{e \in T_{i_h}} w(e)}{w(i_h, i_{h+1})}.$$

One of the most challenging topic in the area is the computation of the mean cover time of a random walk on a graph[2]. This study is linked, in particular, with the complexity of generating a random spanning tree[1, 6, 18]. For a given graph, Proposition 4.2 allows to find an explicit expression for C_i. The above substitution in C_i provides the generating function of the random variable *cover time* which is the length of the random word w chosen from C_i with probability $p(w)$. The mean cover time c_i, which is the expected value for this random variable, can be obtained from the generating function. Another direct technique for computing the mean cover time is by using the automaton recognizing C_i. The automaton $\mathcal{A}_i$ allows in fact to write a linear system of equations on the average "remaining length" of a cover tour from different states, yielding this way the mean cover time. At each state, we can write an equation to compute the expected time to go to the next non visited vertex.

Example 2. Consider the automaton of Figure 2, corresponding to the tree of Figure 1. The mean length of a tour beginning at vertex 1, denoted by c_1, is the same as the average length of a word recognized by the automaton beginning from the initial state to the terminal one. For each state, we can write an equation to compute the average remaining length of a word to reach the terminal state :

$$c_1 = 1 + c_2'$$

where c_2' is the average length of a word from state $(\{1,2\},2)$ to reach the terminal state, which is also the mean cover time of a tour starting at 2 given that vertex 1 is visited. Similarly, at the second (resp. third) level of the automaton, if c_3' (resp. c_4') is the average length of a word from the state $(\{1,2,3\},3)$ (resp. $(\{1,2,3,4\},4)$), then we have

$$c_2' = 3 + c_3', \quad c_3' = 5 + c_4'.$$

Note that the subtree between vertices 1 and 4 is a chain, Section 6.3 deals with computing the cover time of a chain.

Now it is easy to see that

$$c_4' = 1 + \frac{1}{4}c_3' + 3 \times \frac{1}{4}c_5'$$

where c_5' is the average length of a word to go from state $(\{1,2,3,4,5\},5)$ to the terminal one. Note that due to the symmetry of vertices 5, 6 and 7 w.r.t 4, we have $c_5' = c_6' = c_7'$, where c_6' and c_7' are defined in a similar way as c_5'.

Thus, let x_4, y_4, and x_6 be the average lengths of words from the states $(\{1,2,3,4,5\},4)$, $(\{1,2,3,4,5,6\},4)$, and $(\{1,2,3,4,5,6\},6)$, respectively, to the terminal state. Then, we have

$$\begin{aligned} c_5' &= 1 + x_4 \\ x_4 &= 1 + \frac{1}{2}x_6 + \frac{1}{4}(x_4 + 5) + \frac{1}{4}c_5' \\ x_6 &= 1 + y_4 \\ y_4 &= 1 + \frac{1}{2}x_6 + \frac{1}{4}(y_4 + 5). \end{aligned}$$

Solving this system of equations, we get

$$y_4 = 11,\ x_6 = 12,\ x_4 = 17,\ c_5' = 18,\ c_4' = 21,\ c_3' = 26,\ c_2' = 29.$$

Finally, we have the mean cover time starting at vertex 1

$$c_1 = 30.$$

This means that, a random walk starting at 1 will need on the average 30 steps to visit all vertices. By using similar techniques, we find that the mean cover time of a walk starting at vertex 5 :

$$c_5 = c_6 = c_7 = 36.$$

In the case of tree graphs, there are simple relationships between the mean cover time c_i from a giver vertex, and the mean cover times from the neighbouring vertices.

Proposition 5.2 *If i is an internal vertex of the tree T, then*

$$c_i = 1 + \frac{1}{w(i)} \sum_{j \in N(i)} w(i,j)c_j.$$

Proof. A direct probabilistic reasoning establishes the proposition.

It is also possible to apply the above technique using Proposition 4.2 to obtain

$$c_i(z) = \frac{z}{w(i)} \sum_{j \in N(i)} w(i,j) c_j(z)$$

where $c_l(z)$ is the generating function for the length of the cover time from vertex l. The proposition then follows by differentiating the above equation and letting $z = 1$. (we use the fact that $c_l(1) = 1$).

The above proposition does not hold for the leaves. Thus it provides $|I|$ equalities with n unknown mean cover times. Therefore, we can compute any $c_i, i \in I$, in terms of $c_j, j \in F$. It remains henceforth to calculate mean cover times starting at the leaves.

Example 3. Consider again the tree in Section 3. Let us compute c_2, c_3 and c_4. By the above proposition we have

$$c_2 = 1 + \frac{1}{2}c_1 + \frac{1}{2}c_3$$
$$c_3 = 1 + \frac{1}{2}c_2 + \frac{1}{2}c_4$$
$$c_4 = 1 + \frac{1}{4}c_4 + \frac{1}{4}(c_5 + c_6 + c_7)$$

with $c_1 = 30, c_5 = c_6 = c_7 = 36$. The solution of the system is

$$c_2 = \frac{172}{5}, c_3 = \frac{184}{5}, c_4 = \frac{186}{5}.$$

It is clear that the syntactic complexity of C_i is a major difficulty to compute efficiently mean cover times. In very particular cases of graphs, having a strong symmetric structure, it is possible to find a closed expression for this parameter. This is investigated in the next section.

6 Mean cover times for certain classes of graphs

We consider here 3 classes of graphs, having very regular shapes, for which the presented techniques lead to a closed expression. In this investigation, the computation is done only for uniform random walks. In the case of chains, it can be easily extended to non uniform ones. The cycle graphs have been studied in a previous paper[15].

6.1 Complete graphs

The study of the cover time for this class of graphs is very related to the classical problem of *coupon collector* used by statisticians to test the randomness of a sequence of digits[11]. It has been already solved. We use automaton techniques to find the mean cover time. Let K_n, $n \geq 1$, be a complete graph of size n. Taking the advantage of the strong symmetry, we may identify all states of the automaton $\mathcal{A}_i$ of the form (X, j), $j \in X$, for different subsets of vertices X, whenever they have the same cardinality. This is possible if the studied statistics does not make any difference between edges. Thus the automaton reduces to an n-state one, in which a state is identified with k, $1 \leq k \leq n$, where k is the number of already visited vertices. Using this identification, for any state $k, 1 \leq k \leq n-1$, there are two transitions $k \rightarrow k$ and $k \rightarrow k+1$; the first one has the probability $\frac{k-1}{n-1}$ and the second one $\frac{n-k}{n-1}$. Let c_k be the expected number of remaining transitions from the state k to x reach the terminal state n. We have

$$c_k = 1 + \frac{k-1}{n-1} c_k + \frac{n-k}{n-1} c_{k+1}, \quad 1 \leq k \leq n-1, \quad c_n = 0.$$

This yields $c = c_1 = (n-1) \sum_{i=1}^{n-1} \frac{1}{i}$. Therefore the mean cover for k_n from any vertex in $(n-1)H_{n-1}$, which is asymptotically equivalent to $n \ln(n)$.

6.2 Star graphs

A star graph of size n is known for having the least mean cover time in the class ot trees of size n, see [5]. Let S_n be a star graph of size n, $\geq n$. Let c_0 be the mean cover time from the central vertex and let c be the mean cover time from any other vertex. We have obviously $c_0 = c + 1$. The previous technique of simplification may be used to reduce the automaton recognizing the cover tour starting at a leaf. Let $r_1, s_1, r_2, s_2, \ldots, r_k, s_k, \ldots, r_{n-2}, s_{n-2}, r_{n-1}$ be its states, where r_k corresponds to the event that k leaves have been visited and that the current visited vertex is a leaf, and s_k to the fact that k leaves have been visited and the current visited vertex is the center. If c_k denotes the expected number of remaining transitions from the state r_k, we have

$$c_k = 2 + \frac{k}{n-1} + \frac{n-1-k}{n-1}, \; 1 \leq k \leq n-2, \quad c_{n-1} = 0.$$

This yields $c = c_1 = 2(n-1) \sum_{i=1}^{n-2} \frac{1}{i}$. Therefore the mean cover time for a star of size n is $2(n-1)H_{n-2}$, which is asymptotically equivalent to $2n \ln(n)$.

6.3 Chain graphs

The mean cover time for chain graphs or paths have been studied by Brightwell and Winkler[5] who found the minimal value $(n-1)^2$ for the endpoints. They conjectured that the maximum value is $\frac{5(n-1)^2}{4}$ for the midpoint (or midpoints). As we shall see, this conjecture holds only for odd values of n.

We first compute the mean cover time for an endpoint in a chain of size n. Let the vertices be labeled $1, 2, \ldots, n$, where 1 is the starting vertex of the cover tour. According to Proposition 4.2, The cover language C_1 is the same as the hitting language H_{1n}. Thus the mean cover time c_1 and the mean hitting time from 1 to n coincide. The latter has been given by Proposition 5.1. We have therefore

$$c_1 = 1 + 3 + 5 + \ldots + 2n - 3 = (n-1)^2.$$

We distinguish the following two cases :

- n is odd, say $n = 2m + 1$. Let c be the mean cover time from the midpoint. It could be easily shown by using Remark of Section 4, that if x_k is the cover time from a vertex of distance k from the center, then $x_k = c - k^2$.
  We deduce that c is the maximum cover time and that
  $$x_m = c_1 = (n-1)^2 = (c-n)^2.$$
  Thus
  $$c = (n-1)^2 + m^2 = \frac{5(n-1)^2}{4},$$
  which proves the conjecture for the chains of odd length.
- n is even, $n = 2m$. Let c be the mean cover time from the midpoints. As above, the mean cover time x_k from a vertex of distance k from the nearest midpoint is computed easily yielding $x_k = c - k(k+1)$, which proves that c is the maximum mean cover time. Furthermore, we have
  $$c = (n-1)^2 + m(m-1) = (n-1)^2 + (\frac{n}{2} - 1)\frac{n}{2} = \frac{5(n-1)^2 - 1}{4}.$$
  This is slightly different from the conjectured value.

References

1. D.J. Aldous. The random walk construction of uniform spanning trees and uniform labelled trees. *SIAM Journal on Discrete Mathematics*, 3(4):450–465, 1990.
2. D.J. Aldous and J.A. Fill. *Reversible Markov Chains and Random Walks on Graphs*. 1996. Book in preparation.
3. D. Arquès and J. Françon. Arbres bien étiquetés et fractions multicontinues. In B. Courcelle, editor, *Proceedings of 9th Colloquium on Trees in Algebra and Programming*, pages 50–61, Bordeaux, France, March 1984. Cambridge University Press.
4. S. Bhatt and J.Y. Cai. Take a walk, grow a tree. In *29th Annunal IEEE Symposium on Foundations of Computer Science*, pages 469–478, 1988.
5. G. Brightwell and P. Winkler. Extremal cover times for random walks on trees. *Journal of Graph Theory*, 14(5):547–554, 1990.
6. A.Z. Broder. Generating random spanning trees. In *Proc. 30th Ann. IEEE Symp. on Foundations of Computer Science*, pages 442–453, October 1989.
7. A.Z. Broder, A.R. Karlin, P. Raghavan, and E. Upfal. Trading space for time in undirected *s-t* connectivity. In *ACM Symposium on Theory of Computing (STOC)*, pages 543–549, 1989.

8. D. Coppersmith, P. Doyle, P. Raghavan, and M. Snir. Random walks on weighted graphs and applications to on-line algorithms. *Journal of the ACM*, 40(3):421–453, July 1993.
9. M. Dyer, A. Frieze, and R. Kannan. A random polynomial time algorithm for approximating the volume of covex bodies. In *ACM Symposium on Theory of Computing (STOC)*, pages 375–381, 1989.
10. S. Eilenberg. *Automata, Languages, and Machines*, volume A. Academic Press, Newyork, 1974.
11. W. Feller. *An introduction to probability theory and its applications*, volume Vol. 1. 2nd ed. Wiley, New York, 1957.
12. P. Flajolet. Combinatorial aspects of continued fractions. *Discrete Mathematics*, 32:125–161, 1980.
13. A. Israeli and M. Jalfon. Token management schemes and random walks yield self-stabilizing mutual exclusion. In *Proc. of the Ninth Annual Symposium on Principles of Distributed Computing*, pages 119–131, 1990.
14. M. Jerrum and A. Sinclair. Conductance and the rapid mixing property of markov chains: the approximation of the permanent resolved. In *ACM Symposium on Theory of Computing (STOC)*, pages 235–244, 1988.
15. M. Mosbah and N. Saheb. Formal rational fractions and random walks on cycle graphs. Technical Report 1147-96, University of Bordeaux 1, 1996. accepté à FPSAC 97 - Vienne 14-18 juillet 97.
16. M. Mosbah and N. Saheb. Non uniform random spanning trees on weighted graphs. Technical Report 1143-96, University of Bordeaux 1, 1996.
17. P. Tetali and P. Winkler. On a random walk arising in self-stabilizing token management. In *Proceedings of the Tenth Annual ACM Symposium on Principles of Distributed Computing*, pages 273–280, 1991.
18. D.B. Wilson and J.G. Propp. How to get an exact sample from a generic Markov chain and sample a random spanning tree from a directed graph, both within the cover time. In *Proceedings of the Seventh Annual ACM-SIAM Symposium on Discrete Algorithms*, pages 448–457, Atlanta, Georgia, 28–30 January 1996.

Bicliques in Graphs II: Recognizing k-Path Graphs and Underlying Graphs of Line Digraphs

ERICH PRISNER *

Mathematisches Seminar
Universität Hamburg

Abstract. Given an undirected graph G, CHVATAL and EBENEGGER showed that deciding whether there is some loopless digraph D such that G is the underlying graph of the line digraph of D is $\mathcal{NP}$-complete. However, we shall show that the question whether there is such a digraph (with loops allowed) with minimum in-and out degrees not less than 2 can be decided in time $O(|V|^2|E|^2)$ In that case, we show that D is unique modulo reverse, extending previous uniqueness results by VILLAR.
The *k-path graph* $\mathcal{P}_k(H)$ of a graph H has all length-k paths of H as vertices; two such vertices are adjacent in the new graph if their union forms a path or cycle of length $k+1$ in H, and if the edge-intersection of both paths forms a path of length $k-1$. We also show that, given a graph $G = (V, E)$, there is an $O(|V|^4)$-time algorithm that decides whether there is some graph H of minimum degree at least $k+1$ with $G = \mathcal{P}_k(H)$. If it is, we show that k and H are unique, extending previous uniqueness results by XUELIANG LI.
The algorithms are rather similiar and work with the *bicliques*—inclusion-maximal induced complete bipartite subgraphs—of the graphs. Cruical is the fact that underlying graphs of line digraphs, as well as k-path graphs contain only 'few' large bicliques (i.e. bicliques containing $K_{2,2}$).

1 Introduction

Many interconnection networks like the famous (undirected) Kautz, de Brujin, or butterfly graphs seem to consist of a number of complete bipartite graphs, glued together in a certain sophisticated way. Let a *biclique* denote an inclusion-maximal induced complete bipartite subgraphs of a graph. Then bicliques are rather visible in these networks mentioned.

Bicliques play also a fundamental role in certain other graphs. Examples are, as we shall see, underlying graphs of line digraphs, where the well-known line digraph $L(D)$ of a digraph $D = (V, A)$ has A as vertex set, and an arc from

* Mathematisches Seminar, Universität Hamburg Bundesstr. 55, 20146 Hamburg, Germany; supported by the Deutsche Forschungsgemeinschaft under grant no. Pr 324/6-1; part of this research was done at Clemson University, whose hospitality is greatly aknowledged.

xy to zw if and only if $y = z$. Underlying graphs of iterated line digraphs are rather useful for the design of interconnection networks, as has been pointed out in [4]. Kautz, de Bruijn, and butterfly networks arise in this way. Even analysis of these networks can be unified by using the line digraph language. However, the last step of the construction—considering the underlying graph—is somehow artificial and even uneconomical. Therefore one might ask whether there is a similiar way to construct 'good' large interconnection networks from smaller ones without using the bypass over digraphs. One possibility seems to be the so-called 'k-path graph' operator.

We call two length-k paths $x_0, x_1, \ldots, x_k$ and $y_0, y_1, \ldots, y_k$ of an (undirected) graph H *adjacent* whenever their union forms a path or cycle of length $k+1$, and the edge-intersection is a path of length $k-1$. In other words, they are adjacent if (i) $x_i = y_{i+1}$ for $0 \leq i \leq k-1$, or (ii) $x_i = y_{k-1-i}$ for $0 \leq i \leq k-1$, or (iii) $y_i = x_{i+1}$ for $0 \leq i \leq k-1$, or (iv) $y_i = x_{k-1-i}$ for $0 \leq i \leq k-1$. The ***k*-path graph** $\mathcal{P}_k(H)$ of a graph H has all length-k paths of H as vertices, and two vertices are adjacent if the corresponding paths are adjacent by the definition above. Note that what we call $\mathcal{P}_k(H)$ is denoted by $\mathcal{P}_{k+1}(H)$ in [2], [6], and [1], but we adopt the terminology used in [10]. Note also that in most of the papers on k-path graphs, as in [2,6,7,10], the definition is false, although exactly the operator as defined above is intended, since all papers require $\mathcal{P}_k(C_{k+1}) = C_{k+1}$. Since the definition of path graphs is rather similiar to that of iterated line digraphs, the path graphs might be useful for constructing 'good' interconnection networks. For instance, consider the 2-path graph on the left of Figure 1. It is larger than the undirected Kautz and de-Brujin graphs $UK(2,2)$ and $UB(2,2)$, but of same maximum degree and diameter.

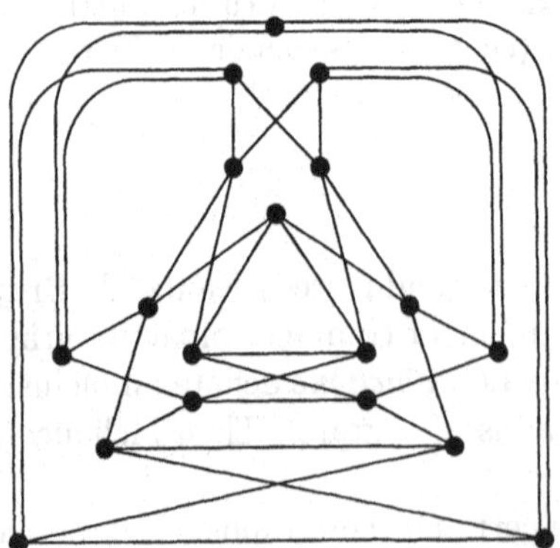

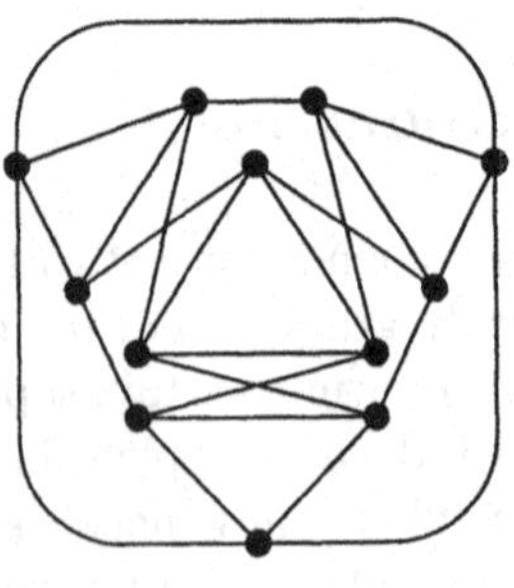

Fig. 1. Two 4-regular graphs of diameter 3

Bicliques arise quite naturally in path graphs or in underlying graphs of line digraphs as follows:

For every length-$(k-1)$ path $x_0, x_1, \ldots, x_{k-1}$ of H, the paths of the form $y, x_0, \ldots, x_{k-1}$ versus the paths of the form $x_0, \ldots, x_{k-1}, z$ form an induced complete bipartite graph in $\mathcal{P}_k(H)$, we denote it by $(x_0, x_1, \ldots, x_{k-1})^*$. The set of all such graphs partitions the edge set of $\mathcal{P}_k(H)$. But not every graph $(x_0, x_1, \ldots, x_{k-1})^*$ is a biclique, and not every biclique has this form.

Every vertex x of D gives rise to some induced complete bipartite graph in $U(L(D))$, whose vertices are the in-going arcs respectively out-going arcs at x, except the possible loop xx. We call this bipartite graph x^*. If V° denotes the vertices stemming from loops of D, then $\{x^*/x \in V(D)\}$ forms a partition of the edge set of $U(L(D)) - V^\circ$. Again, not every x^* is a biclique, and not every biclique in $U(L(D))$ has this form.

Considering these complete bipartite graphs in path graphs and underlying graphs of line digraphs seems rather natural and promising in many contexts. In this paper we will only show their usefullness in the the *recognition* problem. Given a graph G, is there a graph H such that $G = \mathcal{P}_k(H)$, or is there some digraph D such that $G = U(L(D))$. The *reconstruction* problems also involve finding such a H or D. It has been shown in [3] that the recognition of underlying graphs of line digraphs of irreflexive digraphs is $\mathcal{NP}$-complete. In [13] and [6], several uniqueness results have been obtained, provided the minimum degrees of the roots D or H are large enough. In this paper, we shall extend and unify these results. In particular, we shall show that it is possible to recognize k-path graphs of members of some class Γ_k in time $O(|V|^4)$, and that it is possible to recognize underlying graphs of line digraphs of digraphs D with $\delta^-(D), \delta^+(D) \geq 2$ in time $O(|V|^2|E|^2)$. Here Γ_k denotes the class of all graphs H where the start vertex x_0 of every path $x_0, x_1, \ldots, x_t$ of length less than k has at least two neighbors outside the path. Obviously $\Gamma_1 \supseteq \Gamma_2 \supseteq \Gamma_3 \supseteq \cdots$, and $\delta(H) \geq k+1$ implies $H \in \Gamma_k$. Note that all graphs $(P)^*$ of $\mathcal{P}_k(H)$ are bicliques for $H \in \Gamma_k$.

Since the graphs $(x_0, x_1, \ldots, x_{k-1})^*$ in $\mathcal{P}_k(H)$ or x^* in $U(L(D))$ contain quite a lot of the information of H or D, the natural idea is to try to reconstruct the root H respectively D by exposing all these graphs $(x_0, x_1, \ldots, x_{k-1})^*$ or x^*. Our degree restictions on H or D imply that they are actually bicliques, and even contain $K_{2,2}$. We say that a biclique has *type* $\geq k$ if it contains $K_{k,k}$. The overall structure of both our algorithms is as follows:

1. First we list all bicliques of type ≥ 2 of the graph G. Since the number may be exponential (compare [9]), we need a filter where only graphs with few bicliques of type ≥ 2 pass, and where these bicliques can be computed fast. All k-path graphs or underlying graphs of line digraphs should pass.
2. Next we sort the bicliques of type ≥ 2 whether or not they are induced by some length-$(k-1)$ path of H (in the k-path graph case) or some vertex of D (in the line digraph case).
3. Finally we check whether there is some graph H or some digraph D compatible with the information obtained so far.

Step 1 will be treated in Section 2 for both operators. Sections 3 and 4 are devoted to the remainder of the algorithms for k-path graph or underlying graphs of line digraphs, respectively.

2 A local condition

In [9] there has been shown that graphs without induced subgraphs of the form $CP(j) = K_{j,j} - jK_2$ and $H(k,\ell) = kK_1 * \ell K_2$ have 'few' bicliques of type $\geq k$.

Proposition 2.1 *Underlying graphs of line digraphs are $CP(5)-$ and $H(2,2)$-free. If D is irreflexive, then $U(L(D))$ is even $H(2,1)$-free (i.e. diamond-free).* □

Proposition 2.2 *k-path graphs are $CP(4)$-free and $H(1,2)$-free for $k \geq 2$. For $k \geq 3$, they are $H(1,1)$-free (i.e. triangle-free), and for $k \geq 6$, $CP(3)$-free.* □

Thus, by Theorem 5 in [9], underlying graphs of line digraphs have at most $\frac{1}{4}(\frac{|V|}{2})^{20}$ bicliques of type ≥ 2, and k-path graphs have at most $(\frac{|V|}{2})^{16}, (\frac{|V|}{2})^{12}$, or $(\frac{|V|}{2})^{10}$ bicliques of type ≥ 1 for $k = 2, 3 \leq k \leq 5$, and $k \geq 6$, respectively. These are very rough bounds, derived by using nothing else on the structure of these graphs than these two forbidden induced subgraphs. Sharper bounds will be derived in this section.

It is very likely that there exists an algorithm that lists all bicliques (of type $\geq t$) of a graph in time polynomial in the output, similiar as [12] does in the clique case. But even if we had such an algorithm, we would need an a priori condition whether listing all bicliques could be done in polynomial time in terms of G. Here we give such a condition—it gives us a first rough filter on which graphs are really candidates for k-path graphs or underlying graphs of line digraphs.

Consider any edge xy in a graph G. Every biclique $B * C$ of type ≥ 2 that contains the edge xy (say $x \in B, y \in C$) must induce a complete bipartite graph in the graph $G(xy)$ induced by $(N(y) \setminus N(x)) \cup (N(x) \setminus N(y))$ with $B \subseteq N(y) \setminus N(x)$ and $C \subseteq N(x) \setminus N(y)$. On the other hand, if $G(xy)$ can be obtained from the vertex disjoint union of several complete bipartite graphs $B_i * C_i$ with $B_i \subseteq N(y) \setminus N(x)$ and $C_i \subseteq N(x) \setminus N(y)$ by adding some edges inside $N(y) \setminus N(x)$ not connecting vertices inside the same B_i, and some edges inside $N(x) \setminus N(y)$ not connecting vertices inside the same C_i, then the bicliques of type ≥ 2 of G containing xy are just the graphs $(B_i \cup \{x\}) * (C_i \cup \{y\})$. This turns out to be the case for our graphs in question.

Actually, for k-path graphs $G = P_k(H)$, $G(xy)$ is just an induced subgraph of the pattern on the left of Figure 2. We omit the straightforward proof. Therefore, xy is contained in at most 2 bicliques of type ≥ 2 of G, whence G contains at most $2|E|$ such bicliques. For underlying graphs of line digraphs, with a little more effort it can be shown that $G(xy)$ must be an induced of one of the graphs in the middle or on the right of Figur 2.

For simplicity, we subsume both cases under the following pattern: We require for every edge xy of G the following 'neighborhood condition':

(N) The edges between $N(x) \setminus N(y)$ and $N(y) \setminus N(x)$ generate some $\bigcup_{i=1}^{p} K_{r(i),s(i)} \cup mK_1$, with no further edges inside the $K_{r(i),s(i)}$, and with either $p = 2$ and $m = 0$ or $p \leq 1$.

Proposition 2.3 *Checking whether a graph $G = (V, E)$ obeys condition (N), and if it does, computing all its $O(|V||E|)$ bicliques of type ≥ 2 can be done in time $O(|V|^2|E|)$.*

Proof. By the remark above, the bicliques of type ≥ 2 containing xy are generated by the $K_{r(i),s(i)}$ and the other cross edges by adding xy. So, all we have to do is to find these graphs. For every edge xy of G, we proceed as follows: First we compute $N(x) \setminus N(y)$ and $N(y) \setminus N(x)$ in time $O(|V|)$. We choose $a_1 \in N(x) \setminus N(y)$, find all neighbors $b_1, \ldots, b_s$ of a in $N(y) \setminus N(x)$, and all neighbors $a_1, \ldots, a_t$ of b_1 in $N(x) \setminus N(y)$. This requires again time $O(|V|)$. All what remains to check is whether the $a_1, \ldots, a_t$ as well as $b_1, \ldots b_s$ are independent, and whether the neighbors of each a_i in $N(y) \setminus N(x)$ are just $b_1, \ldots, b_s$, and whether the neighbors of each b_j in $N(x) \setminus N(y)$ are just $a_1, \ldots, a_t$. All this can be checked in time $O(\binom{s}{2} + \binom{t}{2} + (s+t)|V|)$. Then we delete $a_1, \ldots, a_t$ in $N(x) \setminus N(y)$ and $b_1, \ldots, b_s$ in $N(y) \setminus N(x)$ and proceed in the remainder. Since all the t add up to $|N(x) \setminus N(y)|$, and the s to $|N(y) \setminus N(x)|$, the total time—for fixed edge xy—is $O(|V|^2)$. Then the total time of the algorithm is $O(|V|^2|E|)$. □

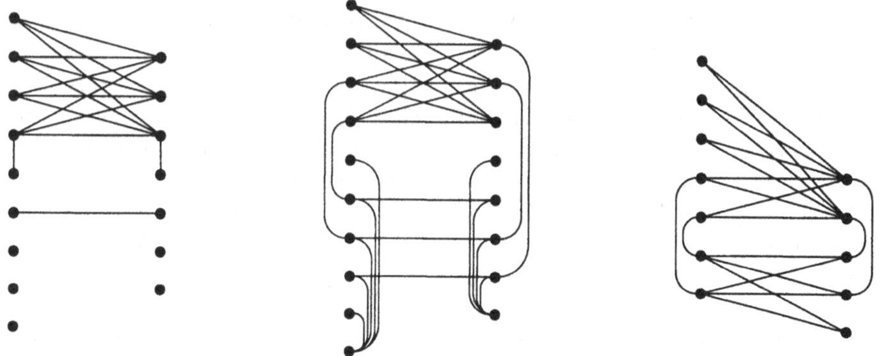

Fig. 2. $N(x) \setminus N(y)$ versus $N(y) \setminus N(x)$ for edges xy in k-path graphs (left) or underlying graphs of line digraphs (middle, and right for digons).

Both graphs in Figure 1 obey condition (N). The left one has 12 bicliques of type ≥ 2, all $K_{2,2}$s, whereas the right one has 9 bicliques, also all $K_{2,2}$s.

3 k-path graphs

3.1 Bicliques of type ≥ 2

Let $G = \mathcal{P}_k(H)$. As noted before, for length-$(k-1)$ paths P in H, the graphs $(P)^*$ induce complete bipartite graphs in G for $k \geq 2$. If $k \leq 3$, then the four length-k subpaths of every length-4 cycle C of G form a $C_4 = K_{2,2}$, a biclique in fact, which we denote by $(C)^{\square}$. It turns out that these two types are the only possibilities for bicliques of type ≥ 2 in G:

Lemma 3.1 *Every induced C_4 in $\mathcal{P}_k(H)$ is either contained in some $(P)^*$ for some path of length $k-1$ of H, or equals $(C)^{\square}$ for some 4-cycle C in H. The second case is only possible for $k = 2, 3$.*

Therefore a graph $(P)^*$ is a biclique in $\mathcal{P}_k(H)$ if it contains $K_{2,2}$.

Lemma 3.2 *Let P_1, P_2 be paths of length $k-1$ in H, and let C_1, C_2 be cycles of length 4 in H.*

a) $(P_1)^*$ and $(P_2)^*$ have one or no common vertex, depending on whether or not $P_1 \cup P_2$ forms a path of length k.
b) $(C_1)^{\square}$ and $(C_2)^{\square}$ have one or no common vertex for $k = 2$, and no common vertex for $k = 3$.
c) $(C_1)^{\square}$ and $(P_1)^*$ have two or no common vertex, depending on whether P_1 is a subpath of C_1 or not.

Having a list of all bicliques of type ≥ 2 for our candidate graph G (see Section 2), our next task is to find out which bicliques of type ≥ 2 stem from paths and which from 4-cycles. Lemma 3.2 is the key for the identification: We find a coloring of the bicliques of type ≥ 2 by two colors such that intersecting bicliques have the same respectively different colors depending on whether they have one respectivley two vertices in common. Such a coloring can be found by breadth first search in the intersection graph of all these bicliques of type ≥ 2, and it is unique if it exists at all. Note that we may require, since we are only interested in graphs H such that $\delta(H) \geq k+1$, that every edge of G lies in some biclique of type ≥ 2.

Every color class forming an edge partition of G, and containing all large bicliques (those containing $K_{2,3}$) may yield the bicliques stemming from paths in H.

For $k \geq 4$ things are easier, since all bicliques of type ≥ 2 stem from paths. That is, in our coloring above we should use only one color. For $k = 3$, the bicliques of one color class should be pairwise vertex-disjoint.

Thus for connected $H \in \Gamma_k$, there remains only an ambiguity for $k = 2$, only if all bicliques are $K_{2,2}$s, and only if the bicliques of both colors partition the edge set. We shall show that these conditions imply $H = K_{3,3}$. Under these conditions, the graphs $(C)^{\square}$ partition the edge set of $\mathcal{P}_k(H)$. Then H must be triangle-free such that every length-3 path lies in a 4-cycle. This implies $\text{diam}(H) \leq 2$. If H is not bipartite, then choose some shortest odd cycle $x_1, x_2, \ldots, x_{2t+1}, x_1$, $t \geq 2$. Then x_1 and x_4 are adjacent by the condition above, thus $x_1, x_4, \ldots, x_{2t+1}, x_1$ would be a shorter odd cycle, a contradiction. Thus H is is bipartite with bipartition $V(H) = V_1 \cup V_2$, then assume there were nonadjacent vertices $x \in V_1, y \in V_2$. This contradicts $\text{diam}(H) \leq 2$, thus H must be complete bipartite. If H contains $K_{3,4}$, then some of the graphs $(x_0x_1)^*$ contain

$K_{2,3}$. On the other hand, the graphs $K_{2,p}$ do not belong to Γ_2. The only remaining graph is $H = K_{3,3}$ But for $H = K_{3,3}$, the set consisting of the graphs $(P)^*$ is isomorphic to the structure consisting of the graphs $(C)^\square$, thus it doesn't matter which one we choose.

3.2 Another characterization of k-path graphs

There is some dualization characterization of 2-path graphs [7], which can be obtained by looking at vertices of H, and the length-2 paths with this vertex as middle vertex. Another dualization is obtained by looking at the length-$(k-1)$ paths in H, and the length-k paths containing it.

To state it for $k = 2$, we need two more definitions: Two induced complete bipartite graphs B_1, B_2 are *on the same side* of a third one, B_3, if $B_1 \cap B_3$ and $B_2 \cap B_3$ both contain just one vertex, and if these two vertices are distinct and nonadjacent. Two length-k paths P_1, P_2 are *on the same side* of a third path P_3 if $P_3 \equiv x_1, x_2, \ldots, x_k$, and $P_i \equiv y_i, x_0, x_1, \ldots x_{k-1}$ for $i = 1, 2$.

Theorem 3.3 *A graph $G = (V, E)$ is a 2-path graph if and only if there are induced complete bipartite subgraphs $(B_i / i \in I)$ partitioning the edge set of E such that*

(1) every two members $B_i, B_j, i \neq j \in I$ have at most one common vertex, and
(2) every vertex lies in exactly two of the graphs B_i, and
(3) the intersection graph of these B_i is a line graph where every B_i has two other graphs B_j, B_p on the same side iff the corresponding length-1 paths P_j, P_p are on the same side of the corresponding P_i.

Proof. If $G = \mathcal{P}_2(H)$, then we take the graphs P^*, for all length-1 paths P of H, as the graphs B_i. Conditions (1) and (3) follow by part (a) of Lemma 3.2. Condition (2) follows since every length-2 path has exactly two subpaths of length 1.

For sufficiency, assume (1) , (2), and (3) are true, let the intersection graph of the B_i be some line graph $\mathcal{P}_1(H)$. We will prove $G = \mathcal{P}_2(H)$ by giving an isomorphism σ between these two graphs. By (2), every vertex x of G lies in two of the bicliques of the cover, say $B_{i(x)}$ and $B_{j(x)}$. The union of the corresponding length-1 paths $P_{i(x)}$ and $P_{j(x)}$ in H is some length-2 subpath of H (since there are no cycles of length 2), which we denote by $\sigma(x)$.

All that remains to show is that σ is a graph isomorphism. The mapping σ is $1-1$, since $x \neq y \in V(G)$ implies $\{i(x), j(x)\} \neq \{i(y), j(y)\}$ by (1), and there is only one way to generate a path of length 2 as union of two length-1 paths. It is surjective, since every length-2 path P is the union of two length-1 paths. The corresponding graphs B_i must contain some common vertex x. Certainly $\sigma(x) = P$. Let finally $x, y \in V(G)$. If $xy \in E(G)$, then w.l.g. $xy \in B_{j(x)} = B_{j(y)}$ by (1) and (2). Then $B_{i(x)}$ and $B_{i(y)}$ are neighbors of $B_{j(x)}$, but on different sides. Therefore, the paths $P_{i(x)}$ and $P_{i(y)}$ have different intersection with $P_{j(x)}$. Thus $P_{i(x)} \cup P_{j(x)} \cup P_{i(y)}$ is a cycle or path of length 3 in H, whence $\sigma(x)$ and

$\sigma(y)$ are adjacent. In the same way, if $\sigma(x)$ and $\sigma(y)$ are adjacent paths in H, then $xy \in E(G)$. □

For higher k, it is not enough to look at this intersection graph, since if the union of two adjacent length-$(k-1)$ paths P_1, P_2 of H forms a length-k cycle, then $(P_1)^*$ and $(P_2)^*$ have no common vertex. However, if $H \in \Gamma_k$, then there are other length-$(k-1)$ paths P_3, P_4 in H such that $(P_1)^*$ and $(P_4)^*$ intersect $(P_3)^*$ in the same bipartition set, and $(P_2)^*$ and $(P_3)^*$ intersect $(P_4)^*$ also in nonadjacent vertices. For a family $(B_i/i \in I)$ of induced complete bipartite subgraphs of a graph G, let $\Psi((B_i/i \in I))$ denote the graph we obtain from the intersection graph of these B_i by adding a vertex from B_i to B_j provided there are members $B_r = U_r * W_r$ and $B_s = U_s * W_s$ such that $B_i \cap U_r \neq \emptyset, U_r \cap U_s \neq \emptyset$, and $U_s \cap B_j \neq \emptyset$.

Using this notion, a similiar characterization of k-path graphs is given, in the full paper. Here we omit it, since its statement is somewhat technical, and since we do not need it. All we need is the following necessary condition:

Theorem 3.4 *If, for $k \geq 2$, a graph $G = (V, E)$ is the k-path graph of some member of Γ_k then there are bicliques $(B_i/i \in I)$ partitioning the edge set of E such that*

(1) every two members $B_i, B_j, i \neq j \in I$ have at most one common vertex, and
(2) every vertex lies in exactly two of the graphs B_i, and
(3) $\Psi(B_i/i \in I)$ is a $\mathcal{P}_{k-1}$-graph.

The proof is similiar to the preceding one.

3.3 The recognition algorithm

The recursive algorithm proceeds as follows.

- We check the local condition (N) of Section 2 and compute all bicliques of type ≥ 2 of G.
- As explained in Subsection 3.1 we find the unique set $(B_j/j \in J)$ of bicliques of G that may stem from length-$(k-1)$ paths.
- Then we compute $\Psi(B_j/j \in J)$. If this equals $\mathcal{P}_{k-1}(H)$ for some $H \in \Gamma_{k-1}$, then H is our only candidate for $\mathcal{P}_k(H) = G$, otherwise G is no k-path graph of some member of Γ_k. Note that for testing whether $\mathcal{P}_k(H) = G$ holds we don't need an expensive isomorphism test. The candidate for the isomorphism σ is already there—for every vertex x of G let $\sigma(x)$ be the union of the paths corresponding to the two B_i containing x. It only must be checked whether σ is an isomorphism (i.e. whether every $\sigma(x)$ is a path, and whether adjacency and nonadjacency is preserved)

Algorithm 3.5 k-PATH GRAPH RECOGNITION (Γ_k)
Instance: *A graph G.*
Question: *Find all $H \in \Gamma_k$ such that $G = \mathcal{P}_k(H)$.*

1. Test whether G obeys the neighborhood condition (N), and if it does, compute all bicliques of type ≥ 2. Test whether every edge lies in at least one and in at most two of these bicliques.
2. Bicolor these bicliques of type ≥ 2 such that intersecting bicliques are colored differently iff the intersection contains only one vertex. Find that color class that
   - contains all large bicliques (i.e. bicliques $\supseteq K_{2,3}$), and
   - partitions the edge set of G, and
   - covers every vertex of G twice,
3. Apply $\mathcal{P}_{k-1}$-GRAPH RECOGNITION(Γ_{k-1}) for the Ψ-graph of the bicliques of that color. If it is $\mathcal{P}_{k-1}(H)$ for some $H \in \Gamma_{k-1}$, then test whether $\mathcal{P}_k(H) = G$ and $H \in \Gamma_k$.

Theorem 3.6 *Algorithm 3.5 is correct. the running time is $O(|V|^4)$, and there is at most one such $H \in \Gamma_k$.*

Proof. The correctness follows by the preceding remarks and Theorem 3.4. For the running time, we use induction on k. For the start, $k = 1$, we use any of the linear-time line graph recognition algorithms in [11] or [5] as 1-PATH GRAPH RECOGNITION. Let now $k \geq 2$, and assume $(k-1)$-PATH GRAPH RECOGNITION (Γ_k) requires time $O(|V|^4)$. Step (1) can be done in time $O(|V|^2|E|)$ by Proposition 2.3. For Step (2), first we order the edges of G (for instance by BFS, in time $O(|E|)$) as $e_1, e_2, \ldots, e_m$ such that each subgraph generated by $e_1, \ldots e_i$ is connected. Then we order the bicliques of type ≥ 2 such that exactly the first $j(i)$ members contain one of $e_1, \ldots, e_i$, for every $1 \leq i \leq |E|$. This assures that each biclique has some common vertex with the union of the bicliques that precede it in the list. If we color the bicliques according to that list, we get uniqueness of the color at each step. Since there are at most $2|E|$ such bicliques, the coloring requires time $O(|E|^2)$. Checking whether a color class obeys the additional conditions can be done during this coloring. Finally, the Ψ-graph of the color class can be computed straightforwardly in time $O(|V|^4)$, and the $\mathcal{P}_{k-1}$-graph test also requires only time $O(|V|^4)$ by the induction hypothesis. □

Let me close this subsection by mentioning that Algorithm k-PATH GRAPH RECOGNITION can be modified to search for arbitrary roots (not necessarily roots in Γ_k). However, then the running time is no longer necessarily polynomial. Details will appear in the full paper.

3.4 The algorithm for k not fixed

Since our algorithms differ only in the call of subroutine in step (3), together they yield some single algorithm, that tests whether for a given graph G there is some integer $k \geq 2$ and some graph $H \in \Gamma_k$ such that $G = \mathcal{P}_k(H)$. We shall see that k and H are unique provided they exist at all.

Although k-path graphs may be line graphs, as can be seen by cycles, we get:

Lemma 3.7 *For $k \geq 2$ and $H \in \Gamma_k$, $\mathcal{P}_k(H)$ is not a line graph.*

Proof. Actually such a graph $\mathcal{P}_k(H)$ is not $K_{1,3}$-free. Choose any length-k path $x_0, x_1, \ldots, x_k$ in H. Since $H \in \Gamma_k$, x_0 has two neighbors y_1, y_2 distinct from $x_1, \ldots, x_{k-1}$, and x_k has two neighbors z_1, z_2 distinct from $x_1, \ldots, x_{k-1}$. We may assume $y_1 \neq x_k$. Then the vertices corresponding to the paths $x_0, x_1, \ldots, x_k$; $y_1, x_0, x_1, \ldots, x_{k-1}$; $x_1, \ldots, x_k, z_1$; and $x_1, \ldots, x_k, z_2$ induce a $K_{1,3}$ in $\mathcal{P}_k(H)$. Therefore this graph can be no line graph. □

Now we proceed as follows: Starting with $G = G_0$, we run Algorithm 3.5 with nonfixed k, compute $G_1 = \Psi(G_0)$, and repeat in this way the algorithm with $G_{i+1} = \Psi(G_i)$ until G_{i+1} is a line graph. Then we have to check backwards.

The algorithm terminates, since $|V(G_{i+1})| \leq |V(G_i)|/2$ as long as none of the conditions is violated—every B_j contains at least 4 vertices, but every vertex lies in exactly two B_j.

Actually, computing the bicliques in the first step could be done easier in all but the first run of the loop, see the full paper.

The overall running time is $O(|V|^4) + O((|V|/2)^4) + \cdots = O(|V|^4)$.

In this way we also obtain an extension of Li's uniqueness result in [6]:

Theorem 3.8 *If $G = \mathcal{P}_k(H_1)$ and $G = \mathcal{P}_j(H_2)$ with $H_1 \in \Gamma_k$ and $H_2 \in \Gamma_j$, then $k = j$ and $H_1 \simeq H_2$.*

The same conclusion holds even without the assumption $H_2 \in \Gamma_j$, see the full paper.

To give examples, we test the graphs of Figure 1.

Coloring the 12 bicliques of the left graph yields one class with three bicliques, and the other with nine. Only the 9 bicliques cover the edge set. The resulting intersection graph is the line graph of the prism $K_3 \times K_2$. Actually, it turns out that $G = \mathcal{P}_2(K_3 \times K_2)$, and this is the only output.

In the graph to the right of Figure 1, we get 6 bicliques of one color, and 3 of the other, thus $k \leq 3$. Only the 6 bicliques cover the edge set, and their intersection graph is the octahedron $\overline{3K_2}$. However, the polarization is 'false' so for every $k \geq 1$, this graph to the right of Figure 1 is no k-path graph of some member of Γ_k. By the slightly more general results of the full paper it can be easily shown that it is no path graph at all.

4 Underlying graphs of line digraphs

As mentioned already, every vertex x of D gives rise to some induced complete bipartite graph x^* in $G = U(L(D))$. For every directed 4-cycle C of D, the vertices of G corresponding to the arcs of C generate a biclique $C_4 = K_{2,2}$ in G, which we denote by $(C)^\square$ again.

Lemma 4.1 *Every induced C_4 in $U(L(D))$ either equals $(C)^\square$ for some directed 4-cycle C in D, or is contained in some x^* with $x \in V(D)$.*

Things get a little more complicated than for k-path graphs, since the intersection pattern of the x^* and $(C)^\square$ is slightly different:

Lemma 4.2 *Let x_1 and x_2 be vertices, and let C_1, C_2 be directed 4-cycles in D.*

a) x_1^* and x_2^* have no, one, or two common vertices, depending on whether or not x_1 and x_2 are joined by no, one, or two arcs in D.
b) $(C_1)^\square$ and $(C_2)^\square$ have no, one, or two common vertices.
c) $(C_1)^\square$ and x_1^* have two or no common vertex, depending on whether C_1 contains x_1 or not.

4.1 Loop vertices and digon edges

We call a vertex x respectively an edge yz of $G = U(L(D))$ a *loop vertex* respectively *digon edge* if the corresponding arcs of D form a loop respectively a digon. Under the assumptions $\delta^-(D), \delta^+(D) \geq 2$, loop vertices do not depend on D, see [13].

Lemma 4.3 *If $\delta^-(D), \delta^+(D) \geq 2$ and if D is weakly connected with more than 5 arcs, then a vertex x of $G = U(L(D))$ stems from a loop if and only if its neighborhood induces a connected complete bipartite graph.*

For the rest of the paper we assume $\delta^-(D), \delta^+(D) \geq 2$. Then all graphs x^* are bicliques of type ≥ 2 of $U(L(D))$ or neighborhoods of loop vertices—we subsume both these types of graphs under the name *candidates*. Our task is to find out which of the candidates actually stem from vertices—they will be colored yellow. Neighborhoods of loop vertices must be yellow.

As VILLAR noted in [13], even under the assumption $\delta^-(D), \delta^+(D) \geq 2$, digon edges cannot be determined. An example is the cube graph, but larger examples could be given as well. However, we shall show in the following that in this case all possible digraphs D are isomorphic.

Digon edges obey the following properties:

- x and y are no loop vertices,
- xy lies in exactly two candidates B_1, B_2,
- every further candidate containing one of x, y must be a C_4 containing one more vertex of B_1 and one more of B_2.

We call xy a *digon impostor* if it obeys these three conditions without being a digon edge.

If a digon impostor x_1x_2 occurs, say x_1 is the arc ab and x_2 the arc bc of D, then there must be arcs cf, fy, cb, ba in D, and $d^-(a) = d^+(a) = d^-(c) = d^+(c) = 2$. Hence every vertex lying in a digon impostor must occur in digon edge too. We define $x_3 = cb$ and $x_4 = ba$. If no directed 4-cycle in D contains both arcs cb and ba, then x_3 lies in only one digon edge or impostor, namely x_2x_3, therefore this must be a digon edge. In the same way, x_1x_4 must be a digon edge, and x_1x_2 is a digon impostor. So assume from now on that there are arcs ad, dc in D with $d = f$ possible. Then all four edges of the induced 4-cycle x_1, x_2, x_3, x_4 are digon edges or digon impostors. Any further nonloop arc beginning or ending at b would reveal the digon edges x_2x_3, x_1x_4, thus we assume in the following that there is no such arc. With $y_4 = ad, y_3 = dc, y_2 = cf$, and $y_1 = fa$, the closed neighborhood of the 4-cycle x_1, x_2, x_3, x_4 must induce the induced cube with vertices $x_1, x_2, x_3, x_4, y_1, y_2, y_3, y_4$, or the cube with one additional (loop) vertex adjacent to x_1, x_2, x_3, and x_4. Since this vertex does not change anything, for simplicity we assume there is no loop bb in D. Except in the case where D is a symmetric C_4 (in which case $U(L(D))$ is the cube graph) there must be more arcs of D incident with d and f. Then either, in case $d = f$, y_1, y_2, y_3, y_4 is no biclique itself but contained in some larger biclique. Then we cannot decide whether x_1x_2 and x_3x_4 are the digon edges, or x_2x_3, x_1x_4, but it doesn't matter anyway, since the situation is symmetrical. So assume in the following $d \neq f$. Then y_1y_2 is contained in the bicliques y_1, y_2, y_3, y_4, y_2, y_1, x_1, x_2, and one further biclique of type ≥ 2, namely f^*. Analogous holds for y_3y_4. If f^* or d^* contains $K_{2,3}$, or if one of the edges y_2y_3, y_1y_4 is contained in just two or more than three bicliques of type ≥ 2, then we can distinguish the edges x_1x_2, x_3x_4 from x_1x_4, x_2x_3. Otherwise there must be arcs fg, gd, dh, hf in D, with $g = h$ possible, and $d^-(d) = d^+(d) = d^-(f) = d^+(f) = 2$. We proceed as above. Except if $U(L(D)) = P_t \times C_4$, eventually we arrive at one of the two situations above: Either we can locate digon edges and digon impostors under our four edges x_1x_2, x_2x_3, x_3x_4, and x_4x_1, or the graph is symmetrical (between x_1x_2, x_3x_4 and x_2x_3, x_4x_1). If $U(L(D)) = P_t \times C_4$, then again there are two possibilities, but both isomorphic.

All this can certainly be implemented to run in time $O(|V|^2|E|^2)$. First we compute the set E_1 of all digon edges or impostors. The three conditions can be checked (for all edges) in overall time $O(|V|^2|E|^2)$. The bottleneck is the third condition, which can be handled with the aid of the $O(|V||E|) \times O(|V||E|)$ matrix where entry a_{ij} indicates the intersection of biclique i of type ≥ 2 with biclique j of type ≥ 2. Those edges of E_1 containing a vertex of degree 1 in (V, E_1) must be digon edges, by the analysis above. We delete them and all its vertices to obtain a subgraph $G_2 \subseteq U(L(D))$. Then we compute all induced 4-cycles in G_2 (in time $O(|E|^2)$), and proceed to investigate the neighborhood of this 4-cycle as far as necessary, which requires again time $O(|V|^2)$ for each 4-cycle.

4.2 The recognition algorithm

Assume now that two candidates of G have nonempty intersection. Assume furthermore that one of them stems from a vertex of D, i.e. has the form x^*. If they have exactly one common vertex, then the other must have the form y^* by Lemma 4.2. If they have exactly two common vertices, then the other must have the form y^* if and only if the common edge is a digon edge, again by Lemma 4.2. Since the intersection graph of the set of these candidates is connected, knowing one x^* means knowing all y^*, since we may require that we know all digon edges. In most cases, we actually know at least one of these graphs x^*, since we may have loop vertices, digon edges, or bicliques containing $K_{2,3}$. Only if all this is not the case, we have to test several hypothesises, namely all bicliques of type ≥ 2 containing any given edge. In this case, actually there may remain an ambiguity. Examples are the cartesian products $C_{2n} \times C_{2m}$, which have two sets of admissible biclique-covers. In the graph $C_4 \times C_4$, we get even three such sets. See subsection 4.3 for further discussion.

The final step necessary is to take the graphs $a^*, b^*, c^*, \ldots$ stemming from vertices of D, and look whether they allow a consistent orientation. That is, we try to orient each such bipartite subgraph from one bipartition set towards the other such that every vertex that lies in two of these graphs is a sink in the one and a source in the other. If that is possible, then the intersection digraph of this family is our root D. Note that the reverse digraph is another root in that case.

Algorithm 4.4 .
Instance: *A graph* G.
Question: *Are there digraphs* D *obeying* $\delta^-(D), \delta^+(D) \geq 2$ *such that* $G = U(L(D))$*?*

1. Test whether G obeys the neighborhood condition (N), and if it does, compute all bicliques B_i of type ≥ 2. Test whether every edge is covered.
2. Compute the set V° of loop vertices (by 4.3).
   Compute an ordering $e_1, e_2, \ldots, e_m$ of the edges of $G \backslash V^\circ$ such that each set $\{e_1, \ldots, e_i\}$ generates a connected graph.
   Compute the set E' of digon edges (as explained above).
3. Color all neighborhoods of loop vertices, all bicliques of type ≥ 2 containing digon edges, and all bicliques containing $K_{2,3}$ yellow.
   For every candidate B_j containing e_1 do:
   - Color B_j yellow.
   - For $i = 2, 3, \ldots m$ do check whether exactly one B_i containing e_i fits with the other yellow B_i, and if it does, color that also yellow.
   - Check whether the yellow B_i have consistent orientations. If, then $G = U(L(D))$ for the intersection digraph of those oriented yellow bicliques.

Theorem 4.5 *Algorithm 4.4 decides in time* $O(|V|^2|E|^2)$*, whether a given graph is the underlying graph of the line digraph of some digraph* D *with both* $\delta^-(D), \delta^+(D) \geq 2$.

We try the algorithm for the two graphs of Figure 1. Both have no loop vertices or digon edges. For the left graph, only the 9 bicliques may stem from vertices of D. However, it is not possible to order them consistently, thus this graph is not an $U(L(D))$. For the right graph, only the six bicliques may stem from vertices of D. Actually, they can be oriented, thus $G = U(L(D))$ with some 6-vertex digraph D.

4.3 Uniqueness

By the algorithm, under the assumption $\delta^-(D), \delta^+(D) \geq 2$, uniqueness (modulo reversal) of the 'root' digraph D with $G = U(L(D))$ follows immediately if the set of graphs x^* is unique. This holds if $G = U(L(D))$ where D is not 2-regular, or contains a loop, or contains a digon (where we only have uniqueness modulo isomorphism).

We want to extend this result. Assume $G = U(L(D_1)) = U(L(D_2))$, where D_1 and D_2 are 2-regular and without loops or digons. Then G is 4-regular without induced $K_{2,3}$. Then the graphs x^* may be different for both cases. That means, G may have more than 2 orientations as underlying graph of the line digraph of a digraph with $\delta^-, \delta+ \geq 2$ There are even graphs allowing 6 such admissible orientations, as $C_4 \times C_4$, for instance. Nevertheless, as it turns out, these orientations must be isomorphic (modulo reversal), and hence D_1 must be isomorphic to D_2 or the reversal of D_2.

Theorem 4.6 *If D_1 and D_2 are digraphs with both minimum in- and out-degree at least 2, then $U(L(D_1)) \simeq U(L(D_2))$ implies that D_1 is isomorphic to D_2 or to its reversal.*

Proof. By the preceding remarks, it suffices to assume that the 4-regular graph G has at least two different partitions of the edge set into C_4s having a consistent orientation. Recall that this means that we can orient the edges of each of these C_4 with two sources and two sinks such that the common vertex (if there is any) of two of these C_4 of the partition is always a source in one and a sink in the other C_4. Obviously, the reversal of a consistent orientation is again consistent. We color the C_4 of the first partition blue, and the C_4 of the second red. We also denote the corresponding consistent 'blue' and 'red' orientations of G as (V, A_B) respectively (V, A_R).

Since the edges of every non-blue C_4 are covered by at least two blue C_4, the blue orientation on this C_4 must contain some directed path of length 2. Therefore, no C_4 is both blue and red.

Every vertex lies in exactly two blue and two red C_4. Moreover, every vertex has both in- and out-degree 2 in both orientations. Since no C_4 is both red and blue, this implies that every vertex has both in- and out-degree 1 in the digraph $J = (V, A_B \cap A_R)$. Then J is the vertex-disjoint union of directed cycles. Note that for every blue or red C_4, exactly two of its edges—obviously opposite edges— are oriented by $A_B \cap A_R$.

Let now $x_0y_0 \in A_R, y_0x_0 \in A_B$. Let $x_0, x_1, \ldots, x_t$ and $y_0, y_1, \ldots, y_r$ be the directed cycles in J containing x_0 respectively y_0. It may be possible that both cycles coincide. Since the edge x_0y_0 must lie in some blue C_4, by the shape of these blue C_4 it must be x_0, y_0, y_1, x_{t-1}. Thus $x_{t-1}y_1 \in A_B$. But the arc $x_{t-1}y_1$ does not occur in J, therefore $x_{t-1}y_1 \notin A_R$, therefore $y_1x_{t-1} \in A_R$. In the same way $y_{r-1}x_1 \in A_R, x_1y_{r-1} \in A_B$.

It follows that if we contract all directed cycles of J, the resulting graph has maximum degree at most 2. Since G is connected, this resulting graph must also be connected, thus it is either a path or a cycle. If it is a cycle, its length must be even. Then we may color the directed cycles of J with colors '1' and '2' such that those with arcs between them have different color. We define a mapping $\Phi : V \to V$ such that $\Phi(x)$ is the out-neighbor of x in J if the directed cycle in which x lies has color 1, and $\Phi(x)$ is the in-neighbor of x in J otherwise. It is fairly easy to see that Φ is an isomorphism between (V, A_B) and (V, A_R). □

References

1. R.E.L. Aldred, M.N. Ellingham, R. Hemminger, P. Jipsen, P_3-isomorphisms for graphs, preprint 1995.
2. H.-J. Broersma, C. Hoede, Path graphs, *J. Graph Theory* 13 (1989) 427-444.
3. V. Chvátal, C. Ebenegger, A note on line digraphs and the directed max-cut problem, *Discrete Applied Math.* 29 (1990) 165-170.
4. M.A. Fiol, J.L.A. Yebra, I. Alegre, Line digraph iterations and the (d,k) digraph problem, *IEEE Transactions on Computers* C-33 (1984) 400-403.
5. P.G.H. Lehot, An optimal algorithm to detect a line graph and output its root graph, *J. Assoc. Comput. Mach.* 21 (1974) 569-575.
6. Xueliang Li, Isomorphisms of P_3-graphs, *J. Graph Theory* 21 (1996) 81-85.
7. H. Li, Y. Lin, On the characterization of path graph, *J. Graph Theory* 17 (1993) 463-466.
8. E. Prisner, Graphs with few cliques, in: *Graph Theory, Combinatorics, and Applications: Proceedings of 7th Quadrennial International Conference on the Theory and Applications of Graphs* (Y. Alavi, A. Schwenk ed.) John Wiley and Sons, Inc. (1995) 945- 956;
9. E. Prisner, Bicliques in graphs I: Bounds on their number, (1996) submitted.
10. E. Prisner, Graph Dynamics, Pitman Research Notes in Mathematics Series 338 (1995), Longman, Essex.
11. N.D. Roussopoulos, A max$\{m,n\}$ algorithm for determining the graph H from its line graph G, *Inform. Process. Lett.* 2 (1973) 108-112.
12. S. Tsukiyama, M. Ide, M. Aiyoshi, I. Shirawaka, A new algorithm for generating all the independent sets, *SIAM J. Computing* 6 (1977) 505-517.
13. J.L. Villar, The underlying graph of a line digraph, *Discrete Applied Math.* 37/38 (1992) 525-538.

Large Networks with Small Diameter

Michael Sampels

Universität Oldenburg, Fachbereich 10, 26111 Oldenburg, Germany
Phone: +49-441-9722-243, Fax: +49-441-9722-242
Email: sampels@informatik.uni-oldenburg.de

Abstract. The construction of large networks with small diameter D for a given maximal degree Δ is a major goal in combinatorial network theory. Using genetic algorithms, together with Cayley graph techniques, new results for this degree/diameter problem can be obtained. A modification of the Todd-Coxeter algorithm yields further results and allows, with Sabidussi's representation theorem, a uniform representation of vertex-symmetric graphs. The paper contains an updated table of the best known (Δ, D)-graphs and a table with the largest known graphs for a given Δ and maximum average distance μ between the nodes.

1 Introduction

The design of efficient interconnection networks (e.g. telecommunication networks, massively parallel computer architectures, or optical networks) mainly focusses on one aim: providing a fast and safe dissemination of a large amount of messages between the nodes. The quality of the network depends on the routing algorithm organizing the transport of information, and on the topology of the network, normally represented by an undirected graph G. The vertices V correspond to the processors or routing chips, and the edges E symbolize the links between the nodes. Not only for first-generation routing algorithms like store-and-forward routing [25] but also for advanced, packet-oriented protocols [13, 27], the distance $d(v, w)$ is highly relevant for the speed of communication between two nodes v and w. Therefore, the construction of large graphs with small diameter $D = \max_{v,w \in V} d(v, w)$ has been of great interest in recent years.

Bermond and Bollobás presented in the survey article [4] various constraints under which the diameter can be analyzed. And probably the most interesting aspect for computer science is to limit the degree Δ of the graph according to hardware constraints on the number of links of a single node. Counting the number of vertices which are at distance $0, 1, 2, \ldots, D$ from a given vertex, it is obvious that there are at most $1+\Delta+\Delta(\Delta-1)+\Delta(\Delta-1)^2+\cdots+\Delta(\Delta-1)^{D-1}$ vertices, thus

$$|V| \leq \begin{cases} \frac{\Delta(\Delta-1)^D-2}{\Delta-2} & \text{if } \Delta \neq 2, \\ 2D+1 & \text{if } \Delta = 2. \end{cases}$$

This bound is due to Moore (circa 1958), and the graphs achieving this bound are called Moore graphs. Besides complete graphs K_n and circles of odd length C_{2n+1}, there are only two further Moore graphs known, the Petersen graph ($\Delta =$

$3, D = 2, |V| = 10$) and the Hoffman-Singleton graph ($\Delta = 7, D = 2, |V| = 50$). There can only be one further non-trivial Moore graph with $\Delta = 57$ and $D = 2$, but it is still unknown whether or not it exists [6].

As a consequence of the preceding results, it is interesting to find graphs near to the Moore bound. But up to now, there have been only three additional graphs which are known to be optimal [5]: ($\Delta = 3, D = 3, |V| = 20$), ($\Delta = 4, D = 2, |V| = 15$), ($\Delta = 5, D = 2, |V| = 24$). For other values, there are essentially three different techniques which have been used for the last two decades to construct large graphs. Products of graphs are defined which combine small graphs to larger graphs by keeping the diameter small [2, 3, 11, 14, 16]. The concept of adding edges and vertices to graphs or trees in order to reduce the diameter is described in [7, 9]. A relatively new approach uses algebraic techniques, mainly Cayley graphs, to construct graphs which are not only large but also vertex-symmetric [8, 17]. Looking the same as viewed from any vertex, Cayley graphs are favorite candidates for interconnection networks, since this property allows simple and economical routing mechanisms [1]. The practical significance of Cayley graphs as interconnection networks was demonstrated by simulation studies in [30], possible applications were considered in [33], and other quality metrics were treated in [29, 31].

Using genetic algorithms in combination with Cayley graphs, we were able to obtain new results for the degree/diameter problem, and present an updated table of (Δ, D)-graphs. Additionally, a table of large graphs with small mean distances $\mu = \frac{1}{|V|\cdot(|V|-1)} \sum_{v,w \in V, v \neq w} d(v, w)$ was set up. The paper is organized as follows: First, Cayley graphs are introduced and a new idea for the representation of vertex-symmetric graphs is postulated. Then, the search heuristics used, which are founded on genetic algorithms, are described. Thereafter, the obtained results are stated and compared with some earlier results. The paper ends with a short case study and a brief perspective on future work.

2 Cayley Graphs and Related Methods

Symmetry is a fundamental virtue in the design of interconnection networks for parallel architectures, because it guarantees feasible routing schemes [23]. With methods founded on Cayley graphs, it is possible to construct networks which are not only vertex-symmetric, but which also have small diameter. Further algebraic techniques allow a uniform representation of vertex-symmetric graphs.

Definition 1. Let Γ be a finite group and S be a symmetric generator set of Γ, i.e. $\langle S \rangle = \Gamma$, $s \in S \Rightarrow s^{-1} \in S$ and $1_\Gamma \notin S$. The Cayley graph $G_S(\Gamma)$ is defined as the undirected graph with vertex set $V = \Gamma$ and edge set $E = \{\{g, h\} \mid g^{-1}h \in S\}$.

Definition 2. A graph $G = (V, E)$ is vertex-symmetric if the group of graph automorphisms $A(G)$ acts transitively on V, i.e. for any two vertices $v, w \in V$ there is a graph automorphism $\alpha \in A(G)$ with $\alpha(v) = w$.

Corollary 3. *Cayley graphs are vertex-symmetric. Let $G_S(\Gamma) = (V, E)$ be the Cayley graph of a group Γ and a symmetric generator set S. For two vertices $v, w \in V = \Gamma$, the mapping $\alpha : x \mapsto wv^{-1}x$ is bijective on Γ as left multiplication with wv^{-1} in Γ. With $\{g, h\} \in E \Rightarrow g^{-1}h \in S \Rightarrow g^{-1}vw^{-1}wv^{-1}h \in S \Rightarrow (wv^{-1}g)^{-1}wv^{-1}h \in S \Rightarrow \{wv^{-1}g, wv^{-1}h\} \in E \Rightarrow \{\alpha(g), \alpha(h)\} \in E$, α is a graph automorphism which maps v to w.* □

It should be pointed out that there are vertex-symmetric graphs which cannot be represented as Cayley graphs, for example the Petersen graph (see [40]). It was shown by Sabidussi [28] that every vertex-symmetric graph can be represented as a Cayley coset graph.

Definition 4. Let Γ be a finite group, Σ a subgroup of Γ, and $S \subset \Gamma$ with $S^{-1} = S$ and $S \cap \Sigma = \emptyset$. The Cayley coset graph $G_S(\Gamma, \Sigma)$ is defined as the undirected graph with vertex set Γ/Σ (left cosets of Σ in Γ) and edge set $E = \{\{g\Sigma, h\Sigma\} \mid g^{-1}h \in \Sigma S \Sigma\}$.

Theorem 5 (Sabidussi's representation theorem). *Let A be the automorphism group of a vertex-symmetric graph G and A_v be the stabilizer subgroup of A for some $v \in V$. Then, G is isomorphic with the Cayley coset graph $G_S(A, A_v)$ where $S = \{\alpha \in A \mid \{v, \alpha(v)\} \in E\}$.*

Proof. See [40].

In the original theorem, the whole automorphism group (and every automorphism α mapping v to one of its neighbors) is needed. For practical reasons, the following refinement is useful.

Theorem 6. *Let G be a vertex-symmetric graph and B a group of automorphisms on G acting transitively on V (not necessarily the whole automorphism group). Let B_v be the stabilizer subgroup of B for some $v \in V$ and $O_1 \cup \cdots \cup O_n = N(v)$ be the orbit-partitioning induced by B_v on the neighbors of v. Let $S = \{\alpha_1, \alpha_1^{-1}, \ldots, \alpha_n, \alpha_n^{-1}\}$ contain one representative automorphism and its inverse for each orbit O_i, i.e. $\alpha_i(v) \in O_i$. Then, $C_S(B, B_v)$ is isomorphic with G.*

Proof. It is easy to see that $S = S^{-1}$ and $S \cap B_v = \emptyset$, because B_v stabilizes v.
$\phi : B_v/B \to V$ given by $\phi(xB_v) = x(v)$, where $xB_v \in B_v/B$, defines a map:
Suppose $xB_v = yB_v$. Then $y = xb$ for some $b \in B_v \Rightarrow \phi(yB_v) = y(v) = (xb)(v) = x(b(v)) = x(v) = \phi(xB_v)$.
ϕ is a graph isomorphism:
ϕ is one-to-one: Suppose $\phi(xB_v) = \phi(yB_v)$, then $x(v) = y(v) \Rightarrow y^{-1}x(v) = v \Rightarrow y^{-1}x \in B_v \Rightarrow x \in yB_v \Rightarrow xB_v = yB_v$.
ϕ is onto: Let w be a vertex of G. Since G is vertex-symmetric and B acts transitively on V, there exists $z \in B$ such that $z(v) = w$. Thus, $\phi(zB_v) = z(v) = w$.
ϕ preserves adjacency of vertices:

(i) $\{xH, yH\} \in E(G_S(B, B_v)) \Rightarrow x^{-1}y \in B_vSB_v \Rightarrow \exists b, c \in B_v, \alpha \in S : x^{-1}y = b\alpha c \Rightarrow b^{-1}x^{-1}yc^{-1} = \alpha \Rightarrow b^{-1}x^{-1}yc^{-1}(v) \in N(v) \Rightarrow yc^{-1}(v) \in N(xb(v)) \Rightarrow y(v) \in N(x(v)) \Rightarrow \{x(v), y(v)\} \in E(G) \Rightarrow \{\phi(xH), \phi(yH)\} \in E(G)$.
(ii) $\{x, y\} \in E(G) \Rightarrow \exists \alpha \in B : \alpha(x) = v$ and $\{v, \alpha(y)\} \in E(G)$. As $\alpha(y) \in N(v)$, say $\in O_i$, it exists $\alpha_i \in S, \beta \in B_v$ such that $\alpha(y) = \beta\alpha_i(v)$ and $\beta\alpha_i \in B_vBB_v$. Thus, $\{B_v, \beta\alpha_i B_v\} \in E(C_S(B, B_v)) \Rightarrow \{\alpha^{-1}B_v, \alpha^{-1}\beta\alpha_i B_v\} \in E(C_S(B, B_v))$. $\phi(\alpha^{-1}B_v) = \alpha^{-1}(v) = x$ and $\phi(\alpha^{-1}\beta\alpha_i B_v) = \alpha^{-1}(\beta\alpha_i(v)) = \alpha^{-1}(\alpha(y)) = y$.

□

In order to construct new large graphs with given degree and small diameter or mean distance two techniques proved worthwhile. On the one hand, we used a special family of groups: semidirect products of cyclic groups. On the other hand, we worked with abstract presentations of groups, which allows principally with Theorem 6 to analyze every vertex-symmetric graph.

Definition 7. Let $\mathbb{Z}_n$ be the cyclic group of integers with addition modulo n. Every group automorphism of $\mathbb{Z}_n$ can be represented by a unit of the ring $\mathbb{Z}_n$. If the multiplicative order of the unit $a \in \mathbb{Z}_n$ divides m, a semidirect product of $\mathbb{Z}_m$ with $\mathbb{Z}_n$ can be defined by $\mathbb{Z}_m \times_a \mathbb{Z}_n = \{[x, y] \mid x \in \mathbb{Z}_m, y \in \mathbb{Z}_n\}$ and $[x, y][u, v] = [x + u \bmod m, ya^u + v \bmod n]$.

Definition 8. Let $S = \{a, b, \ldots\}$ be a generator set of a group Γ. A set of relations $r_i(a, b, \ldots) = 1_\Gamma$ $(i = 1, 2, \ldots, s)$ is called an abstract presentation of Γ if every relation satisfied by the generators is an algebraic consequence of these particular relations.

The construction of the multiplication table of an abstractly presented group can be done with the Todd-Coxeter algorithm [37], whose original intention was the systematic enumeration of cosets of a subgroup in the given abstract group. The idea of the algorithm is to enumerate the products $\{a, b, \ldots, aa, ab, \ldots\}$ while checking the given relations among them, until no further products can be defined. There are several methods for defining new products and checking them (for an overview see [24]). We improved a breadth-first-search method founded on [19] in that it allows a simultaneous computation of the diameter of the corresponding Cayley graph. This method turned out to be the fastest strategy, in comparison to others, for the examined groups [38].

Example 1. The automorphism group of the Petersen graph has an order of 120 [40]. With Theorem 6, it is sufficient to determine a transitively acting (on V) subgroup of the whole automorphism group. We choose the group $B = \langle a = (1,6)(2,8,5,9)(3,10,4,7), b = (1,2,3,4,5)(6,7,8,9,10)\rangle$ with $|B| = 20$ (see Fig. 1). The abstract presentation of B is $\langle a, b \mid a^4 = b^5 = b^{-1}a^{-1}b^2a = 1_B\rangle$. The vertex 1 is stabilized by a^2, thus $B_1 = \langle a^2\rangle$. The Cayley coset graph $C_{\{a,a^{-1},b,b^{-1}\}}(B, B_1)$ can be seen from the graph in Fig. 1 (right side) by identifying the cosets as nodes with the induced edges (as undirected edges).

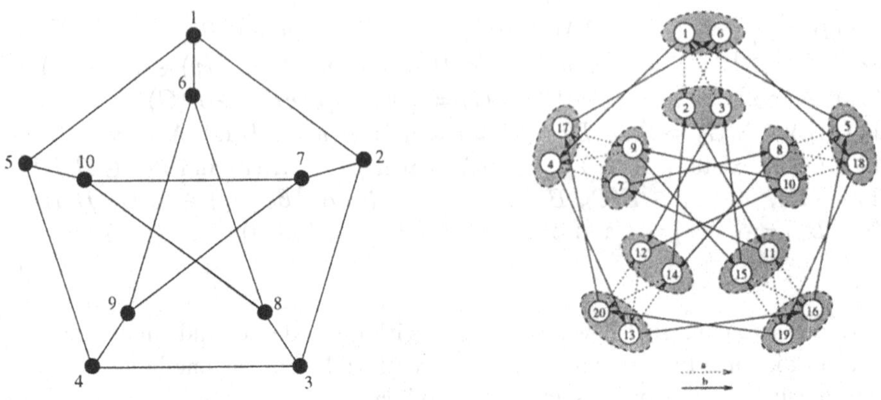

Fig. 1. Petersen graph and its representation as Cayley coset graph

3 Genetic Algorithms and Parameter Tuning

The idea of genetic algorithms (GAs) is to base an optimization algorithm on the natural evolution process with three basic operators [22]: selection of individuals due to a fitness function, recombination of new individuals from the selected ones, and mutation of individuals. GAs were successfully used in graph theory for approximations of NP-complete problems (e.g. TSP), but they also delivered new results for more structural questions about Ramsey problems [18]. The main task of suiting GAs to a specific problem is to find adequate realizations of the basic operators and to adjust the parameters (e.g. maximal number of generations, number of individuals per generation) for the given optimization problem. The basic principle of the algorithm used here is shown in Fig. 2.

Different versions of the single procedures of the GA were tested by some case studies with known groups. The number of graphs which had to be generated until the optimum was found was compared with statistical methods. With the results, the GA could be refined and the parameters could be trimmed.

`select(`n`)`: This function chooses n generator sets of the current generation. With a uniformly distributed selection, each generator set is chosen with equal probability. With a fitness-distributed selection, the probability that a generator set is selected is proportional to its fitness (here Δ or μ of the corresponding graph). At a confidence level of 0.05, we could not discover a statistically valid difference between both techniques, and we used fitness-distributed selection as default.

`crossover(`n`)`: This function generates one new generator set T from n generator sets $S_1, \ldots, S_n$ from the actual generation by selecting Δ single generators at random. This can be done once again with uniform probability or according to the fitness of the individuals. At a confidence level of 0.05, we could not obtain a statistically evident difference between these mechanisms, also. We used uniform probability as default.

`mutate`: A given generator set T can be manipulated by replacing $n \leq |T|$

```
init;                       // Initialize randomly a start population with
                            // p generator sets S_1,...,S_p of order Δ
fitness;                    // Compute D or μ for each C_{S_i}(⟨S_i⟩)
for i = 1 to #generations do
    for j = 1 to #mutated do
        select(1);          // Choose one S_i from the actual generation
        crossover(1);       // Generate a copy T of S_i
        mutate;             // Mutate T and add it to the actual generation
    od;
    for j = 1 to #offspring do
        select(2);          // Choose S_i, S_j from the actual generation
        crossover(2);       // Recombine S_i and S_j to one new generator set T
        mutate;             // Mutate T and add it to the actual generation
    od;
    for j = 1 to #new_generator_sets do
        generate;           // Add a new, random generator set to the
                            // actual generation
    od;
    fitness;                // Compute D or μ for each new graph
    kill;                   // Delete the worst graphs of the generation
od;
```

Fig. 2. GA used in the search for large Cayley graphs with small diameter or mean distance and degree Δ

generators with new ones while retaining the symmetry of T. Our result was that replacing only one generator and its inverse is at least 2.5 times better than if we would replace more generators (confidence level: 0.05).

`fitness` and `kill`: At the end of each iteration, the new generator sets are assessed by the diameter or mean distance of the corresponding Cayley graphs. We checked three strategies how to build the new generation from the old one: (i) kill the worst individuals, (ii) kill the oldest individuals, (iii) kill the individuals randomly. The first mode turned out to be at least three times better than the others (confidence level: 0.05).

Parameters: With various statistical analyses, the parameters of the GA were trimmed, and we achieved the best results with: *#generations* = max. 100, *#mutated* = 5, *#offspring* = 2, *#new_generator_sets* = 2. Keeping the number of individuals per generation at 10 with the function `kill` proved best.

A much deeper statistical analysis of a greater portion of the experiments can be found in [21]. Fig. 3 depicts the main comparison between the optimized GA and a pure random search for both, the degree/diameter optimization and the optimization of mean distances. The figures show that we could speed up the search up to a factor of 5, which warrants the use of GAs. Indeed, we found no negative example where the GA is worse than random search by the degree/diameter optimization. Regarding the mean distance problem, there are certain examples (e.g. example (h) in Fig. 3), which can be put down to the fact that all optimizations and parameter trimming were done in orientation towards the degree/diameter problem.

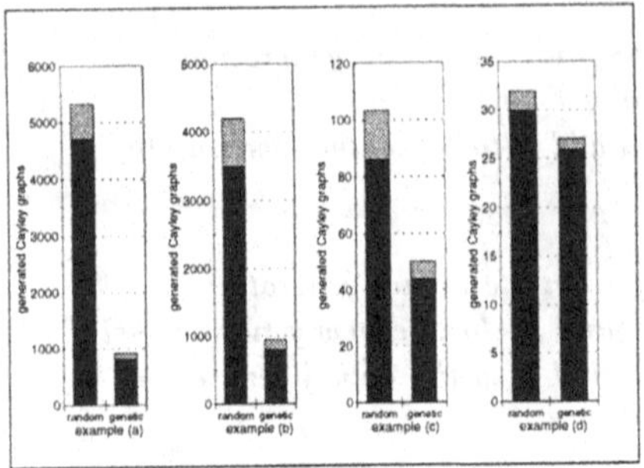

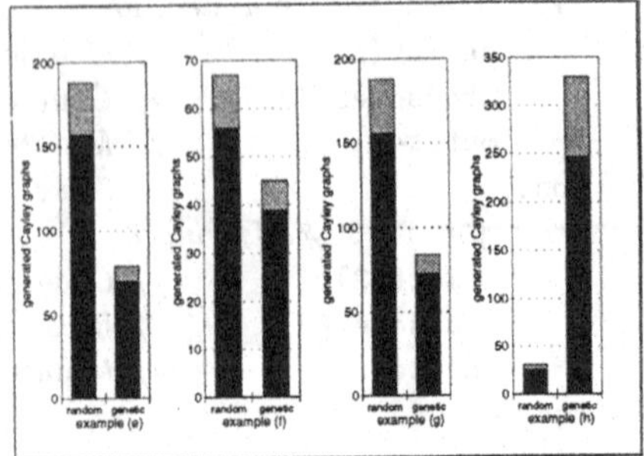

Fig. 3. Comparison between the GA and random search for degree/diameter (left) and mean distance optimization (right) (confidence level: 0.05)

4 Search Field and Selection of Groups

Besides the problem of finding a good generator set S in a given group Γ, a main problem is deciding which groups to search through. It is obvious that abelian groups produce no good results, because for any two generators a, b we can obtain at most only three vertices at distance 2 from one given vertex: aa, bb, and $ab = ba$, whereas it is possible to get four vertices at distance 2 in a non-abelian group. This disadvantage even increases for larger distances. Nevertheless, some lower bounds can be obtained from abelian groups [31]. We examined all group classes used in [17], but only semidirect products of cyclic groups (see Def. 7) yielded new results. To optimize the value for a (Δ, D)-pair, we searched randomly through semidirect products of orders from 100% to 110% of the order of the best known (Δ, D)-graph. If a first check of a group produced good results, it was further investigated by the GA.

Furthermore, we searched through a list of all groups with order less than 1 000, which was compiled in a library of the algebra tool GAP [35]. We were able to find those Cayley graphs being already in the table, but we got no better results. This manifests the quality of the graphs in the table with order less than 1 000.

For larger graphs, the following technique was useful: We took a quite good vertex-symmetric graph for a given (Δ, D)-pair and calculated the abstract presentation of the underlying group. Then, we modified the abstract presentation of the group by exchanging one relation with a new one and checked it with the modified Todd-Coxeter algorithm. If the corresponding Cayley graph was near to an entry in the table, we started the GA on this group.

5 Results

Fig. 4 shows an up-to-date table of the largest known graphs for a given degree and diameter. Almost every paper on this topic [4, 5, 8, 10, 11, 16, 17, 26, 36] contained an updated version of this table. Some examples of past advances are listed in Fig. 5.

$\Delta \setminus D$	2	3	4	5	6	7	8	9	10
3	*10*	*20*	38	70	130	184	320	540	938
4	*15*	41	95	364	740	1 155	**3 080**	7 550	**17 604**
5	*24*	70	**210**	**546**	2 754	**5 500**	**16 956**	**52 768**	**145 880**
6	32	108	**375**	**1 395**	7 860	**19 065**	**74 256**	**278 046**	**954 480**
7	*50*	144	**672**	2 756	**11 110**	**50 020**	**216 160**	**953 586**	**5 243 030**
8	57	**253**	1 081	4 895	39 396	127 134	**660 765**	**2 943 720**	**7 739 472**
9	74	585	**1 536**	7 752	75 198	264 024	**1 355 424**	**5 094 726**	**19 873 350**
10	91	650	2 211	**12 642**	133 500	**556 803**	**3 696 600**	**9 910 080**	**47 129 712**

Fig. 4. Largest known graphs for a given degree Δ and diameter D (new results in **bold**, optimal results in *italics*)

Excluding the 33 new values discovered in this work, since 1995, some new results have been obtained independently by P. Hafner [20] and O. Wohlmuth [39].[1] Comellas [12] and Delorme [15] collect new results for the degree/diameter problem and present them on the web. We include a table with the generators of the new records as appendix.

A table with the largest known graphs for a given degree and bounded mean distance – to our knowledge – is set up for the first time in Fig. 6. Detailed presentations of these graphs can be found in [32].

The practical relevance of the construction of large graphs is exemplary demonstrated with two case studies in Fig. 7. An advanced wormhole routing protocol [34] was implemented, and the network latency was compared for increasing message injection rates (Poisson distribution of number of messages, uniform distribution of communication pairs) on different topologies.

It can always be concluded (confidence level 0.05) that the graphs constructed here can handle a higher load of random message traffic, and that for a fixed

[1] The ($\Delta = 6, D = 3, |V| = 108$)-graph constructed by Wohlmuth and the $(8, 4, 1\,081)$-, $(8, 5, 4\,895)$-, $(9, 5, 7\,752)$-, $(10, 4, 2\,211)$-graphs by Hafner, which are listed in Fig. 4 are still records for the degree/diameter problem. Hafner also found new records for graphs with $\Delta > 10$.

(Δ, D)	1981 [4]	1984 [16]	1986 [5]	1987 [10]	1992 [8]	1995[1] [11]	1994 [17]	this work
(4, 8)	1 230	1 872	1 872	2 673	2 943	3 025	3 025	3 080
(4, 10)	2 560	7 000	13 056	13 056	15 657	–	16 555	17 604
(5, 4)	170	174	174	174	182	186	186	210
(5, 5)	386	532	532	532	532	532	532	546
(7, 6)	8 024	8 998	10 546	10 546	10 554	10 566	10 566	11 110
(8, 3)	192	200	200	200	203	220	234	253

Fig. 5. Selected advances in the degree/diameter problem over the last fifteen years

$\Delta \setminus \mu$	≤ 2	≤ 3	≤ 4	≤ 5	≤ 6	≤ 7	≤ 8	≤ 9	≤ 10
3	*12*	*30*	72	136	272	602	1 026	2 054	4 082
4	20	66	203	620	1 806	5 418	15 500	43 380	
5	24	110	486	1 830	7 104	26 960			
6	40	189	936	4 446	20 271				
7	50	294	1 650	9 020	50 020				
8	60	440	2 628	[2]16 555					
9	72	558	3 942	28 476					
10	84	756	5 720	48 006					

Fig. 6. Largest known graphs for a given degree Δ and maximal mean distance μ (optimal results in *italics*)

injection rate, the message dissemination is faster than on standard topologies like binary hypercubes or rectangular meshes.

6 Conclusions and Future Work

By the use of genetic algorithms in connection with algebraic techniques, we were able to get 33 new results for the degree/diameter problem. This shows, in addition to [18], that the use of GAs in graph theory is not limited to standard optimization problems like TSP. They can be successfully used for more

[1] Though published earlier, the table in [17] is better than that in [11].

[2] This graph was constructed by Hafner [20].

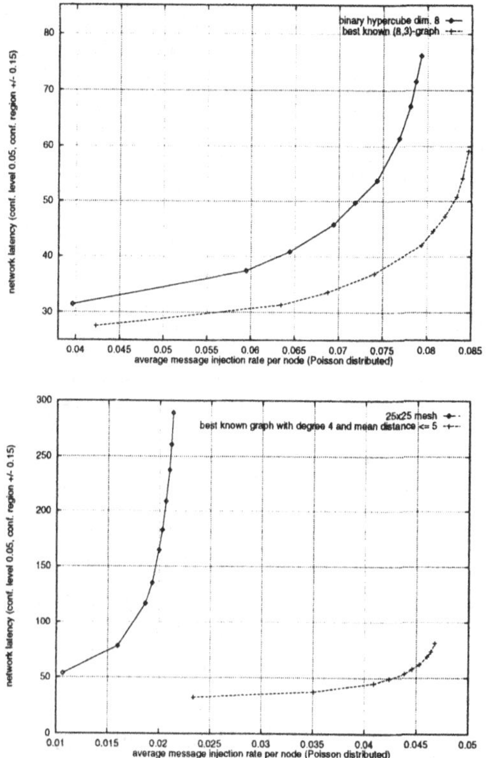

Fig. 7. Case study of random message traffic at different loads: $(8,3)$-graph (253 nodes) versus binary hypercube H_8 (256 nodes) (top) and $(4, \mu \leq 5)$-graph (620 nodes) versus 25×25 mesh (625 nodes) (bottom)

structural problems. It seems as though the class of Cayley graphs is exhausted for graphs of order less than 1 000. But with more computational power, it is very likely that better results with Cayley graphs of order above 1 000 may be achieved in the future.

Moreover, the class of Cayley coset graphs, which covers all vertex-symmetric graphs, has to be checked. This is feasible with the refinement of Sabidussi's representation theorem. We have already started experimenting in this area, but have not obtained new results, yet.

Acknowledgements

The author would like to thank Berthold Hagmann for his implementation of the genetic algorithms, Lutz Twele for the modification of the Todd-Coxeter algorithm, and Torsten Müller for the wormhole routing simulator. Many thanks also to Paul Hafner for useful discussions and suggestions.

Appendix: Groups and Generators for selected graphs

(Δ, D)	order	group	generators (order) or presentation
$(4,8)$	3 080	presentation	$ba^{-1}b^4a^{-1}baba^{-2} = a^4ba^2b^{-1}a^2b^{-1}a^2b = a^4b^4a^{-2}b^3 = 1$
$(4,10)$	17 604	$\mathbb{Z}_{36} \times_{38} \mathbb{Z}_{489}$	[35,421](36), [4,254](9), [1,139](36), [32,67](9)
$(5,4)$	210	index 14 subgr. in $\mathbb{Z}_{42} \times_{39} \mathbb{Z}_{70}$	[28,14](15), [7,18](6), [21,14](2), [14,56](15), [35,48](6)
$(5,5)$	546	$\mathbb{Z}_6 \times_{69} \mathbb{Z}_{91}$	[2,74](21), [3,69](2), [4,10](21), [3,86](2), [3,67](2)
$(5,7)$	5 500	$\mathbb{Z}_{20} \times_{28} \mathbb{Z}_{275}$	[3,72](20), [19,120](20), [10,132](2), [17,139](20), [1,215](20)
$(5,8)$	16 956	$\mathbb{Z}_{12} \times_{578} \mathbb{Z}_{1413}$	[7,450](12), [10,167](18), [6,585](2), [5,1404](12), [2,277](18)
$(5,9)$	52 768	$\mathbb{Z}_{32} \times_{1231} \mathbb{Z}_{1649}$	[2,137](16), [29,1450](32), [16,85](2), [30,638](16), [3,121](32)
$(5,10)$	145 880	$\mathbb{Z}_{40} \times_{1728} \mathbb{Z}_{3647}$	[36,1402](70), [21,957](40), [20,2541](2), [4,1706](70), [19,110](40)
$(6,4)$	375	presentation	$a^5 = (ab)^3 = ab^2ab^{-1}ab^{-1} = ca^{-1}b^2a^{-2} = 1$
$(6,5)$	1 395	$\mathbb{Z}_{45} \times_{9} \mathbb{Z}_{31}$	[11,1](45), [39,8](15), [10,13](9), [34,11](45), [6,29](15), [35,16](9)
$(6,7)$	19 065	$\mathbb{Z}_{15} \times_{324} \mathbb{Z}_{1271}$	[5,1077](123), [14,1196](15), [8,40](15), [10,727](123), [1,151](15), [7,139](15)
$(6,8)$	74 256	$\mathbb{Z}_{48} \times_{1255} \mathbb{Z}_{1547}$	[43,450](48), [31,267](48), [28,635](12), [5,601](48), [17,1158](48), [20,1163](12)
$(6,9)$	278 046	$\mathbb{Z}_{54} \times_{788} \mathbb{Z}_{5149}$	[50,143](27), [39,1382](18), [5,4774](54), [4,1042](27), [15,3646](18), [49,1091](54)
$(6,10)$	954 480	$\mathbb{Z}_{80} \times_{8645} \mathbb{Z}_{11931}$	[43,6754](80), [36,3112](60), [19,1914](80), [37,778](80), [44,3995](60), [61,8319](80)
$(7,4)$	672	$\mathbb{Z}_6 \times_{39} \mathbb{Z}_{112}$	[2,73](48), [5,54](6), [5,71](12), [3,42](2), [4,23](48), [1,22](6), [1,31](12)
$(7,6)$	11 110	$\mathbb{Z}_{10} \times_{791} \mathbb{Z}_{1111}$	[1,631](10), [2,850](55), [6,255](55), [5,303](2), [9,554](10), [8,811](55), [4,823](55)
$(7,7)$	50 020	$\mathbb{Z}_{20} \times_{125} \mathbb{Z}_{2501}$	[9,1614](20), [13,659](20), [3,2023](20), [10,2196](2), [11,1014](20), [7,1051](20), [17,29](20)
$(7,8)$	216 160	$\mathbb{Z}_{32} \times_{1007} \mathbb{Z}_{6755}$	[7,6477](32), [13,6227](32), [15,6378](32), [16,2030](2), [25,4881](32), [19,6689](32), [17,764](32)

(Δ, D)	order	group	generators (order) or presentation
$(7,9)$	953 586	$\mathbb{Z}_{54} \times_{16387} \mathbb{Z}_{17659}$	[32,2018](27), [41,1372](54), [52,1818](27), [27,1560](2), [22,12912](27), [13,3806](54), [2,36](27)
$(7,10)$	5 243 030	$\mathbb{Z}_{130} \times_{380} \mathbb{Z}_{40331}$	[60,26296](403), [84,6032](65), [4,17745](65), [65,19034](2), [70,9633](403), [46,33119](65), [126,12358](65)
$(8,3)$	253	$\mathbb{Z}_{11} \times_{3} \mathbb{Z}_{23}$	[2,0](11), [2,13](11), [3,1](11), [4,22](11), [9,0](11), [9,19](11), [8,17](11), [7,2](11)
$(8,8)$	660 765	$\mathbb{Z}_{105} \times_{1299} \mathbb{Z}_{6293}$	[74,5328](105), [57,4049](35), [69,2830](35), [49,1753](435), [31,360](105), [48,4295](35), [36,2770](35), [56,190](435)
$(8,9)$	2 943 720	$\mathbb{Z}_{72} \times_{17344} \mathbb{Z}_{40885}$	[56,21163](765), [65,32979](72), [52,28332](1530), [23,37340](72), [16,10117](765), [7,13919](72), [20,2438](1530), [49,13330](72)
$(8,10)$	7 739 472	$\mathbb{Z}_{312} \times_{16253} \mathbb{Z}_{24806}$	[104,22051](74418), [225,575](104), [230,4842](156), [59,13725](312), [208,2755](74418), [87,17941](104), [82,2682](156), [253,4493](312)
$(9,4)$	1 536	presentation	$e^2 = dbeb^{-1}d^{-1}e = ac^2ad^2 = (ad)^3 = ad^{-1}eda^{-1}e = ded^{-1}b^{-1}eb = ab^{-1}da^{-1}de = ebcec^{-1}b^{-1} = cd^{-1}edc^{-1}e = acbdb^{-1}e = badb^{-2}ad^{-1} = 1$
$(9,8)$	1 355 424	$\mathbb{Z}_{168} \times_{523} \mathbb{Z}_{8068}$	[49,4508](24), [51,7732](56), [141,2259](56), [88,5901](84), [84,4594](2), [119,6492](24), [117,2516](56), [27,3267](56), [80,487](84)
$(9,9)$	5 094 726	$\mathbb{Z}_{294} \times_{2331} \mathbb{Z}_{17329}$	[288,12473](19747), [32,2016](147), [143,8415](294), [265,9895](294), [147,2494](2), [6,16543](19747), [262,13615](147), [151,6302](294), [29,4959](294)
$(9,10)$	19 873 350	$\mathbb{Z}_{1050} \times_{7496} \mathbb{Z}_{18927}$	[496,5873](4725), [231,18443](50), [791,3475](150), [766,4763](4725), [525,7948](2), [554,5647](4725), [819,2324](50), [259,7129](150), [284,223](4725)

(Δ, D)	order	group	generators (order) or presentation
$(10,5)$	12 642	$\mathbb{Z}_{42} \times_{124} \mathbb{Z}_{301}$	[18,24](7), [31,123](42), [8,292](21), [22,207](21), [37,42](42), [24,291](7), [11,226](42), [34,4](21), [20,222](21), [5,14](42)
$(10,7)$	556 803	$\mathbb{Z}_{117} \times_{35} \mathbb{Z}_{4759}$	[79,2789](117), [34,2008](117), [36,4255](13), [93,3212](39), [68,3596](117), [38,1416](117), [83,2737](117), [81,3140](13), [24,1309](39), [49,2561](117)
$(10,8)$	3 696 600	$\mathbb{Z}_{120} \times_{24107} \mathbb{Z}_{30805}$	[86,935](60), [41,26529](120), [88,29978](15), [23,4082](120), [108,12955](10), [34,25700](60), [79,12908](120), [32,24782](15), [97,1796](120), [12,7145](10)
$(10,9)$	9 910 080	$\mathbb{Z}_{216} \times_{23007} \mathbb{Z}_{45880}$	[179,1239](216), [52,35690](108), [205,29836](216), [133,5885](216), [194,16531](216), [37,31687](216), [164,550](108), [11,6532](216), [83,24925](216), [22,301](216)
$(10,10)$	47 129 712	$\mathbb{Z}_{1776} \times_{23819} \mathbb{Z}_{26537}$	[421,18797](1776), [649,20946](1776), [1472,24312](1887), [1376,17320](1887), [1265,16954](1776), [1355,21715](1776), [1127,3382](1776), [304,13326](1887), [400,22970](1887), [511,2471](1776)

References

1. Bruce W. Arden and K. Wendy Tang. Representations and routing for Cayley graphs. *IEEE Transactions on Communications*, 39(11):1533–1537, November 1991.
2. J. C. Bermond, C. Delorme, and G. Farhi. Large graphs with given degree and diameter III. *Annals of Discrete Mathematics*, 13:23–32, 1982.
3. J. C. Bermond, C. Delorme, and G. Farhi. Large graphs with given degree and diameter. II. *Journal of Combinatorial Theory, Series B*, 36:32–48, 1984.
4. Jean-Claude Bermond and Béla Bollobás. The diameter of graphs — A survey. *Congressus Numerantium*, 3:3–27, 1981.
5. Jean-Claude Bermond, Charles Delorme, and J.-J. Quisquater. Strategies for interconnection networks, some results from graph theory. *Journal of Parallel and Distributed Computing*, 3:433–449, 1986.
6. Norman Biggs. *Algebraic Graph Theory*. Cambridge University Press, 2nd edition, 1993.
7. Shahid H. Bokhari and A. D. Raza. Reducing the diameters of computer networks. *IEEE Transcations on Computers*, C-35(8):757–761, August 1986.

8. Lowell Campbell et al. Small diameter symmetric networks from linear groups. *IEEE Transactions on Computers*, 41(2):218–220, 1992.
9. Alan G. Chalmers and Steve Gregory. Constructing minimum path configurations for multiprocessor systems. *Parallel Computing*, 19:343–355, 1993.
10. Fan R. K. Chung. Diameters of graphs: Old and new results. *Congressus Numerantium*, 60:295–317, 1987.
11. F. Comellas and J. Gómez. New large graphs with given degree and diameter. In Y. Alavi and A. Schwenk, editors, *Graph Theory, Combinatorics and Algorithms*, volume 1, pages 221–233, New York, 1995. John Wiley & Sons, Inc.
12. Francesc Comellas. comellas@mat.upc.es.
13. Robert Cypher, Friedhelm Meyer auf der Heide, Christian Scheideler, and Berthold Vöcking. Universal algorithms for store-and-forward and wormhole routing. In *Proceedings of the 26th ACM-STOC*, pages 356–365, 1996.
14. C. Delorme. Examples of products giving large graphs with given degree and diameter. *Discrete Applied Mathematics*, 37/38:157–167, 1992.
15. Charles Delorme. cd@lri.fr.
16. Charles Delorme and G. Farhi. Large graphs with given degree and diameter – part I. *IEEE Transactions on Computers*, C-33(9):857–860, September 1984.
17. Michael J. Dinneen and Paul R. Hafner. New results for the degree/diameter problem. *Networks*, 24:359–367, 1994.
18. Geoffrey Exoo. Applying optimization algorithms to Ramsey problems. In Yousef Alavi, editor, *Graph Theory, Combinatorics, Algorithms and Applications*, pages 175–179. John Wiley & Sons, Inc, New York, 1991.
19. H. Felsch. Programmierung der Restklassenabzählung einer Gruppe nach Untergruppen. *Numerische Mathematik*, 3:250–256, 1961.
20. P. Hafner. Large Cayley graphs and digraphs for the degree/diameter problem: an update. (in preparation), 1997.
21. Berthold Hagmann. Optimierungstechniken zum Entwurf günstiger Netztopologien mittels Cayley-Graphen. Diploma thesis, Universität Oldenburg, 1997.
22. J. H. Holland. *Adaptation in Natural and Artificial Systems.* University of Michigan Press, Ann Arbor, 1975.
23. S. Lakshmivarahan, Jung-Sing Jwo, and S. K. Dhall. Symmetry in interconnection networks based on Cayley graphs of permutation groups: A survey. *Parallel Computing*, 19:361–407, 1993.
24. John Leech. Coset enumeration. In Michael D. Atkinson, editor, *Computational Group Theory*, pages 3–18. Academic Press, 1984.
25. F. Thomson Leighton. *Parallel Algorithms and Architectures.* Morgan Kaufmann Publishers, San Mateo, California, 1992.
26. Gerard Memmi and Yves Raillard. Some new results about the (d, k) graph problem. *IEEE Transactions on Computers*, C-31(8):784–791, August 1982.
27. Friedhelm Meyer auf der Heide and Berthold Vöcking. A packet routing protocol for arbitrary networks. In E. W. Mayr and C. Puech, editors, *Proceedings of the 12th Annual Symposium on Theoretical Aspects of Computer Science (STACS '95)*, LNCS 900, pages 291–302, 1995.
28. Gert Sabidussi. Vertex-transitive graphs. *Monatshefte für Mathematik*, 68:426–438, 1964.
29. Michael Sampels. Algebraic constructions of efficient systolic architectures. In *Proceedings of the 2nd International Conference on Massively Parallel Computing Systems (MPCS '96)*, pages 15–22. IEEE Computer Society Press, 1996.

30. Michael Sampels. Cayley graphs as interconnection networks: A case study. In *Proceedings of the 7th International Workshop on Parallel Processing by Cellular Automata and Arrays (PARCELLA '96)*, pages 67–76, Berlin, 1996. Akademie-Verlag.
31. Michael Sampels. Massively parallel architectures and systolic communication. In *Proceedings of the 5th Euromicro Workshop on Parallel and Distributed Processing (PDP '97)*, pages 322–329. IEEE Computer Society Press, 1997.
32. Michael Sampels. Representation of vertex-symmetric interconnection networks. In *Proceedings of the 2nd International Conference on Parallel Processing & Applied Mathematics.* (to appear), 1997.
33. Michael Sampels and Stefan Schöf. Massively parallel architectures for parallel discrete event simulation. In *Proceedings of the 8th European Simulation Symposium (ESS '96)*, volume 2, pages 374–378. SCS, 1996.
34. Christian Scheideler and Berthold Vöcking. Universal continuous routing strategies. In *Proceedings of the 8th ACM-SPAA*, pages 142–151. ACM, 1996.
35. Martin Schönert. GAP - groups, algorithms and programming. Lehrstuhl D für Mathematik, RWTH Aachen, 1995.
36. Robert M. Storwick. Improved construction techniques for (d,k) graphs. *IEEE Transactions on Computers*, C-19:1214–1216, December 1970.
37. J. A. Todd and H. S. M. Coxeter. A practical method for enumerating cosets of a finite abstract group. *Proceeding of the Edinburgh Mathematical Society*, 5(2):26–34, 1936.
38. Lutz Twele. Effiziente Implementierung des Todd-Coxeter Algorithmus im Hinblick auf Grad/Durchmesser-Optimierung von knotentransitiven Graphen. Diploma thesis, Universität Oldenburg, 1997.
39. Otto Wohlmuth. A new dense group graph discovered by an evolutionary approach. In *Paralleles und Verteiltes Rechnen, Beiträge zum 4. Workshop über wissenschaftliches Rechnen.* Shaker Verlag, 1996.
40. H. P. Yap. *Some topics in graph theory.* London Mathematical Society Lecture Note Series 108. Cambridge University Press, 1986.

The Bounded Tree-Width Problem of Context-Free Graph Languages

K. Skodinis *

University of Passau,
94030 Passau, Germany

Abstract. We show that the following (equivalent) problems are P-complete:

1. Does a given confluent NCE graph grammar only generate graphs of bounded tree-width? and
2. is the graph language generated by a given confluent NCE graph grammar an HR language?

This settles the complexity of these important problems on graph grammars.

1 Introduction

Graph grammars can be seen as the extension of context-free grammars from strings to graphs. They are based on node replacement (NR graph grammars) [Nag80,JR80a,JR80b,Cou87,Eng89,EKR91] or on edge replacement (HR graph grammars) [Cou90,Hab92]. A graph grammar consists of a finite set of productions, which are used to expand graphs or hypergraphs by repeated replacements of nodes or hyperedges. The language of a graph grammar is the set of all terminal labeled graphs derivable from some axiom. The most powerful NR graph grammars are the edNCE graph grammars which generate directed graphs with labeled edges.

A graph grammar is confluent if simultaneously applicable productions can be applied in any order. Since the environment of an hyperedge replacement is static, it is easy to see that HR graph grammars are confluent. NR graph grammars are not necessarily confluent, and there are graph languages, e.g. the set of all graphs over some alphabets, which cannot be generated by confluent NR graph grammars. It is well known that confluence of NR graph grammars is decidable in polynomial time [Kau85]. Confluent NR graph grammars and HR graph grammars are often called context-free.

HR graph languages are a proper subclass of confluent NR graph languages [ER90,CER93]. The complement graph of binary trees are an example. Although HR graph languages are quite restrictive they are very useful in the theory of graph grammars. One reason is that they contain only graphs of bounded tree-width, which contrasts with confluent NR graph languages. It is well known

* The work of the author was supported by the Deutsche Forschungsgemeinschaft (DFG) grant Br-835-7-1

that many generally NP-hard graph problems become solvable in polynomial time if the tree-width is bounded [ALS91,AP89]. The tree-width of a graph is the minimum width over all its tree-decompositions (see for example [Bod96]).

Hence from the algorithmic point of view, the question whether or not a graph language can be generated by an HR graph grammar is important. It has been shown that confluent NR graph languages of bounded degree are HR graph languages [Bra91,EH91,EH94]. Courcelle [Cou95] came up with a characterization. He has shown that a confluent NR graph grammar $\mathcal{G}$ can be generated by an HR graph grammar if and only if the graph language of $\mathcal{G}$ has a bounded tree-width, or if and only if there is an integer n, such that graph grammar $\mathcal{G}$ does not generate a graph containing a complete bipartite subgraph $K_{n,n}$.

In this paper we consider confluent NR graph grammars generating undirected graphs without edge labels (C-NCE graph grammars) and we examine the complexity of the problem above. It is known that it is decidable [Cou95]. We show that it is P-complete. The proof is based on a pumping argument as it is known e.g. for the context-free string languages. A similar argument has been used in [SW97]. We prove an even more general lemma: It is P-complete whether or not a general NCE graph grammar $\mathcal{G}$ generates graphs containing complete bipartite graphs $K_{n,n}$ of arbitrarily large size n as subgraphs .

Hence, for a confluent NCE graph grammar $\mathcal{G}$ the problem whether or not a graph language of $\mathcal{G}$ has bounded tree-width and the problem whether or not a graph language of $\mathcal{G}$ can be generated by an HR graph grammar are P-complete. This complexity result can be generalized to (general) confluent NR (C-edNCE) graph grammars.

2 Preliminaries

In this section we recall the basic notions concerning graphs and node-replacement graph grammars. We deal with undirected, node labeled graphs. The approach can be extended to directed graphs with node and edge labels.

Definition 1 (Graphs). Let Σ be a finite set of node labels. A graph over Σ is a system $G = (V, E, \phi)$, where

1. V is a finite set of nodes,
2. $E \subseteq \{\{u,v\} \mid u, v \in V, u \neq v\}$ is a finite set of edges, and
3. $\phi : V \to \Sigma$ is a node labeling function.

A node labeled by $a \in \Sigma$ is called an a-node.

Next we define the substitution of node u in graph G by the graph R. This substitution mechanism is used in derivation steps of NCE graph grammars.

Definition 2 (Substitutions). Let G and R be two graphs over Σ. Let $C \subseteq \Sigma \times V_R$, be an embedding relation for R, and let u be a node from G. The graph $G[u/_C R]$ is obtained by replacing node u by R with respect to C as follows (see Fig. 1).

1. Let J be the union of G (without node u and its incident edges) and a copy of R which is disjoint with G,
2. For each edge $\{v, u\}$ from G add an edge $\{v, w\}$ to the set of edges of J if and only if $(\phi(v), w) \in C$. The resulting graph is $G[u/_C R]$.

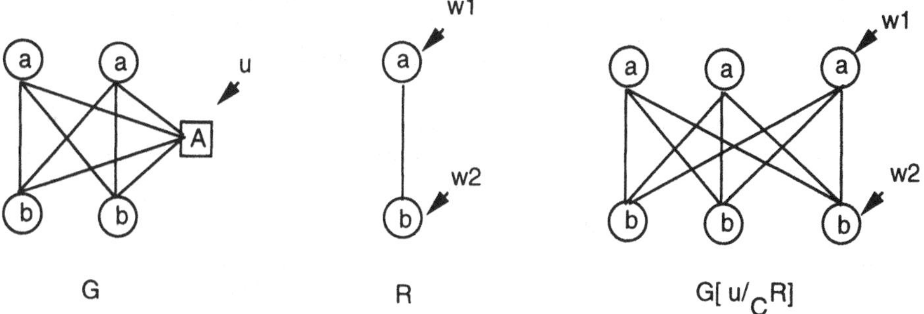

Fig. 1. Graphs G, R, and $G[u/_C R]$ for $C = \{(b, w_1), (a, w_2)\}$.

Thus $G[u/_C R]$ is obtained substituting u by R and establishing connections according to C.

Definition 3 (NCE graph grammars). An NCE (neighborhood controlled embedding) graph grammar is a system $\mathcal{G} = (\Sigma, \Delta, S, P)$, where

1. Δ and Σ are alphabets with $\Delta \subseteq \Sigma$ as the set of terminal node labels, and $\Sigma - \Delta$ as the set of nonterminal node labels, respectively,
2. S is a graph over Σ, the axiom of $\mathcal{G}$, and
3. P is a finite set of productions. Each production is a triple (A, R, C), where A is a nonterminal node label from $\Sigma - \Delta$ (the left-hand side), R is a graph over Σ (the right-hand side), and C is an embedding relation for R.

The next definition shows how NCE graph grammars define sets of graphs by derivations.

Definition 4 (Derivations and languages). Let $\mathcal{G}$ be an NCE graph grammar and let G and H be graphs over Σ. We say that G directly derives H in $\mathcal{G}$, denoted by $G \Longrightarrow_p H$ *(or just* $G \Longrightarrow H$*)* if there is an A-node u in G and a production $p = (A, R, C)$, such that $H = G[u/_C R]$.

The language of an NCE graph grammar $\mathcal{G}$ is $L(\mathcal{G}) = \{G \mid S \stackrel{*}{\Longrightarrow} G\}$, where S is the axiom of $\mathcal{G}$ and $\stackrel{*}{\Longrightarrow}$ is the transitive and reflexive closure of $\Longrightarrow$.

An NCE graph grammar is *confluent* (C-NCE) if for every graph G derivable from the axiom of $\mathcal{G}$, all nonterminal nodes u, v in G, and all productions $(\phi(u), H, D), (\phi(v), J, F)$ in $\mathcal{G}$:

$$G[u/_D H][v/_F J] = G[v/_F J][u/_D H].$$

In confluent graph grammars the order in which the productions are applied is irrelevant for the resulting graph.
An NCE graph grammar is *boundary* (B-NCE) if the axiom and the right hand-side of each production do not contain adjacent nonterminal nodes.
It is *linear* (L-NCE) if the axiom and the right hand-side of each production have at most one nonterminal node.
The linear, boundary and confluent NR graph languages form a proper hierarchie [EL89,CER93]. HR graph languages are a proper subclass of the boundary NR graph languages [ER90,CER93] and incomparable to the linear NR graph languages.

An eNCE graph grammar is an extension of NCE, with terminal and nonterminal node labels as well as terminal and nonterminal edge labels. If additionally directions are allowed then we denote such graph grammars by edNCE.
edNCE and eNCE graph grammars can generate so-called *blocking edges*. These are nonterminal edges incident to two terminal nodes. Dealing with blocking edges is complicated. If the elimination of the blocking edges is possible, it is of high complexity, see [SW95].

In parallel to context-free string grammars a graph grammar is called *reduced* if every label A is reachable from the axiom S and can generate a terminal graph. Using the classical techniques it can be seen that every NCE (C-NCE, B-NCE, L-NCE) graph grammar can be transformed into an equivalent reduced NCE (C-NCE, B-NCE, L-NCE) graph grammar in polynomial time.

For the complexity problems on NCE graph grammars we define the size of a graph grammar as the size of the string obtained when writing down the grammar in the usual way.

Next we recall the notions of the subgraph-derivation, ancestor relation, descendant relation, and edge-preserving derivation from [SW97].

Definition 5 (Subgraph derivations). A graph H' is directly subgraph derivable from some graph G, denoted by $G \longmapsto H'$ (or $G \longmapsto_p H'$), if there is a graph H, such that $G \Longrightarrow_p H$ and H' is a subgraph of H.

We say $G \overset{*}{\longmapsto} H$ is a *subgraph derivation*, where $\overset{*}{\longmapsto}$ is the transitive and reflexive closure of $\longmapsto$; in that case, we also say that H is *subgraph derivable* from G.
If $H_1, \ldots, H_n$ is a sequence of graphs, such that for all i, either $H_i \mapsto H_{i+1}$ or H_{i+1} is a subgraph of H_i, then we also write $H_1 \mapsto H_2 \mapsto \cdots \mapsto H_n$.

Definition 6 (Ancestor- and descendant relation). Let $G \Longrightarrow_p H$ be a direct derivation step performed with the application of production $p = (A, R, C)$ to an A-node u of G.

1. Node u is the direct ancestor of all nodes and edges of R in H, and all nodes and edges of R are direct descendants of u.
2. Let v be a B-neighbor of u in G and $\{v, u\}$ be an edge in G. Edge $\{v, u\}$ is a direct ancestor of every edge $\{v, w\}$ in H, $w \in R$, and all edges between v and nodes of R in H are direct descendants of edge $\{v, u\}$.

The direct ancestor- and descendant relations can be extended to the case of direct subgraph derivation steps $G \longmapsto_p H$. The reflexive and transitive closure of the direct ancestor- and direct descendant relations are called ***ancestor relation*** and *descendant relation*, respectively.

Definition 7 (Edge-preserving derivations). Let $\mathcal{G}$ be an NCE graph grammar and let G be a graph with two adjacent nodes u and v with labels A and B. We say that there is an edge-preserving (A, B)-derivation in $\mathcal{G}$ if G can derive a graph H which has an edge incident to a terminal labeled descendant node of u and a terminal labeled descendant node of v.

We also need some further results from [SW97].

Lemma 8. *Let $\mathcal{G}$ be an NCE graph grammar and $G \longmapsto H$ a direct subgraph derivation step of $\mathcal{G}$. Let G' and H' be subgraphs of G and H, respectively, such that every node and every edge of H' has at least one ancestor in G'. Then either H' is a subgraph of G' or there is a subgraph derivation step $G' \longmapsto H'$ of $\mathcal{G}$.*

Lemma 9. *Given an NCE graph grammar $\mathcal{G}$ and two graphs G, H. If the number of nodes and edges of G and H is bounded by some k, then the question whether or not $G \overset{*}{\longmapsto} H$ holds is nondeterministically decidable in logarithmic space in the size of $\mathcal{G}$. If, moreover, a partial function $d : V_H \to V_G$ is given, then it can be decided whether the following holds in the subgraph derivation $G \mapsto^* H$: if $d(u) = v$ then u is a descendant of v.*

Lemma 9 implies that for a given NCE graph grammar $\mathcal{G}$ it is nondeterministically decidable in logarithmic space whether or not there is an edge-preserving (A, B)-derivation in $\mathcal{G}$.

3 The bounded tree-width problem

At first we recall the definition of the tree-decomposition and the tree-width of a given graph G, see [Klo94].

Definition 10. Let $G = (V, E)$ be a graph. A tree-decomposition of G is a pair (X, T), where X is a collection $X = \{X_i \mid i \in I\}$ and $T = (I, F)$ is a tree with one node for each subset of X, such that

1. $\cup_{i \in I} X_i = V$,
2. for every edge $\{v, u\}$ of G there is a subset $X_i \in X$ which contains both v and u
3. for every node v of G the set of nodes $\{i \mid v \in X_i\}$ forms a subtree of T.

The width of a tree-decomposition (X, T) is $\max_{i \in I}(|X_i| - 1)$. The tree-width of G is the minimum width over all its tree-decompositions.

The *bounded tree-width problem* of a given confluent NCE graph grammar $\mathcal{G}$ is the question, whether or not there exists an integer n, such that the tree-width of all graphs in $L(\mathcal{G})$ is less than n. In this section we analyze the *bounded tree-width problem* of a given confluent NCE graph grammar.

A graph $G = (V \overset{\bullet}{\cup} V', E)$ with node set $V \overset{\bullet}{\cup} V'$, where $|V| = |V'| = n$ and edge set $E = \{\{v_1, v_2\} \mid v_1 \in V, v_2 \in V'\}$, is called a complete bipartite graph $K_{n,n}$. n is the size of $K_{n,n}$.

The following theorem of Courcelle characterizes the relationship between confluent NCE and HR graph languages [Cou95].

Theorem 11 (Courcelle). *Let $\mathcal{G}$ be a confluent graph grammar.*

1. *The following conditions are equivalent.*
   (a) *$L(\mathcal{G})$ is HR,*
   (b) *$L(\mathcal{G})$ is of bounded tree-width,*
   (c) *there is an integer n, such that complete bipartite subgraph $K_{n,n}$ is not a subgraph of any graph $G \in L(\mathcal{G})$,*
   (d) *$L(\mathcal{G})$ is sparse.*
2. *One can decide whether or not these conditions hold.*

Because of Theorem 11 above we prefer the following definition for the bounded tree-width problem of a confluent graph grammar.

Problem: Bounded tree-width problem.

Instance: A confluent NCE graph grammar $\mathcal{G}$.

Question: Is there an integer n, such that all complete bipartite subgraphs of graphs in $L(\mathcal{G})$ have size less than n?

We will show that the bounded tree-width problem of confluent NCE graph grammars (and hence the problem whether or not the conditions of Theorem 11 above hold) is P-complete. In order to do this we prove a more general lemma concerning reduced NCE graph grammars.

Lemma 12. *Let $\mathcal{G}$ be a reduced NCE graph grammar with axiom S. $L(\mathcal{G})$ contains a complete bipartite subgraph $K_{n,n}$ of arbitrary size n if and only if*

1. *$S \overset{*}{\longmapsto} \hat{G}_1 \overset{*}{\longmapsto} \hat{G}_2 \overset{*}{\longmapsto} \hat{G}_3 \overset{*}{\longmapsto} \hat{G}_4$ is a subgraph derivation, where graphs $\hat{G}_i$, $1 \leq i \leq 4$, are defined as in Fig. 2 for some labels A', A, B, C, D, such that*
   (a) *node u_1 is descendant of u, and nodes v_1, v_2 are descendants of v,*
   (b) *node v_{21} is descendant of v_2, and nodes v_{11}, v_{12} are descendants of v_1,*
   (c) *node v_{211} is descendant of v_{21}, node v_{111} is descendant of v_{11}, and nodes v_{121}, v_{122} are descendants of v_{12},*
   (d) *there exists an edge-preserving (A, B)-derivation in $\mathcal{G}$.*

   or
2. *$S \overset{*}{\longmapsto} \hat{H}_1 \overset{*}{\longmapsto} \hat{H}_2 \overset{*}{\longmapsto} \hat{H}_3 \overset{*}{\longmapsto} \hat{H}_4$ is a subgraph derivation, where graphs $\hat{H}_i$, $1 \leq i \leq 4$, are defined as in Fig. 2 for some labels A', A, B, C, D, E, such that*

(a) *node u_1 is descendant of u, node v_1 is descendant of v, and nodes w_1, w_2 are descendants of w,*
(b) *node w_{11} is descendant of w_1 and nodes v_{11}, v_{12} are descendants of v_1,*
(c) *node w_{111} is descendant of w_{11}, node v_{111} is descendant of v_{11}, and nodes v_{121}, v_{122} are descendants of v_{12},*
(d) *there exists an edge-preserving (A, B)-derivation in $\mathcal{G}$.*

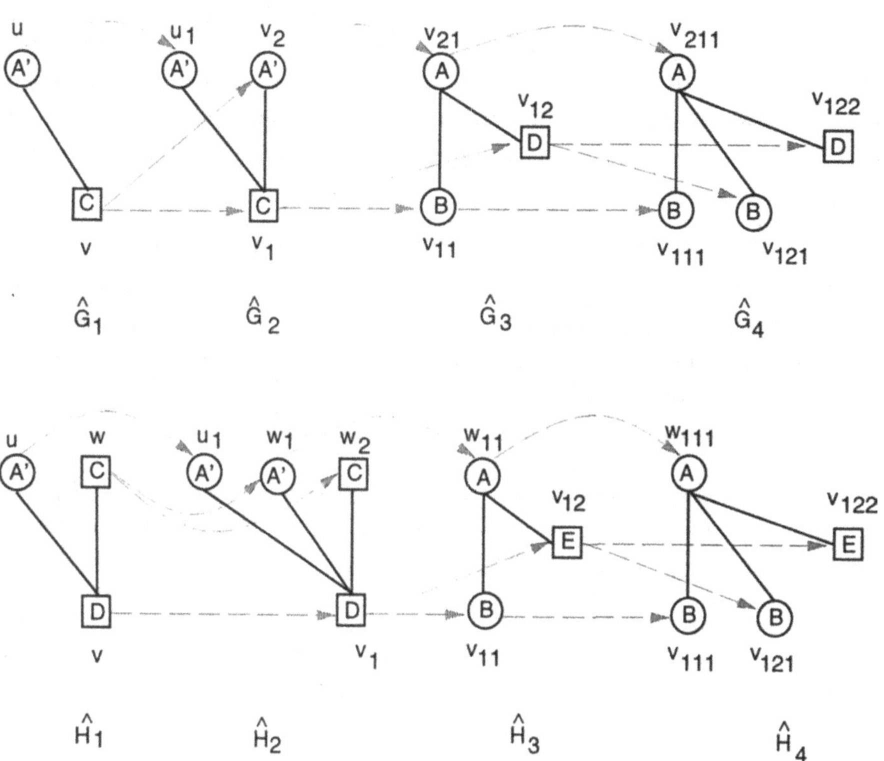

Fig. 2. The dotted arrows indicate the node descendant relation.

Proof. "If-case:" Assume there exists a subgraph derivation $S \overset{*}{\longmapsto} \hat{G}_1 \overset{*}{\longmapsto} \hat{G}_2 \overset{*}{\longmapsto} \hat{G}_3 \overset{*}{\longmapsto} \hat{G}_4$ (or $S \overset{*}{\longmapsto} \hat{H}_1 \overset{*}{\longmapsto} \hat{H}_2 \overset{*}{\longmapsto} \hat{H}_3 \overset{*}{\longmapsto} \hat{H}_4$, respectively). Using a pumping argument it is not difficult to show that there exists a subgraph derivation $S \overset{*}{\longmapsto} H$, such that H contains a complete bipartite subgraph $K_{n,n}$ with n A-nodes and n B-nodes, for arbitrarily large n. By the existence of an edge-preserving (A, B)-derivation and by the reduction of the graph grammar $\mathcal{G}$ there is a derivation of a terminal graph G containing a complete bipartite subgraph with n terminal a-nodes and n terminal b-nodes for some terminal labels a, b.

"Only-if-case:" Let c be the maximum number of nodes in the right hand-sides of the productions. Assume there is a derivation $S = G_1 \Rightarrow G_2 \Rightarrow \cdots \Rightarrow G_k$, where $G_k \in L(\mathcal{G})$ contains a complete bipartite graph $K_{n,n}$ of size $n = |\Delta|^2 \cdot c^{4(|\Delta|+1)|\Delta|^{|\Delta|+3}}$.

Since the number of node labels is bounded by $|\Delta|$ there is a complete bipartite subgraph $K_{m,m} = (V \dot{\cup} V', E)$, such that all nodes of V have the label a and all nodes of V' have the label b. Then $m \geq n/|\Delta|^2 \geq c^{4(|\Delta|+1)|\Delta|^{|\Delta|+3}}$.

The proof will be done in two steps. The first is easy. We mark all ancestors of $K_{m,m}$ in each G_i, $1 \leq i \leq k$. Let H_i be the subgraph of G_i, $1 \leq i \leq k$, induced by the marked nodes. By Lemma 8 $S = H_1 \longmapsto \cdots \longmapsto H_k = K_{m,m}$ is a subgraph derivation by $\mathcal{G}$. In the second and complicated step we successively delete nodes and edges from the graphs $H_1, \ldots, H_k$ until subgraph derivation $S = H_1 \longmapsto \cdots \longmapsto H_k = K_{m,m}$ contains one of the desired subgraph derivations $S \overset{*}{\longmapsto} \hat{G}_1 \overset{*}{\longmapsto} \hat{G}_2 \overset{*}{\longmapsto} \hat{G}_3 \overset{*}{\longmapsto} \hat{G}_4$, or $S \overset{*}{\longmapsto} \hat{H}_1 \overset{*}{\longmapsto} \hat{H}_2 \overset{*}{\longmapsto} \hat{H}_3 \overset{*}{\longmapsto} \hat{H}_4$.

Without loss of generality we assume that axiom S consists of a single node and that all graphs H_i, $1 \leq i \leq k$, are pairwise disjoint. Let $S = H_1 \longmapsto \cdots \longmapsto H_k = K_{m,m}$ be the subgraph derivation obtained in the first step. Let $V_i \dot{\cup} V_i'$ be an arbitrary partition of the nodes in H_i, such that all nodes of V_i are ancestors of a-nodes and all nodes of V_i' are ancestors of b-nodes in $K_{m,m}$. It is easy to see that the ancestor relation guarantees that the subgraph of H_i induced by $V_i \dot{\cup} V_i'$ contains the complete bipartite subgraph $(V_i \dot{\cup} V_i', E')$, where E' is the set of the edges of H_i between nodes of V_i and nodes of V_i'.

Before we start with the second step we recursively define for each H_i, $1 \leq i \leq k-1$, the so-called *generator nodes* a-g_i and b-g_i. At the beginning the generator nodes a-g_1, b-g_1 are identical with axiom S. The generator node a-g_i of H_i for $i = 2, \ldots, k-1$, is one of the descendant nodes of generator node a-g_{i-1} in H_{i-1} with the most descendant a-nodes in $K_{m,m}$. The selection between two nodes with the same number of descendant nodes in $K_{m,m}$ is done in an arbitrary manner. Similarly, the generator node b-g_i of H_i is one of the descendant nodes of generator node b-g_{i-1} in H_{i-1} with the most descendant b-nodes in $K_{m,m}$. The following observation is of importance.

(*) If generator nodes a-g_i and b-g_i are identical in H_i, it is possible that generator nodes a-g_{i+1} and b-g_{i+1} are also identical in H_{i+1}. On the other hand if a-g_i and b-g_i are different in H_i nodes a-g_{i+1} and b-g_{i+1} are always different in H_{i+1}.

A subgraph derivation step $H_i \longmapsto H_{i+1}$ is called *a-growing* (*b-growing*) if generator node a-g_i (b-g_i) of H_i has more descendant a-nodes in $K_{m,m}$ than generator node a-g_{i+1} (b-g_{i+1}) of H_{i+1}; otherwise it is called *a-stagnating* (*b-stagnating*). Obviously only *a-growing* and *b-growing* increase the number of descendant nodes of the generator nodes in a subgraph derivation step. Because the maximal number of nodes in the right-hand sides of the productions is bounded by c, our subgraph derivation $H_1 \longmapsto \ldots \longmapsto H_n$ has at least

$$\log_c c^{4(|\Delta|+1)|\Delta|^{|\Delta|+3}} = 4(|\Delta|+1)|\Delta|^{|\Delta|+3}$$

a-growing subgraph derivation steps and the same number of *b-growing* subgraph derivation steps.

For $i = 1, \ldots, k-1$ we will delete nodes and edges from $H_i, H_{i+1}, \cdots, H_k$ as follows.
Let V_i, V_i' be the actual partition of the nodes in $H_i - \{a\text{-}g_i, b\text{-}g_i\}$, such that all nodes of V_i are ancestors of a-nodes and all nodes of V_i' are ancestors of b-nodes in $K_{m,m}$. At the start we have $V_1 = V_1' = \emptyset$.

1. For each (non-generator) node v of V_i (v' of V_i') in H_i we select exactly one of its descendants w (w') in H_{i+1}, which has a descendant a-node (b-node) in $K_{m,m}$. All other descendant nodes of v (v') (and their incident edges) in H_{i+1} are removed.
   We define $V_{i+1} := \{w \mid v \in V_i, w \text{ is a descendant node of } v \text{ in } H_{i+1}\}$ and $V_{i+1}' := \{w' \mid v' \in V_i, w \text{ is a descendant node of } v' \text{ in } H_{i+1}\}$.
2. We choose (if existing) one arbitrary direct descendant node u (u') of the generator node a-g_i (b-g_i) in H_{i+1}, which is not the a-generator (b-generator) node in H_{i+1}, such that u (u') has a descendant a-node (b-node) in $K_{m,m}$. We remove all edges between u and nodes of V_{i+1}' as well as all edges between u' and nodes of V_{i+1} constructed in 1. We define $V_{i+1} := V_{i+1} \cup \{u\}$ and $V_{i+1}' := V_{i+1}' \cup \{u'\}$. All other direct descendant nodes of a-g_i and b-g_i (and their incident edges) in H_{i+1}, except the generator nodes a-g_{i+1} and b-g_{i+1} are removed from H_{i+1}.
   Notice that a-g_i and b-g_i may be identical in H_i.

Observe that the subgraph induced by $V_i \cup V_i'$ in H_i is complete bipartite. Moreover all nodes of V_i are connected to node $b - g_i$, and all nodes of V_i' to node $a - g_i$.

Now there remains a subgraph derivation $H_1 \longmapsto \cdots \longmapsto H_k$ with at least $4(|\Delta|+1)|\Delta|^{|\Delta|+3}$ *a-growing* and at least $4(|\Delta|+1)|\Delta|^{|\Delta|+3}$ *b-growing* subgraph derivation steps, such that each *a-growing* step generates exactly one new node in V_{i+1} and each *b-growing* step exactly generates one new node in V_{i+1}'.

By (*), there is some l, such that the first part $H_1 \longmapsto \cdots \longmapsto H_l$ of the subgraph derivation contains only graphs whose a- and b-generator nodes are identical, whereas the second part $H_{l+1} \longmapsto \cdots \longmapsto H_{n-1}$ contains only graphs whose a- and b-generator nodes are different. There are four cases:

Case 1: the first part has at least $2(|\Delta|+1)|\Delta|^{|\Delta|+3}$ *a-growing* and at least $2(|\Delta|+1)|\Delta|^{|\Delta|+3}$ *b-growing* subgraph derivation steps,
Case 2: the first part has at least $2(|\Delta|+1)|\Delta|^{|\Delta|+3}$ *a-growing* and the second part at least $2(|\Delta|+1)|\Delta|^{|\Delta|+3}$ *b-growing* subgraph derivation steps,
Case 3: (similarly to Case 2) the first part has at least $2(|\Delta|+1)|\Delta|^{|\Delta|+3}$ *b-growing* and the second part at least $2(|\Delta|+1)|\Delta|^{|\Delta|+3}$ *a-growing* subgraph derivation steps, or
Case 4: the second part has at least $2(|\Delta|+1)|\Delta|^{|\Delta|+3}$ *a-growing* and at least $2(|\Delta|+1)|\Delta|^{|\Delta|+3}$ *b-growing* subgraph derivation steps.

Notice that the number of *a-growing* and *b-growing* subgraph derivation steps chosen before $(2(|\Delta|+1)|\Delta|^{|\Delta|+3})$ is completely needed only in Case 4. For the other cases a smaller number is sufficient.

Case 1: The first part $H_1 \longmapsto \cdots \longmapsto H_l$ of the subgraph derivation has at least $2(|\Delta|+1)|\Delta|^{|\Delta|+3}$ *a-growing* and at least $2(|\Delta|+1)|\Delta|^{|\Delta|+3}$ *b-growing* subgraph derivation steps.

It is easy to see that there is an index j, such that either

- $H_1 \longmapsto \cdots \longmapsto H_j$ has at least $(|\Delta|+1)|\Delta|^{|\Delta|+3}$ *a-growing* subgraph derivation steps and $H_{j+1} \longmapsto \cdots \longmapsto H_l$ has at least $(|\Delta|+1)|\Delta|^{|\Delta|+3}$ *b-growing* subgraph derivation steps or
- $H_1 \longmapsto \cdots \longmapsto H_j$ has at least $(|\Delta|+1)|\Delta|^{|\Delta|+3}$ *b-growing* subgraph derivation steps and $H_{j+1} \longmapsto \cdots \longmapsto H_l$ has at least $(|\Delta|+1)|\Delta|^{|\Delta|+3}$ *a-growing* subgraph derivation steps.

Without loss of generality we assume that the first alternative holds. At the beginning we remove

1. all nodes (and their incident edges) of V_j' from H_j,
2. all ancestor nodes of V_j' (and their incident edges) from $H_1, \ldots H_{j-1}$, except the generator nodes,
3. all descendant nodes of V_j' (and their incident edges) from $H_{j+1}, \ldots H_l$,
4. all nodes of V_i (and their incident edges),$i = j+1, \ldots, l$, from H_i which are not descendants of nodes of V_j.

Obviously the number of *a-growing* subgraph derivation steps in $H_1 \longmapsto \cdots \longmapsto H_j$ and the number of *b-growing* subgraph derivation steps in $H_{j+1} \longmapsto \cdots \longmapsto H_l$ are unchanged.

Next we enumerate the non-generator nodes in the graphs $H_1, \ldots, H_j$ starting with index 1 as follows. A node whose direct ancestor node is a generator node gets the least unused index. A node whose ancestor node in the previous graph is not a generator node gets the same index as its ancestor node. In the same way we enumerate the non-generator nodes in the graphs $H_{j+1}, \ldots, H_l$ but starting with index $1'$ (see also Fig. 3).

We now select certain graphs $H_{x_1}, \ldots, H_{x_{|\Delta|+1}}$, in the subgraph derivation $H_1 \longmapsto \cdots \longmapsto H_j$, such that two of them contain the desired graphs $\hat{G}_1, \hat{G}_2$. In a similar way we select certain graphs $H_{y_1}, \ldots, H_{y_{|\Delta|+1}}$ in the subgraph derivation $H_{j+1} \longmapsto \cdots \longmapsto H_l$, such that two of them contain the other graphs $\hat{G}_3, \hat{G}_4$. In order to do so we continuously *remove* graphs. Initially, all graphs $H_1, \ldots, H_l$ are present. At the beginning we remove all graphs H_i, $1 < i \leq l$, which have the same number of nodes as H_{i-1}. That means we consider only *a-growing* and *b-growing* subgraph derivation steps. The number of *a-growing* subgraph derivation steps in $H_1 \longmapsto \cdots \longmapsto H_j$ is at least

$$(|\Delta|+1)|\Delta|^{|\Delta|+2} \text{ (actually } (|\Delta|+1)|\Delta|^{|\Delta|+3})$$

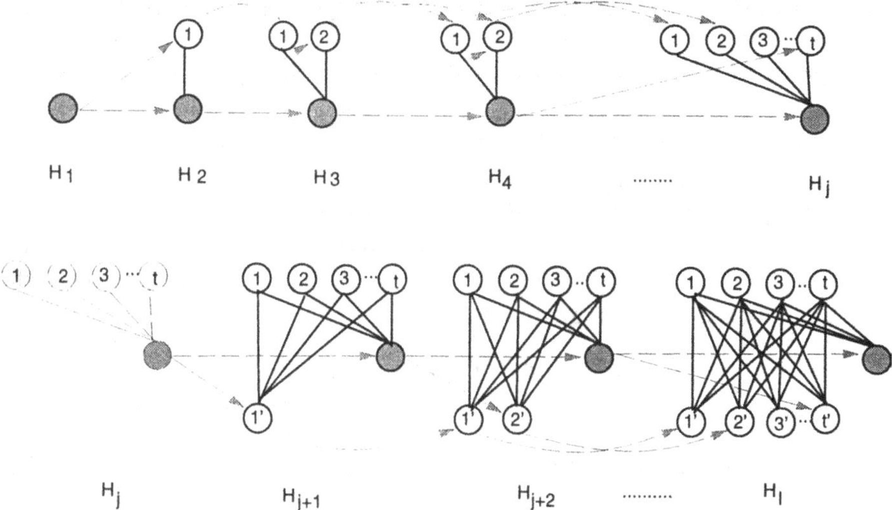

Fig. 3. The dotted arrows indicate the node descendant relation. The dark nodes are the generator nodes.

and the number of *b-growing* subgraph derivation steps in $H_{j+1} \longmapsto \cdots \longmapsto H_l$ is at least

$$(|\Delta| + 1)|\Delta|^{|\Delta|+3}.$$

Each *a-growing* and *b-growing* subgraph derivation step increases the number of nodes in the graphs by one.

Next we describe the selection of graphs $H_{x_1}, \ldots, H_{x_{|\Delta|+1}}$ in the obtained subgraph derivation. We determine the node label that is most frequently used by the generator node a-g_i. For example, let C be this node label, then we remove all graphs where the generator node is not labeled by C. At least

$$(|\Delta| + 1)|\Delta|^{|\Delta|+1}$$

graphs remain where all generator nodes have label C.

Let H_f be the first graph in the obtained rest sequence. Assume H_f has c_f numbered nodes. We determine the node label which is most frequently used by the node with index c_f (notice that every graph in the sequence with index greater than f has at least $c_f + 1$ numbered nodes). Let this node be an A'-node. Then we remove all graphs in which the node with index c_f is not labeled by A'. There remain at least

$$(|\Delta| + 1)|\Delta|^{|\Delta|}$$

graphs in which all generator nodes have label C, all nodes with index c_f are A'-nodes and descendant nodes of the generator node of H_f. Let H_{x_1} be the first graph in the sequence. We select this graph.
Assume H_{x_1} has c_1 numbered nodes. Notice that (1) all graphs in the sequence

with index greater than x_1 contain a node with number c_1 and (2) every node with index higher than c_1 is a descendant node of the generator node a-g_{x_1}. We remove H_{x_1} from the sequence. There remain at least

$$(|\Delta|+1)|\Delta|^{|\Delta|}-1 \geq |\Delta||\Delta|^{|\Delta|}$$

graphs in which all generator nodes have label C, all nodes with index c_1 are A'-nodes and descendant nodes of the generator node of H_{x_1}.

Let H_f be the first graph in the obtained rest sequence. We repeat the procedure above and we select the graph H_{x_2} in the same way as we selected the graph H_{x_1} before. There remain at least

$$|\Delta||\Delta|^{|\Delta|-1}-1 \geq (|\Delta|-1)|\Delta|^{|\Delta|-1}$$

graphs.

After $|\Delta|+1$ iterations we select $|\Delta|+1$ graphs $H_{x_1},\cdots,H_{x_{|\Delta|+1}}$. Note that there are more than enough graphs for $|\Delta|+1$ iterations. Since there are only $|\Delta|$ different node labels, there have to exist two subgraphs H_{x_i}, H_{x_j}, $x_i < x_j$, such that H_{x_i} contains some subgraph $\hat{G}_1$ and H_{x_j} contains some subgraph $\hat{G}_2$ as defined in Fig. 2, such that both satisfy the conditions necessary for case 1 of Lemma 12.

Now we describe the selection of graphs $H_{y_1},\ldots,H_{y_{|\Delta|+1}}$ in the subgraph derivation $H_{j+1} \longmapsto \cdots \longmapsto H_l$. Assume that graph H_{x_j} has c_{x_j} numbered nodes. We determine in the subgraph derivation $H_{j+1} \longmapsto \cdots \longmapsto H_l$ the node label pair, which is most frequently used by the b-generator node and by the node having index c_{x_j} in H_i, $j+1 \leq i \leq l$. For example, let D, A be this node label pair, then we remove all graphs in the subgraph derivation $H_{j+1} \longmapsto \cdots \longmapsto H_l$ in which the b-generator node is not labeled by D or the node with index c_{x_j} is not labeled by A. There remain at least

$$(|\Delta|+1)|\Delta|^{|\Delta|+1}$$

graphs in which all generator nodes have label C and all nodes with index c_{x_j} have label A. In the same way as before we can find two subgraphs H_{y_i}, H_{y_j}, $y_i < y_j$, in the subgraph derivation $H_{j+1} \longmapsto \cdots \longmapsto H_l$, such that H_{y_i} contains some subgraph $\hat{G}_3$ and H_{y_j} contains some subgraph $\hat{G}_4$ as defined in Fig. 2.

Obviously, the a-generator node of H_{x_j} is an ancestor node of the b-generator node of H_{y_i} and the node with index c_{x_j} in graph H_{y_i} is a descendant of the node with index c_{x_j} in H_{x_j}. The other required descendant relations and edge-preserving (A, B)-derivations follow immediately by the selection process. This completes the proof for this case.

Case 2 and Case 3: The proof is analogous to case 1.

Case 4: The second part $H_{l+1} \longmapsto \cdots \longmapsto H_k$ has at least $2(|\Delta|+1)|\Delta|^{|\Delta|+3}$ *a-growing* and at least $2(|\Delta|+1)|\Delta|^{|\Delta|+3}$ *b-growing* subgraph derivation steps.

The proof is similar to case 1. Here we split the subgraph derivation in two parts, $H_{l+1} \longmapsto \cdots \longmapsto H_j$ and $H_{j+1} \longmapsto \cdots \longmapsto H_k$ for some index j. In the first part we find two subgraphs H_{x_i}, H_{x_j}, $x_i < x_j$, such that H_{y_i} contains some subgraph $\hat{H}_1$ and H_{x_j} contains some subgraph $\hat{H}_2$ as defined in Fig. 2. In the second part we find two subgraphs H_{y_i}, H_{y_j}, $y_i < y_j$, such that H_{y_i} contains some subgraph $\hat{H}_3$ and H_{x_j} contains some subgraph $\hat{H}_4$ as defined in Fig. 2.

By Lemmas 9 and 12 and from the fact that every NCE graph grammar can be transformed in polynomial time into an equivalent reduced NCE graph grammar, the problem whether or not $L(\mathcal{G})$ contains complete bipartite subgraphs $K_{n,n}$ of an arbitrarily large size n belongs to P.
On the other hand using standard techniques it is easy to see that the problem is at least as hard as the emptiness problem of context-free string grammars, and hence P-hard.

Theorem 13. *Let $\mathcal{G}$ be an NCE graph grammar. The problem whether or not $L(\mathcal{G})$ contains complete bipartite subgraphs $K_{n,n}$ of an arbitrarily large size n is log-space complete for P.*

Corollary 14. *Let $\mathcal{G}$ be a confluent NCE graph grammar. The bounded tree-width problem of $\mathcal{G}$ is log-space complete for P.*

4 Conclusions

In this paper we have examined the bounded tree-width problem of a given confluent NCE graph grammar. The problem is equivalent to the generation of large complete bipartite graphs. We have shown the P-completeness. This result can be generalized to context-free nonblocking confluent eNCE (edNCE) graph grammars. The only difference are multiple edges. The proofs are based on the same idea.

If the NR graph grammars contain blocking edges the following complexity result can be easily obtained by Lemma 12 and by the emptiness problem of confluent edNCE graph grammars from [SW95]:
The bounded tree-width problem is DEXPTIME-complete for confluent and boundary edNCE and PSPACE-complete for linear edNCE graph grammars. Moreover the problem whether or not the language of a given edNCE graph grammar contains complete bipartite subgraphs $K_{n,n}$ of an arbitrarily large size n is undecidable.

5 Acknowledgments

I am grateful to Prof. Dr. F.J. Brandenburg for many suggestions. I thank an anonymous referee for pointing out an error in Lemma 12 and his comments on an earlier version of this paper.

References

[ALS91] S. Arnborg, J. Lagergren, D. Seese. Easy problems for tree-decomposable graphs. *J. of Algorithms*, 12:308–340, 1991.

[AP89] S. Arnborg, A. Proskurowski. Linear time algorithms on graphs with bounded tree-width. *LNCS*, 317:105–119, 1989.

[Bra87] F.J. Brandenburg. On partially ordered graph grammars. *LNCS*, 291: 99–111, 1987.

[Bra91] F.J. Brandenburg. The equivalence of boundary and confluent graph grammars on graph languages of bounded degree. *LNCS*, 488:312–322, 1991.

[Bod96] H . L. Bodlaender. A linear time algorithm for finding tree-decompositions of small treewidth. In *SIAM J. Computing*, 25:1305–1317, 1996.

[Cou87] B. Courcelle An axiomatic definition of context-free rewriting and its application to NLC graph grammars. *Theoret. Comput. Sci.*, 55:141–181, 1987.

[Cou90] B. Courcelle. The monadic second-order logic of graphs I: Recognizable sets of finite graphs. *Inform. and Comput.*, 85:12–75, 1990.

[Cou95] B. Courcelle Structural properties of context-free sets of graphs generated by vertex replacement. *Inform. and Comput.*, 116:275–293, 1995.

[CER93] B. Courcelle, J. Engelfriet, and G. Rozenberg. Handle-rewriting hypergraph grammars. *J. Comput. System Sci.*, 46:218–270, 1993.

[Eng89] J. Engelfriet. Context-free NCE graph grammars. *LNCS*, 380:148–161,1989.

[EH91] J. Engelfriet and L. M. Heyker. The string generating power of context-free hypergraph graph grammars. *J. Comput. System Sci.*, 43:328–360, 1991.

[EH94] J. Engelfriet and L. M. Heyker. Hypergraph languages of bounded degree. *J. Comput. System Sci.*, 48:58–89, 1994.

[EL89] J. Engelfriet and G. Leih. Linear graph grammars: Power and complexity. *Inform. and Comput.*, 81:88–121, 1989.

[ELW90] J. Engelfriet, G. Leih, and E. Welzl. Boundary graph grammars with dynamic edge relabeling. *J. Comput. System Sci.*, 40:307–345, 1990.

[ER90] J. Engelfriet, G. Rozenberg. A comparison of boundary graph grammars and context-free hypergraph grammars. *Inform. and Comput.*, 84:163–206, 1990.

[EKR91] H. Ehrig, H.J. Kreowski, and G. Rozenberg. *Proceedings of Graph-Grammars and Their Application to Computer Science '90*, Vol. 532 of *LNCS*. Springer-Verlag, 1991.

[ENRR87] H. Ehrig, M. Nagl, A. Rosenfeld, and G. Rozenberg. *Proceedings of Graph-Grammars and Their Application to Computer Science '86*, Vol. 291 of *LNCS*. Springer-Verlag, 1987.

[Hab92] A. Habel. *Hyperedge Replacement: Grammars and Languages*, Vol. 643 of *LNCS*, Springer-Verlag, 1992.

[JR80a] D. Janssens and G. Rozenberg. On the structure of node label controlled graph languages. *Inform. Sci.*, 20:191–216, 1980.

[JR80b] D. Janssens and G. Rozenberg. Restrictions, extensions, and variations of NLC grammars. *Inform. Sci.*, 20:217–244, 1980.

[Kau85] M. Kaul. Syntaxanalyse von Graphen bei Präzedenz-Graphgrammatiken. Dissertation, Universität Osnabrück, Osnabrück, Germany, 1985.

[Klo94] T. Kloks. *Treewidth. Computations and Approximations*. Vol. 842 of *LNCS*, Springer-Verlag, 1994.

[Nag76] M. Nagl. Formal languages of labelled graphs. *Computing* 16:113–137, 1976.

[Nag80] M. Nagl. A tutorial and bibliographical survey of graph grammars. *LNCS*, 73:70–126, 1980.

[SW95] K. Skodinis and E. Wanke. Emptiness problems of eNCE graph languages. *J. Comput. System Sci.*, 51:472–485, 1995.

[SW97] K. Skodinis and E. Wanke. The Bounded Degree Problem for eNCE Graph Grammars. *Inform. and Comput.*, 135:15–35, 1997.

Structured Programs have Small Tree-Width and Good Register Allocation (extended abstract)

Mikkel Thorup

Department of Computer Science, University of Copenhagen, Universitetsparken 1 2100 Kbh. Ø, Denmark. (mthorup@diku.dk, http://www.diku.dk/~mthorup)

Abstract. The register allocation problem for an imperative program is often modelled as the coloring problem of the interference graph of the control-flow graph of the program. The interference graph of a flow graph G is the intersection graph of some connected subgraphs of G. These connected subgraphs represent the lives, or life times, of variables, so the coloring problem models that two variables with overlapping life times should be in different registers. For general programs with unrestricted gotos, the interference graph can be any graph, and hence we cannot in general color within a factor $O(n^{\varepsilon})$ from optimality unless NP=P.
It is shown that if a graph has tree-width k, we can efficiently color any intersection graph of connected subgraphs within a factor $(\lfloor k/2 \rfloor + 1)$ from optimality. Moreover, it is shown that structured ($\equiv$ goto-free) programs, including, for example, short circuit evaluations and multiple exits from loops, have tree-width at most 6. Thus, for every structured program, we can do register allocation efficiently within a factor 4 from optimality, regardless of how many registers are needed.
The bounded tree-decomposition may be derived directly from the parsing of a structured programs, and it implies that the many techniques for bounded tree-width may now be applied in control-flow analysis, solving problems in linear time that are NP-hard, or even P-space hard, for general graphs.

1 Introduction

The *register allocation problem* for an imperative program P is usually modelled as the coloring problem of the interference graph I of the control-flow graph G of P [2, 15, 16, 21]. The *control-flow graph* G is a digraph representing the flow of control between program points in the execution of the program P (see Figure 1, page 6). The orientation is, however, unimportant in our context, and we will hence perceive G as undirected. The *life time of a variable* is a connected subgraph of G, and the *interference graph* I is the intersection graph of the set X of all life times of variables in the program P. Thus adjacency in I denotes intersecting life times of variables. Consequently, our problem of coloring the interference graph I models that two variables with overlapping life times should be in different registers.

It should be noted that even with a good coloring, we may still run short of physical registers, in which case we have an additional ***spilling problem*** of simulating desired registers with the physical registers by copying to and from memory locations. The coloring then limits the amount of memory locations needed. The spilling problem is not addressed in this paper.

For general programs with unrestricted gotos, the control-flow graph can be any graph, and so can the interference graph. Hence for some fixed $\varepsilon > 0$, we cannot in polynomial time color within a factor $O(n^{\varepsilon})$ from optimality unless NP=P [30]. The current best approximation factor is $O(n(\log\log n)/(\log n)^3)$ [26]. However, in this paper we show that for structured ($\equiv$ goto-free) programs[1], including, for example, short circuit evaluations and multiple exits from loops, we can do register allocation in polynomial time within a factor 4 from optimality.

Recently Kannan and Proebsting [27] showed that if the control-flow graph of a program is series-parallel, we can color the interference graph within a factor 2 from optimality. The relevance of series-parallelism follows from the well-known fact (see e.g. [2]) that many of the structured language constructs, such as if-then-else, and while-loops, allows programs to be recursively sub-divided into basic blocks with a single entry and a single exit point. Such a recursive sub-division immediately gives a series-parallel decomposition of the flow-graph (see e.g. [33]). However, even within structured languages, there are well-known exceptions to the sub-division into basic-blocks/series-parallelism. For example [27] points to short circuit evaluation where the evaluation of a boolean expression is terminated as soon as the correct answer is found, e.g. if $e = x_1 \vee \cdots \vee x_n$ is true, then e is only evaluated till the first true x_i is found. Other exceptions include loops with multiple exits/breaks and programs/functions with multiple stop-/return-statements. In [27] the problem of dealing with these exceptions is suggested. So far, however, there has been no approach suggesting that the control-flow graphs of general structured programs should be any simpler than general graphs. The concept of reducibility [2, pp. 606–607] of control-flow graphs is associated with structured programs, but reducibility only refers to the orientation of the edges (any acyclic orientation is reducible). In our case, we are only interested in the underlying undirected graph, so the requirement of reducibility does not limit the class of graphs considered.

Here we address the problem of Kannan and Proebsting by showing that bounded tree-width, as defined in [35], captures all the above mentioned exceptions.

Theorem 1. *Assuming short circuit evaluation,*

- *Goto-free Algol [31] and Pascal [39] programs have control-flow graphs of tree-width* ≤ 3.
- *All Modula-2 [40] programs have control-flow graphs of tree-width* ≤ 5.
- *Goto-free C [28] programs have control-flow graphs of tree-width* ≤ 6.

---

[1] Structured programs is not an well-defined agreed-upon term (see e.g. [18, 20, 29, 31, 32]). In this paper, we are referring specifically to the aspect of being goto-free.

Without short circuit evaluation, each of the above widths drops by one[2]*. Control-flow graphs with tree-decompositions of the above widths are derived directly (linear time, small constants) from the parse trees of the programs.*

The reason for the gap between Algol/Pascal and Modula-2 is that Modula-2 has loops with multiple exits and multiple returns from functions. The further gap to C is due to C's continue-statement jumping to the beginning of a loop.

Series-parallel graphs are graphs of tree-width 2, and here we generalize the technique from [27] for series-parallel graphs to show

Theorem 2. *Given a tree-decomposition of width k of a graph G, we can efficiently color the intersection graph I of any set X of connected subgraphs of G within a factor $(\lfloor k/2 \rfloor + 1)$ from optimality. If n is the number of nodes in G and ω is the maximal number of subgraphs from X intersecting a single vertex in G, the coloring is done in time $O(k\omega n + \omega^{2.5} n)$. Also, we can color I within a factor $(k+1)$ from optimality in time $O(k\omega n)$.*

Note that for $k = 2$, we get the factor 2 from [27]. Also note that the tree-width 1 graphs are the forests for which we get a factor 1. Hence follows the colorability of chordal graphs [23, 24]. Combining Theorems 2 and 1, we get

Corollary 3. *In time $O(\omega n + \omega^{2.5} n)$, the register allocation problem can be solved within the following factors from optimality:*

- *2 (2) for Algol/Pascal,*
- *3 (3) for Modula-2, and*
- *4 (3) for C.*

The parenthesized numbers are without short circuit evaluation. If we only want to spend time $O(\omega n)$, we can get the factors: 4 (3) for Algol/Pascal, 6 (5) for Modula-2, and 7 (6) for C.

The bounded tree-width of structured programs implies that the many techniques for bounded tree-width may now be applied in control-flow analysis, solving problems in linear time that are NP-hard [4, 5, 11, 17], or even P-space hard [9], for general graphs.

It should be noted that our work on the register allocation problem does not follow the usual pattern of deriving fast algorithms for graphs of bounded tree-width. First of all we are studying intersection graphs of connected subgraphs rather than the graph itself. Second, the coloring problem we consider is NP-complete for any $k > 1$. The NP-completeness follows from the fact that a cycle has tree-width 2, and for a cycle, the coloring problem for intersection graphs is known to be NP-complete [22].

The paper is divided as follows. Section 2 addresses Theorem 1. Section 3 proves Theorem 2. Finally, Section 4 gives a more detailed comparison with previous register allocation, and puts the results in a broarder perspective.

[2] Standard Pascal and Algol does not have short circuit evaluation while standard Modula-2 and C does have short circuit evaluation.

For space reasons, several important issues are deferred to the full version of this paper [37]. As mentioned previously, the full version contains the proof of Theorem 5. Moreover, it showns that simple listings are preserved under the standard optimizations mentioned in [2]. Further, it contains a simple linear time heuristics for finding a good tree-decomposition directly from the three-address code generated, as in [2], from a structured program. Also, it presents algorithms for efficiently computing the minimum separators of any listing. All these algorithms are important for the integration of our approach with many existing compilers.

For basic definitions for programs, grammars, and control-flow graphs, the reader is referred to [2]. Concerning bounded tree-width, we shall use the following definition:

Definition 4. Given a graph G, a ($\leq k$)-*complex listing* is a listing of the vertices of G such that for every vertex $v \in V$, there is a set S_v of at most k of the vertices preceding v in the listing, whose deletion from G separates v in G from all the vertices preceding v in the listing. In this case, we say that G is ($\leq k$)-*complex*. The set S_v is referred to as the *separator* of v in the listing.

Note above that for a given listing and a vertex v, there is a unique minimal choice of the separator S_v; namely as the set of preceding vertices that can be reached by a path from v with no interior vertices preceding v in the listing.

Theorem 5 [19]. *A graph is k-complex if and only if it has tree-width k (as defined in in [35]). Moreover there are linear transformations between k-complex listings and tree-decompositions of width k.*

In the rest of this paper, we will only talk about bounded tree-width in terms of Definition 4.

2 Simplicity of structured programs

In this section, we will address Theorem 1. Our first tool, is the following simple lemma:

Lemma 6. *From a ($\leq k$)-complex listing L of a graph G, we can derive a ($\leq k$)-complex listing of G with an edge $\{v, w\}$ contracted.*

Proof: By symmetry, we may assume v comes before w. Then, we contract the edge $\{v, w\}$ identifying both end-points with v, and deleting w from L. ∎

We will argue the correctness of Theorem 1 by first studying a Modula-2 [40] inspired toy-language STRUCTURED, illustrating all the essential problems of finding good listing of control-flow graphs for structured programs. Afterwards we will outline how the complexity of STRUCTURED carries over to `goto`-free Pascal, Modula-2, and `goto`-free C.

A program in STRUCTURED starts with the key word `program`, then comes the statement of the program, and finally comes the key word `margorp`. Of statements, we have sequences of statements separated by `;`, the conditional statements: `if-then-else-fi` and `if-then-else-fi`, a general loop structure `loop-pool`, an `exit`-statement terminating the nearest surrounding loop, and a `stop`-statement ending the execution of a program. If an `exit`-statement is not surrounded by a loop, we view it as a `stop`-statement. Finally, we have "atomic" statements consisting of single lower-case letters `s`, `t`,... These should be thought of as representing statements such as assignments that do not affect the flow of control of the program, hence which are irrelevant to finding a good listing. We are assuming the standard that control-flow graphs are generated separately for the main program and for procedures, i.e. we are not doing inter-procedural analysis. Hence our atomic statements may also represent procedure calls.

For boolean expressions, we have the connectives `or`, and `and`, both of which are evaluated with short circuit evaluation. Also, we have "atomic" boolean expressions represented by single lower-case letters `a`, `b`,... These should be thought of as representing constants or program variables.

In Figure 1 is given an example of a control-flow graph for a STRUCTURED program. Typically control-flow graphs are depicted with fewer nodes, that is, typically there is not a node for every key word in the program. However, key words are removed by contraction of edges, and by Lemma 6, contractions do not increase the complexity.

We will now describe the order in which to visit the words in a STRUCTURED program so that it corresponds to a (≤ 5)-complex listing. Generally we just follow the structure of the program in a top-down fashion, visiting all key words of a composite structure before we descend into its different parts. Thus, for a program `program` S `margorp`, we first visit the two key words `program` and `margorp`, and then recursively, we visit the contents of the statement S.

For a composite statement of the form

S`;`S', we first visit `;`, and then we visit S and S' recursively.

`loop` S `pool`, we first visit `loop` and `pool`, and then we visit S recursively.

`if` B `then` S `else` S' `fi`, we first visit `if`, `fi`, `then`, and `else`, and then we visit B, S, and S' recursively.

`if` B `then` S `fi`, we first visit `if`, `fi`, and `then`, and then we visit B and S recursively.

For a composite boolean expression of the form

B `or` B', we first visit `or`, and then we visit B, and B' recursively.

B `and` B', we first visit `and`, and then we visit B, and B' recursively.

An example of the described visit sequence is given in Figure 1.

Theorem 7. *All control-flow graphs derived from STRUCTURED are (≤ 5)-complex.*

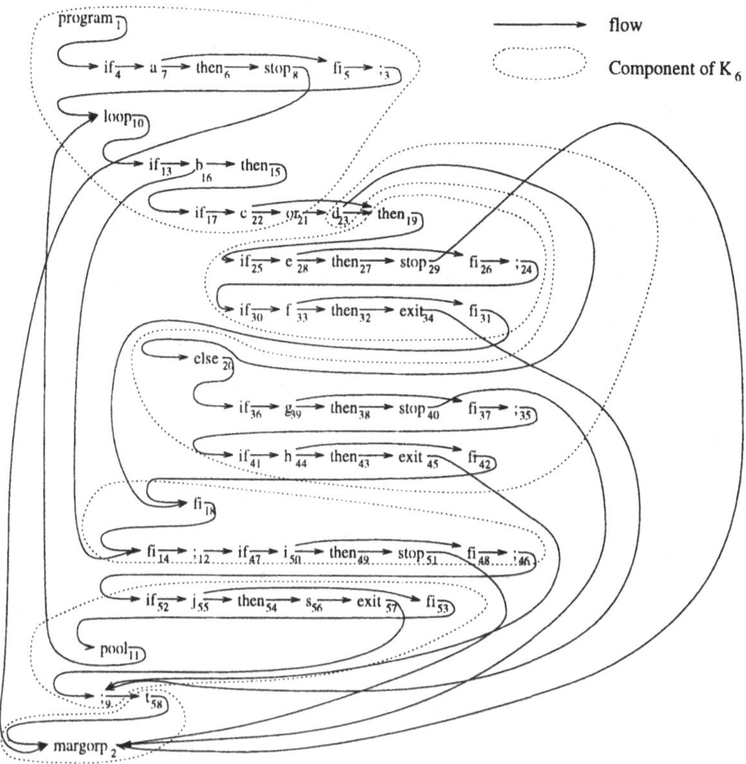

Fig. 1. The 5-complex control-flow graph of a full program produced by FLOW. It is shown how the edges can be contracted so as to derive a clique of size 6, which is trivially 5-complex. Hence the original control-flow graph is (≥ 5)-complex. The (≤ 5)-complex listing produced by FLOW demonstrates that it is exactly 5-complex.

Proof: We prove the theorem by showing that the above described visit sequence always corresponds to a (≤ 5)-complex listing.

Note that for a statements S, the sub-graph of the control flow graph corresponding to S has only the following potential neighbors:

in the word preceeding S.
out the word succeding S.
exit the word succeding the nearest surrounding loop, or `margorp` if there is no surrounding loop.
stop the end of a program `margorp`.

Similarly for a boolean expression B, the potential neighbors of the sub-graph corresponding to B are

in the word preceeding B.

true the word we jump to if B evaluates to true.
false the word we jump to if B evaluates to false.

Following the recursive structure of our visiting sequence, we will now show both (1) that the all the neighbors of a statement S or boolean expression B are visited before we start visiting words in S or B, and (2) that all words get a separator of size at most 5.

Given a program $\texttt{program}\,S\,\texttt{margorp}$, we first visit `program` and `margorp` that get separators $\emptyset$ and $\{\texttt{margorp}\}$, respectively. The neighbors of S are then $in = \texttt{program}$ and $out = exit = stop = \texttt{margorp}$, all of which have been visited before we start visiting S.

For a statement S of the form $S'\,;\,S''$, we first visit ;. Since none of the words in S' or S'' have been visited, ; gets separator $\{in, out, exit, stop\}$. Now ; separates S' from out and S'' from in. Hence the neighbors of S' and S'' are the same as those of S except that $out' = in'' =$;. Since all the neighbors of S where visited before we started visiting S, it follows that the neighbors of S' and S'' are visited before we start visiting S' and S''.

Now consider the case of a loop statement $S = \texttt{loop}\,S'\,\texttt{pool}$, where first we visit `loop` and `pool`. Then $\{in, out, exit, stop\}$ is a separator for `loop` and $\{\texttt{loop}, out, exit, stop\}$ is a separator for `pool`. For S' we then get the neighbors $in' = \texttt{loop}$, $out' = \texttt{pool}$, $exit' = out$, and $stop' = stop$.

Next consider a conditional statement $S = \texttt{if}\,B'\,\texttt{then}\,S''\,\texttt{else}\,S'''\,\texttt{fi}$. Recall that we first visit `if`, `fi`, `then`, and `else`, and then we visit B', S'', and S''' recursively. Thus `if` gets the separator $\{in, out, exit, stop\}$, `fi` gets the separator $\{\texttt{if}, out, exit, stop\}$, `then` gets the separator $\{\texttt{if}, \texttt{fi}, exit, stop\}$, and `else` gets the separator $\{\texttt{if}, \texttt{fi}, \texttt{then}, exit, stop\}$. For B' we get the neighbors $in' = \texttt{fi}$, $true' = \texttt{then}$, and $false' = \texttt{else}$. For S'', we get the neighbors $in'' = \texttt{then}$, $out'' = \texttt{fi}$, $exit'' = exit$, and $stop'' = stop$. The neighbors of S''' are the same as those of S'' except that $in''' = \texttt{else}$. The `if-then-fi` version is done similarly.

Concerning statements consisting of a single word, the separator of `exit` is $\{in, exit\}$, the separator of `stop` is $\{in, stop\}$, and the separator of atomic statements `t`, `s`, .. is $\{in, out\}$.

For a boolean expression $B = B'\,\mathsf{or}\,B''$, we first visit or, which gets separator $\{in, true, false\}$. The neighbors of B' and B'' are like those of B except that $false' = in'' = \mathsf{or}$. The case of $B = B'\,\mathsf{and}\,B''$ is similar except that we get $true' = in'' = \mathsf{and}$. The separator of an atomic boolean a, b, .. is $\{in, true, false\}$.
■

Note in the above proof that only word that gets a separator of size 5 is `else`. Unfortunately, a separator of size 5 cannot be avoided in general. The example in figure 1 shows that there are are programs written in STRUCTURED that do not have (≤ 4)-complex listings, hence that the bound in Theorem 7 cannot be improved.

We will now show how to adapt the proof of Theorem 7 so as to prove the claims in Theorem 1. First, in Theorem 1 it was claimed that the complexity decreases by one if we do not have short circuit evaluation. Recall from the proof of Theorem 7 that only the word `else` gets a separators of size 5. Suppose we do not have short circuit evaluation. The immediate consequence is that for a boolean expression B, the neighbors *true* and *false* both become identified with the word succeeding B in the program. Now consider a statement $S =$ **if** B' **then** S'' **else** S''' `fi`. Let B' be of the form Cb meaning that b is the last word in B'. Without short circuit evaluation, we know that the evaluation of B' pass through b. Now change the visiting sequence so that we first visit `if`, `fi`, b, `then`, and `else`, and then visit C, S'', and S''' recursively. The separators for `if` and `fi` are unchanged, but we get the separator $\{\texttt{if}, \texttt{fi}, \mathit{exit}, \mathit{stop}\}$ for b and $\{b, \texttt{fi}, \mathit{exit}, \mathit{stop}\}$ for both `then` and `else`, that is, no word gets a separator of size > 4.

We will now argue that the programming language Modula-2 has the same complexity as STRUCTURED. First note that Modula-2 does have all the structural (flow-affecting) statements from STRUCTURED. The `stop`-statement from STRUCTURED is not directly available, but in Modula-2 functions we can have multiple return-statements and these have the effect of the stop-statement on the control-flow graphs for functions. Thus, the complexity of Modula-2 is no less than that of STRUCTURED. We now need to show that the structural statements of Modula-2 that are not in STRUCTURED do not increase the complexity. Of these we have `while`-statements, `repeat`-statements and `case`-statements. The `while`-statements and `repeat`-statements are just special cases of our general loop. If a `case`-statement is evaluated by visiting the cases one by one until the right one is found, the control-flow of a `case`-statement is equivalent to that of a a combination of `if-then-else-fi`-statements. Alternatively, based on the value of the case selector, we may be able to jump directly to the relevant case. However, as is described in the full version of this paper [37], even such multi-branching does not increase the complexity. This completes our argument that Modula-2 has the same complexity as STRUCTURED.

Next we turn our attention to Pascal and C. First note that these programming languages do not use key-words like `fi` explicitly terminating an `if-then-else`-statement. This difference is, however, inconsequential, for we can always remove a `fi` from the control-flow graph by contracting the outgoing edge to the succeeding word. By Lemma 6, such a contraction cannot increase the complexity.

We will now argue that the complexity of `goto`-free Pascal is two less than that of Modula-2. The difference is due to the Pascal's lack of a general `loop`-statement with multiple `exits`, and lack of multiple `stop/return`-statements. Referring to the proof of Theorem 7, this removes both the *exit* and the *stop* from the potential neighborhood of statements. Consequently the complexity is reduced by two except that we now have to deal directly with `while`-loops and `repeat`-loops. In the discussion of Modula-2, these loops where just considered special cases of the general loop, but now we will show that they are strictly simpler as they have only one pre-determined exit. Consider a loop of the form $S =$ `while` B' `do` S'' `elihw`, where the neighborhood of S is *in* and *out*. Following the usual pattern, we first visit `while`, `do`, and `elihw`, and then we visit B' and

S'' recursively. Thereby we get the separators $\{in, out\}$ for `while`, $\{$`while`$, out\}$ for `do`, and $\{$`do`, `while`$\}$ for `elihw`. Thus, all the key-words of a `while`-loop get separators of size 2. For B' the neighborhood becomes $in' =$ `while`, $true' =$ `do`, and $false' = out$. For S'' the neighborhood becomes $in'' =$ `do` and $out'' =$ `while`. A `repeat`-loop may be treated identically, and hence we conclude that the complexity of Pascal is reduced by two relative to that of STRUCTURED.

In Theorem 1, we claimed that the complexity of `goto`-free C is one higher than that of Modula-2. The difference is that for loops in C, besides a `break`-statement corresponding to the above `exit`-statement, we have a `continue`-statement bringing the control back to the beginning of the loop. Consequently, the neighborhood of statements is augmented with a *continue* being the beginning of the nearest surrounding loop. The separators then have to be increased accordingly. This completes our outline of a proof of Theorem 1.

Robustness with respect to other optimizations In the full version of this paper [37], it is shown that simple listings are preserved under the standard optimizations mentioned in [2]. For most optimizations, we just note that they can be done preserving the structural statements of a program. This holds for: redundant-instruction elimination, algebraic simplifications, use of machine idioms, elimination of global common subexpression, code motion, copy propagation, and elimination of induction variables. The only optimization from [2] that has to be done after the translation into object code, such as three-address code, is flow-of-control optimization. However, it turns out that standard flow-of-control optimization essentially can be viewed as contraction of edges, which by by Lemma 6 cannot increase the complexity.

3 Coloring the intersection graphs of k-complex graphs

Let $G = (V, E)$ be a graph and $v_1, \ldots, v_n$ a k-complex listing of its vertices. Moreover, let X be a set of connected sub-graphs of G, called *variables*, and let the *interference graph* I be the intersection graph over X. That is, two variables $v, w \in X$ are adjacent in I if and only if they they intersect in G. In this section, we study the problem of coloring I given that k is a constant. Our colors will be numbers $1, 2, \ldots$, and our aim is to minimize the maximum color used.

We use $\chi(I)$ to denote the chromatic number of I, i.e. the minimal number of colors needed to color I. For each v_i, set $P_{v_i} = \{v_1, \ldots, v_{i-1}\}$. Moreover, let S_{v_i} denote the set of $w \in P_{v_i}$ that can be reached by a path from v_i with no interior vertices in P_{v_i}. By Definition 4, $|S_{v_i}| \leq k$. Let X_{v_i} be the set of variables $x \in X$ containing v. Note that X_{v_i} is a clique in I. Set $\omega = \max |X_v| \leq \chi(I)$. Finally, let $X^*_{v_i}$ denote $X_{v_i} \cup \bigcup_{w \in S_{v_i}} X_w$.

Lemma 8. *If $x \in X_v$ does not intersect P_v but intersects a variable y intersecting P_v, then y intersects S_v.*

Proof: There exists a path in $x \cup y$ from v to a vertex in P_v. The first vertex u in this path which is in P_v is in S_v. Since x does not intersect P_v, $u \in y$. ∎

The following generalizes an algorithm from [27] for series-parallel graphs ($k \leq 2$):

Algorithm A: Colors I with at most $(k+1)\omega \leq (k+1)\chi(I)$ colors.

A.1.For $i := 1, \ldots, n$,

A.1.1. Color the uncolored variables $x \in X_{v_i}$ with the smallest colors not used by variables intersecting S_{v_i}.

Correctness: From Lemma 8 it follows that the algorithm produces a proper coloring of I, i.e. that no two intersecting variables get the same color. When coloring the uncolored variables in X_{v_i}, the largest color used is at most the total number of colors used in $X^*_{v_i}$. Moreover $X^*_{v_i}$ is the union of at most $k+1$ optimally colored sets; namely the cliques $X_w, w \in S_{v_i} \cup \{v_i\}$. Thus the largest color used is $\leq (k+1)\omega$. ∎

Note that we color I without constructing I. The algorithm is trivially implemented to run in time $O(n\omega k)$. This matches the time bound of the second algorithm from Theorem 2.

A *biclique* is a graph whose vertex set is partitioned into two cliques. Then

Lemma 9 [27]. *Let $G = (V_1 \cup V_2, E)$ be a biclique on n vertices with the induced subgraphs on V_1 and V_2 being cliques. Then there is an $O(n^{2.5})$ algorithm that optimally colors G.* ∎

Using the same idea as in [27], we get an improved algorithm using at most $k\chi(I)$ colors if we modify step A.1.1 as follows. If $S_{v_i} \neq \emptyset$, choose any $p(v_i) \in S_{v_i}$ and color the biclique $I|(X_{v_i} \cup X_{p(v_i)})$ using Lemma 9. Rename the colors so that the coloring of $X_{p(v_i)}$ is not changed, and such that the new colors for X_{v_i} are the smallest not used in any $X_w, w \in S_{v_i}$. If $S_{v_i} = \emptyset$ then $X^*_{v_i} = X_{v_i}$ is one optimally colored set, and if $S_{v_i} \neq \emptyset$, $X^*_{v_i}$ is the union of at most k optimally colored sets; namely the biclique $X_{v_i} \cup X_{p(v_i)}$ and the cliques $X_w, w \in S_{v_i} \setminus \{p(v_i)\}$. Thus the modified algorithm uses at most $k\chi(I)$ colors. In [27], $k = 2$, so the change brings their approximation factor down from 3 to 2, which is their main result.

We will now get further down to $(\lfloor k/2 \rfloor + 1)\chi(I)$ colors by carefully choosing the $p(v_i) \in S_{v_i}$. Let $p(v_i) = \perp$ denote that $p(v_i)$ is undefined. Note that any graph F with edges $\{v, p(v)\}$, where $p(v) \in S_v$, is acyclic since $v_h \in S_{v_i}$ implies $h < i$. Hence F is a forest.

Algorithm B: Colors I with at most $(\lfloor k/2 \rfloor + 1)\chi(I)$ colors.

B.1.For $i := 1, \ldots, n$,

B.1.1. Let M_i be a maximal matching in the forest on S_{v_i} with edges $\{w, p(w)\} \in S^2_{v_i}$.

B.1.2. If M_i is perfect ($\bigcup M_i = S_{v_i}$) then

B.1.2.1. $p(v_i) := \perp$.

B.1.2.2. Color the uncolored variables $x \in X_{v_i}$ with the smallest colors not used by variables intersecting S_{v_i}.

B.1.3. If M_i is imperfect then

B.1.3.1. Choose $p(v_i)$ from $S_{v_i} \setminus \bigcup_i M_i$.

B.1.3.2. Color the biclique $I|(X_{v_i} \cup X_{p(v_i)})$ using Lemma 9. Rename the colors so that the coloring of $X_{p(v_i)}$ is not changed, and such that the new colors for X_{v_i} are the smallest not used in any $X_w, w \in S_{v_i}$.

Correctness: First note the loop invariant that $X_v \cup X_{p(v)}$ is an optimally colored biclique for all $p(v) \neq \perp$. For any $W \subseteq V$, let $\#(W)$ denote $|W| - |M|$ where M is a maximal matching in the forest on W with edges $\{w, p(w)\} \in W^2$. Then $\bigcup_{w \in W} X_w$ is the union of at most $\#(W)$ optimally colored sets; namely the $|M|$ bicliques $X_w \cup X_{p(w)}, \{w, p(w)\} \in M$, and the $|W| - 2|M|$ cliques $X_w, w \in W \setminus \bigcup M$. By induction on i we will show $\#(S_{v_i} \cup \{v_i\}) \leq \lfloor k/2 \rfloor + 1$.

If M_i is perfect, $\#(S_{v_i}) = |S_{v_i}|/2$, so $\#(S_{v_i} \cup \{v_i\}) \leq \lfloor k/2 \rfloor + 1$. Note that $S_{v_1} = \emptyset$, so this covers the base case of our induction.

For the case where M_i is imperfect, let h is the largest index such that $v_h \in S_{v_i}$. Then $S_{v_i} \subseteq S_{v_h} \cup \{v_h\}$. Hence $\#(S_{v_i}) \leq \#(S_{v_h} \cup \{v_h\})$. By induction, $\#(S_{v_h} \cup \{v_h\}) \leq \lfloor k/2 \rfloor + 1$. Moreover, $M_i \cup \{\{v_i, p(v_i)\}\}$ is a matching in $S_{v_i} \cup \{v_i\}$, so $\#(S_{v_i} \cup \{v_i\}) = \#(S_{v_i})$. That is,

$$\#(S_{v_i} \cup \{v_i\}) = \#(S_{v_i}) \leq \#(S_{v_h} \cup \{v_h\}) \leq \lfloor k/2 \rfloor + 1.$$

This completes the induction, thus proving that the algorithm uses at most $(\lfloor k/2 \rfloor + 1)\chi(I)$ colors. ■

Proof of Theorem 2: First note that it only takes linear time to find a maximal matching M in a forest F, as in step B.1.1. Greedily pick for M any leaf incident edge $\{v, w\}$, and recurse on $F \setminus \{v, w\}$. The leaves are found by keeping track of the degrees. Then Theorem 2 follows from Lemma 9 and the correctness of Algorithms A and B. ■

It should be noted that our improvement from k to $\lfloor k/2 \rfloor + 1$ is not an improvement for the case $k = 2$ studied in [27], but it is an improvement for any $k > 2$.

4 Concluding remarks

So far, we have only addressed structured programs of imperative languages. However, one could imagine that even goto-users have structured thoughts [29], hence that also the control-flow-graphs of their programs have simple listings/bounded tree-width. For variable k, the problem of deciding the tree-width is NP-hard [3]. For fixed k, however, there are linear time algorithms [8]. Also, for variable k, there has been work done on polynomial approximating algorithms [10]. Finally, in the full version of this paper [37] is presented a heuristics that deals directly with three-address code that may contain any number of programmer supplied gotos. Our derivation of simple listings from syntax, as described in Section 2, is, however, much simpler than the general approaches to tree-width, so the general advice following from this paper is: you help not only yourself and your fellow humans [20, 32], but also the optimizer, by not using goto's, thus making the structure explicit from the syntax.

Concerning functional programming languages, the ML Kit with Regions [6, 38, Tofte *pc*] compiles Standard ML into goto-free blocks with structured statements of the form treated in this paper. Thus the techniques of this paper are not limited to imperative languages.

Detailed comparison with previous register allocators:

• The classic approach [15, 16] to register allocation via graph coloring uses the scheme: let x be a variable/vertex in the interference graph I. Color $I \setminus \{x\}$ recursively. Color x with the least color not used by any neighboring variable in I. Typically x is chosen to be of low degree, but this does not in itself lead to any guarantees for the quality of the produced coloring.

Consider the order of which our $(k+1)$-approximation algorithm (Algorithm A, Section 3) colors the vertices. The correctness proof implies that if we choose the x in the classic scheme following this order reversely, the largest degree encountered becomes at most $(k+1)\omega - 1$. Here ω is the maximal number of variables live at any single point of the control-flow graph. Hence, given our ordering of the variables, the classic coloring scheme will use at most $(k+1)\omega$ colors, and since at least ω colors are needed, this is at most a factor $(k+1)$ from optimality. Thus our new register allocation algorithms can be seen as clever ways of running the classic register allocation, giving a good worst-case performance. In practice, we may, of course, often get a lot closer to optimality.

• The interference graph I may be of size quadratic in the number of variables and its construction is considered a main obstacle for coloring based register allocation. Hence, for space reasons, many heuristics aim at only having parts of the I constructed at any time [14, 25, 34]. For our $(\lfloor k/2 \rfloor + 1)$-approximation algorithm (Algorithm B, Section 3) the biggest sub-graphs considered are of size $O(\omega^2)$. From the k-complex listing, for each variable, our $(k+1)$-approximation algorithm identifies a small set of potential colored neighbors, but it never checks if they are actual neighbors, that is, it never checks for any two variables whether they actually interfere. Thus our $(k+1)$-approximation algorithm does not produce any part of the interference graph!

• In [25] they color straight line code optimally. By definition the control-flow graph of straight line code is a single path which is 1-complex and is hence also optimally colored by our $(\lfloor k/2 \rfloor + 1)$-approximation algorithm. In [25] they try to use this for the coloring of the straight line code in an innermost loop - which is assumed to be executed most frequently. Good coloring of the innermost loops is also the concern in [14]. However, as observed in [14], if there is more than one innermost loop, the coloring of one may negatively effect the possibility of coloring the other. The variables of different innermost loops may interfere non-trivially, so we cannot just address them independently. Now, suppose that all the most critical parts, like innermost loops, have been marked. Our approximation algorithms can then be used to first find a good coloring of the variables in the critical parts, and afterwards color the rest of the variables with different colors.

• Our algorithms are generalizations of those in [27] for series parallel control-flow graphs. If the control-flow graph is not series parallel, [27] suggest heuristics for removing a minimal set of edges so that the graph becomes a series-parallel. The removed "exception" edges requires special treatment. If the program execution goes through an exception edge, all register values may have to be reassigned. Note that with our approach we have no exceptions: the tree-width may vary, but this only affects the quality of the coloring; no special action needs to be taken.

In [27] they mention short circuit evaluation as a prime example of an obstruction to series parallelism. For example, goto-free Pascal without short circuit evaluation has series parallel control-flow graphs, but if we allow short circuit evaluation we may get exceptions to series parallelism. In fact, short circuit evaluation alone may give rise to arbitrarily many exceptions to series parallelism. However, the tree-width only grows from 2 to 3. As stated in Corollary 3, for our $(\lfloor k/2 \rfloor + 1)$-approximation algorithm, this change does not give a worse approximation factor!

Implications

• With reference to Theorem 1 and 5, it seems that tree-width of control-flow graphs offers a well-defined mathematical measure for how structured programs, or programming languages, are. Tree-width is an established measure for the computational complexity of graphs, and now it turns out to capture aspects of structured programming [18].

• Our result suggests banning gotos for the sake of optimization. Gotos have long been considered harmful to the readability of programs for humans [20, 32], and further gotos may obstruct bounded tree-width. Wirth's move from Pascal [39] to Modula-2 [40] is an excellent example of what can be done. In Modula-2 there are no gotos, but to reconcile the programmers, the language have been enriched with some extra exit structures - multiple exits from loops and multiple returns from functions. As a consequence, we get a tree-decomposition of width at most 5 as a free side-effect of the parsing of any Modula-2 program.

• A substantial theory of tree-width has developed, and we contribute to this theory by showing that not only are graphs of bounded tree-width computationally simple, but so are their intersection graphs. Concretely we color intersection graphs of graphs with tree-width k within a factor $O(k)$ from optimality, while for some ε, coloring within a factor $O(n^{\varepsilon})$ from optimality is NP-hard for general graphs [30]. This is the first concrete result demonstrating the computational simplicity of intersection graphs of graphs of bounded tree-width.

• For bounded tree-width, many linear time algorithms are known for problems that are otherwise NP-complete [4, 5, 11, 17] or PSPACE-complete [9]. As a consequence of Theorem 1 together with Theorem 5, this understanding may now be applied in control-flow graph analysis. The constants bounding the tree-width are truly small (≤ 6), allowing us to develop algorithms working well in both theory and practice. A first concrete application of the bounded tree-width result of this paper within control-flow graph analysis is presented in [1].

References

1. S. ALSTRUP, P.W. LAURIDSEN, AND M. THORUP, Generalized dominators for structured programs. In *Proceedings of the 3rd Static Analysis Symposium, LNCS 1145*, pages 42–51, 1996.
2. A.V. AHO, R. SETHI, AND J.D. ULLMAN, *Compilers: Principles, Techniques, and Tools*, Addison-Wesley, Reading, Mass., 1986.
3. S. ARNBORG, D.G. CORNEIL, AND A. PROSKOROWSKI, Complexity of Finding Embeddings in a k-Tree, *SIAM J. Alg. Disc. Meth.* **8** (1987) 277–284.
4. S. ARNBORG, J. LAGERGREN, AND D. SEESE, Easy problems for tree-decomposable graphs, *J. Algorithms* **12** (1991) 308–340.
5. S. ARNBORG AND A. PROSKOROWSKI, Linear time algorithms for NP-hard problems restricted to partial k-trees, *SIAM J. Alg. Disc. Meth.* **23** (1989) 11–24.
6. L. BIRKDAL, M. TOFTE, AND M. VEJLSTRUP, From region inference to von Neyman Machines via region representation inference. *in* "Proc. POPL'96," pp. 171–183, 1996.
7. H.L. BODLAENDER, A Tourist Guide Through Treewidth, *Acta Cybernetica* **11** (1993) 1–23.
8. H.L. BODLAENDER, A Linear Time Algorithm for Finding Tree-Decompositions of Small Treewidth, *in* "Proc. 25th STOC," pp. 226–234, 1993.
9. H.L. BODLAENDER, Complexity of Path Forming Games, *Theor. Comp. Sc.* **110** (1993) 215–245.
10. H.L. BODLAENDER, J.R. GILBERT, H. HAFSTEINSSON, AND T. KLOKS, Approximating Treewidth, Pathwidth, Frontsize, and Shortest Elimination Tree, *J. Algorithms* **18**, 2 (1995) 221–237.
11. R.B. BORIE, R.G. PARKER, AND C.A. TOVEY, Automatic Generation of Linear-Time Algorithms from Predicate Calculus Descriptions of Problems on Recursively Constructed Graph Families, *Algorithmica* **7** (1992) 555–581.
12. P. BRIGGS, Register allocation via graph coloring, PhD Thesis, Rice University, 1992.
13. P. BRIGGS, K.D. COOPER, K. KENNEDY, AND L. TORCZON, Coloring heuristics for register allocation, *in* "Proc. SIGPLAN'89 Conf. Programming Language Design and Implementation," pp. 275–284, 1989.
14. D. CALLAHAN AND B. KOBLENZ, Register allocation via hierarchical graph coloring, *in* "Proc. SIGPLAN'91 Conf. Programming Language Design and Implementation," pp. 192–203, 1991.
15. G.J. CHAITIN, Register Allocation and Spilling via Graph Coloring, *in* "Proc. SIGPLAN'82 Symp. Compiler Construction," pp. 98–105, 1982.
16. G.J. CHAITIN, M.A. AUSLANDER, A.K. CHANDRA, J. COCKE, M.E. HOPLINS, AND P.W. MARKSTEIN, Register allocation via graph coloring, *Computer Languages* **6** (1981) 47–57.
17. B. COURCELLE AND M. MOSBAH, Monadic second-order evaluations on tree-decomposable graphs, **6** (1993) 49–82.
18. O.J. DAHL, E.W. DIJKSTRA, AND C.A.R. HOARE, *Structured Programming*, Academic Press, London, 1972.
19. N. DENDRIS, L. KIROUSIS, AND D. THILKOS, Fugitive-serach gamses on graphs and related parameters, *Theor. Comp. Sc.* **172** (1997) 233-254.
20. E.W. DIJKSTRA, Go To Statement Considered Harmful, *Comm. ACM* **11**, 3 (1968) 147–148.

21. A.P. Ershov, Reduction of the problem of memory allocation in programming to the problem of colouring the vertices of a graph, *Doklady Academii Nauk SSSR* **142**, 4 (1962) 785–787. English version in *Soviet Mathematics* **3** (1962) 163–165.
22. M.R. Garey, D.S. Johnson, G.L. Miller, and C.H. Papadimitriou, The Complexity of Coloring Circular Arcs and Chords, *SIAM J. Alg. Discr. Meth.* **1**, 2 (1980) 216–227.
23. F. Gavril, Algorithms for Minimum Coloring, Maximum Clique, Minimum Covers by Cliques, and Maximum Independent Set of Chordal Graphs, *SIAM J. Comp.* **1**, 2 (1972) 180–187.
24. F. Gavril, The Intersection Graph of Subtrees in Trees Are Exactly the Chordal Graphs, *J. Comb. Th. Ser. B* **16** (1974) 47–56.
25. R. Gupta, M.L. Soffa, and T. Steele, Register allocation via clique separators, *in* "Proc. SIGPLAN'89 Conf. Programming Language Design and Implementation," pp. 264–274, 1989.
26. M. M. Halldórsson, A Still Better Performance Guarantee for Approximate Graph Coloring, *Inf. Proc. Lett.* **45** (1993) 19–23.
27. S. Kannan and T. Proebsting, Register Allocation in Structured Programs, *in* "Proc. 6th SODA," pp. 360–368, 1995.
28. B.R. Kernighan and D.M. Ritchie, *The C Programming Language*, Prentice-Hall, New Jersey, 1978.
29. D.E. Knuth, Structured Programming with Go To Statements, *ACM Computing Surveys* **6**, 4 (1974) 261–301.
30. C. Lund and M. Yannakakis, On the Hardness of Approximating Minimization Problems, *J. ACM* **41** (1994) 960–981.
31. P. Naur, Revised Report on the Algorithmic Language Algol 60, *Comm. ACM* **6**, 1 (1963) 1–17.
32. P. Naur, Go To Statements and Good Algol Style, *BIT* **3**, 3 (1963) 204–208.
33. T. Nishizeki, K. Takamizawa, and N. Saito, Algorithms for detecting series-parallel graphs and D-charts, *Trans. Inst. Elect. Commun. Eng. Japan* **59**, 3 (1976) 259–260.
34. C. Norris and L.L. Pollock, Register Allocation over the Program Dependence Graph, *in* "Proc. SIGPLAN'94 Conf. Programming Language Design and Implementation," pp. 266–277, 1994.
35. N. Robertson and P.D. Seymour, Graph Minors I: Excluding a Forest, *J. Comb. Th. Ser. B* **35** (1983) 39–61.
36. N. Robertson and P.D. Seymour, Graph Minors XIII: The Disjoint Paths Problem. *J. Comb. Th. Ser. B* **63** (1995) 65–110.
37. M. Thorup, Structured Programs have Small Tree-Width and Good Register Allocation. Latest full version: `http://www.diku.dk/~mthorup/PAPERS/register.ps.gz`.
38. M. Tofte and J-P. Talpin, Implementing the call-by-value lambda-calculus using a stack of regions. *in* "Proc. POPL'94," pp. 188–201, 1994.
39. N. Wirth, The Programming Language PASCAL, *Acta Informatica* **1** (1971), 35–63.
40. N. Wirth, *Programming in Modula-2 (3rd corr. ed.)*, Springer-Verlag, Berlin, New York, 1985.

A Measure of Parallelization for the Lexicographically First Maximal Subgraph Problems

Ryuhei Uehara

Center for Information Science, Tokyo Woman's Christian University,
uehara@twcu.ac.jp

Abstract. A maximum directed tree size (MDTS) is defined by the maximum number of the vertices of a directed tree on the directed acyclic graph of a given undirected graph. The MDTS of a graph *measures* the parallelization for the lexicographically first maximal subgraph (LFMS) problems. That is, the complexity of the problems on a graph family $\mathcal{G}$ gradually increases as the value measured on each graph in the family grows; (1) if the MDTS of each graph in $\mathcal{G}$ is $O(\log^k n)$, the lexicographically first maximal independent set problem on $\mathcal{G}$ is in NC^{k+1}, and the LFMS problem for π is in NC^{k+s}, where π is a property on graphs such that π is nontrivial, hereditary, and NC^{s-1} testable; (2) both problems above are P-complete if the MDTS of each graph in $\mathcal{G}$ is cn^{ϵ}. It is worth remarking that the problem to compute the MDTS is in NC^2. This is important in the sense that a "measure" means only if measuring the complexity of a problem is easier than solving the problem.
Key words: Analysis of algorithms, P-completeness, NC algorithms, the lexicographically first maximal independent set problem, the lexicographically first maximal subgraph problems.

1 Introduction

The maximal independent set (MIS) problem is a typical maximality problem, that is to find a maximal vertex-induced subgraph that satisfies a specified graph property. Since Karp and Wigderson showed that the MIS problem is in the class NC [12], much work has been devoted to the study of parallel complexity of maximality problems [14, 1, 7, 6, 4, 17, 18]. On the other hand, the lexicographically first maximal independent set (LFMIS) problem is a typical P-complete problem [3], and P-completeness of the lexicographically first maximal subgraph (LFMS) problems for some properties π was shown [15, 17] (see also [8] for a comprehensive reference). As noticed by Iwama and Iwamoto in [10], one of approaches to making clear the boundary between the classes NC and P is to find (sufficient) conditions for problems to be in NC or to be P-complete. In [10], they produced a new problem on graphs, called α-connectivity, whose complexity gradually increases as the value of α grows. Our approach is slightly different from theirs. We provide a "measure" of parallelization for the LFMIS and LFMS problems. That is, the complexity of the problems on a family of graphs gradually increases as the value measured on each graph in the family grows. The measure is the *maximum directed tree size* (MDTS) of G, which is defined by the maximum number of the vertices of a directed tree on the directed acyclic graph of G. The problem to compute the MDTS is in NC^2. This result is important since a "measure" means to us only if measuring the complexity of a problem is easier than solving the problem. We define two graph families $\mathcal{G}_{\log^k n}$ and $\mathcal{G}_{\text{poly}}$ as follows:

$\mathcal{G}_{\log^k n}$ contains every graph whose MDTS is $O(\log^k n)$ for some positive integer k, and

$\mathcal{G}_{\text{poly}}$ contains every graph whose MDTS is at most cn^ϵ for any fixed positive constant c and ϵ.

Then the results in this paper are the following: The LFMIS problem on $\mathcal{G}_{\log^k n}$ is in NC^{k+1}, and the problem on $\mathcal{G}_{\text{poly}}$ is P-complete. Let π be a property on graphs such that π is nontrivial, hereditary on vertex-induced subgraphs, satisfied by all independent edges, and NC^{s-1} testable. Then, the LFMS problem for π on $\mathcal{G}_{\log^k n}$ is in NC^{k+s}, and the problem on $\mathcal{G}_{\text{poly}}$ is P-complete.

2 Preliminaries

We will deal only with graphs and digraphs without loops or multiple edges. Throughout this paper, unless stated otherwise, G always denotes the input (undirected) graph, $V = \{0, 1, \cdots, n-1\}$ and E denote the set of vertices and edges in G, respectively. Without loss of generality, we assume that G is connected. The *neighborhood* of a vertex v in G, denoted $N_G(v)$, is the set of vertices in G adjacent to v. The *degree* of a vertex v in G is $|N_G(v)|$, and denoted by $d_G(v)$. Vertices of degree 0 are called *isolated vertices*. For $U \subseteq V$, $N_G(U)$ is $\cup_{u \in U} N_G(u)$, and $G[U]$ is the graph (U, F), where $F = \{\{u,v\} \mid u, v \in U \text{ and } \{u,v\} \in E\}$. For a graph G, $|G|$ denotes the number of vertices in G.

Let $X = \{x_1, x_2, \cdots, x_k\}$ and $Y = \{y_1, y_2, \cdots, y_h\}$ be any subsets of V. (Throughout this paper, we assume that sets are always sorted, that is, $x_i < x_j$ and $y_i < y_j$ for each $1 \le i < j \le k, h$.) Let $<$ be the total ordering on the sets X and Y defined as follows: $X < Y$ if and only if $x_i = y_i$ for every i with $1 \le i \le k$, or there is an index $i' \ge 1$ such that $x_j = y_j$ for every $1 \le j < i'$, and $x_{i'} < y_{i'}$.

A subset U of V is called an *independent set* if $G[U]$ only contains isolated vertices. A *maximal independent set* (MIS) in G is an independent set that is not properly contained in another independent set. The MIS problem is to find, given a graph G, an MIS in G. The *lexicographically first maximal independent set* (LFMIS) in G is the MIS I in G such that $I < J$ for every MIS J in G. The LFMIS problem is to find, given a graph G, the LFMIS in G.

For $G = (V, E)$, $\vec{G} = (V, \vec{E})$ is defined by the directed acyclic graph obtained from G by replacing each edge $\{u, v\}$ by the arc $(\min\{u,v\}, \max\{u,v\})$. (It is easy to show that the resulting graph $\vec{G}$ is acyclic for any graph G.) The *neighborhood* of a vertex v in $\vec{G}$, denoted $N_{\vec{G}}(v)$, is the set of vertices u in $\vec{G}$ such that $(v, u) \in \vec{E}$. The *outdegree* of a vertex v in $\vec{G}$, denoted by $d^+_{\vec{G}}(v)$, is $|N_{\vec{G}}(v)|$, and the *indegree* of a vertex v in $\vec{G}$, denoted by $\mathrm{d}^-{}_{\vec{G}}(v)$, is $|N_G(v)| - |N_{\vec{G}}(v)|$. We mention that any directed acyclic graph $\vec{G}$ has at least one vertex v with $\mathrm{d}^-{}_{\vec{G}}(v) = 0$, and at least one vertex u with $\mathrm{d}^+{}_{\vec{G}}(u) = 0$ (see [9, Theorem 16.2]).

Lemma 1. *Let v be a vertex with $\mathrm{d}^-{}_{\vec{G}}(v) = 0$. Then v is in the LFMIS of G.*

Proof. Let I be the LFMIS of G. To derive a contradiction, assume that $v \notin I$. Let $U = N_G(v) \cap I$. Then, by the maximality of I and the assumption, $U \neq \emptyset$. Let $J = I - U + \{v\}$. Clearly, J is an independent set, and it is not difficult to see that there is a maximal independent set K such that $J \subseteq K$, consequently, $I - U + \{v\} \subseteq K$. For each u in U, since $\mathrm{d}^-{}_{\vec{G}}(v) = 0$, $u > v$ holds. This implies that $K < I$, which contradicts that I is the LFMIS of G. □

Now, we define the measure of the parallelization for the lexicographically first maximal problems. A *directed tree* T in G is the edge-induced subgraph of G such that T is a rooted tree, and every path from the root to each leaf is a directed path on $\vec{T}$. The *maximum directed tree size* (MDTS) of G, denoted by $\mathrm{MDTS}(G)$, is defined by the maximum number of vertices of a directed tree in G.

Lemma 2. *Let T be a directed tree in G with $|T| = \mathrm{MDTS}(G)$. Then $\mathrm{d}^-{}_{\vec{G}}(r) = 0$, where r is the root of T.*

Proof. To derive contradictions, assume that $\mathrm{d}^-{}_{\vec{G}}(r) > 0$. Then there is a vertex s with (s, r) in $\vec{G}$. First we assume that s is in T. Then the directed path from r to s with the arc (s, r) makes a cycle, which contradicts the acyclicness of $\vec{G}$. Thus, s is not in T. Then, adding s to T, we get the directed tree properly containing T. This contradicts that $|T| = \mathrm{MDTS}(G)$. □

We use P to denote the class of all polynomial time computable problems, and NC^k to denote the class of all problems computable by a uniform polynomial size circuit family of depth $O(\log^k n)$. The class NC is defined by $\cup NC^k$ (for further details, refer to [3, 11]). Although the P-completeness is defined via NC^1-reducibility in [3], we use the log space reducibility simply as in [15]: A function F_0 is said to be *P-complete* if F_0 is in P and for each F in P there are log space computable functions f and g such that $F(x) = g(F_0(f(x)))$ for all inputs. It is well known that the MIS problem is in NC [12, 1, 14, 7, 6, 13], and the LFMIS problem is one of the fundamental P-complete problems [15, 8].

Recall that the EREW PRAM is the parallel model where the processors operate synchronously and share a common memory, but no two of them are allowed simultaneous access to a memory cell (whether the access is for reading or for writing in that cell). The CRCW PRAM differs from the EREW PRAM in that both simultaneous reading and simultaneous writing to the same cell are allowed; in case of simultaneous writing, the processor with lowest index succeeds. It is well known [11] that a problem is in NC^{k+1} if it is solved in $O(\log^k n)$ time using a polynomial number of processors on a CRCW or EREW PRAM. As the input representation of G, to express $\vec{G}$, we assume that each vertex v has two lists of the edges incident to it, one is the list of the arcs to v, and the other is the list of the arcs from v.

Lemma 3 [2]. *The problem to compute* MDTS(G) *is in* NC^2.

Proof. For each vertex v, let r_v be the number of vertices reachable from v on $\vec{G}$. It is not difficult to see that MDTS(G) is equal to the maximum number of r_v for each v in G. That is, the following algorithm computes MDTS(G):

1: In parallel, compute r_v for each vertex v in G.
2: Compute $\max\{r_v \mid v \text{ is in } G\}$.

The classic naive parallel algorithm can compute the first step as a kind of graph reachability problem. It computes the transitive closure of the adjacency Boolean matrix of G, and r_v is given by the number of 1s of the vth row (the details are discussed in [5, 16]). Thus, the first step is computed by using $O(\log n)$ time and n^3 processors on a CRCW PRAM. Clearly, the resources are sufficient to compute the second step. Thus, the problem is in NC^2. □

3 Lexicographically first maximal independent set problem

3.1 An algorithm for the LFMIS problem

The parallel algorithm for the LFMIS problem is very simple:

1: Set $G' := G$, and $I := \emptyset$.
2: While G' has at least one vertex, do the following step: In parallel, for each vertex v with $d^-_{\vec{G'}}(v) = 0$, add v into I, and delete v and every vertex in $N_{\vec{G'}}(v)$ from G'.
3: Output I.

Lemma 4. *Let v and v' be any vertices with $d^-_{\vec{G'}}(v) = d^-_{\vec{G'}}(v') = 0$. Then $v \notin N_{\vec{G'}}(v')$.*

Proof. Assume that $v \in N_{\vec{G}}(v')$. Then (v', v) is the arc in G', which contradicts $d^-_{\vec{G'}}(v) = 0$. □

Lemma 4 shows that every vertex v with $d^-_{\vec{G'}}(v) = 0$ never fail to be added into I in step 2. That is, such a vertex is not deleted from G' as the neighbor of another vertex.

Theorem 5. *The algorithm outputs the LFMIS of G.*

Proof. We show the theorem by induction on the number of iteration. Let I_G be the LFMIS of G, and I_i be the contents of the variable I after the ith iteration. Let D_i be the set of vertices that are deleted from G' as the neighbors of the vertices in I_i in step 2. Then, it is sufficient to show that I_i is a subset of I_G,

$I_{i+1} - I_i \neq \emptyset$, and every vertex in D_i is not in I_G for each $i \geq 1$. Lemma 1 and 4 imply that I_1 is a subset of I_G, and D_1 is a set of the vertices not in I_G. We consider the ith iteration with $i > 1$. Let I' be the set of vertices v in G' with $d^-_{\vec{G'}}(v) = 0$. Since $\vec{G'}$ is acyclic, $I' \neq \emptyset$, or $I_{i+1} - I_i \neq \emptyset$. Next we show that every vertex v in I' is not only in the LFMIS of G', but also in I_G. To derive a contradiction, we assume that v is a vertex in $I' - I_G$. Let $U = N_G(v) \cap I_G$. Then, by the maximality of I_G and the assumption, $U \neq \emptyset$. Let u is a vertex in U. If u is in I_{i-1}, v must be deleted from G' as a neighbor of u since $v \notin I_{i-1}$, or $u < v$. On the other hand, u is not in D_{i-1} since u is in I_G. Thus, u must be in G', or U is the set of vertices in G'. Let $J = I_G - U + \{v\}$. Clearly, J is an independent set of G, and there is a maximal independent set K such that $J \subseteq K$, consequently, $I_G - U + \{v\} \subseteq K$. For each u in U, since $d^-_{\vec{G'}}(v) = 0$ and u is in G', $u > v$ holds. This implies that $K < I_G$, which contradicts that I_G is the LFMIS of G. Thus, I_i is the subset of I_G, and clearly, every vertex in D_i is not in I_G. This completes the proof. □

For the time complexity of the algorithm, we have the following theorem:

Theorem 6. *Let $\mathcal{G}$ be the family of graphs whose MDTS are bounded above by $t(n)$. Then the LFMIS problem on $\mathcal{G}$ can be solved in $O(t(n))$ time using $O(n+m)$ processors on an EREW PRAM.*

Proof. We first show the number of the iterations of the while-loop. Fix an iteration. Let T be a directed tree in G' with $|T| = \text{MDTS}(G')$, and r be the root of T. Then, by Lemma 2, $d^-_{\vec{G'}}(r) = 0$. Thus, r will be deleted from G' in step 2, consequently, MDTS(G') will decrease at least one through step 2. Since MDTS(G') $= 0$ if and only if G' contains no vertices, the number of the iterations of the while-loop is bounded above by $t(n)$. The implementation of the algorithm on an EREW PRAM is easy, and omitted here. □

3.2 *P*-completeness of the LFMIS problem

It is well known that the LFMIS problem is *P*-complete [15, 8]. In other words, the LFMIS problem on $\mathcal{G}$ is *P*-complete, where $\mathcal{G}$ is the family of all graphs whose MDTS are bounded above by n. Using this observation, we have the following theorem.

Theorem 7. *The LFMIS problem on $\mathcal{G}_{\text{poly}}$ is P-complete under the log space reductions.*

Proof. We reduce the general LFMIS problem to the restricted one. For a given graph $G(V, E)$ with n vertices, we construct a new graph $G'(V', E')$ that consists of $f(n)$ copies of G and an additional vertex, where $f(n) = \left(\frac{n+1}{c}\right)^{\frac{1}{\epsilon}}$. The role of the additional vertex will be explained later on. Intuitively, the vertex $in+j$ in the ith copy of G corresponds to the vertex j in G for each i with $0 \leq i \leq f(n) - 1$ and j with $0 \leq j \leq n-1$. More precisely, $V' = \{0, 1, \cdots, nf(n)\}$, and E' consists of $\{\{in + j_0, in + j_1\} \mid \{j_0, j_1\} \in E\}$ for each i with $0 \leq i \leq f(n) - 1$, and j_0, j_1 with $0 \leq j_0, j_1 \leq n-1$. The additional vertex $nf(n)$ joints each copy of G in the following way: There exists at least one vertex v in G such that $d^{+}_{\vec{G}}(v) = 0$ since $\vec{G}$ is acyclic. For the vertex v, E' also consists of $\{in+v, nf(n)\}$ for each i with $0 \leq i \leq f(n)-1$. Clearly, this is the log space reduction, and G' is connected. Now, MDTS(G') is at most $n+1$ and the number of vertices of G' is $n' = nf(n)+1$. Thus, for the resulting graph G' with n' vertices, MDTS(G') $\leq n+1 = c(f(n))^{\epsilon} \leq c(nf(n)+1)^{\epsilon} = cn'^{\epsilon}$. This completes the proof. □

Summarizing up Lemma 3, Theorem 6 and 7, we obtain the following corollary:

Corollary 8. *1. The problem to compute MDTS on any graph family is in NC^2.*
2. The LFMIS problem on $\mathcal{G}_{\log^k n}$ is in NC^{k+1}, and the problem on $\mathcal{G}_{\mathrm{poly}}$ is P-complete.

4 Lexicographically first maximal subgraph problem for a local property

Miyano showed that the lexicographically first maximal subgraph (LFMS) problem for a hereditary property is *P*-complete [15], and Shoudai and Miyano showed that maximal subgraph problem for a local property is in *NC* [18]. Using their techniques, we can generalize the results in the previous section.

Let π be a property on graphs. The problem we consider in this section is the lexicographically first maximal subgraph (LFMS) problem for π that is stated as follows: given a graph $G = (V, E)$, find the lexicographically first maximal subset U of V whose vertex-induced subgraph satisfies π. We say that π is *nontrivial* on a graph family $\mathcal{G}$ if infinitely many graphs in $\mathcal{G}$ satisfy π and some graph in $\mathcal{G}$ violates π, and π is *hereditary* on vertex-induced subgraphs if all vertex-induced subgraphs also satisfy π. We say that a graph $G = (V, E)$ is a *minimal graph violating* π with respect to vertices if G violates π and the vertex-induced subgraph $G[U]$ satisfies π for every subset U of V with $U \neq V$.

Theorem 9. *Let π be a property on graphs satisfying the following conditions;*

- *π is nontrivial,*
- *π is hereditary on vertex-induced subgraphs,*
- *π is satisfied by all independent edges, and*
- *π is NC^{s-1} testable for some positive integer s.*

Then the LFMS problem for π on $\mathcal{G}_{\log^k n}$ is in NC^{k+s}, and the problem on $\mathcal{G}_{\mathrm{poly}}$ is P-complete.

Proof. First, we consider the algorithm for the LFMS problem. For subsets W and U of vertices with $W \cap U = \emptyset$, let $E^W_U = \{\{v, w\} \mid v \neq w, v, w \in W$ and there is $u \in U$ such that $\{v, u\} \in E$ and $\{w, u\} \in E\}$. Then let $H^W_U = (W, E[W] \cup E^W_U)$, where $E[W] = \{\{u,v\} \in E \mid u,v \in W\}$. Then the algorithm is described as follows:

```
1. begin /* G = (V, E) is an input */
2.   W := V; U := ∅;
3.   while W ≠ ∅ do begin
4.     I := {v | d⁻_{H_U^W}(v) = 0};
5.     U := U ∪ I;
6.     W := W − I;
7.     W := W − {w ∈ W | G[U ∪ {w}] violates π};
8.   end
9.   output U;
10. end
```

Since π is hereditary and satisfied by all independent edges, π is satisfied by all independent set. By the definition of H_U^W, for each vertex in U, at most one of its neighbors is set into U from W in an iteration. Using the same techniques in the proof of Theorem 5 and the proof of [18, Theorem 3], we can show the correctness of the algorithm. To evaluate the number of the iterations, we use the potential function argument. The potential function $\phi(W,U)$ is defined by MDTS(H_U^W). Let W_i and U_i, respectively, be the contents of the variables W and U before the ith iteration. Then, it is easy to see that $\phi(W_1,U_1) = \phi(V,\emptyset) =$ MDTS(G), and $\phi(W_i,U_i) > 0$ if and only if $W_i \neq \emptyset$. We now show that $\phi(W_i,U_i) - \phi(W_{i+1},U_{i+1}) \geq 1$ for each $i > 0$ by induction. Lemma 2 shows that $\phi(W_1,U_1) - \phi(W_2,U_2) \geq 1$, since each root of a tree T with $|T| =$ MDTS(G) is set into I_1 in step 4 and deleted from W_1 in step 6. Here, fix an iteration $i > 1$ of the while-loop. We assume that $\phi(W_i,U_i) - \phi(W_{i+1},U_{i+1}) < 1$, and derive contradictions. Let T be a directed tree in $H_{U_{i+1}}^{W_{i+1}}$ with $|T| =$ MDTS($H_{U_{i+1}}^{W_{i+1}}$). We first assume that T contains no edge in $E_{U_{i+1}}^{W_{i+1}}$. Then T must be the directed tree in $H_{U_i}^{W_i}$. This implies that $\phi(W_i,U_i) - \phi(W_{i+1},U_{i+1}) = 0$, and $|T| =$ MDTS($H_{U_i}^{W_i}$). Thus, Lemma 2 shows that the root of T has no indegree in $H_{U_i}^{W_i}$, or the root must be set into I_i and deleted from W_i in the ith loop. This contradicts that the root of T exists in the $(i+1)$st loop. Thus, T contains at least one edge in $E_{U_{i+1}}^{W_{i+1}}$. We first assume that T contains only one edge $e = \{v,w\}$ in $E_{U_{i+1}}^{W_{i+1}}$. By construction of $E_{U_{i+1}}^{W_{i+1}}$, there is a vertex $u \in U$ such that $\{v,u\} \in E$ and $\{w,u\} \in E$. Two cases arise. First, u is set into U in the ith step. Then T with u must be a directed tree in $H_{U_i}^{W_i}$. This contradicts the assumption that $\phi(W_i,U_i) - \phi(W_{i+1},U_{i+1}) < 1$. Second, u is set into U in i'th step with $i' < i$. Then T must be a directed tree in $H_{U_i}^{W_i}$. This also contradicts the assumption. Thus $\phi(W_i,U_i) - \phi(W_{i+1},U_{i+1}) \geq 1$ when T contains only one edge in $E_{U_{i+1}}^{W_{i+1}}$. Using a simple induction on the number of the edges, we can get $\phi(W_i,U_i) - \phi(W_{i+1},U_{i+1}) \geq 1$ when T contains two or more edges in $E_{U_{i+1}}^{W_{i+1}}$. Thus, the number of the iterations is bounded above by $\phi(W_1,U_1) =$ MDTS(G). Hence, the algorithm solves the LFMS problem for π

in $O(T(n)\text{MDTS}(G))$ time using polynomial number processors on an EREW PRAM for any input graph G, where $T(n)$ is the time to test whether a given graph satisfies π with polynomial number processors on an EREW PRAM. This completes the proof of the former half of the theorem.

The LFMS problem on graphs (with no restrictions) for π is P-complete under the log space reductions [15, Theorem 5]. Using the same log space reduction in Theorem 7, we can show the P-completeness of the LFMS problem for π on $\mathcal{G}_{\text{poly}}$. This completes the proof of the latter half of the theorem. □

5 Concluding Remarks

Toda noticed that we can use the maximum length of a directed path in $\vec{G}$ as another measure [19]. We can derive the same results using this measure as using the MDTS.

The further work is to completely characterize the class NC^k using the value measured. To do this, we must show that the LFMIS problem on the graphs with value k is not only in NC^{k+1}, but also NC^{k+1}-complete.

Acknowledgment

I indebted to Professor Seinosuke Toda and Professor Zhi-Zhong Chen for many helpful comments. My thanks also to Dr. Koichi Yamazaki for discussions and encouragement. I also wish to thank the referees for pointing out an error in an earlier draft.

References

1. N. Alon, L. Babai, and A. Itai. A Fast and Simple Randomized Parallel Algorithm for the Maximal Independent Set Problem. *Journal of Algorithms*, 7:567–583, 1986.
2. Z.-Z. Chen. Personal communication. 1997.
3. S.A. Cook. A Taxonomy of Problems with Fast Parallel Algorithms. *Information and Control*, 64:2–22, 1985.
4. K. Diks, O. Garrido, and A. Lingas. Parallel algorithms for finding maximal k-dependent sets and maximal f-matchings. In *ISA '91 Algorithms*, pages 385–395. Lecture Notes in Computer Science Vol. 557, Springer-Verlag, 1991.
5. H. Gazit and L. Miller. An Improved Parallel Algorithm That Computes the BFS Numbering of a Directed Graph. *Information Processing Letters*, 28:61–65, 1988.
6. M. Goldberg and T. Spencer. A New Parallel Algorithm for the Maximal Independent Set Problem. *SIAM Journal on Computing*, 18(2):419–427, 1989.
7. M. Goldberg and T. Spencer. Constructing a Maximal Independent Set in Parallel. *SIAM J. Disc. Math.*, 2(3):322–328, 1989.
8. R. Greenlaw, H.J. Hoover, and W.L. Ruzzo. *Limits to Parallel Computation*. Oxford University Press, 1995.
9. F. Harary. *Graph Theory*. Addison-Wesley, 1972.
10. K. Iwama and C. Iwamoto. α-Connectivity: A Gradually Nonparallel Graph Problem. *Journal of Algorithms*, 20:526–544, 1996.

11. R.M. Karp and V. Ramachandran. Parallel Algorithms for Shared-Memory Machines. In J. van Leeuwen, editor, *The Handbook of Theoretical Computer Science, vol. I: Algorithms and Complexity*, pages 870–941. MIT Press, 1990.
12. R.M. Karp and A. Wigderson. A Fast Parallel Algorithm for the Maximal Independent Set Problem. *Journal of American Computing Machinery*, 32(4):762–773, 1985.
13. D.C. Kozen. *The Design and Analysis of Algorithms*. Springer-Verlag, 1992.
14. M. Luby. A Simple Parallel Algorithm for the Maximal Independent Set Problem. *SIAM Journal on Computing*, 15(4):1036–1053, 1986.
15. S. Miyano. The Lexicographically First Maximal Subgraph Problems: P-Completeness and *NC* Algorithms. *Mathematical Systems Theory*, 22:47–73, 1989.
16. C.H. Papadimitriou. *Computational Complexity*. Addison-Wesley Publishing Company, 1994.
17. D. Pearson and V.V. Vazirani. Efficient Sequential and Parallel Algorithms for Maximal Bipartite Sets. *Journal of Algorithms*, 14:171–179, 1993.
18. T. Shoudai and S. Miyano. Using Maximal Independent Sets to Solve Problems in Parallel. *Theoretical Computer Science*, 148:57–65, 1995.
19. S. Toda. Personal communication. 1997.

Make your Enemies Transparent

Tanja Vos and **Doaitse Swierstra**

Utrecht University, Department of Computer science
P.O. Box 80.089, 3508 TB Utrecht, The Netherlands
e-mail: {tanja, doaitse}@cs.ruu.nl.

Abstract. We design a distributed program that sorts a connected network of processes in an unstable environment. We show two techniques that enable us to factor out dynamic changes to the topology of the network in the specification of the program.

keywords: distributed programs, unstable environments, UNITY, convergence, transparency law.

1 Introduction

Our starting point is a **connected network of processes** in which: every process has a unique label which is used to identify its address; every process has a unique local variable, which can store a data value; a process can only communicate with a process to which it is connected; a process can only communicate with one other process at the same time.

Moreover, we have an **unstable environment**, in which enemies lurk to tamper with the configuration of the network. In particular, we consider two kinds of enemies. Firstly, *external agents* that can change the data values of the processes. Secondly, *daemons* that can deactivate and re-activate communication links.

The primary result of this paper is a technique for factoring these enemies out of the specification of a problem. Consequently, we can discard certain environmental unstabilities and concentrate on the actual problem which is solved by the program that is designed. To illustrate these techniques, we shall design a distributed program which **sorts** the network. That is a program that does not alter the multi-set of the processes values, and (according to some predefined orderings, $\prec_p$ and $\prec_v$, on the labels and the data values respectively) will bring the network in a state in which, for any pair of processes, the ordering ($\prec_p$) on the labels of these processes is reflected in the ordering ($\prec_v$) on the data values that reside at these processes.

2 Preliminaries

The results in this paper, are formalised with UNITY, a programming logic developed to support the design and verification of distributed programs. The basic foundations of UNITY can be found in [CM88]; several extensions regarding compositionality and convergence can be found in [Pra95]. One of the extensions in [Pra95] is concerned with self-stabilisation and convergence of programs. Roughly speaking, a self-stabilising program is a program which is capable of recovering from arbitrary transient failures. Obviously such a property is very useful, although the requirement to allow arbitrary failures may be too strong. A more restricted form of self-stabilisation, called convergence, allows a program to recover only from certain failures. Self-stabilisation and convergence are considered to be essential properties for distributed programs and evidently very significant for our purposes.

This section informally introduces some concepts needed to understand the rest of this paper, for a more formal treatment the reader is referred to [Pra95,VSP96].

A UNITY program Π is a quadruple $(R_\Pi, W_\Pi, I_\Pi, A_\Pi)$, where R_Π and W_Π are the sets of variables Π reads and writes respectively, I_Π is the initial condition, and A_Π is the set of actions of Π. A program execution is infinite, in each step of the execution an action is selected[1] and executed.

In UNITY, single-action progress properties are defined by the **ensures** relation. For a program Π, p **ensures** q means that:

- for all actions $a \in A_\Pi$, if p is true at some point of the computation and q is not, then after executing a, p remains true or q becomes true.
- there exists an action a in A_Π, such that if p is true at some point of the computation and q is not, then after executing a, q becomes true. Consequently, eventually q becomes true.

A predicate p is *confined* by a set of variables V, denoted by $p\mathsf{C}V$, if p does not restrict the value of any variable outside V (e.g. $x < y$ is confined by $\{x, y\}$ but not by $\{y\}$).

A predicate J is *stable* in a program Π (denoted ${}_\Pi\vdash \circlearrowright J$), if once J holds during the execution of Π, it will remain to hold forever. Note that J is an invariant (denoted ${}_\Pi\vdash \Box J$) if J is stable and holds initially.

A program Π *converges* from p to q ($p\mathsf{C}W_\Pi$ and $q\mathsf{C}W_\Pi$) under the stability of J (denoted by $J\ {}_\Pi\vdash p \rightsquigarrow q$), if, given that ${}_\Pi\vdash \circlearrowright J$, the program Π started in p shall eventually find itself in a situation where q holds and will remain to hold. Intuitively, a program Π for which this holds can recover from failures which preserve the validity of p and the stability of J.

Furthermore, $|S|$ is used to denote the cardinality of a set S; N_0 denotes the natural numbers including zero; we assume the reader is familiar with the meanings of anti-symmetric, transitive and total orderings; for a binary relation R on a finite set A ($|A| = N$), we define the *transitive closure* of R (denoted by R^{tr}) by induction as in [RW92], i.e. $R^{tr} = R_N$, where:

[1] Selection is nondeterministically, with fairness condition that every action is selected infinitely often

$(x, y) \in R_1 = (x, y) \in R$
$(x, y) \in R_n = (x, y) \in R_{n-1} \vee (\exists z \in A : (x, z) \in R_{n-1} \wedge (z, y) \in R_{n-1})$

Finally, all results in this paper are mechanically verified using the the theorem proving environment HOL [GM93]. Since mechanical verification is beyond the scope of this paper, the reader is referred to [VSP96] for an intensive treatment of these verification activities.

3 Formal Specification of what the Program is to do

A network of processes is modelled by a triple (P, C, D), where

- P is a set containing the labels of all processes in the network. Consequently, since every process has a unique label this implies that $|P|$ equals the number of processes in the network.
- C is the topology of the network, $C \subseteq P \times P$. For the sake of clarity, we shall call elements of this set *communication links*. The communication links that are active (i.e. up) and over which communication can actually take place, shall be called *connections*. Note that the set of communication links C is static, the set of connections is dynamic, and the latter is always $\subseteq$-ed in C.
- (P, C) is considered to be an undirected graph, so there is at least one process, the number of processes is finite and $(p, q) \in C \Rightarrow (q, p) \in C$.
- The distribution of the local *variables* that store the data value that resides at a process, is given by a function D that maps a process-label to the *local variable* of that process[2]. The initial distribution of the *data values* in the network shall be denoted by the function I, which maps a process-label to the *data value* which is initially stored in the local variable of that process.

The values of a program's variables at a given moment reflect the *state* of the program. Consequently, a state of the program is represented by a function from variables to the universe of values that these variables can have. As a result, in every state s the distribution of data values is given by $(s \circ D)$[3].

A triple (P, C, D) which is in compliance with the constraints above shall be denoted by **Network**(P, C, D). Such a network is defined to be sorted in some state s, when the following property[4] is satisfied:

$$\mathbf{Sorted}_{(P,D,\prec_p,\prec_v)}\ s = \forall i, j \in P : i \prec_p j \Rightarrow ((s \circ D)\, i) \prec_v ((s \circ D)\, j) \quad (1)$$

where $\prec_p$ and $\prec_v$ are orderings on the process labels and the data values respectively.

[2] Since every process has a unique variable: $\forall i, j \in P : i \neq j \Rightarrow (D\, i) \neq (D\, j)$.
[3] So, if s_0 is the initial state of the program, $I = (s_0 \circ D)$.
[4] $\mathbf{Sorted}_{(P,D,\prec_p,\prec_v)}$ is called a *state-predicate*, i.e. a predicate of type: *state* $\rightarrow$ *bool*

Let us start with the assumption that $\prec_v$ is a total order and determine which properties the ordering $\prec_p$ must satisfy under this assumption. First, it must be anti-symmetric. Consider a network in which every process contains a different data value, that is for all states s: $\forall i,j \in P :(i \neq j) \Rightarrow ((s \circ D)\, i) \neq ((s \circ D)\, j)$. Suppose that $\prec_p$ is not anti-symmetric, so it is possible that there are processes i and j such that $(i \neq j) \wedge (i \prec_p j) \wedge (j \prec_p i)$ holds. From $(i \neq j)$ we deduce that $((s \circ D)\, i) \neq ((s \circ D)\, j)$, and from the anti-symmetric property of $\prec_v$ we derive that $((s \circ D)\, i) \prec_v ((s \circ D)\, j)$ and $((s \circ D)\, j) \prec_v ((s \circ D)\, i)$ cannot both hold at the same time. Consequently, the network cannot be sorted and thus $\prec_p$ must be anti-symmetric. Second, $\prec_p$ must be transitive, for again consider a network in which every process contains a different data value. Suppose that $\prec_p$ is not transitive, so it is possible that there are i, j and k $(i \neq j \neq k)$ such that $i \prec_p j$ and $j \prec_p k$ and $k \prec_p i$ simultaneously hold. Again it can be concluded from the anti-symmetry property of $\prec_v$ that it is impossible to sort the network. Whether $\prec_p$ is reflexive, anti-reflexive, or neither makes no difference. Because if $i \prec_p i$ holds, (1) is valid since $\prec_v$ is reflexive; and if $\neg(i \prec_p i)$ holds then (1) is trivially valid. Finally, we assume that $\prec_p$ is non-empty, since if $\prec_p = \emptyset$ then (1) is a tautology and consequently there is no use in constructing a sorting program.

Until now, we have made the following assumptions about the environment in which the program will operate. First, we assumed an unstable environment in which external agents can change the data values of the processes, and daemons can tamper with the status (i.e. up or down) of the communication links. Second, we assumed a network of processes, in which two processes can only compare their data values if they have a connection (i.e. active communication link) between them, and a process can only compare its value with one other value at the same time. In order to sort all the data values of a network, a sufficient number of these values have to be compared (i.e. a sufficient number of connections must be present) and appropriate actions must be taken according to the result of this comparison. Of course the way these values are compared and which actions will be taken accordingly are matters of *how* the program will achieve *what* it is to do, and must not be part of the specification. But conditions on the environment in which the program is required to operate are matters of *what* the program is to do, viz. under what circumstances it must fulfil its requirements. Although, at this stage, we may not make any assumptions whatsoever on how the program will achieve the requested results, nevertheless we have to take full account of the limitations the environment imposes upon the possible implementations of the program, by specifying and if necessary strengthening the properties of this environment. It is obvious that, in this case, we have to curtail the set of possible failures that can change the environment, and settle for convergence instead of self-stabilisation. As it happens, we cannot allow arbitrary communication links to fail. Consider for example the network in Figure 1, where the labels (written above the processes) are characters, and the data values (written inside the processes) are numbers. As a result of an environmental failure, the communication link between processes b and c is down. Let $\prec_p$ and $\prec_v$ be the lexicographic order on characters and the less-than-or-equal ($\leq$) order on numbers respectively.

There exists no implementation whatsoever which does not alter the multi-set of the processes's values and can sort this network, since:

- the values of processes a, c, d and e are already sorted according to Definition (1), so these processes shall not undertake any action[5]
- the same holds for the processes a and b
- process b cannot compare its value with any of processes c, d and e, so nothing will be done by process b either.

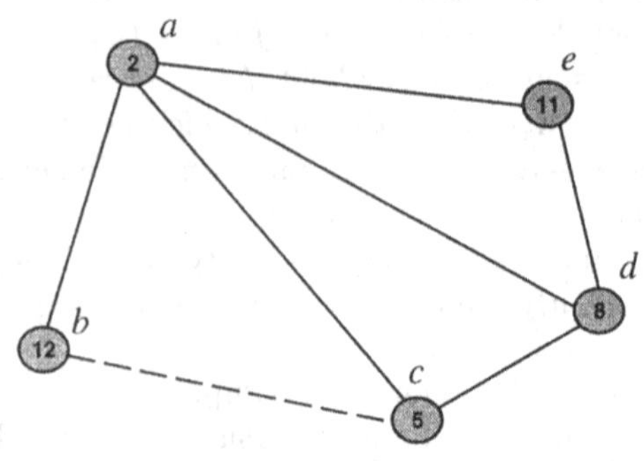

Fig. 1. A network which cannot be sorted.

Consequently, we must formalise a condition which states when there are still enough connections left for any implementation to sort the network. This condition then imposes a restriction on the set of failures from which the convergent program can recover. Moreover, we need to make the status (i.e. up or down) of the available communication links (i.e. C) dependent of the state such that the presence of connections becomes dynamic. To establish the latter, we introduce variables $\mathbf{aC}\ i\ j$ for each $(i, j) \in C$ of type boolean, and model link failures as follows:

$(s \circ \mathbf{aC})\ i\ j = \mathbf{true}$, if link $(i, j) \in C$ is up (i.e. an active Connection) in state s
$(s \circ \mathbf{aC})\ i\ j = \mathbf{false}$, if link $(i, j) \in C$ is down in state s

For all states s we define: $\mathbf{AC}\ s = \{(i, j) \mid (i, j) \in C \wedge (s \circ \mathbf{aC})\ i\ j\}$

A minimal and sufficient condition on the connections must imply that if the network is not yet sorted, then there must always be processes which recognise

[5] Note that by concluding that no action will be undertaken, we do *not* make assumptions about the eventual implementation. We merely use the fact that a convergent program is being designed, which means that if the program finds itself in the required situation it will stay there.

that their values are not sorted; in other words, among the pairs of connected processes whose labels are ordered by $\prec_p$, there must at least be one pair whose values are out-of-order. For, if this condition is satisfied, it will always be possible to undertake some action if the network is not yet sorted, and situations as sketched in the example above cannot arise. Before this condition is formalised, first the definition of a **W**rong **P**air of processes is given.

$$\mathbf{WP}^{(i,j)}\ s = i,j \in P \wedge (i \prec_p j) \wedge \neg(((s \circ D)\, i) \prec_v ((s \circ D)\, j)) \tag{2}$$

Obviously,

$$\neg\mathbf{Sorted}_{(P,D,\prec_p,\prec_v)}\ s \Leftrightarrow \exists i,j \in P : \mathbf{WP}^{(i,j)}\ s \tag{3}$$

Thus the condition, which we are looking for must satisfy in state s

$$\begin{array}{l}\mathbf{SufficientConnections}_{(\mathbf{AC},\prec_p)}\ s \Rightarrow \\ \quad \neg\mathbf{Sorted}_{(P,D,\prec_p,\prec_v)}\ s \Rightarrow \exists u,v \in P : \mathbf{WP}^{(u,v)}\ s \wedge (u,v) \in (\mathbf{AC}\ s)\end{array} \tag{4}$$

Now, we state that the following definition of $\mathbf{SufficientConnections}_{(\mathbf{AC},\prec_p)}$ satisfies(4), the proof of which shall be given below.

$$\mathbf{SufficientConnections}_{(\mathbf{AC},\prec_p)}\ s = (\prec_p \subseteq (\prec_p \cap (s \circ \mathbf{AC}))^{tr}) \tag{5}$$

In order to verify that Definition (5) of $\mathbf{SufficientConnections}_{(\mathbf{AC},\prec_p)}$ satisfies (4), assume that for all states s $(\prec_p \subseteq (\prec_p \cap (\mathbf{AC}\, s))^{tr})$ and $\neg\mathbf{Sorted}_{(P,D,\prec_p,\prec_v)}\, s$ hold.

From (3) and the second assumption we can deduce that there exist a $i,j \in P$, such that $\mathbf{WP}^{(i,j)}\ s$ (and thus $i \prec_p j$). Consequently, the first assumption tells us that $(i,j) \in (\prec_p \cap (\mathbf{AC}\ s))^{tr}$, i.e. $(i,j) \in (\prec_p \cap (\mathbf{AC}\ s))_N$, where $N = |P|$. We now prove that for all processes $i,j \in P$:

$$\mathbf{WP}^{(i,j)}\ s \wedge (i,j) \in (\prec_p \cap (\mathbf{AC}\ s))_N \Rightarrow (\exists u,v \in P : \mathbf{WP}^{(u,v)}\ s \wedge (u,v) \in (\mathbf{AC}\ s))$$

by induction on N.

INDUCTION BASE: case 1

Assume $\mathbf{WP}^{(i,j)}\ s$ and $(i,j) \in (\prec_p \cap (\mathbf{AC}\ s))_1$. Rewriting the second assumption gives us $(i,j) \in (\prec_p \cap (\mathbf{AC}\, s))$, i.e. $(i,j) \in (\mathbf{AC}\, s)$, which together with the first assumption establishes this case.

INDUCTION HYPOTHESIS: for all $M < N$ and for all processes $i,j \in P$:

$$\mathbf{WP}^{(i,j)} s \wedge (i,j) \in (\prec_p \cap (\mathbf{AC} s))_M \Rightarrow (\exists u,v \in P : \mathbf{WP}^{(u,v)} s \wedge (u,v) \in (\mathbf{AC} s))$$

INDUCTION STEP: case N

Assume $\mathbf{WP}^{(i,j)}\,s$ and $(i,j) \in (\prec_p \cap (\mathbf{AC}\,s))_N$. The second assumption gives us two cases:

- $(i,j) \in (\prec_p \cap (\mathbf{AC}\,s))_{N-1}$, this case is trivially proven by the **IH**.

or,

- $\exists k \in P : (i,k) \in (\prec_p \cap (\mathbf{AC}\,s))_{N-1} \wedge (k,j) \in (\prec_p \cap (\mathbf{AC}\,s))_{N-1}$, since $\mathbf{WP}^{(i,j)}\,s$, and thus $\neg((s \circ D) i) \prec_v ((s \circ D) j)$, we have, due to the transitivity of $\prec_v$, that $\neg((s \circ D)\,i) \prec_v ((s \circ D)\,k) \vee \neg((s \circ D)\,k) \prec_v ((s \circ D)\,j)$, again the **IH** proves this case.

□

Now, we are almost ready to construct the formal specification of what the program is to do. We have defined what the program must establish, i.e. sort a particular kind of network of processes, we have defined under which conditions it must achieve that, i.e. there is a total ordering on the processes data values; there is an anti-symmetric and transitive relation on the processes labels; and there is a restriction upon the failures that may occur (5). There is, however, one, important but obvious, thing that must be embodied in the specification of the sorting program. During the activity of sorting the network, we want the distribution of the values among the processes to remain a permutation of the initial distribution (i.e. the one with which the program started). If we do not require this, a simple program which just assigns the same value to all processes would achieve, by reflexitivity of $\prec_v$, a sorted network (according to Definition (1)). Obviously, this is not what we want. Consequently, we need a definition of permutation:

$$\begin{aligned}\mathbf{Permutation}_{(P,D,I)}\ s\ &= \exists f \in P \rightarrow P : (\mathbf{Bijection} f)\\ &\qquad \wedge (\forall i \in P : ((s \circ D)\,i) = (I(f\,i)))\end{aligned}$$

The formal specification[6] of the program in terms of the convergence operator can be found in Figure 2. Note that I is not a program variable, but a *proof variable*, i.e. introduced to reason about the program (namely the permutation-part). Consequently, $(D = I)$ is not an initial *program* condition, but an initial *specification* condition, which, intuitively, makes the specification state that:

IF *the program is started in some state s and I is a snapshot of the distribution of data values in that state s, and* $\mathbf{SufficientConnections}_{(\mathbf{AC},\prec_p)}$ *and* $\mathbf{Permutation}_{(P,D,I)}$ *are stable in the program,*

THEN *the program will eventually find itself in a situation (i,e, a state) in which the network is sorted.*

[6] Be aware of the overloading. $(D = I)$ means $(\lambda s.\forall i \in P : (s \circ D)\,i = I\,i)$, **Total** $(\prec_v, D)$, means $(\forall s : \mathbf{Total}\,(\prec_v, \{(s \circ D) i \mid i \in P\}))$, and $\wedge$ works on state-predicates.

Specification 1.

$$\frac{\begin{array}{c}\mathbf{Network}(P,C,D) \wedge \mathbf{Total}(\prec_v, D) \wedge (\prec_p \neq \emptyset)\\ \mathbf{AntiSymmetric}(\prec_p, P) \wedge \mathbf{Transitive}(\prec_p, P)\end{array}}{\begin{array}{l}(\ {}_{\mathsf{Sort}}\vdash \Box\mathbf{Permutation}_{(P,D,I)}) \wedge \\ (\ {}_{\mathsf{Sort}}\vdash \Box\mathbf{SufficientConnections}_{(\mathbf{AC},\prec_p)}) \wedge \\ (\mathbf{Permutation}_{(P,D,I)} \wedge \mathbf{SufficientConnections}_{(\mathbf{AC},\prec_p)} \\ \qquad\qquad\qquad {}_{\mathsf{Sort}}\vdash (D = I) \leadsto \mathbf{Sorted}_{(P,D,\prec_p,\prec_v)})\end{array}}$$

Fig. 2. Formal specification.

Actually, there are two different techniques we have used to deal our enemies.

External agents that tamper with data values are made latent by using the convergence operator and the specification condition $(D = I)$. For, if, during the execution of the program, some agent tampers with data values, we interpret the resulting state s as a new initial state (and thus I as a snapshot of the distribution of data values in that state s). Again the convergence operator guarantees that the network will eventually get sorted and stay sorted.

To make the daemons that disable or enable communication links transparent, this technique cannot be used, since the predicate **SufficientConnections** is not confined by the write variables of the sorting program[7], and thus cannot be put in conjunction with the initialisation-condition $(D = I)$. These daemons are made transparent by making the status of communications links dependent of the state of the program, by specifying that the program can only converge to the required situation if **SufficientConnections** is stable in the program, and, moreover, by stating that **SufficientConnections** holds in the initial state (i.e. it is an invariant). For, if, during the execution of the program, some daemon deactivates a communication link in such a way that **SufficientConnections** still holds the specification tells us that the network will eventually get sorted and remain sorted.

3.1 Refine the Specification

Program development in UNITY [CM88] consists of refining (i.e. adding detail) to specifications until the latter are solely expressed in terms of **ensures**. Subsequently, behaviour of individual actions of the program can be extracted.

The whole refinement of the progress-part of Specification 1, can be found in the full paper [VSP96]. Here, only the solution strategy[8] and the results are presented. The validity of $\mathbf{Network}(P,C,D)$, $\mathbf{Transitive}(\prec_p, P)$, $\mathbf{Total}(\prec_v, D)$ and $\mathbf{AntiSymmetric}(\prec_p, P)$ shall be implicitly assumed from now on. The

[7] Evidently, the write variables of the program shall be $\{D\,i \mid i \in P\}$

[8] That is, the strategy we want our program to employ in order to satisfy the specification.

solution strategy[9] which shall be used to refine Specification 1, is one that reduces the number of wrong pairs of processes.

$$\mathbf{WPs}_{(P,D,\prec_p,\prec_v)} s = \{(i,j) | i,j \in P \wedge (i \prec_p j) \wedge \neg(((s \circ D)\, i) \prec_v ((s \circ D)\, j))\}$$

In other words, during the execution of a program – which uses this strategy to sort a network – progress is ensured since the number of wrong pairs of processes reduces. In a sorted network there are no wrong pairs of processes (i.e. $|\mathbf{WPs}_{(P,D,\prec_p,\prec_v)} s| = 0$). Consequently, since $|\mathbf{WPs}_{(P,D,\prec_p,\prec_v)} s|$ is always a value from N_0, the less-than ($<$) is known to be a well-founded relation on N_0, and since the value of $|\mathbf{WPs}_{(P,D,\prec_p,\prec_v)} s|$ reduces during the execution of a program that exploits our solution strategy, the network shall eventually get sorted. Refining the progress-part of our specification according to this proposed solution strategy, leaves us with[10]:

$$\begin{array}{l} \forall m > 0 : \forall i,j \in P : \\ \quad (\lambda\, s.(|\mathbf{WPs}_{(P,D,\prec_p,\prec_v)} s| = m) \wedge \mathbf{WP}^{(i,j)}\, s \wedge (i,j) \in (\mathbf{AC}\, s)) \\ \qquad \textsf{ensures} \\ \quad (\lambda\, s.|\mathbf{WPs}_{(P,D,\prec_p,\prec_v)} s| < m) \end{array}$$

3.2 Construct a Program Which Satisfies this Refined Specification

Considering the properties of the network – principally the property that a process can only communicate with one other process at the same time – it is evident that the only thing two connected processes can do is compare their values and swap them if they are out-of-order with respect to the processes labels. The resulting program is presented in Figure 3. Note that the program performs a topological sort on the directed acyclic graph $G_{\prec_p} = (P, \{(u,v) \mid u \prec_p v\}$.

```
prog    Sort
read    {D i | i ∈ P} ∪{aC i j | (i, j) ∈ C}
write   {D i | i ∈ P}
init    SufficientConnections(AC,≺p)
assign  ‖i, j : (i, j ∈ P) ∧ (i ≺p j) ∧ (i, j) ∈ C ::
           if(aC i j) is true in current state
           then(D i), (D j) :=min≺v(D i, D j), max≺v(D i, D j)
        (*else skip*)
```

Fig. 3. The sorting program

[9] There are other possible solution strategies, see for example [CM88].

[10] Note that ◯and ensures work on state-predicates, hence the lambda-constructions.

3.3 Prove that the Program Satisfies the Specification

The main theorem that had to be verified in order to show that the program meets its specification, was that the program employs the solution strategy which was introduced in Section 3.1. That is, we must prove that *if* the program finds itself in some state s in which holds that there still exist wrong pairs of processes, *then if* t is the state in which the program results after swapping the data values of some connected wrong pair (which exists because of **SufficientConnections**), *then* the number of wrong pairs in state t is less than the number of wrong pairs in state s. More formally, given the following definition of swapping to values during the transition from s to t:

$$\begin{aligned}\textbf{Swapped}_{(s,t,P,D,\prec_p,\prec_v)} = {} & \exists i,j \in P : (i \prec_p j) \wedge \neg(((s \circ D)\, i) \prec_v ((s \circ D)\, j)) \\ & \wedge (\forall k \in P : (k \neq i \wedge k \neq j) \Rightarrow (s{\circ}D)\, k = (t{\circ}D)\, k) \\ & \wedge (t{\circ}D)\, j = (s{\circ}D)\, i \wedge (t{\circ}D)\, i = (s{\circ}D)\, j\end{aligned}$$

Then the main theorem that we had to prove was:

$$\frac{m \neq 0 \wedge |\mathbf{WPs}_{(P,D,\prec_p,\prec_v)} s| = m \wedge \mathbf{Swapped}_{(s,t,P,D,\prec_p,\prec_v)}}{|\mathbf{WPs}_{(P,D,\prec_p,\prec_v)} t| < m}$$

4 Parallel Composition and the Transparency Law

A consequence of the absence of ordering in the execution of a UNITY program is that the parallel composition of two programs can be modelled by simply merging the variables and actions of both programs. In UNITY parallel composition is denoted by $[\!]$. So $W_{P[\!]Q}$ is the union of W_P and W_Q. Moreover, two programs are called *write-disjoint* (denoted by $P \div Q$), if $(W_P \cap W_Q = \emptyset)$. The following is a very important law for write-disjoint programs:

$$\frac{P \div Q \ \wedge\ ({}_Q\vdash \circlearrowleft J)\ \wedge\ (J\ {}_P\vdash p \rightsquigarrow q)}{J\ {}_{P[\!]Q}\vdash p \rightsquigarrow q} \quad \textbf{(Transparency Law)}$$

Intuitively, if P and Q are write disjoint, then Q cannot write P's variables. So, if Q does not destroy the P's safety property (i.e. J), then Q cannot destroy P's progress (i.e. $J\ {}_P\vdash p \rightsquigarrow q$), since predicates p and q are confined by the write variables of P.

5 Making the Daemons Explicit

The second technique mentioned at the end of Section 3 resulted in implicit daemons that made communication links fail, by making the status of the links dependent on the state of the program. In this section we shall make these daemons explicit, in order to show that the technique is suitable for making these daemons tacit. The UNITY program modelling the daemons is given in Figure 4.

```
prog   Daemon
read   {aC i j | (i, j) ∈ C}
write  {aC i j | (i, j) ∈ C}
init   SufficientConnections(AC,≺p)
assign ‖i, j : (i, j ∈ P) ∧ (i, j) ∈ C ::
          if SufficientConnections(AC−{(i,j)},≺p) is true in current state
          then aC i j := ¬aC i j
         (*else skip*)
```

Fig. 4. The daemon

Consequently, we have to show that, if we execute our program Sort within this inimical environment, then the network still eventually gets sorted. So, we have to show that the following is satisfied:

Specification 2.

$$\frac{\begin{array}{c}\mathbf{Network}(P,C,D) \wedge \mathbf{Total}(\prec_v, V) \wedge (\prec_p \neq \emptyset)\\ \mathbf{AntiSymmetric}(\prec_p, P) \wedge \mathbf{Transitive}(\prec_p, P)\end{array}}{\begin{array}{l}(\ {}_{\mathsf{Sort}\|\mathsf{Daemon}}\vdash \Box\mathbf{Permutation}_{(P,D,I)})\wedge\\ (\ {}_{\mathsf{Sort}\|\mathsf{Daemon}}\vdash \Box\mathbf{SufficientConnections}_{(\mathbf{AC},\prec_p)})\wedge\\ (\mathbf{Permutation}_{(P,D,I)} \wedge \mathbf{SufficientConnections}_{(\mathbf{AC},\prec_p)}\\ \qquad\qquad {}_{\mathsf{Sort}\|\mathsf{Daemon}}\vdash (D = I) \rightsquigarrow \mathbf{Sorted}_{(P,D,\prec_p,\prec_v)})\end{array}}$$

Using the Transparency law on this specification leaves us with the following requirements for Sort, Daemon and their composition:

$$(\ {}_{\mathsf{Sort}\|\mathsf{Daemon}}\vdash \Box\mathbf{Permutation}_{(P,D,I)}) \wedge (\ {}_{\mathsf{Sort}\|\mathsf{Daemon}}\vdash \Box\mathbf{SufficientConnections}_{(\mathbf{AC},\prec_p)})$$

$$\wedge$$

$$\mathsf{Sort}\div\mathsf{Daemon} \wedge\ {}_{\mathsf{Daemon}}\vdash \circlearrowleft \mathbf{Permutation}_{(P,D,I)} \wedge \mathbf{SufficientConnections}_{(\mathbf{AC},\prec_p)}$$

$$\wedge$$

$$\mathbf{Permutation}_{(P,D,I)} \wedge \mathbf{SufficientConnections}_{(\mathbf{AC},\prec_p)}$$

$${}_{\mathsf{Sort}}\vdash (D = I) \rightsquigarrow \mathbf{Sorted}_{(P,D,\prec_p,\prec_v)}$$

Evidently, these requirements are satisfied (the last was already stated in Section 3.3). Consequently, our technique succecfully made our enemies transparent.

Acknowledgements The encouragement and guidance of Dr. I.S.W.B Prasetya are gratefully acknowledged.

References

[CM88] K.M. Chandy and J. Misra. *Parallel Program Design - A Foundation.* Addison-Wesley, 1988.

[GM93] Mike J.C. Gordon and Tom F. Melham. *Introduction to HOL.* Cambridge University Press, 1993.

[Pra95] W Prasetya. *Mechanically Supported Design of Self-stabilizing Algorithms.* PhD thesis, UU, Oct 1995.

[RW92] K.A. Ross and C. R. B. Wright. *Discrete Mathematics.* Prentice-Hall, 1992.

[VSP96] T.E.J. Vos, S.D. Swierstra, and I.S.W.B. Prasetya. *Formal Methods and Mechanical Verification Applied to the development of a Convergent Distributed Sorting Program.* Technical report, UU, 1996.

Optimal Fault-Tolerant ATM-Routings for Biconnected Graphs

Koichi Wada[1], Wei Chen[1], Yupin Luo[2] and Kimio Kawaguchi[3]

[1] Nagoya Institute of Technology,
Gokiso-cho, Syowa-ku, Nagoya 466, JAPAN
email:(wada,chen)@elcom.nitech.ac.jp

[2] Department of Automation, Tsinghua University,
Beijing 100084, P.R.China
email:luo@iris.au.tsinghua.edu.cn

[3] Osaka Institute of Technology,
1-79-1 Kitayama, Hirakata 573-01, JAPAN
email:kawaguci@is.oit.ac.jp

Abstract. We study the problem of designing fault-tolerant virtual path layouts for an ATM network which is a biconnected network of n processors in the surviving route graph model. The surviving route graph for a graph G, a routing ρ and a set of faults F is a directed graph consisting of nonfaulty nodes with a directed edge from a node x to a node y iff there are no faults on the route from x to y. The diameter of the surviving route graph could be one of the fault-tolerance measures for the graph G and the routing ρ. When a routing is considered as a virtual path layout, we can discuss the fault tolerance of virtual path layouts in the ATM network. In this paper, we show that we construct three routings for any biconnected graph such that the diameter of the surviving route graphs is optimal and they satisfy some desirable properties of virtual path layouts in ATM networks.

1 Introduction

Consider a communication network or an undirected graph G in which a limited number of link and/or node faults F might occur. A *routing* ρ for a graph defines at most one path called *route* for each ordered pair of nodes. We assume that it must be chosen without knowing which components might be faulty.

Given a graph G, a routing ρ and a set of faults F, the *surviving route graph* $R(G,\rho)/F$ is defined to be a directed graph consisting of all nonfaulty nodes in G, with a directed edge from a node x to a node y iff the route from x to y is intact. The diameter of the surviving route graph (denoted by $D(R(G,\rho)/F)$) could be one of the fault-tolerance measures for the graph G and the routing ρ [2, 4]. Many results have been obtained for the diameter of the surviving route graph [6, 12, 15, 17].

The Asynchronous Transfer Mode(ATM) is a promising technology for high-speed networking[13]. ATM is based on relatively small fixed size packets called cells. Cells are routed through a layout of virtual paths(VPs), as well as sequences of such virtual paths, called virtual channels(VCs). If we consider virtual path layouts as routings in the surviving route model, we can discuss the fault-tolerance of ATM networks in the same model. [3] has given a graph-theoretical model that captures the characteristics of ATM networks, especially virtual paths. The virtual path layouts in ATM networks must have the following properties[3]. (1) The number of VPs which is contained in any VC should be small. (2) The number of virtual paths passing through an edge (called *edge-load*) should be as least as possible. (3) The number of occupied entries in the VP routing table should be low enough in the network. In that model, some virtual path layout problems are treated for trees [3, 9], chains [8] and complete graphs with faulty links [7].

When we consider the fault tolerance of virtual path layouts in ATM networks, in order that routings satisfy the constraints stated above, the diameter of the surviving route graph should be small(for (1)), the maximum of the number of routes defined for a node (called *route degree of the routing*) and the total number of routes defined in the routing should be as least as possible, because the size of routing tables is dominated by the route degree of the routing and the edge-load of the routing is dependent on the total number of routes and is bounded to it. And furthermore, if there is an edge between two nodes for which a virtual path must be defined, the virtual path should be defined as the edge(we call such routings *edge-routings*). Edge-routings decrease their edge-load and the sizes of the routing tables.

If the diameter of the surviving route graph is minimal over all routings on a given graph, the routing is said to be *optimal.* It has been shown that an optimal routing ρ can be constructed for any biconnected graph [18]. However, from the viewpoint of virtual path layouts in ATM-networks, the routing ρ is far from the satisfaction. The total number of routes defined in ρ is $n(n-1)$ and the route degree of ρ is $n-1$. Furthermore, ρ is not even an edge-routing.

In this paper, we improve the routing ρ and we construct the following routings which satisfy the ATM-routing requirements.

1. an edge-routing λ in which the total number of routes is $O(n \log n)$ and $D(R(G,\lambda)/\{f\}) \leq 2$ for any fault f.
2. an edge-routing π in which the total number of routes is $O(n)$ and $D(R(G,\pi)/\{f\}) \leq 2$ for any fault f.
3. an edge-routing ν in which the total number of routes is $O(n)$, the route degree is $O(\sqrt{n})$ and $D(R(G,\nu)/\{f\}) \leq 3$ for any fault f.

The routings λ and π are optimal. Because as long as faults are assumed to occur in a network, the diameter of the surviving route graph is more than one. The difference between λ and π is that λ has the property that for any node pair x and y, $\lambda(x,y) = \lambda(y,x)$ (we call such a routing *bidirectional*) but π is not bidirectional. In order to make the routing λ bidirectional, $O(n \log n)$ routes are needed in it. On the other hand, if we remove the condition that routings are bidirectional, $O(n)$ routes are sufficient to obtain an optimal routing. We also show that the routing ν is optimal if we consider all the routings in which the total number of routes is $O(n)$ and the route degree is $O(\sqrt{n})$. That is, if we consider such routings, the diameter of the surviving route graph is at least three.

2 Preliminary

In this section, we give definitions and terminology. We refer readers to [10] for basic graph terminology.

Unless otherwise stated, we deal with an undirected graph $G = (V, E)$ that corresponds to an ATM network. For a node v of G, $N_G(v) = \{u|(v,u) \in E\}$ and $deg_G(v) = |N_G(v)|$. $deg_G(v)$ is called *degree* of v and if G is apparent it is simply denoted by $deg(v)$. A graph G is *k-connected* if there exist k node-disjoint paths between every pair of distinct nodes in G. Usually 2-connected graphs are called *biconnected graphs.* The *distance* between nodes x and y in G is the length of the shortest path between x and y and is denoted by $dis_G(x,y)$. The *diameter* of G is the maximum of $dis_G(x,y)$ over all pairs of nodes in G and is denoted by $D(G)$. Let $P(u,v)$ and $P(v,w)$ be a path from u to v and a path from v to w, respectively. In general, even if both $P(u,v)$ and $P(v,w)$ are simple, the concatenation of $P(u,v)$ and $P(v,w)$ is not always simple. Thus we consider two kinds of concatenation: one is a usual concatenation (denoted by $P(u,v)\cdot P(v,w)$) and the other is a special concatenation (denoted by $P(u,v) \odot P(v,w)$), which is defined as the the shortest path from u to w in the graph $P(u,v) \cup P(v,w)$ to make the concatenated path simple.

Let $G = (V, E)$ be a graph and let x and y be nodes of G. Define $P_G(x,y)$ to be the set of all simple paths from the node x to the node y in G, and $P(G)$ to be the set of all simple paths in G. A *routing* is a partial function $\rho : V \times V \to P(G)$ such that $\rho(x,y) \in P_G(x,y)(x \neq y)$. The path specified to be $\rho(x,y)$ is called the *route from x to y*. The length of the route $\rho(x,y)$ is denoted by $|\rho(x,y)|$. For the routes $\rho(x_{i-1},x_i)(1 \leq i \leq p)$, define $[\rho(x_0,x_1),\rho(x_1,x_2),\ldots,\rho(x_{p-1},x_p)]$ to be $\rho(x_0,x_1)\cdot\rho(x_1,x_2)\cdot\ldots\cdot\rho(x_{p-1},x_p)(p \geq 1)$. We call $[\rho(x_0,x_1),\rho(x_1,x_2),\ldots,\rho(x_{p-1},x_p)]$ a *route sequence of length p from x_0 to x_p*.

For a graph $G = (V, E)$, let $F \subseteq V \cup E$ be a set of nodes and edges called a set of *faults.* We call $F \cap V(= F_V)$ and $F \cap E(= F_E)$ the set of *node faults* and the set of *edge faults*, respectively. If an object such as a route or a route sequence does not contain any element of F, the object is said to be *fault free.*

For a graph $G = (V, E)$, a routing ρ on G and a set of faults $F(= F_V \cup F_E)$, the *surviving route graph*, $R(G,\rho)/F$, is a directed graph with node set $V - F_V$ and edge set $E(G,\rho,F) = \{< x,y > |\rho(x,y) \; is \; defined \; and \; fault \; free\}$. In what follows, unless confusion arises we use notations for directed graphs as the same ones for undirected graphs.

In the surviving route graph $R(G,\rho)/F$, when $F = \phi$ the graph is called the route graph. In the route graph, the outdegree of a node v is called *the route degree of a node v* and the maximum of the route degree of all nodes is called *the route degree of the routing* ρ. If the number of (directed) edges in the route graph is m, the routing ρ is called *m-route-routing* or simply *m-routing.*

For a graph $G = (V, E)$ and a routing ρ, if for any edge (x, y) in G, $\rho(x, y)$ is defined and is assigned to the edge, ρ is called *edge-routing.*

A routing ρ is a *bidirectional* routing if $\rho(x, y) = \rho(y, x)$ for any node pair (x, y) in the domain of ρ. If a routing is not bidirectional, it is called *unidirectional.* Note that if the routing ρ is bidirectional, the surviving route graph $R(G, \rho)/F$ can be represented as an undirected graph.

Given a graph G and a routing property P, A routing ρ on G is *optimal* with respect to P if $max_{F s.t. |F| \leq k}$ $(D(R(G, \rho)/F))$ is minimal over all routings on G satisfying P. Note that from the definition of the optimality, if $D(R(G, \rho)/F)$ is 2 for any set of faults F such that $|F| \leq k$, the routing is obviously optimal with respect to any property. If the property P is known, we simply call the routing is optimal.

3 s-t Numbering

An s-t numbering for a biconnected graph is developed in the linear time algorithm for testing planarity of a graph [5] and it is used to solve several graph problems in linear time such as bipartition of biconnected graphs [16] and 2-path tree problem [11]. In the construction of the optimal routing for biconnected graphs, the s-t numbering plays an important role [18], and we also use it to construct the optimal ATM-routings shown in this paper.

Given an edge (s, t) of a biconnected graph $G = (V, E)$, a bijective function $g : V \rightarrow \{1, 2, \ldots, |V| = n\}$ is called an *s-t numbering* if the following conditions are satisfied:

- $g(s) = 1$, $g(t) = n$ and
- Every node $v \in V - \{s, t\}$ has two adjacent nodes u and w such that $g(u) < g(v) < g(w)$.

For a node v in G and an s-t numbering g, we define two paths $P_I[v, t]$ and $P_D[v, s]$ as follows:

(1) $P_I[v, t] = (v_0(= v), v_1, \ldots, v_k(= t))$, where $v \neq s$ and $g(v_i) = max\{g(u) | u \in N_G(v_{i-1})\} (1 \leq i \leq k)$ and

(2) $P_D[v, s] = (v_0(= v), v_1, \ldots, v_k(= s))$, where $v \neq t$ and $g(v_i) = min\{g(u) | u \in N_G(v_{i-1})\} (1 \leq i \leq k)$.

Note that if (v, s) and (v, t) are in E, $P_D[v, s] = (v, s)$ and $P_I[v, t] = (v, t)$ from the definition.

From the definition of the s-t numbering, two paths $P_I[v, t]$ and $P_D[v, s]$ are well defined and $P_I[x, t]$ and $P_D[x, s]$ are node-disjoint for any node $x(\neq s, t)$, which can be easily derived.

4 Optimal Routing ρ

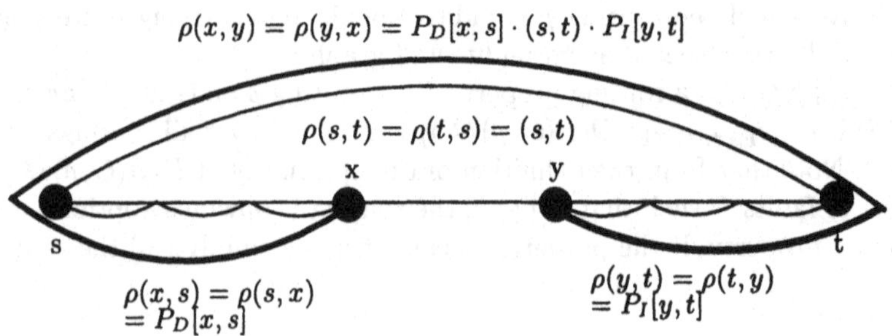

Fig. 1. The optimal routing ρ.

It is shown in [18] an optimal routing ρ can be constructed for any biconnected graph. Fig. 1 shows the routing ρ for a biconnected graph whose nodes are s-t numbered. In Fig. 1, nodes are located from left to right with the s-t numbering. The routing ρ is not only optimal for any biconnected graph but also can be extended to an optimal routing for any connected graph[18]. However from the viewpont of ATM-routings, ρ does not satisfy several desirable properties. In particular, since the routing ρ is defined as a total function, ρ is an $n(n-1)$-routing and the route degree of ρ is $n-1$. Furthermore, the routing ρ is not an edge-routing. In fact, if ρ is modified to an edge-routing, ρ can not be optimal. Consider a biconnected graph $G = (V, E)$ in which $(u, v) \in E$, $(u, s) \in E$ and $(v, t) \in E$ and $deg(u) = deg(v) = 2$. If ρ is an edge-routing, $\rho(u, v)$ must be the edge (u, v). Also for any z such that $g(s) < g(z) < g(u)$ the route $\rho(z, u)$ contains the edge (u, v) and for any w such that $g(v) < g(w) < g(t)$ the route $\rho(w, v)$ contains the edge (u, v). Thus if the edge (u, v) becomes faulty, there does not exist a node x such that both $\rho(u, x)$ and $\rho(x, v)$ are fault free and the distance between u and v in the surviving route graph is 3 by using the fault free routes $\rho(u, s)$, $\rho(s, t)$ and $\rho(v, t)$.

In the following three sections, we will construct three optimal routings for biconnected graphs with desirable properties as ATM-routings. Unlike ρ, these three routings are edge-routings. The first one is a bidirectional $O(n \log n)$-routing. In some ATM model [3], it is assumed that routings are bidirectional. Although it is not necessary that routings should be bidirectional for ATM-routings, the difference of constructing optimal routings between bidirectional routings and unidirectional ones seems to be fairly large and therefore it is theoretically interesting whether there exists an optimal and bidirectional routing with $o(n^2)$ routes or not. For example, we can easily construct an optimal routing for hypercube graphs if it is not bidirectional [4]. However, we have shown that it is very difficult to construct an optimal and bidirectional routing for hypercube graphs [12].

Next, we show that we can construct an optimal $O(n)$-routing, if the assumption that routings are bidirectional is removed. This routing is also optimal from the viewpoint of the number of routes because the route graph is not connected if the routing has routes fewer than n.

The last routing is more desirable as ATM-routings. The routing is not only an optimal and bidirectional $O(n)$-routing but also its route degree is $O(\sqrt{n})$. On the other hand, the first two routings must have $\Omega(n)$ route degree.

5 Optimal and Bidirectional O(n log n)-Routing

In this section, we construct an optimal edge-routing λ. The routing λ is bidirectional $O(n \log n)$-routing.

It is easily shown that we can not construct any optimal and bidirectional edge-routing for the 4-node cycle graph. See [12]. Thus in what follows, we assume that the number of nodes is at least 5.

We define the routing λ for a biconnected graph $G = (V, E)$ with n nodes and m edges on which an s-t numbering g is given. Let s' be a node such that $g(s') = 2$ and let t' be a node such that $g(t') = n-1$. The nodes of $V - \{s, t, s', t'\}$ are denoted by $b_0, \ldots, b_{q-1}, a_1, \ldots, a_p$, with the s-t numbering from left to right, where $p + q = n - 4$ and p is the least number p' such that $p' \geq \log(n - 4 - p')$. Note that $p = O(\log n)$ and the index j of b_j can be represented with p-bit number $j = j_p \ldots j_1$.

routing λ

If$(x, y) \in E$ then $\lambda(x, y) = \lambda(y, x) := (x, y)$.
Otherwise,

1. For $x \neq t$, $\lambda(s, x) = \lambda(x, s) := P_D[x, s]$.
2. For $x \neq s$, $\lambda(x, t) = \lambda(t, x) := P_I[x, t]$.
3. $\lambda(s', t') = \lambda(t', s') := (s', s) \cdot (s, t) \cdot (t, t')$.
4. For $x \neq s, t, t'$,

$$\lambda(s', x) = \lambda(x, s') := \begin{cases} (x, s) \cdot (s, s') & \text{if } (x, s) \in E \\ P_I[x, t] \cdot (t, s) \cdot (s, s') & \text{otherwise.} \end{cases}$$

5. For $x \neq s, s', t$,

$$\lambda(x, t') = \lambda(t', x) := \begin{cases} (x, t) \cdot (t, t') & \text{if } (x, t) \in E \\ P_D[x, s] \cdot (s, t) \cdot (t, t') & \text{otherwise.} \end{cases}$$

6. For i and $j (1 \leq i < j \leq p)$, $\lambda(a_i, a_j) = \lambda(a_j, a_i) := P_D[a_i, s] \cdot (s, t) \cdot P_I[a_j, t]$.

7. For $i(1 \leq i \leq p)$ and $j(0 \leq j \leq q-1)$,
   let $j_p \ldots j_1$ be the p-bit number of j, then

$$\lambda(a_i, b_j) = \lambda(b_j, a_i) := \begin{cases} P_D[b_j, s] \cdot (s,t) \cdot P_I[a_i, t] & \text{if } j_i = 0 \\ P_I[b_j, t] \odot P_I[a_i, t] & \text{if } j_i = 1. \end{cases}$$

8. For $i(1 \leq i \leq q-1)$, $\lambda(b_0, b_i) = \lambda(b_i, b_0) := P_D[b_i, s] \odot P_D[b_0, s]$.

Note that the notation $\odot$ for the path concatenation is effective in 7. and 8., because for example in 7. if $P_I[b_j, t]$ and $P_I[a_i, t]$ have common nodes except t, $P_I[b_j, t] \odot P_I[a_i, t]$ is defined as the shortest path in $P_I[b_j, t] \cup P_I[a_i, t]$.

It is easily verified that λ is a bidirectional edge-routing. It is also checked that λ is an $O(m+n \log n)$-routing, and its route degree is $O(n)$, because $p = O(\log n)$ and all the route between b_i and b_j are not defined. The number of routes in λ can be reduced to $O(n \log n)$ because for any biconnected graph with n nodes and m edges, the number of edges can be reduced to $O(n)$ with preserving the biconnectivity [14].

Now we prove that λ is an optimal routing. Intuitively, λ is based on the routing ρ [18] and in order to reduce the number of routes and to make the routing λ an optimal edge-routing, s' ,t' and a_i are introduced. We make use of a_i so that two nodes between which the route is not defined are connected with distance two(see Lemma 2). Also the route between $s'(t')$ and $x(\neq s, s', t', t)$ in the above definition 4.(5.) are defined to make λ an optimal edge-routing.

For a node or an edge $w \in V \cup E$ and two nodes x and y such that $g(x) < g(y)$, if one of the following conditions is satisfied, we write $w \in [x, y]$.

1. $w \in V$ and $g(x) \leq g(w) \leq g(y)$ or
2. $w = (w_1, w_2) \in E$, $g(x) \leq min(g(w_1), g(w_2))$ and $max(g(w_1), g(w_2)) \leq g(y)$.

The following is a key lemma to show that routings are optimal.

Lemma 1. *Let $G = (V, E)$ be a biconnected graph whose nodes are s-t numbered with g. For a routing σ and two nodes $x_i (i = 1, 2)$ in G such that $g(x_1) < g(x_2)$, if $\sigma(x_i, s)$ and $\sigma(s, x_i)$ are defined with elements only in $[s, x_i]$, $\sigma(x_i, t)$ and $\sigma(t, x_i)$ are defined with elements only in $[x_i, t]$ and paths $P(x_1, x_2)$ and $P(x_2, x_1)$ consist of elements only in $[s, x_1]$ and $[x_2, t]$ then at least one of the two route sequences and the path $[\sigma(x_1, s), \sigma(s, x_2)]$, $[\sigma(x_1, t), \sigma(t, x_2)]$, and $P(x_1, x_2)$ (or $\sigma(x_2, s) \cdot \sigma(s, x_1)$, $\sigma(x_2, t) \cdot \sigma(t, x_1)$, and $P(x_2, x_1)$) is fault-free for any fault f in G.*

Proof. We only show the case from x_1 to x_2, since the other case can be proved similarly. Note that when this lemma is considered, it is sufficient to consider faulty nodes and faulty edges only in $[s, x_1]$, $[x_1, x_2]$ and $[x_2, t]$. Because other edges(such as (e_1, e_2) ($e_1 \in [s, x_1]$ and $e_2 \in [x_2, t]$)) are never uesd to define routes. If f is the edge (s, t) or $f \in [s, x_1]$, $[\sigma(x_1, t), \sigma(t, x_2)]$ is fault free. If $f \in [x_2, t]$, then $[\sigma(x_1, s), \sigma(s, x_2)]$ is fault free. Otherwise($f \in [x_1, x_2]$), the path $P(x_1, x_2)$ does not contain the fault f from the definition of $P(x_1, x_2)$.

The following lemma shows that there is a node via which b_i and b_j are connected with distance 2 if the fault is in $[b_i, b_j]$.

Lemma 2. *For the routing λ and two nodes b_i and b_j $(0 \le i < j \le q-1)$, there is a node a_k such that $\lambda(b_i, a_k) = P_D[b_i, s] \cdot (s,t) \cdot P_I[a_k, t]$ and $\lambda(a_k, b_j) = P_I[b_j, t] \odot P_I[a_k, t]$.*

Proof. Let $i_1 \dots i_p$ and $j_1 \dots j_p$ be the p-bit representation of i and j, respectively. Since $i_1 \dots i_p < j_1 \dots j_p$, there exists a bit position k such that $i_k = 0$ and $j_k = 1$. From the definition of λ, the lemma follows.

We are ready to prove that λ is optimal.

Theorem 3. *Let G be a biconnected graph with $n(\ge 5)$ nodes. Then $D(R(G,\lambda)/\{f\}) \le 2$ for any fault f in G.*

Proof. Let $R = R(G,\lambda)/\{f\}$. Let x and y be any pair of distinct nonfaulty nodes in G.

(1) Suppose that $x = s$ and $y = t$. If f is not the edge (s,t) then $dis_R(s,t) = 1$. Otherwise, the route sequence $[\lambda(s,s'), \lambda(s',t)] = (s,s') \cdot P_I[s',t]$ can not contain f. Thus, $dis_R(s,t) \le 2$.

In the remaining cases, we must consider the two subcases whether $(x,y) \notin E$ or $(x,y) \in E$.

(2) Suppose that $x \ne s$ and $y = t$. If $(x,t) \notin E$, there are two node-disjoint route sequences $\lambda(x,t) = P_I[x,t]$ and $[\lambda(x,s), \lambda(s,t)] = P_D[x,s] \cdot (s,t)$. Since at most one of the route sequences can be faulty by one fault, $dis_R(x,t) \le 2$. In the case that $(x,t) \in E$, since $P_I[x,t] = (x,t)$, the same way can be used.

(3) For the case that $x = s$ and $y \ne t$, it can be proved similar to the case (2).

(4) Suppose that $x = s'$ and $y = t'$. For the case that $(s',t') \notin E$, at least one of the route sequences $[\lambda(s',s), \lambda(s,t')]$, $[\lambda(s',t), \lambda(t,t')]$ and the route $\lambda(s',t')$ is fault free because λ (and s' and t') satisfy the conditions of Lemma 1. Thus, $dis_R(s',t') \le 2$.

In the case that $(s',t') \in E$, if $f \ne (s',t')$ then $dis_R(s',t') = 1$. Otherwise$(f = (s',t'))$, since $n \ge 5$ there is a node z such that $g(s') < g(z) < g(t')$. From the definition of λ, both $\lambda(s',z)$ and $\lambda(z,t')$ are fault free independently of the condition whether there is an edge between z and s (or t) or not. Thus, $dis_R(s',t') \le 2$.

(5) Suppose that $x = s'$ and $y \ne t, t'$. The case that $(s',y) \notin E$ can be proved with Lemma 1 similar to the case (4). The difference is that $\lambda(y,s')$ is differently defined according to the condition whether $(y,s) \in E$ or not: $P_I[y,t] \cdot (t,s) \cdot (s,s')$(if $(y,s) \notin E$) and $(y,s) \cdot (s,s')$(if $(y,s) \in E$). However, this route also satisfies the condition of Lemma 1, that is this route does not contain any element in $[s',y]$.

We consider the case that $(s',y) \in E$. If $f \ne (s',y)$ it is obvious. Else there is a node z such that $(y,z) \in E$ and $g(y) < g(z)$. If $z \ne t$ then $\lambda(z,s') = P_I[z,t] \cdot (t,s) \cdot (s,s')$ is fault free. Since $\lambda(y,z) = (y,z)$, $dis_R(s',y) \le 2$. If

$z = t$, then $\lambda(y,t') = (y,t)\cdot(t,t')$ is fault free. Since $\lambda(t',s')$ is also fault free, $dis_R(s',y) \le 2$.

(6) The case that $x \ne s, s'$ and $y = t'$ can be similarly proved if the role of s' is interchanged with that of t' in the proof of (5).

(7) Suppose that $x = a_i$ and $y = a_j (1 \le i < j \le p)$ Since the case that $(a_i,a_j) \notin E$ can be also proved with Lemma 1 similar to the case (4), we consider the case $(a_i,a_j) \in E$. Since if $f \ne (a_i,a_j)$ it is obvious, we consider the case that $f = (a_i,a_j)$. In this case there exists a node z such that $(a_j,z) \in E$ and $g(a_j) < g(z)$. Similar to the case (5), we have $dis_R(a_i,a_j) \le 2$.

(8) Suppose that $x = b_i (0 \le i \le q-1)$ and $y = a_j (1 \le j \le p)$. For the case that $(b_i,a_j) \notin E$, Lemma 1 is used. When this case is applied to Lemma 1, as the path between b_i and a_j the route sequence $[\lambda(b_i,b_0),\lambda(b_0,a_j)]$(if $i \ne 0$) or the route $\lambda(b_0,a_j)$(if $i = 0$) is considered instead of the route $\lambda(b_i,a_j)$. The route $\lambda(b_i,a_j)(i \ne 0)$ may contain elements in $[b_i,a_j]$ when j-th bit of the p-bit number i is equal to 1. However the route sequence $[\lambda(b_i,b_0),\lambda(b_0,a_j)]$(if $i \ne 0$) and the route $\lambda(b_0,a_j)$(if $i = 0$) does not contain any element in $[b_i,a_j]$. Thus, $dis_R(b_i,a_j) \le 2$. For the case that $(b_i,a_j) \in E$ and $f = (b_i,a_j)$, it can be proved similar to the $(b_i,a_j) \notin E$ by using the route sequence $[\lambda(b_i,b_0),\lambda(b_0,a_j)]$(if $i \ne 0$) and the route $\lambda(b_0,a_j)$(if $i = 0$).

(9) The case that $x = b_i$ and $y = b_j (0 \le i < j \le q-1)$. can be proved similar to the case (8) by using Lemma 1 and 2.

6 Optimal O(n)-Routing

In this section, we show that we can construct an optimal $O(n)$-routing π if we remove the condition that routings are bidirectional.

In λ of the preceding section, since $\lambda(a_i,b_j)$ and $\lambda(b_j,a_i)$ must be equal, that is, the route is assigned to one of the two paths $P_D[b_j,s]\cdot(s,t)\cdot P_I[a_i,t]$ and $P_I[b_j,t] \odot P_I[a_i,t]$, $O(\log n)$ a_i are necessary for a fixed j in order that Lemma 2 holds. However, if the routing is not bidirectional, such a_i's are not needed and s' (or t') can play such roles instead of a_i's by assigning $\pi(s',x)$ and $\pi(x,s')$(or $\pi(t',x)$ and $\pi(x,t')$) to different paths.

For the case of a 4-node cycle graph, we can define an optimal edge-routing since a 4-node cycle graph is isomorphic to the 2-dimensional hypercube C_2 and it is shown that an optimal and edge-routing can be defined for C_2[4]. Thus, the routing π is assumed to be defined for a biconnected graph $G = (V,E)$ with $n(\ge 5)$ nodes. Also we can assume that $|E| = O(n)$.

routing π

If$(x,y) \in E$ then $\pi(x,y) = \pi(y,x) := (x,y)$.
Otherwise,

1. For $x \ne t$, $\pi(s,x) = \pi(x,s) := P_D[x,s]$.
2. For $x \ne s$, $\pi(x,t) = \pi(t,x) := P_I[x,t]$.

3. For $x \neq s, t$,

$$\pi(s', x) := \begin{cases} (x, s) \cdot (s, s') & \text{if } (x, s) \in E \\ (s', s) \cdot (s, t) \cdot P_I[x, t] & \text{otherwise.} \end{cases}$$

4. For $x \neq s, t, t'$, $\pi(x, s') := P_D[x, s] \odot (s, s')$.
5. For $x \neq s, t$,

$$\pi(t', x) := \begin{cases} (t', t) \cdot (t, x) & \text{if } (x, t) \in E \\ (t', t) \cdot (t, s) \cdot P_D[x, s] & \text{otherwise.} \end{cases}$$

6. For $x \neq s, s', t$, $\pi(x, t') := P_I[x, t] \odot (t, t')$.

It is easily shown that π is an edge-routing and an $O(n)$-routing but a unidirectional one. Note that since π is not bidirectional, when we prove the optimality of π, we must consider the any ordered pair x and y of nonfaulty nodes.

The following lemma is easily verified from the definition.

Lemma 4. *Let x and y be two nodes such that $g(x) < g(y)$, $x \neq s$ and $y \neq t$. If the fault is in $[x, y]$, the route sequences $[\pi(x, s'), \pi(s', y)]$ and $[\pi(y, t'), \pi(t', x)]$ are fault free*[4].

Theorem 5. *Let G be a biconnected graph with $n(\geq 5)$ nodes. Then $D(R(G, \pi)/\{f\}) \leq 2$ for any fault f in G.*

Proof. Let $R = R(G, \pi)/\{f\}$. Let x and y be any pair of distinct nonfaulty nodes in G.

(1) $x = s$ and $y = t$, (2) $x \neq s$ and $y = t$ and (3) $x = s$ and $y \neq t$ are proved same as that of Theorem 3. Since unidirectional routes are not used in these proofs, the cases from y to x can be also proved.

(4) Suppose that $x = s'$ and $y = t'$. For the case that $(s', t') \notin E$, it can be shown similar to the case (4) in the proof of Theorem 3 with Lemma 1. And since unidirectional routes are not used in the proof, the case from y to x is also proved.

In the case that $(s', t') \in E$, if $f \neq (s', t')$ then $dis_R(s', t') = dis_R(t', s') = 1$. Otherwise($f = (s', t')$), since $n \geq 5$ there is a node z such that $g(s') < g(z) < g(t')$. From the definition of π, both $\pi(s', z)$ and $\pi(z, t')$ (from s' to t') and both $\pi(t', z)$ and $\pi(z, s')$ (from t' to s') are fault free. Thus, $dis_R(s', t') \leq 2$ and $dis_R(t', s') \leq 2$.

(5-1) The case from $x = s'$ to $y \neq t, t'$ is considered. The case that $(s', y) \notin E$ can be proved with Lemma 1 similar to the case (5) in the proof of Theorem 3. We consider that $(s', y) \in E$. If $f \neq (s', y)$ then it is obvious. Else there is a node z such that $(y, z) \in E$ and $g(y) < g(z)$. If $z \neq t$ then $\pi(s', z) = (s's) \cdot (s, t) \cdot P_I[z, t]$ is fault free. Since $\pi(y, z) = (y, z)$, $dis_R(s', y) \leq 2$. If $z = t$, then $\pi(t', y) = (t', t) \cdot (t, z)$ is fault free. Since $\pi(s', t')$ is also fault free, $dis_R(s', y) \leq 2$.

[4] If $x = s'$ or $y = t'$ then the route $\pi(s', y)$ or $\pi(t', x)$ is fault free.

(5-2) The case from $y \neq t, t'$ to $x = s'$ is considered. The both cases $(s', y) \notin E$ and $(s', y) \in E$ can be proved with Lemma 1 similar to the case (5) in the proof of Theorem 3.

(6) The case from $x \neq s, t'$ to $y = t'$ and that from $y = t'$ to $x \neq s, t'$ can be similarly proved if the role of s' is interchanged with that of t' in the proof of (5-1) and (5-2), respectively.

(7) The case that $x \neq s, s', t', t$ and $y \neq s, s', t', t$ can be similarly proved with Lemma 1 and 4

7 Optimal O(n)-Routing with Route Degree o(n)

Although the number of routes defined in the routings λ or π is $o(n^2)$, the route degree of λ or π is $n-1$. In this section, we construct an $O(n)$-routing ν with route degree $O(\sqrt{n})$ such that the diameter of the surviving route graph is three for any one fault. We also show that the routing is optimal in $O(n)$-routings with route degree $O(\sqrt{n})$.

Theorem 6. *Let a routing σ on G be an $O(n)$-routing. If the route degree of σ is at most $o(n)$, $D(R((G,\sigma)/\{f\}) \geq 3$ for any fault f .*

Proof. We show that this statement holds even if there is no fault in G. Assume that the diameter of the route graph is at most 2. The least number of edges of a graph of n nodes with degree k and diameter 2 is $\Omega(n^2/k)$ [1]. Since the route degree of σ is $o(n)$, the number of edges in the route graph must be $\omega(n)$. It is a contradiction.

We construct an $O(n)$-routing ν with route degree $O(\sqrt{n})$ which attains the lower bound in Theorem 6. The routing ν is a hierarchical one based on the optimal routing ρ in [18].

We assume that an s-t numbering g is defined on a biconnected graph $G = (V, E)(|V| \geq 5)$. We divide the nodes of G into $\ell = \lfloor n/k \rfloor$ sections of size k each except the last section. Note that the last section contains at most $2k-1$ nodes. For each section denoted by $V_i(1 \leq i \leq \ell)$, the least numbered node and the largest numbered node are denoted by s_i and t_i, respectively. Note that $s = s_1$ and $t = t_k$.

routing ν

1. For $x \in V_i(1 \leq i \leq \ell)$ such that $g(s_i) < g(x) < g(t_i)$,
   $\nu(x, s_i) = \nu(s_i, x) := P_D[x, s] \odot P_D[s_i, s]$ and
   $\nu(x, t_i) = \nu(t_i, x) := P_I[x, t] \odot P_I[t_i, t]$, and
   $\nu(x, s_{i+1}) = \nu(s_{i+1}, x) := P_D[x, s] \odot P_D[s_{i+1}, s]$(if $i < \ell$) and
   $\nu(x, t_{i-1}) = \nu(t_{i-1}, x) := P_I[x, t] \odot P_I[t_{i-1}, t]$(if $1 < i$).
2. For $i, j(1 \leq i < j \leq \ell)$,
   $\nu(s_i, s_j) = \nu(s_j, s_i) := P_D[s_i, s] \odot P_D[s_j, s]$ and
   $\nu(t_i, t_j) = \nu(t_j, t_i) := P_I[t_i, t] \odot P_I[t_j, t]$.

3. For $i, j (1 \le i \le j \le \ell)$,

$$\nu(s_i, t_j) = \nu(t_j, s_i) := \begin{cases} (s_1, t_\ell) & \text{if } i = 1 \text{ and } j = \ell \\ P_D[s_1, t_j] & \text{if } i = 1 \text{ and } j \neq \ell \\ P_I[s_i, t_\ell] & \text{if } i \neq 1 \text{ and } j = \ell \\ P_D[s_i, s] \cdot (s, t) \cdot P_I[t_j, t] & \text{otherwise.} \end{cases}$$

4. For x and y such that the routes $\nu(x, y) = \nu(y, x)$ are defined in 1.–3., if $(x, y) \in E$ then $\nu(x, y) = \nu(y, x)$ is changed to (x, y).

The notation $\odot$ for the path concatenation is used in 1. and 2., because for example s_i and s_j are not connected by using elements only in $[s_i, s_j]$ in general and s_i and s_j are connected via s similar to the routing λ.

It is easily verified that ν is a bidirectional edge-routing and an $O(n + k^2)$-routing with route degree $O(k)$. Thus, if $k = \lfloor \sqrt{n} \rfloor$ then ν is an $O(n)$-routing with route degree $O(\sqrt{n})$.

Theorem 7. *Let G be a biconnected graph with at least 5 nodes. Then $D(R(G, \nu)/\{f\}) \le 3$ for any fault f in G.*

Proof. Let $R = R(G, \nu)/\{f\}$. Let x and y be any pair of distinct nonfaulty nodes in G.

(1) Suppose that $x = s$ and $y = t$. If f is not the edge (s, t) then $dis_R(s, t) = 1$. Otherwise, for any $z \in \{s_i, t_i\} (1 < i < \ell)$, the route sequence $[\nu(s, z), \nu(z, t)]$ can not contain f. Thus, $dis_R(s, t) \le 2$.

(2-1) Suppose that $x = s$ and $y \in \{s_j, t_j\} (1 \le j \le \ell$ and $y \neq t)$ or $x \in \{s_i, t_i\} (1 \le i \le \ell$ and $x \neq s)$ and $y = t$. For the former case, since there are two node-disjoint route sequences $\nu(s, y)$ and $[\nu(s, t), \nu(t, y)]$, $dis_R(x, y) \le 2$. The latter case can be proved similarly.

(2-2) Suppose that $x \in \{s_i, t_i\}$ and $y \in \{s_j, t_j\}$ $(1 \le i \le j \le \ell$, $x \neq s$ and $y \neq t)$.

If $f = (s, t)$ or $f \in [s, x]$, $\nu(x, t = t_\ell)$ and $\nu(y, t = t_\ell)$ are fault free. Thus, $dis_R(x, y) \le 2$. Similarly, if $f \in [y, t]$ since $\nu(x, s = s_1)$ and $\nu(y, s = s_1)$ are fault free $dis_R(x, y) \le 2$. Otherwise ($f \in [x, y]$), $\nu(x, s_1)$, $\nu(s_1, t_\ell)$ and $\nu(y, t_\ell)$ are fault free. Thus, $dis_R(x, y) \le 3$.

(3) Suppose that $x \in V_i - \{s_i, t_i\}$and $y \in V_j - \{s_j, t_j\}$ $(1 \le i \le j \le \ell)$ such that $g(x) < g(y)$.

If $f = (s, t)$ or $f \in [s, x]$, since $\nu(x, t_i)$ and $\nu(y, t_j)$ are fault free, $dis_R(x, t_i) \le 1$ and $dis_R(y, t_j) \le 1$. Also since $\nu(t_i, t_j)$ does not contain the fault, $dis_R(x, y) \le 3$. Similarly, $dis_R(x, y) \le 3$ holds for the case that $f \in [y, t]$. Otherwise($f \in [x, y]$), $\nu(x, s_i)$, $\nu(s_i, t_j)$ and $\nu(y, t_j)$ are fault free. Thus, $dis_R(x, y) \le 3$.

(4) Otherwise, $x \in V_i$ and $y \in V_j$ $(1 \le i \le j \le \ell)$ such that either $x \in \{s_i, t_i\}$ or $y \in \{s_j, t_j\}$. The cases that $x = s_i$ and $g(s_j) < g(y) < g(t_j)$ and that $g(s_i) < g(x) < g(t_i)$ and $y = t_j$ can be proved similar to the case (3). For the case that $x = t_j$ and $g(s_j) < g(y) < g(t_j)$, except that $f \in [x, y]$ we can prove similar o the case (3). In the case that $f \in [x, y]$, if $1 < i$ and $i < \ell$ then $\nu(x, t_{i-1})$, $\nu(t_{i-1}, s_{j+1})$ and $\nu(y, s_{j+1})$ are fault free. Thus, $dis_R(x, y) \le 3$. If $i = 1$ or $i = \ell$,

then $\nu(x = s_1, t_\ell)$, $\nu(t_\ell, t_j)$ and $\nu(y, t_j)$(if $i = 1$) and $\nu(x = s_i, s_1)$, $\nu(s_1, t_\ell)$ and $\nu(y, t_\ell)$(if $i = \ell$) are fault free, respectively. Thus, $dis_R(x, y) \leq 3$.

The case that $g(s_i) < g(x) < g(t_i)$ and $y = t_j$ can be treated symmetrically.

8 Concluding Remarks

We have shown three optimal edge-routings with desirable properties as ATM-routings. The last routing ν is most desirable among them, because the route degree is $O(\sqrt{n})$. Thus, it is an interesting open question whether or not there exists an edge-routing such that the route degree is $O(\sqrt{n})$ and the diameter of the surviving route graph is 2. In order to construct such routings, the number of routes defined in them must be $\Theta(n^{1.5})$. It is also an interesting question whether or not there is an optimal and bidirectional edge-routing with $O(n)$ routes.

Acknowledgement This research was supported in part by a Scientific Research Grant-In-Aid from the Ministry of Education, Science and Culture, Japan, and by The Telecommunications Advancement Foundation.

References

1. B.Bollobás: Extremal graph theory, *Academic Press*, 172(1978).
2. A.Broder, D.Dolev, M.Fischer and B.Simons: "Efficient fault tolerant routing in network," *Information and Computation*75,52–64(1987).
3. I.Cidon, O.Gerstel and S.Zaks: "A scalable approach to routing in ATM networks," *Proc. 8th International Workshop on Distributed Algorithms*, LNCS 859, 209–222 (1994).
4. D.Dolev, J.Halpern , B.Simons and H.Strong: "A new look at fault tolerant routing," *Information and Computation*72,180–196(1987).
5. S. Evens: Graph algorithms, *Computer Science Press*, Potomac, Maryland(1979).
6. P.Feldman: "Fault tolerance of minimal path routing in a network,*in Proc. 17th ACM STOC*,pp.327–334(1985).
7. L.Gąsieniec, E.Kranakis, D. Krizanc and A.Pelc: "Minimizing Congestion of Layouts for ATM Networks with Faulty Links," *Proc. The 21st International Symposium on Mathematical Foundations of Computer Science*, LNCS 1113, 372–381(1996).
8. O.Gerstel, A. Wool and S.Zaks: "Optimal Layouts on a Chain ATM Network," *Proc. The 3rd Annual European Symposium on Algorithms*, LNCS 979, 508–522 (1995).
9. O.Gerstel and S.Zaks: "The Virtual Path Layout Problem in Fast Networks," *Proc. 13th ACM Symposium on Principles of Distributed Computing*, 235–243 (1994).
10. F.Harary, Graph theory, *Addison-Wesley*, Reading, MA(1969).
11. A. Itai and M. Rodeh: "The multi-tree approach to reliability in distributed networks," *Information and Computation*79,43–59(1988).
12. K.Kawaguchi and K.Wada: "New results in graph routing," *Information and Computation*, 106, 2, 203–233 (1993).
13. J.Y.Le Boudec: "The Asynchronous Transfer Mode:A Tutorial," *Computer Networks and ISDN Systems*, 24, 279–309 (1992).

14. H.Nagamochi and T.Ibaraki : "A linear-time algorithm for finding a sparse k-connected spanning subgraph of a k-connected graph," *Algorithmica*, 7, 5/6, 583–596 (1992).
15. D.Peleg and B.Simons: "On fault tolerant routing in general graph," *Information and Computation*74,33–49(1987).
16. H.Suzuki, N.Takahashi and T.Nishizeki: "A linear algorithm for bipartition of bi-connected graphs," *Information Processing Letters*, 33, 5, 227–231 (1990).
17. K.Wada and K.Kawaguchi: "Efficient fault-tolerant fixed routings on $(k+1)$-connected digraphs," *Discrete Applied Mathematics*, 37/38, 539–552 (1992).
18. K.Wada, Y.Luo and K.Kawaguchi: "Optimal Fault-tolerant routings for Connected Graphs," *Information Processing Letters*, 41, 3, 169–174 (1992).

List of Participants

Paola Alimonti
alimon@dis.uniroma1.it

Luitpold Babel
babel@statistik.tu-muenchen.de

Georg Baier
baier@first.gmd.de

Oliver Bastert
bastert@statistik.tu-muenchen.de

Stefan Baumann
stefanB@statistik.tu-muenchen.de

Hans Bodlaender
hansb@cs.ruu.nl

Stephan Brandt
brandt@math.fu-berlin.de

Paola Campadelli
campadelli@mc.dsi.unimi.it

Claudia Maria Clò
clo@di.unipi.it

Christian Capelle
capelle@lirmm.fr

Derek Corneil
corneil@cs.toronto.edu

Elias Dahlhaus
dahlhaus@informatik.uni-koeln.de

Fabrizio d'Amore
damore@dis.uniroma1.it

Daniele Degiorgi
degiorgi@inf.ethz.ch

Miriam Di Ianni
diianni@dsi.uniroma1.it

Feodor Dragan
dragan@informatik.uni-rostock.de

Thomas Erlebach
erlebach@informatik.tu-muenchen.de

Marcelo Feighelstein
marcelof@cs.technion.ac.il

Stefan Felsner
felsner@inf.fu-berlin.de

Andreas Fest
fest@math.tu-berlin.de

Ulrich Fuchs
fuchs@math.fu-berlin.de

Ewgenij Gawrilow
gawrilow@math.tu-berlin.de

Giampaolo Greco
greco@dsi.uniroma1.it

Jens Gustedt
gustedt@math.tu-berlin.de

Michel Habib
habib@lirmm.fr

Dagmar Handke
Dagmar.Handke@uni-konstanz.de

Stephan Hartmann
hartmann@math.tu-berlin.de

Christoph Helmberg
helmberg@zib.de

Petr Hliněný
hlineny@kam.ms.mff.cuni.cz

Olaf Jahn
jahno@math.tu-berlin.de

Ojvind Johansson
ojvind@nada.kth.se

Lefteris Kirousis
kirousis@ceid.upatras.gr

Ton Kloks
kloks@math.utwente.nl

Ekkehard Köhler
ekoehler@math.tu-berlin.de

Klaus Kriegel
kriegel@inf.fu-berlin.de

Arfst Ludwig
lnd@lsb.de

Mariel Lüdecke
luedecke@statistik.tu-muenchen.de

Alberto Marchetti-Spaccamela
marchetti@dis.uniroma1.it

Ernst Mayr
mayr@informatik.tu-muenchen.de

Rolf H. Möhring
moehring@math.tu-berlin.de

Mohamed Mosbah
mosbah@labri.u-bordeaux.fr

Michel Morvan
morvan@litp.ibp.fr

Haiko Müller
rmueller@wiwi.hu-berlin.de

Rudolf Müller
hm@minet.uni-jena.de

Matthias Müller-Hannemann
mhannema@math.tu-berlin.de

Manfred Nagl
nagl@i3.informatik.rwth-aachen.de

Gabriele Neyer
neyer@inf.ethz.ch

Hartmut Noltemeier
noltemei@informatik.uni-wuerzburg.de

Martin Oellrich
oellrich@math.tu-berlin.de

Roberto Posenato
posenato@dsi.unimi.it

Erich Prisner
ms6a013@math.uni-hamburg.de

Michael Sampels
sampels@informatik.uni-oldenburg.de

Wolfram Schlickenrieder
schlicke@math.tu-berlin.de

Rainer Schrader
schrader@zpr.uni-koeln.de

Andreas Schulz
schulz@math.tu-berlin.de

Alexander Schwartz
schwartz@math.tu-berlin.de

Konstantin Skodinis
skodinis@fmi.uni-passau.de

Martin Skutella
skutella@math.tu-berlin.de

Hermann Stolle
Hermann.Stolle@zib.de

Frederik Stork
stork@math.tu-berlin.de

Jan Arne Telle
telle@ii.uib.no

Dimitrios Thilikos
sedthilk@cs.ruu.nl

Mikkel Thorup
mthorup@diku.dk

Gottfried Tinhofer
gottin@statistik.tu-muenchen.de

Ryuhei Uehara
uehara@twcu.ac.jp

Marc Uetz
uetz@math.tu-berlin.de

Tanja Vos
tanja@cs.ruu.nl

Koichi Wada
wada@elcom.nitech.ac.jp

Dorothea Wagner
Dorothea.Wagner@uni-konstanz.de

Peter Widmayer
widmayer@inf.ethz.ch

David P. Williamson
dpw@watson.ibm.com

Authors' Index

List of WG Proceedings

WG'75 U. Pape (Ed.): *Graphen-Sprachen und Algorithmen auf Graphen.* 1. Fachtagung Graphentheoret. Konzepte der Informatik, Hanser, Munich, 1976, 236 pages, ISBN 3-446-12215-X

WG'76 H. Noltemeier (Ed.): *Graphen, Algorithmen, Datenstrukturen.* Proc. of WG'76, Graphtheoretic Concepts in Computer Science, Hanser, Munich, 1977, 336 pages, ISBN 3-446-12330-4.

WG'77 J. Mühlbacher (Ed.): *Datenstrukturen, Graphen, Algorithmen.* Proc. of WG'77, Hanser, Munich, 1978, 368 pages, ISBN 3-446-12526-3.

WG'78 M. Nagl and H.-J. Schneider (Eds.): *Graphs, Data Structures, Algorithms.* Proc. of WG'78, Hanser, Munich, 1979, 320 pages, ISBN 3-446-12748-3.

WG'79 U. Pape (Ed.): *Discrete Structures and Algorithms.* Proc. of WG'79, Hanser, Munich, 1980, 270 pages, ISBN 3-446-13135-3.

WG'80 H. Noltemeier (Ed.): *Graphtheoretic Concepts in Computer Science.* Proc. of WG'80, Lecture Notes in Computer Science 100, Springer-Verlag, Berlin, 1981, 403 pages, ISBN 0-387-10291-4

WG'81 J. Mühlbacher (Ed.):*Proc. of the 7th Conf. Graphtheoretic Concepts in Computer Science (WG'81).* Hanser, Munich, 1982, 355 pages, ISBN 3-446-13538-3.

WG'82 H.-J. Schneider and H. Göttler (Eds.): *Proc. of the 8th Conf. Graphtheoretic Concepts in Computer Science (WG'82).* Hanser, Munich, 1983, 280 pages, ISBN 3-446-13778-5.

WG'83 M. Nagl and J. Perl (Eds.): *Proc. WG'83, Workshop on Graphtheoretic Concepts in Computer Science.* Trauner, Linz, 1984, 397 pages, ISBN 3-853-20311-6

WG'84 U. Pape (Ed.): *Proc. WG'84, Workshop on Graphtheoretic Concepts in Computer Science.* Trauner, Linz, 1985, 381 pages, ISBN 3-853-20334-5.

WG'85 H. Noltemeier (Ed.): *Graphtheoretic Concepts in Computer Science.* Proc. WG'85, Trauner, Linz, 1986, 443 pages, ISBN 3-853-20357-4.

WG'86 G. Tinhofer and G. Schmidt (Eds.): *Graph-Theoretic Concepts in Computer Science.* Proc. WG'86, Lecture Notes in Computer Science 246, Springer-Verlag, Berlin, 1987, 305 pages, ISBN 0-387-17218-1.

WG'87 H. Göttler and H.-J. Schneider (Eds.): *Graph-Theoretic Concepts in Computer Science.* Proc. WG'87, Lecture Notes in Computer Science 314, Springer-Verlag, Berlin, 1988, 254 pages, ISBN 0-387-19422-3

WG'88 J. van Leeuwen (Ed.): *Graph-Theoretic Concepts in Computer Science.* Proc. WG'88, Lecture Notes in Computer Science 344, Springer-Verlag, Berlin, 1989, 457 pages, ISBN 0-387-50728-0.

WG'89 M. Nagl (Ed.): *Graph-Theoretic Concepts in Computer Science.* Proc. WG'89, Lecture Notes in Computer Science 411, Springer-Verlag, Berlin, 1990, 374 pages, ISBN 0-387-52292-1.

WG'90 R. Möhring (Ed.): *Graph-Theoretic Concepts in Computer Science.* Proc. WG'90, Lecture Notes in Computer Science 484, Springer-Verlag, Berlin, 1991, 360 pages, ISBN 0-387-53832-1.

WG'91 G. Schmidt and R. Berghammer (Eds.): *Graph-Theoretic Concepts in Computer Science.* Proc. WG'91, Lecture Notes in Computer Science 570, Springer-Verlag, Berlin, 1992, 253 pages, ISBN 0-387-55121-2.

WG'92 E. W. Mayr (Ed.): *Graph-Theoretic Concepts in Computer Science.* Proc. WG'92, Lecture Notes in Computer Science 657, Springer-Verlag, Berlin, 1993, 350 pages, ISBN 0-387-56402-0.

WG'93 J. van Leeuwen (Ed.): *Graph-Theoretic Concepts in Computer Science.* Proc. WG'93, Lecture Notes in Computer Science 790, Springer Verlag, Berlin, 1994, 431 pages, ISBN 0-387-57889-4.

WG'94 E. W. Mayr, G. Schmidt and G. Tinhofer (Eds.): *Graph-Theoretic Concepts in Computer Science.* Proc. WG'94, Lecture Notes in Computer Science 903, Springer-Verlag, Berlin, 1995, 414 pages, ISBN 3-540-59071-4.

WG'95 M. Nagl (Ed.): *Graph-Theoretic Concepts in Computer Science.* Proc. WG'95, Lecture Notes in Computer Science 1017, Springer-Verlag, Berlin, 1995, 406 pages, ISBN 3-540-60618-1.

WG'96 F. d'Amore, P. G. Franciosa and A. Marchetti-Spaccamela (Eds.): *Graph-Theoretic Concepts in Computer Science.* Proc. WG'96, Lecture Notes in Computer Science 1197, Springer-Verlag, Berlin, 1997, 410 pages, ISBN 3-540-62559-3

WG'97 R. H. Möhring (Ed.): *Graph-Theoretic Concepts in Computer Science.* Proc. WG'97, Lecture Notes in Computer Science, Springer-Verlag, Berlin, this volume.

Vol. 1302: P. Van Hentenryck (Ed.), Static Analysis. Proceedings, 1997. X, 413 pages. 1997.

Vol. 1303: G. Brewka, C. Habel, B. Nebel (Eds.), KI-97: Advances in Artificial Intelligence. Proceedings, 1997. XI, 413 pages. 1997. (Subseries LNAI).

Vol. 1304: W. Luk, P.Y.K. Cheung, M. Glesner (Eds.), Field-Programmable Logic and Applications. Proceedings, 1997. XI, 503 pages. 1997.

Vol. 1305: D. Corne, J.L. Shapiro (Eds.), Evolutionary Computing. Proceedings, 1997. X, 307 pages. 1997.

Vol. 1306: C. Leung (Ed.), Visual Information Systems. X, 274 pages. 1997.

Vol. 1307: R. Kompe, Prosody in Speech Understanding Systems. XIX, 357 pages. 1997. (Subseries LNAI).

Vol. 1308: A. Hameurlain, A M. Tjoa (Eds.), Database and Expert Systems Applications. Proceedings, 1997. XVII, 688 pages. 1997.

Vol. 1309: R. Steinmetz, L.C. Wolf (Eds.), Interactive Distributed Multimedia Systems and Telecommunication Services. Proceedings, 1997. XIII, 466 pages. 1997.

Vol. 1310: A. Del Bimbo (Ed.), Image Analysis and Processing. Proceedings, 1997. Volume I. XXII, 722 pages. 1997.

Vol. 1311: A. Del Bimbo (Ed.), Image Analysis and Processing. Proceedings, 1997. Volume II. XXII, 794 pages. 1997.

Vol. 1312: A. Geppert, M. Berndtsson (Eds.), Rules in Database Systems. Proceedings, 1997. VII, 214 pages. 1997.

Vol. 1313: J. Fitzgerald, C.B. Jones, P. Lucas (Eds.), FME '97: Industrial Applications and Strengthened Foundations of Formal Methods. Proceedings, 1997. XIII, 685 pages. 1997.

Vol. 1314: S. Muggleton (Ed.), Inductive Logic Programming. Proceedings, 1996. VIII, 397 pages. 1997. (Subseries LNAI).

Vol. 1315: G. Sommer, J.J. Koenderink (Eds.), Algebraic Frames for the Perception-Action Cycle. Proceedings, 1997. VIII, 395 pages. 1997.

Vol. 1316: M. Li, A. Maruoka (Eds.), Algorithmic Learning Theory. Proceedings, 1997. XI, 461 pages. 1997. (Subseries LNAI).

Vol. 1317: M. Leman (Ed.), Music, Gestalt, and Computing. IX, 524 pages. 1997. (Subseries LNAI).

Vol. 1318: R. Hirschfeld (Ed.), Financial Cryptography. Proceedings, 1997. XI, 409 pages. 1997.

Vol. 1319: E. Plaza, R. Benjamins (Eds.), Knowledge Acquisition, Modeling and Management. Proceedings, 1997. XI, 389 pages. 1997. (Subseries LNAI).

Vol. 1320: M. Mavronicolas, P. Tsigas (Eds.), Distributed Algorithms. Proceedings, 1997. X, 333 pages. 1997.

Vol. 1321: M. Lenzerini (Ed.), AI*IA 97: Advances in Artificial Intelligence. Proceedings, 1997. XII, 459 pages. 1997. (Subseries LNAI).

Vol. 1322: H. Hußmann, Formal Foundations for Software Engineering Methods. X, 286 pages. 1997.

Vol. 1323: E. Costa, A. Cardoso (Eds.), Progress in Artificial Intelligence. Proceedings, 1997. XIV, 393 pages. 1997. (Subseries LNAI).

Vol. 1324: C. Peters, C. Thanos (Eds.), Research and Advanced Technology for Digital Libraries. Proceedings, 1997. X, 423 pages. 1997.

Vol. 1325: Z.W. Raś, A. Skowron (Eds.), Foundations of Intelligent Systems. Proceedings, 1997. XI, 630 pages. 1997. (Subseries LNAI).

Vol. 1326: C. Nicholas, J. Mayfield (Eds.), Intelligent Hypertext. XIV, 182 pages. 1997.

Vol. 1327: W. Gerstner, A. Germond, M. Hasler, J.-D. Nicoud (Eds.), Artificial Neural Networks – ICANN '97. Proceedings, 1997. XIX, 1274 pages. 1997.

Vol. 1328: C. Retoré (Ed.), Logical Aspects of Computational Linguistics. Proceedings, 1996. VIII, 435 pages. 1997. (Subseries LNAI).

Vol. 1329: S.C. Hirtle, A.U. Frank (Eds.), Spatial Information Theory. Proceedings, 1997. XIV, 511 pages. 1997.

Vol. 1330: G. Smolka (Ed.), Principles and Practice of Constraint Programming – CP 97. Proceedings, 1997. XII, 563 pages. 1997.

Vol. 1331: D. W. Embley, R. C. Goldstein (Eds.), Conceptual Modeling – ER '97. Proceedings, 1997. XV, 479 pages. 1997.

Vol. 1332: M. Bubak, J. Dongarra, J. Waśniewski (Eds.), Recent Advances in Parallel Virtual Machine and Message Passing Interface. Proceedings, 1997. XV, 518 pages. 1997.

Vol. 1333: F. Pichler. R.Moreno-Díaz (Eds.), Computer Aided Systems Theory – EUROCAST'97. Proceedings, 1997. XII, 626 pages. 1997.

Vol. 1334: Y. Han, T. Okamoto, S. Qing (Eds.), Information and Communications Security. Proceedings, 1997. X, 484 pages. 1997.

Vol. 1335: R.H. Möhring (Ed.), Graph-Theoretic Concepts in Computer Science. Proceedings, 1997. X, 376 pages. 1997.

Vol. 1336: C. Polychronopoulos, K. Joe, K. Araki, M. Amamiya (Eds.), High Performance Computing. Proceedings, 1997. XII, 416 pages. 1997.

Vol. 1337: C. Freksa, M. Jantzen, R. Valk (Eds.), Foundations of Computer Science. XII, 515 pages. 1997.

Vol. 1338: F. Plášil, K.G. Jeffery (Eds.), SOFSEM'97: Theory and Practice of Informatics. Proceedings, 1997. XIV, 571 pages. 1997.

Vol. 1339: N.A. Murshed, F. Bortolozzi (Eds.), Advances in Document Image Analysis. Proceedings, 1997. IX, 345 pages. 1997.

Vol. 1340: M. van Kreveld, J. Nievergelt, T. Roos, P. Widmayer (Eds.), Algorithmic Foundations of Geographic Information Systems. XIV, 287 pages. 1997.

Vol. 1341: F. Bry, R. Ramakrishnan, K. Ramamohanarao (Eds.), Deductive and Object-Oriented Databases. Proceedings, 1997. XIV, 430 pages. 1997.

Vol. 1342: A. Sattar (Ed.), Advanced Topics in Artificial Intelligence. Proceedings, 1997. XVIII, 516 pages. 1997. (Subseries LNAI).

Vol. 1344: C. Ausnit-Hood, K.A. Johnson, R.G. Pettit, IV, S.B. Opdahl (Eds.), Ada 95 – Quality and Style. XV, 292 pages. 1997.

Lecture Notes in Computer Science

For information about Vols. 1–1265

please contact your bookseller or Springer-Verlag

Vol. 1266: D.B. Leake, E. Plaza (Eds.), Case-Based Reasoning Research and Development. Proceedings, 1997. XIII, 648 pages. 1997 (Subseries LNAI).

Vol. 1267: E. Biham (Ed.), Fast Software Encryption. Proceedings, 1997. VIII, 289 pages. 1997.

Vol. 1268: W. Kluge (Ed.), Implementation of Functional Languages. Proceedings, 1996. XI, 284 pages. 1997.

Vol. 1269: J. Rolim (Ed.), Randomization and Approximation Techniques in Computer Science. Proceedings, 1997. VIII, 227 pages. 1997.

Vol. 1270: V. Varadharajan, J. Pieprzyk, Y. Mu (Eds.), Information Security and Privacy. Proceedings, 1997. XI, 337 pages. 1997.

Vol. 1271: C. Small, P. Douglas, R. Johnson, P. King, N. Martin (Eds.), Advances in Databases. Proceedings, 1997. XI, 233 pages. 1997.

Vol. 1272: F. Dehne, A. Rau-Chaplin, J.-R. Sack, R. Tamassia (Eds.), Algorithms and Data Structures. Proceedings, 1997. X, 476 pages. 1997.

Vol. 1273: P. Antsaklis, W. Kohn, A. Nerode, S. Sastry (Eds.), Hybrid Systems IV. X, 405 pages. 1997.

Vol. 1274: T. Masuda, Y. Masunaga, M. Tsukamoto (Eds.), Worldwide Computing and Its Applications. Proceedings, 1997. XVI, 443 pages. 1997.

Vol. 1275: E.L. Gunter, A. Felty (Eds.), Theorem Proving in Higher Order Logics. Proceedings, 1997. VIII, 339 pages. 1997.

Vol. 1276: T. Jiang, D.T. Lee (Eds.), Computing and Combinatorics. Proceedings, 1997. XI, 522 pages. 1997.

Vol. 1277: V. Malyshkin (Ed.), Parallel Computing Technologies. Proceedings, 1997. XII, 455 pages. 1997.

Vol. 1278: R. Hofestädt, T. Lengauer, M. Löffler, D. Schomburg (Eds.), Bioinformatics. Proceedings, 1996. XI, 222 pages. 1997.

Vol. 1279: B. S. Chlebus, L. Czaja (Eds.), Fundamentals of Computation Theory. Proceedings, 1997. XI, 475 pages. 1997.

Vol. 1280: X. Liu, P. Cohen, M. Berthold (Eds.), Advances in Intelligent Data Analysis. Proceedings, 1997. XII, 621 pages. 1997.

Vol. 1281: M. Abadi, T. Ito (Eds.), Theoretical Aspects of Computer Software. Proceedings, 1997. XI, 639 pages. 1997.

Vol. 1282: D. Garlan, D. Le Métayer (Eds.), Coordination Languages and Models. Proceedings, 1997. X, 435 pages. 1997.

Vol. 1283: M. Müller-Olm, Modular Compiler Verification. XV, 250 pages. 1997.

Vol. 1284: R. Burkard, G. Woeginger (Eds.), Algorithms — ESA '97. Proceedings, 1997. XI, 515 pages. 1997.

Vol. 1285: X. Jao, J.-H. Kim, T. Furuhashi (Eds.), Simulated Evolution and Learning. Proceedings, 1996. VIII, 231 pages. 1997. (Subseries LNAI).

Vol. 1286: C. Zhang, D. Lukose (Eds.), Multi-Agent Systems. Proceedings, 1996. VII, 195 pages. 1997. (Subseries LNAI).

Vol. 1287: T. Kropf (Ed.), Formal Hardware Verification. XII, 367 pages. 1997.

Vol. 1288: M. Schneider, Spatial Data Types for Database Systems. XIII, 275 pages. 1997.

Vol. 1289: G. Gottlob, A. Leitsch, D. Mundici (Eds.), Computational Logic and Proof Theory. Proceedings, 1997. VIII, 348 pages. 1997.

Vol. 1290: E. Moggi, G. Rosolini (Eds.), Category Theory and Computer Science. Proceedings, 1997. VII, 313 pages. 1997.

Vol. 1291: D.G. Feitelson, L. Rudolph (Eds.), Job Scheduling Strategies for Parallel Processing. Proceedings, 1997. VII, 299 pages. 1997.

Vol. 1292: H. Glaser, P. Hartel, H. Kuchen (Eds.), Programming Languages: Implementations, Logigs, and Programs. Proceedings, 1997. XI, 425 pages. 1997.

Vol. 1293: C. Nicholas, D. Wood (Eds.), Principles of Document Processing. Proceedings, 1996. XI, 195 pages. 1997.

Vol. 1294: B.S. Kaliski Jr. (Ed.), Advances in Cryptology — CRYPTO '97. Proceedings, 1997. XII, 539 pages. 1997.

Vol. 1295: I. Prívara, P. Ružička (Eds.), Mathematical Foundations of Computer Science 1997. Proceedings, 1997. X, 519 pages. 1997.

Vol. 1296: G. Sommer, K. Daniilidis, J. Pauli (Eds.), Computer Analysis of Images and Patterns. Proceedings, 1997. XIII, 737 pages. 1997.

Vol. 1297: N. Lavrač, S. Džeroski (Eds.), Inductive Logic Programming. Proceedings, 1997. VIII, 309 pages. 1997. (Subseries LNAI).

Vol. 1298: M. Hanus, J. Heering, K. Meinke (Eds.), Algebraic and Logic Programming. Proceedings, 1997. X, 286 pages. 1997.

Vol. 1299: M.T. Pazienza (Ed.), Information Extraction. Proceedings, 1997. IX, 213 pages. 1997. (Subseries LNAI).

Vol. 1300: C. Lengauer, M. Griebl, S. Gorlatch (Eds.), Euro-Par'97 Parallel Processing. Proceedings, 1997. XXX, 1379 pages. 1997.

Vol. 1301: M. Jazayeri, H. Schauer (Eds.), Software Engineering - ESEC/FSE'97. Proceedings, 1997. XIII, 532 pages. 1997.